BUILDING CONSTRUCTION
— AND —
DESIGN

BUILDING CONSTRUCTION
— AND —
DESIGN

JAMES AMBROSE
University of Southern California

SPRINGER SCIENCE+BUSINESS MEDIA, LLC

Originally published by Van Nostrand Reinhold Company Inc. in 1992

Library of Congress Catalog Card Number 92-5094
ISBN 978-1-4615-6585-7 ISBN 978-1-4615-6583-3 (eBook)
DOI 10.1007/978-1-4615-6583-3

Published by Van Nostrand Reinhold
115 Fifth Avenue
New York, New York 10003

Chapman and Hall
2-6 Boundary Row
London, SE1 8HN, England

Thomas Nelson Australia
102 Dodds Street
South Melbourne 3205
Victoria, Australia

Nelson Canada
1120 Birchmount Road
Scarborough, Ontario M1K 5G4, Canada

16 15 14 13 12 11 10 9 8 7 6 5 4 3 2 1

Library of Congress Cataloging-in-Publication Data

Ambrose, James E.
Building construction and design / James Ambrose.
p. cm.
Includes index.
ISBN 978-1-4615-6585-7
1. Building. 2. Architecture. I. Title.
TH145.A58 1992
690—dc20 92-5094
CIP

Contents

Preface

This book addresses, for the student of architecture, a subject that is of interest to a wide range of people with various relationships to the designing and constructing of buildings. There is at any given time a great mass of information on these topics which may be accessed for application to the many tasks of designers, builders, suppliers, and others. The enormity of this information resource is at once reassuring to those who regularly encounter needs for it, and overwhelming to just about everyone who needs to figure out how to use it. It is especially important for the young designer to be aware of and familiar with the entire framework and of the place of design within it.

This book relates the topic of building construction to the basic problems of building design. The basic assumption here is that the need to design precedes the need to build, and that real concern for how to build comes from a desire to build something. This is the normal process of development for designers and others who start from the point of desiring a building and then proceed to determine what it should be. Intense concern for specific consideration of building materials, systems, and details of construction thus emerges at a later stage of design, typically after the general form, size, and essential nature of the building are already proposed.

A major problem with using the great mass of available information about building construction is that it is largely not oriented to the purposes of education or design; thus it is not user-friendly to the student or apprentice in architecture, or to any others who do not already have a broad grasp of what buildings and building construction are all about. Rich as they may be as information sources, *Sweets Catalog Files*, the *CSI Spec-Data System*, and *Architectural Graphic Standards* are not friendly to inexperienced users. You have to already know the trade lingo, what you are looking for, and pretty much why you are looking for it to make effective use of these resources.

The latest, most extensive, most detailed information about building construction is largely produced by persons engaged in the manufacturing, supplying, specifying, purchasing, regulating, financing, or insuring of building materials and products, and in their applications to making buildings.

The information presented is often slanted towards the specific concerns of those who produce it. Thus it is to be expected that materials forthcoming from manufacturers and suppliers are shaped to the purpose of promoting sales of their products, that insurers and financiers have strong economic concerns, and that specification writers and enforcers of building codes are somewhat paranoid about specific language and terminology that is legally-binding.

This book attempts to be user-friendly to the person who is basically more interested in buildings in general and somewhat less in the specific concerns for their construction. The basic purposes here are to develop a general view of buildings and the problems of their design as they relate to the eventual need to construct them. Highly specific, detailed information, such as how to attach gypsum drywall to wood studs, can be pursued through appropriate sources, once the real need for the information is established. The need for having the drywall, for attaching it to the studs, for having wood studs—and indeed, for having the wall in the first place—need to precede the search for the specific information.

Buildings are complex and the topic of their construction is correspondingly extensive. In order to cut down the size of this presentation, the topic here is limited to concerns for basic issues and fundamental types of construction elements. Detailed information is given only for sample elements in most discussions, with references cited for more complete information. The three sources mentioned earlier are indeed such types of references, and are rich with detailed information. A major purpose of this book is to prepare readers for appreciating and being able to utilize such sources to obtain information that is supportive of building design work.

Much of the work that appears here was first developed in the writing of four books that treated this topic in a partial manner; each book developing a selected portion of the whole topic of building construction. The topic concentrations of those books, as represented by their subtitles, are: Enclosure Systems, Interior Systems, Site and Below-Grade Systems, and Service Systems. As most people who work in building design are concerned primarily with only limited parts of the whole problem, those partial topic developments relate more closely to the limited working needs of many designers.

While much of the work presented here is collected from the four shorter volumes, the need here is to develop the *whole* topic of building design for construction which brings in many additional issues not treated in the shorter books. On the other hand, the greater concern here for the total picture brings some need for less full development of individual topics. Thus this book is both more and less than simply a collection of the work in the four previous books. They have their selected purposes, and so does this broader work.

My views of the need for this book and my ideas for its contents have emerged from many years of experience as a building designer and a teacher. I am grateful to all my former students, coworkers, and others whose feedback of frustrations have shaped those views and ideas. I am also grateful to the many people at Van Nostrand Reinhold who have helped to bring my rough materials into being as an actual book; particularly my editors Everett Smethurst and Wendy Lochner as well as Cindy Zigmund and Alberta Gordon.

I am also grateful to various organizations who have permitted the use of materials from their publications, as noted throughout the book.

Finally, I am grateful to the members of my family who steadfastly support and tolerate me in my home office working situation. For this book, I am particularly grateful for the assistance of my wife, Peggy, and my son, Jeff.

James Ambrose
Westlake Village, California

BUILDING CONSTRUCTION — AND — DESIGN

1

Introduction

This book deals with the construction of buildings. The topic is viewed broadly, and includes the determination of what is to be built as well as the actual work of building. Consideration is also given to the producing and supplying of the materials and products used for the construction. Considered as such, the topic includes the work and interests of many people and a very large number of factors that influence the final form and detail of the end product: the constructed building.

If fully developed so broadly, the topic of building construction has the potential for producing an enormous treatment. In fact, considerable material is presented here, but there is no intention to provide a totally comprehensive source of information. Many well-organized and reasonably available sources of information exist, and there is no intention to fully duplicate them here.

The basic intent here is to explain and instruct, for the purpose of helping readers to gain a broad and fundamental view of the topic. For those who wish to pursue the general topic or any part of it in greater depth, guidance to many other sources is provided.

In developing the materials in this book, a basic viewpoint taken is that of the building designer. Limited treatment is given to problems involving the management of construction work and the manufacture and marketing of materials and products, although some understanding of these concerns is necessary for intelligent design work. A similar view is taken of concerns for construction financing, contracts and litigation, insurance, building code administration, and the myriad of services and businesses that relate to the necessary work of producing buildings.

The critical focus here is on the concerns of the prime designer of the building. For a traditional view, the prime designer is considered to be an architect and the design work is assumed to be accomplished within the functioning of a professional architectural practice. However, for all but the simplest of projects, the design work is certain to involve many people, with most performing some fraction of the total design work. The designed building is thus finally achieved as an aggregate of many decisions made by individuals as contributions to the work.

This is not essentially a book on how to manage a design practice. The basic concern here is for the design work itself, and what is fundamental, significant, and useful to the completion of successful designs. Concentration is on the designed object, but with a basic orientation on the object as viewed by the designer.

Building construction is primarily dealt with here as it occurs at present in the United States. This does not necessarily affect basic concerns for many fundamental issues, such as how to keep rain from coming through a roof. It does, however, limit the scope of treatment of construction materials, systems, and processes and of specific design standards and criteria.

1.1 OPERATION OF THE BUILDING PROCESS

In the United States, how buildings get designed, how the materials, products, and equipment for their construction are produced, and how the buildings get built are all strongly related to the general nature of the free enterprise system. Controls of many kinds exist, but all of the processes are subject to manipulation by creative entrepreneurs. Market competition and the tensions developed by contentions between groups with opposed views or purposes make for a dynamic state of activities. Building designers, specifications writers, building materials purchasers, and building construction contractors must all learn to function in this highly charged, continuously changing atmosphere.

What gets built and how it gets built are largely based on our individual and collective experiences. Buildings must usually endure over some time span, and it is natural to derive evaluations from observation of the endurance and general performance of the buildings of the past. This often results in the development of some attachment to old familiar ways and a reluctance to accept change. This conservative attitude is most notable among those persons with major concern for the financing, insuring, and legal ordinance enforcement activities related to building. New and better ways always present some attraction to designers, but the need for assurance and reliability of performance is strongly affected by the successes and failures of the past. (See Figure 1.1.)

With regard to who does what and how they relate to each other, all of the people and organizations involved in the building process operate in a variety of ways. The basic process itself, however, usually includes the following fundamental activities:

1. Someone wants a building, has some specific usage in mind for it, is willing to pay for it, and will own or sell it when it is built.
2. Someone must determine how the building will look, how its parts will be arranged, how it will be built, what equipment will be installed, and—in general—what the finished building, ready for occupancy, will be.
3. Someone must produce the required materials, products, and equipment for the construction of the building.
4. And, finally, someone must build it.

In the case of pioneer settlers, all four efforts may be done by a single person. In modern society, however, the individual activities as described are likely to be done by separate persons or groups. In any event, all the work must be done, whoever does it.

The discussions in this book deal principally with the finished construction and only minimally with the construction process and the arrangements for its operation. Nevertheless, it is good to have some knowledge of the ordinary processes and the roles of the various participants, as the end product is often strongly affected by the process, by its management, and by all those who are involved.

1.2 PARTICIPANTS IN THE BUILDING PROCESS

While many other ways of operating are possible, the following is a traditional form of operation of the building process.

1. Someone wants a building and retains the services of a building design specialist (read: architect), and, if the advice is taken, for the preparation of the necessary documentation of the building design to permit construction to occur.

Figure 1.1 Late nineteenth-century building, still expressing many features of the old masonry wall-bearing construction, although its basic structure is steel frame. Auditorium Hotel and Office Building, Chicago, by Adler and Sullivan. Eighty years later, the Sears Tower looms over it, clearly expressing the curtain wall covered steel frame and its basic lateral resistive system—a bundled tube form.

2. The architect first prepares preliminary drawings and models to explain the proposed design of the building to the potential owner. Once the general nature of the design is accepted, the architect then proceeds to prepare the detailed drawings and written specifications as required to completely describe the desired construction.
3. Because modern buildings are quite complex, it is likely that the architect will need help from various other specialists, such as the following:

   Structural engineers—to design the foundations and other elements of the building's load-resisting structure for response to effects of gravity, wind, earthquakes, and building usage.

Civil engineers and landscape designers—to develop the site grading, site drainage, site constructions, plantings, and other elements necessary for the development of the building's site and surroundings.

Mechanical engineers—to design the systems for plumbing and for heating, ventilating, and air conditioning (called HVAC) for the building.

Electrical engineers—to design the systems for electrical power and for powered illumination of the building.

These specialists may be outside consultants or may be employees in a single organization with the architect. For special problems, or simply for large complex projects, the list of specialists may be longer, including persons with expertise in interior design, sound control, vertical transportation, fire safety, and so on.

4. Various companies produce materials, products, and equipment for buildings, which they market directly to builders or provide through suppliers of building materials or services.
5. Organizations engage in construction of buildings and offer their services for contract, bidding for the work as it is described by the architect and the other specialists. *General contractors* engage the work on the whole building project but much of the work is often done by the individual, specialty *subcontractors,* who do limited parts of the work, such as, erection of the steel frame or installation of the plumbing.

This is a common, historically traditional form for the production of buildings. However, there are many possible variations, depending somewhat on the nature, size, or location of the building project. Participation by developers, construction managers, or companies that package several of the activities described may significantly alter the processes, the forms of contractual agreements, or other aspects of the organization or operation of the work. Still, the work that needs to be done is generally the same.

In addition to the participants described in the preceding discussions, there are typically many other persons with some interest or involvement in the building process. These may not contribute directly to the design work, the production of materials, or the construction work, but nevertheless have some concern for the work. In some situations they may be sources of major influence on some decisions that affect the end product. Some of the participants of this nature include the following:

Financing organizations that make loans to the owner, the builder, and maybe the architect—before fees are collected.

Insurance companies that insure the building when it is done, the contractors and workers during the construction, and the liability of the architect and other designers.

Building code writers and enforcers, who develop many design and construction requirements, issue permits for construction and occupancy, and generally survey the work for compliance with the law.

Industry, professional, and trade union organizations that may set standards or otherwise influence the work to promote and protect the interests of their members.

And lawyers—representing all parties and interests and handling negotiations for every contractual arrangement, as well as any disputes.

Finally, a group that must be acknowledged are the persons who write the specifications that are part of the construction contract. This group—and their organization—the Construction Specifications Institute (CSI)—exerts influence on the general organization of the activities of building design and construction and the production and marketing of building materials and products. They are the originators and custodians of the Masterformat system of classification that is used industry-wide.

In many essential ways the quality of building construction is basically controlled by the specifications, and not by the drawings used to describe the construction. The drawings control form and dimensions and the arrangements of parts of the construction, but the actual materials and products used and the quality of the work is controlled by the specifications. This is a situation not widely understood nor appreciated by building designers, many of who have little to do with the actual preparation of the written specifications.

Construction of buildings is big business in the United States, and the number of people and organizations interested and involved is enormous. While the focus in this book is on design, the designer must work with a realistic awareness of the many sources of constraint on the design decisions and end product.

Considerable effort is expended by various people in the form of some striving against traditional constraints. Designers, builders, investors, and others are constantly looking for ways to reduce costs, save time, simplify or otherwise improve various processes, and generally improve upon what and how we build. Many changes come out of this effort—not always to the benefit of improving the end products. General economic and time pressures on both design and construction work, for example, have resulted in a steady erosion of the quality of craftsmanship in both areas. Designers work hastily to meet deadlines; careful and deliberate workers are soon replaced by others who are capable of slapping things together faster.

Quality hand-crafted work is pretty much relegated to incidental, decorative elements. The general construction is mostly achieved with machine-produced parts. The emerging major participants in the building design/build process of the future may well be the computer and its industrial robots. Craft—if still in the construction dictionary—will take on a new meaning.

1.3 ORGANIZATION OF THE BUILDING INDUSTRY

The building industry in the United States operates essentially as a free enterprise system, and is not actually "organized" as such. Nevertheless, there are many component organizations that interact in the overall activity of building design and construction; many of which are themselves quite organized and which represent fairly established and tightly defined units of the building industry. Notable units of this kind are described in the preceding section.

The business of building mostly organizes around the general investment and ownership of buildings and the processes of financing, designing, and construction. These processes are sometimes quite formally defined, as for institutional buildings—especially those owned by governments. For commercial projects—such as tract housing and shopping centers—the process is more freely formed in response to market conditions and operations.

The strongest forms of organization are those controlled by law; including administration of building codes and zoning ordinances. Regulations for fire safety, health standards, worker safety, and facilities for the physically handicapped impose such restrictions. So do laws that deal with property ownership, contracts, required bonding, and other matters that affect the general business relationships and interactions of the people in the building business.

In a large sense, however, the way we do business is controlled by habits and customs; good old "business as usual." It is smart to know the letter of the law; but highly practical to know how things really work in the everyday operations of the business. It takes involvement and experience to obtain that know-how.

1.4 REGULATION OF THE BUILDING INDUSTRY

Governmental regulation of building construction is well established through building codes and other standards, as well as by the general laws that affect people and their relationships. Beyond this, however, there are many forms of self-regulation, established by the various groups in the industry.

All the major professional groups have their codes of ethics, identified with the major professional organizations. Trade unions and industry organizations abound; with just about any identifiable group having their own organization. (The Single Ply Roofing Institute, The Jute Carpet Backing Council, The Flat Glass Jobbers Association, The Insect Wire Screening Bureau, for some examples.) While many of these groups exist for self-promotion, they also usually establish some form of regulation for their members; many of which become basic industry standards for designers and builders. Some major ones of this kind have achieved the stature of being recognized by building codes as basic references; examples being the ACI Code (American Concrete Institute) and the AISC Specification (American Institute of Steel Construction).

There are also many organizations that exist primarily for the purpose of developing standards which others use. This includes the publishers of the national model codes (UBC, BOCA, SSBC) and organizations such as ASTM, ANSI, and NFPA.

In one way or another, just about anything done for the design or construction of a building is subject to some controlling regulations that have some stature for recognition. These may vary from strong laws to simple recommendations, but usually carry some force of encouragement for their use.

1.5 ROLE OF THE BUILDING DESIGNER

As the discussion in the preceding section points out, there are various possibilities for the employment or the contractual situation of the building designers (see Figure 1.2). While the specific details of such arrangements may strongly affect the nature of any legal conditions of the design work there is a great deal of the character of the work that should not be so conditioned.

Design must be done responsibly and competently within the reasonable limits of the designers' capabilities, experience, and resources. If any conditions make this unlikely to be able to be achieved, it is the responsibility of the designers to make

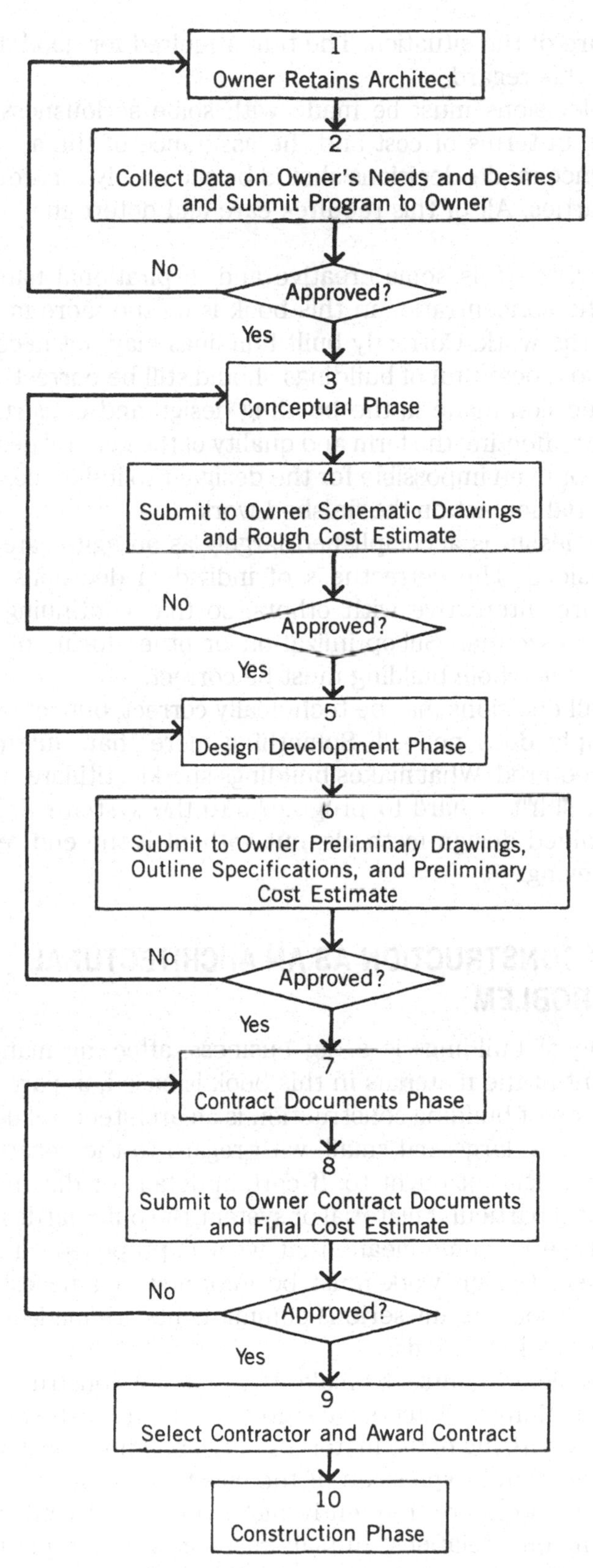

Figure 1.2 Steps in the basic building design and build process. Reprinted from *Building Engineering and Systems Design*, 2nd ed., with permission of the publisher, Van Nostrand Reinhold.

others aware of the situation. The time required for good design work is one major concern in this regard.

Design decisions must be made with some seriousness, befitting their relative importance in terms of cost and the assurance of the adequacy of the completed building. Once made, decisions should be accurately recorded and communicated to affected parties. All of this requires care and determined effort on the part of the designers.

Hopefully, there is some creative and inspirational nature to the design work, although the concentration in this book is on the more mundane and businesslike aspects of the work. Correctly built buildings may not necessarily be beautiful, but even the most beautiful of buildings should still be correctly built.

Every decision made in the building design and construction process has some potential for affecting the form and quality of the general design. Many decisions may be difficult or even impossible for the designer to influence, but this does not reduce their potential impact on the finished work.

Building design is accomplished largely as an aggregate of many individual, detailed decisions. The correctness of individual decisions is important, but many decisions are interactive with others, so that continuing coordination of design activities is essential. Suboptimization or other forms of partial success are not acceptable; the whole building must be correct.

Finally, all decisions may be technically correct, but somehow the overall building design simply does not jell. Something more than automated, efficient decision making is required. What makes buildings sparkle, titillate, inspire, or generally uplift the human spirit is hard to program into the systems design, value analysis, and computer-aided design methods, but without it the end results often seem hardly worth achieving.

1.6 BUILDING CONSTRUCTION AS AN ARCHITECTURAL DESIGN PROBLEM

The making of buildings is a big business, affecting many people. However, the development of the materials in this book is based, in part, on the attitude that the determination of building construction is an architectural design problem. The solving of problems, large and small, with regard to the construction work affects the design—or at least it ought to. If certain details or dimensions are not feasible to achieve with a particular material or system, the building designers should accept the facts, or prove by some means that what is proposed for the design can truly be accomplished. Design work must be informed, not necessarily in early sketching stages, but as soon as any serious commitments are made and the detailed development of the work proceeds.

It is possible, of course, to make the building construction itself a major design determinant. Many architects have great concern for the proper use of, revealing of, and even dramatizing of the materials of the building construction. However, respect for correctness and expression of the construction is one thing; reverence for it is another. Architects are frequently more concerned with form, space, appearance, illusion, symbolic relations, human responses and aspirations, or other matters; caring somewhat less about the expression of the construction or the revealing of the building's functioning elements.

There is no intention to advocate any particular design philosophy in this book. Proper construction is considered essentially as a simple, pragmatic matter. It is considered to have a reasonable value and to be something that deserves serious attention—but it is not essential that it command dominance in all architectural design decisions.

For most designers the actual work of design usually is a very personal matter. Launching points, decision sequences, and the relative weights of different values vary considerably from designer to designer as well as from project to project with a single designer. Approaching the subject of building construction as an architectural design issue simply means that concerns for the construction emerge as necessities in the process of design. The building designer does not set out basically to produce some construction, but rather encounters construction considerations in the process of designing a building. First comes the building, with all of its planning concerns—then the problems of making it.

1.7 HOW TO USE THIS BOOK

The basic purpose of this book is to help the inexperienced designer, whether in a school design class or in an apprentice position in a design office. However, anyone who is interested in the topic of building construction and its effects on building design should be able to derive some use from this book.

There is no single method for the most effective use of the materials in this book. Readers are sure to vary considerably with regard to interests, needs, and backgrounds. While the materials have been arranged with some logical sequence in mind, it is anticipated that the typical reader will take up the materials in random fashion. This is to be expected all the more if the reader is seeking some help in the progress of some design work.

We do not mean to discourage any readers from taking up the book materials in the sequence presented; this is likely to make most sense, if the reader is essentially uninformed in the general topic. Development of a fluent "construction" vocabulary is also to be generally encouraged, as the discipline of proper terminology is quite essential for accurate communication in the building design and construction business.

Finally, for many readers, this book may serve a principal function in pointing out sources for additional information.

2

Building Design Issues and Criteria

The design of buildings is a progressive effort in problem solving. Some problems are big and call for extensive study, careful analysis, creative innovation, and sage judgment. But mostly, problems are small, relatively simple, and time-consuming only due to their large number.

Problems relating to building construction influence design decisions at all stages of the design process. The design work at all times must be done in anticipation of the need to achieve the constructed work. Choices of form, dimensions, and spatial arrangement in even the most preliminary design work should be made with some consciousness of the implications in terms of the structural requirements and other issues relating to the building construction. This can be an inhibiting factor for the free, roaming mind of the imaginative designer, but ignoring it for long results in a lot of really silly architecture.

This chapter presents discussions of some of the principal issues in building design that relate most directly and importantly to the final development of the building construction. These issues include concerns that are imposed on the designer—such as building code requirements—and many that are simple, logical concerns for the design of any building.

Logical concerns include those for the functional usage of the building by its occupants, for appearance, for historic continuity, for social significance, and so on. The obtaining of correctly constructed buildings is important, but it is certainly not all that building design is about.

Design criteria generally emerge from a combination of past experience and some reference to knowledgeable and authoritative sources. While a conservative, experience-based continuity is to be encouraged, progress can often be achieved only by defying some entrenched attitudes and long-standing habits. Many of yesterdays rebels and screwballs are todays innovative heroes.

2.1 GENERAL ISSUES

The design of a building must relate to the specific needs of the users of the building. This should be the primary basis for the development of design requirements. However, in any situation, there are also various general sources of influence which may be detached from the specific conditions of the work, but which may be capable of having a major impact on the design. This section deals briefly with some of these general sources.

A major influence may be the personal attitudes, values and design philosophies of the designers. If the finished work is to have any inspirational character, it must come primarily from the creative efforts of the designers. It is to be hoped that the inspired design is also functional, economical, and otherwise generally correct.

Dollar cost is a major factor in most building design work. Buildings represent major investments, whether for a single wage earner or for some large corporation. Control of costs is a major constraint in most design situations, and should never be entirely out of the mind of the designers. Most building designs start out with a "budget" that provides an overall envelope within which the designer must make many individual decisions. A decision might be made to splurge a little here (for really good carpet), but will probably imply the need to save a little there (simplify the roof structure, etc.).

During the design work costs may be dealt with in various ways. Some references to current prices and market conditions may be made at different times, either for comparative evaluation of alternatives or to prepare an estimate of total costs. In a more general way, some overall principles may be applied to the work as a customary means for achieving cost reduction. Examples of the latter are the repetitive use of single elements versus maximum variety, the general simplification of forms, striving for ordered sets in plan arrangements and assemblages, and the use of standard products versus custom-made items.

Real costs, however, are those that the builders must deal with at the time of the construction work. These costs can usually only be educated guesses (called *cost estimates*) when made during the early design stages prior to the actual start of construction.

Time is also important. Design work is almost always done within some time constraints, including those that anticipate the completion of the construction and the occupancy of the building. Owners understandably desire to have their buildings for use as soon as possible, and a leisurely pace for the design work is not likely to soothe the persons who are paying the design fees.

For investment properties, with construction that normally requires considerable time, an extra pressure may be exerted on the design work through use of the fast-track method. This procedure consists of an overlapping of the design and build activities by starting of the construction work while some final design is still being completed. The fast-track method requires that some portions of the design—notably those involving early stages of the construction—be firmly set in the preliminary design stage. Major changes in later stages of design are understandably more difficult in this situation.

Time is money in many ways. Interest on loans, taxes, rents, and regular payrolls go on, whether any progress in design or construction occurs or not. For many dedicated designers, a particular design may never be finished; it is simply a matter of being forced to quit. Tightly scheduled design work implies the necessity for a lot of hard decisions about when to quit.

For the sake of good design, time constraints should be made to bear primarily on an increase of efficiency in those aspects of the work that are less critical to creative design. These aspects include time spent in gathering information, in the making of drawings, in communications, and so on.

Concerns for the safety of building users and for the public in general are the source for most of the criteria in building codes. The two major concerns for safety—fire and structural collapse—have major impacts on the selection of construction materials and systems.

In the past, design for safety was dealt with mostly by simply assuring compliance with the minimum requirements of the legally enforceable building codes and zoning ordinances. As in other situations, however, code minimums are not necessarily related to optimal performance. With the accumulation of better data from research and building performance, and the development of computer-aided design procedures allowing for more intelligent design, the code requirements have receded as design standards. In due time the codes will probably have less "thou shalts" and "thou shalt nots," and will establish criteria relating to specifically regulated performance values.

There is nothing new about a view of building code criteria as being less than optimal. The basic purpose of codes is to protect the public by assuring minimum adequacy. The law cannot force everyone to use the best methods of construction. If a building owner wants more than the bare minimum in any regard, and is willing to pay for it, simple compliance with code minimums is probably not enough.

An emerging concern in recent times is that for hazardous materials in buildings. A growing list of materials is steadily evolving from a status of not desired to one of outright condemned. At any given time there are materials and products that were only recently widely used and are now condemned, and others that are still used but subjected to major challenges. Lead-based paints moved along the route to condemnation long ago. Asbestos and formaldehyde are more recent additions.

Objectionable materials sometimes emerge as popular for certain useful purposes. They may even establish a priority status over alternatives, become widely used, and represent major investments for some companies. Getting them on a condemned list may not be easy, even with considerable evidence of objectionable performance. Building codes sometimes help in this matter, but are usually restricted to acting only after the fact, when a final condemnation has been unequivocally established by others.

Major influences on use of building materials and details of the building construction have been developed from recent concerns for energy use. The passive functions of a building relating to thermal control, use of daylight for interior illumination, and general relations to the sun have become the sources for major criteria in design selections. These concerns fluctuate in relative importance over time, in different regions, and with individual designers or owners, but tend generally to have some pervasive influence on building product development.

Concerns for passive response and energy conservation impact the greatest on design of the building enclosure systems. Choices for elements of the construction for roofs and exterior walls almost always include consideration of these matters.

At any time, in any palce, for any particular building, other issues may be the sources of major criteria for the building design and may exert a major influence on the choice of materials, systems, details, or general methods of construction. Popular building products and systems become so mostly for practical reasons: they do the job acceptably and are economical. When some unusual or special requirements dominate the value system, however, the usual solutions may not be indicated.

2.2 Design Tasks

An early task of the building designer, of course, is the clear definition of the design problem. For construction concerns, the "problem" may emerge largely from data relating to the location of the building site, the dates of commencement and completion of the construction work, stated needs and desires of the owner, and many inherent requirements implied by the specific type, size, and form of the building. The design problem must be stated for the purposes of the designers, but should also be presented in a manner that is understandable by others, including the owner. Common values and goals should be established very early in the design process.

At this stage of design it is important to establish the basic sources for design criteria and data. Many design decisions will be subject to the judgment of the designers, but very few will be made with freedom from imposed constraints.

All applicable building codes and zoning ordinances should be established, together with any special regulations pertinent to the usage (occupancy) or location of the building. Codes are, of course, usually quite minimal in nature; typically developed to protect the public by assuring some barely adequate minimum level of the construction. Work that aspires to have optimal performance or higher quality in general should not be designed to the limits of code-required minimums.

With design requirements reasonably defined, the design work itself begins with preliminary determinations of basic forms, major control dimensions, and general arrangements of fundamental components of the building. This preliminary work sets up the general solution in broad terms and anticipates the more detailed definitions in later stages of the design. In many regards, the preliminary design is the most difficult stage of the design work. Bad judgments at this stage can lead to some unreal situations later on. Unrealistic assumptions regarding capabilities of materials or systems, plan areas or dimensions required for construction elements, or other practical matters may result in unsolvable problems at later stages of the design work when details must be clearly developed. Extensive design experience is most critical to the earliest stages of design.

Once the general design is cast in broad terms by the preliminary design, the work proceeds to a definitive stage. Precision and completeness of details and specific properties and identities of all materials and components must be developed. Choices must be made from ranges of possible alternatives. Some comparative analyses may be required to substantiate design decisions.

Ideally, comparative evaluations should be made with neutrally prepared data. Realistically, choices must largely be made with information obtained from manufacturers and suppliers and with judgments based on the designer's personal experience. Conflicting claims for products and personal bias from good or bad experiences often cloud the issues.

Most building designers are not also builders, and have largely only secondary experiences and some degree of detachment from the actual business of doing the construction. An outside source of aid that may be highly useful at the definitive stage of design is from persons experienced in actual construction work. They may also have personal biases, but the inputs can still be of value.

The basic purpose of the definitive stage of design is to solidify the choices of major elements of the construction and to set critical control dimensions. Once this work is accomplished, the final stage of design proceeds to the completion of the design in documented form—in a combination of the construction drawings and written specifications. This documentation is the basis for the contracts for the construction work, and in a sense it is the final product of the design work.

Although a great amount of detailed work must be carefully done at this stage, it is the design work that is most readily accomplished. This is largely because it is work that is best supported by sources of information. Assuming that the choices made in the earlier stages have not set up any unsolvable situations, the necessary details mostly fall into place. Major usage of existing materials and products will assure this all the more.

When a proposed design involves innovative forms, use of new materials or systems, or extensions of old materials or systems to new uses, all stages of design will be somewhat more difficult. Designers who work on the cutting edge of developing new technologies and design concepts must generally invest considerable greater effort in the design work. This becomes especially critical in the final stages of design, when specific data may be difficult or even impossible to find.

As stated previously, the functional termination of the design work is largely represented by the completion of the construction documentation in the form of drawings and specifications. Construction drawings are basically graphic, but also typically contain extensive data in the form of dimensions and written notes. The specifications are almost entirely written, and as such are not so much the domain of most architectural designers, who tend to be highly trained in graphic work and not often very much oriented to numeric or written work.

The specifications are quite important in the control of the quality of the construction. The drawings may assure proper form and dimensions, but the correctness and quality of materials and products and the general soundness of the construction work is largely controlled by the specifications. Once the progress of the work moves the action out of the hands of the designers and into the field, control will be hard to maintain without a complete and very well written set of specifications.

This book is not about how to write specifications, although it deals with many of the factors relating to them. The emphasis here is on graphic presentation, because it is the best means for learning to visualize the development of the building construction, both in its parts and in their associations in the whole building.

A highly useful graphic element for direct presentation of the form and detail of construction is the cut section. Three types of cut section are used extensively in explaining construction of the building examples in Chapter 10. These are the building plans (horizontal sections, cut at about 3 ft above the floor), the full building section (like the plan, but cut vertically across the building), and the detailed section of some segment of the construction, showing all the parts as clearly as possible.

2.3 DESIGN CRITERIA

As has been discussed in the preceding sections, much of the hard data and many issues for design derive from sources external to the actual building design problem. Documentation of such material is extensive in the forms of building codes, research reports, handbooks, and the standards accepted for general use by various industry and professional groups. For any of the issues previously mentioned, some form of documentation exists, typically containing data and recommendations. A major problem for designers is that of deciding what to use from this mass of documentation, especially when some conflicts develop between different sources or the attempts to satisfy conflicting value systems (code-minimum adequacy versus maximum user satisfaction; inexpensive versus optimal performance, etc.).

Building codes, particularly the code that has legal jurisdiction for the building, cannot be ignored. Industry standards must be acknowledged if particular materials or products are used. However, dealing with many issues requires some selection by designers of a particular source for data and design criteria.

Concerns for energy use were strongly advocated only by a few persons and by

Figure 2.1 Upper: The Monsanto "House of the Future" at Disneyland in Anaheim, California. Molded plastic modules reinforced with glass fiber form the roof, floor, and walls of this experimental house that opened for display in June, 1957. Developed by Monsanto Chemical Company in cooperation with architects Hamilton and Goody of Cambridge, MA, following preliminary design and research at the Massachusetts Institute of Technology. Lower photo shows a cutaway section of the floor structure. Building form and detail clearly express the material and the structure.

some groups with small influence before major events forced concern on the general public. Only after the public response became undeniable was major attention obtained from the building industry and design professionals as a whole. Even now, conflicting positions are held, and some judgment must be made as to what is acceptable or optimal for design criteria or performance evaluation.

With experience any designer eventually acquires a personal set of design criteria that reflects some personal values, some specific good and bad experiences, a personal knowledge base, and individual responses to current trends and emerging issues. This adds up to some personal approach to design that may be very individual or may generally conform to popular trends.

The materials presented in this book undoubtedly reflect some of the author's personal values and attitudes and the general accumulation of experiences as a teacher, writer and design professional. However, it is not the intention here to advocate any particular attitude about building design other than a desire for reasonable competency and responsibility. It is inevitable that any presentation of materials on building construction that is limited by the need for brevity requires selections as to what to include. It is also necessary in showing illustrations of construction to show some particular details, which are typically only one of a wide range of possibilities. What is shown and what is not may well be considered to define some preferences.

In many cases the choices made for illustrations here are quite arbitrary. The main purpose often is simply to illustrate the general form of construction in a complete manner. In some cases the range of alternatives may be discussed. Mostly, however, the reader should accept that building construction is potentially much more diverse (for example, see Figure 2.1), and good design tolerates wider choices, than most of those shown here.

3

Sources for Design Information

Work in building design brings with it a continuing involvement in a search for information. This search is not limited to one for information about building construction, but if the designer is required to carry the work through the stages of producing construction drawings, then seeking answers to questions about details of the construction and about product availability and usage in general will constitute a major effort.

There is a great mass of information about buildings and the materials and products used for their construction. Almost more critical than its volume is the fact that it is constantly growing and changing. Even in the day of the computer, it is a monumental task to try to keep on top of it and to obtain the latest information on any topic. New products are developed; shifting market conditions affect prices and the relative competitiveness of products; codes and design standards are revised; and research results in slow but steady modification of experience-based practices.

The professional designer is expected to keep on top of all of this to some degree, but it is a real challenge. What is generally necessary is for a person to become familiar with the significant sources of information that relate to particular types of questions. While questions about building construction have a potential for infinite variety, they generally break down into a few basic types. In similar fashion, the useful sources of information also group into a few major categories.

Experience brings with it a developed sense of awareness of the typical kinds of questions that one can expect to encounter in various design situations. Accomplishing the work also generally available sources of information. Individual designers will have some particular situation regarding the types of questions they find most difficult, the information sources they have available, and the general means of access they are able to use (local library, computer-accessible data base, etc.).

The essential purpose of this chapter is to describe various types of questions that typically arise in building design work and to discuss the most useful sources of information. Some of the best sources may be difficult for some individuals to access. The specific operation of an effective information-gathering effort depends considerably on the designer's working environment.

3.1 GENERAL SOURCES OF INFORMATION

There are many publishers of information about buildings and many types of publications. Publishers vary from government-sponsored research groups (with reasonably unbiased views) to the manufacturers and distributors of building materials and products (understandably biased in favor of their commercial interests).

For a simplified view, publishers may be grouped into three main categories, as follows:

1. Mostly unbiased groups (*bias* referring to the favoring of choices for commercial purposes), with the general interests of the public or the working designer in mind. These groups include the publishers and administrators of building codes, research groups that are not industry sponsored, and professional groups (AIA, ASCE, CSI, etc.).
2. General publishers, concerned with selling their publications, but usually with no interest in promoting particular views, products, design philosophies, construction processes, and so on.
3. Largely biased groups, including individual manufacturers and the distributing outlets for their materials and products. This category also includes industry-wide organizations, such as the AISC and ACI, although much of their efforts may be in the interests of setting industry-wide standards for quality and safety controls.

Useful, and even necessary, design information is forthcoming from all of these groups. Understandable bias must be recognized and reasonably dealt with. You simply cannot expect the American Institute for Steel Construction (AISC) to provide you with a generally unbiased analysis of the criteria for selecting building materials; they exist primarily for the purpose of promoting the use of steel! Ask them about concrete—and ask the concrete people about wood—and ask the wood people about steel.

What is necessary is to use all available sources for whatever they can provide. For example, the single, most comprehensive source of information about commercially produced materials, products, and equipment—indispensable to building designers—is *Sweets Catalog Files*, consisting primarily of collections of industry-supplied information brochures, largely *advertising* brochures. You really have to use them for design work, but you must keep in mind the nature of the sources.

Publications must be considered from many points of view. First, and certainly quite critical, is their timeliness. For concerns of a general nature (for example, the basic forms of brick masonry), this may be of less importance. However, for the development of the construction details or writing of specifications, it is vital to have up-to-the-minute information regarding product availability and prices, code requirements, construction practices, and so on.

Books from general publishers (such as this one) require considerable time for writing, editing and publishing, and must usually stay on sale for a number of years to pay back the high costs of production and marketing. They can be good sources for general information, but cannot be relied on for up-to-the-minute data (see Figure 3.1).

A second concern for publications is the nature of their contents in terms of topic scope and type of design information. Most publications are generally narrow in both of these regards, typically covering a very limited topic (for example, steel structures). They also frequently have a limited development of the topic (for example, investigative theory and structural design computations, but no construction details, specifications, construction assemblage procedures, etc.).

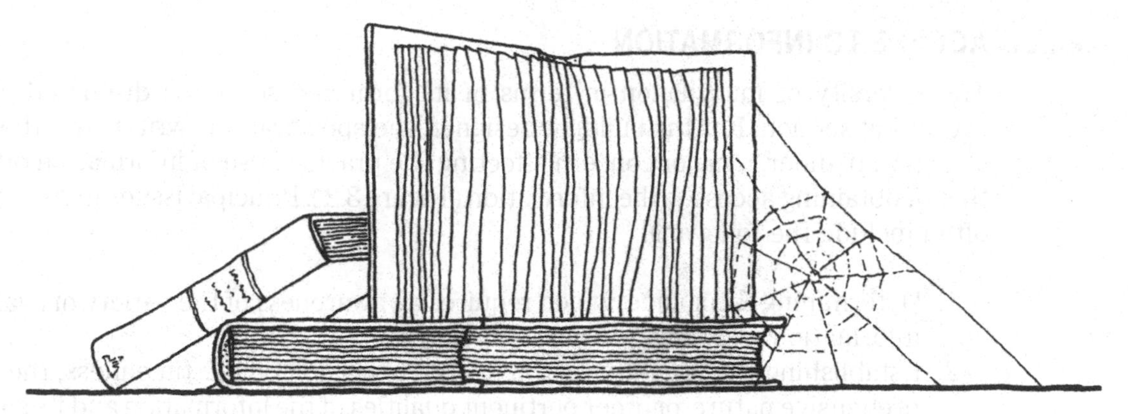

Figure 3.1 Information—printed and published—ages rapidly.

Narrow scope may not necessarily be construed as totally negative. A book that is narrow in the manner indicated in the parenthetic comments in the preceding paragraph is *Steel Structures: Behavior and Design*, by Salmon and Johnson (Harper & Row, 1980), which runs to 1007 pages in the second edition. Narrow as it may be, it is an excellent text for the general theory and design of steel structures. What it is not is a general reference for design usage considerations for steel in buildings; but then it does not claim to be that.

It is a challenge to maintain a complete, up-to-date collection of all the books on any technical subject of significant scope. Large libraries have difficulty doing this, so individual collectors with limited funds may well expect considerable frustration with it. High cost, short shelf life, and the lack of good sources for information about new work (not just advertisements) all add to the problem.

There are, of course, other sources of information besides books. Working professional designers usually subscribe to or have access to one or more periodicals (magazines, newsletters, journals) that deal with their fields of interest. These can provide news about current activities, events, and developments, and sometimes reviews of new publications. Many periodicals have extensive advertisements, which can be leads to information about products and services.

As with books, the sheer number and total cost of all the periodicals in any defined area is overwhelming to individual users. Some are published weekly, and some very important news sources in the construction industry are published *daily*. You could obtain all of these publications, but it would be a full-time job to try to read them.

A reasonable goal is to develop a general awareness of all the significant news sources and some feasible means for accessing those that you feel are vital to your work. A major first step in this regard is to carefully analyze the scope and definable limits of your personal involvements and concerns, using the forms of identity used by publishers and the building industry.

Another approach to selective reduction of the useful sources of information is to formulate a list of the specific questions that routinely put you into an information search. This can yield a personal checklist for evaluation of sources.

With some identification of the useful sources of information established, the problem remains as to how to gain access to them. The general problems of obtaining access to information on building construction are discussed in the next section. A description of various sources for information on building construction is given at the end of this chapter.

3.2 ACCESS TO INFORMATION

The diversity of information in terms of its form and source is discussed in the preceding section. For the building designer, the specifications writer, and the construction planner, a major concern affecting the practical use of information often is that of obtaining access to the information (Figure 3.2). Principal issues in this regard often include the following:

1. Determining the source; which requires an awareness of the variety of available information and the accessible sources.
2. Establishing the lack of bias, the degree of reliability, the timeliness, the comprehensive nature, or other pertinent qualities of the information and its source.
3. Obtaining access to the information at the time it is needed for a task at hand.

The easiest information to access is that which is within arm's reach: in a file or in a book on the shelf, or, in these times, at the beck and call of the computer, requiring only a few strokes of the keyboard. This assumes, of course, that you know the information is there, and only need to do what is necessary to call it up.

At the other end, the hardest information to access is that which you do not know exists. To obtain that kind of information you have to go after it like a researcher, which first requires a very clear definition of your purposes and needs. After that, it becomes a test of your abilities to snoop and pry.

Special difficulties are involved when information is available only at a price. The price may be in dollars, in some commitment, or in a permanent place on a junk mail list. These matters may not affect the quality or value of the information, but often affect the practicality of accessing it.

Both the methods used to access information and the particular forms of information that are readily accessible depend on the working conditions of the individual information user. For persons who work in an organization (office, school, government department, etc.), the facilities and operation conditions in the organization will establish the system for information handling. If the organization has a library or a central computer data base, these become primary components for information use.

Figure 3.2 Access to information is not always easy to obtain.

If a person works with many others who perform similar tasks, information is likely to be shared on some cooperative basis. If an individual within an organization serves an essentially unique role in the organization, sharing may be with persons in other organizations. In any event, whatever function one serves in the general processes of designing and making buildings, it is unlikely to be so unique that it is not performed by many others. There is thus considerable potential for sharing, whether on an interoffice basis or in a worldwide network.

If you don't have access to a good technical library, cannot afford to subscribe to data-sharing networks, and work on your own (alone or as the only person doing what you do in a group), you may very likely have a major problem with accessing the information. If you require many kinds of information, you probably cannot establish your own unshared inventory of materials because of the cost and attention required to amass it and maintain it. This is not necessarily a hopeless situation, but it most likely is one that requires careful consideration.

Every building designer has some specific situation regarding the working conditions that affect needs for and use of information. With experience, needs for information can to some extent be anticipated with sufficient lead time so as not to impede the process and schedule of the design work. This is not always easy, however, even when design work is of a repetitive nature.

In any design work, some situations arise where some new piece of information is suddenly needed on short notice. Still the general work of design is largely predictable (not necessarily the outcome, but the problems and the process) and very many questions can be anticipated in general form.

With experience, therefore, comes some developed procedure or design methodology that incorporates typical questions that designers may expect to have to answer. Knowing these questions ahead of time, the practical designer will organize some procedure for identifying major needs for information and some strategy for accessing critical information well ahead of the time it is needed for the design work.

Basic reasons for using information in the building design process are discussed in the next section. A general discussion of information sources is given in Sec. 3.1.

3.3 USE OF INFORMATION FOR DESIGN

Input of information is required at all stages of design (see Figure 3.3). Some considerable information must be assembled before even the most preliminary design work can start. Much of the information needed at this stage, however, comes from design program requirements as generated by analyses of needs or desires of the owners, investors, or users. Data may be required to quantify some of the requirements as they progress from simple general goals to specific detailed objectives.

Some typical problems that require support of information are imaged by the following questions:

1. What materials or types of construction may be considered for a particular building?
   Information sources: Building codes; local usages and availability; usage appropriate to the building type (traditionally as well as in recent work).
2. What is a reasonable expectation for the construction cost?

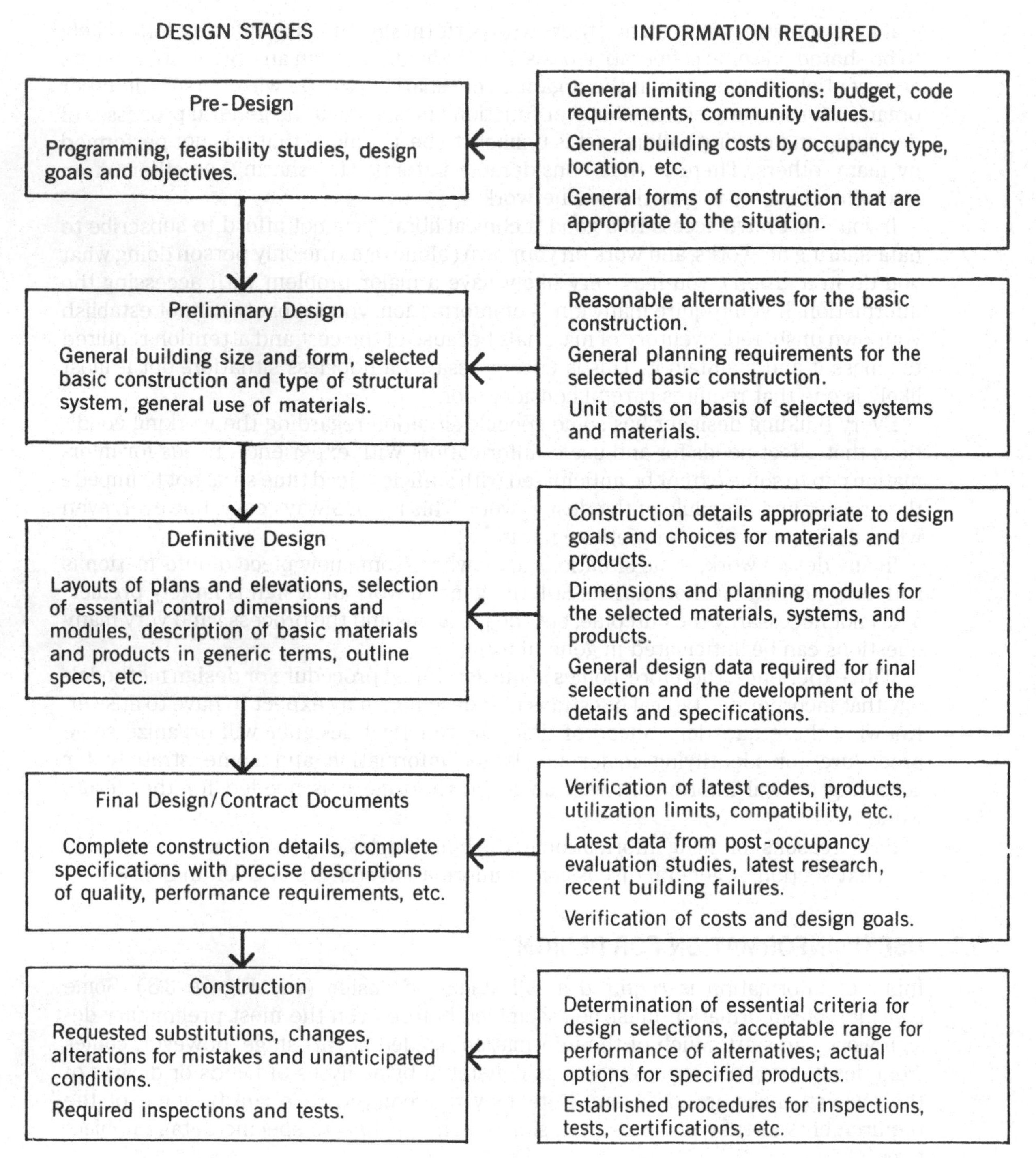

DESIGN STAGES
INFORMATION REQUIRED
Pre-Design
Programming, feasibility studies, design goals and objectives.
General limiting conditions: budget, code requirements, community values.
General building costs by occupancy type, location, etc.
General forms of construction that are appropriate to the situation.
Preliminary Design
General building size and form, selected basic construction and type of structural system, general use of materials.
Reasonable alternatives for the basic construction.
General planning requirements for the selected basic construction.
Unit costs on basis of selected systems and materials.
Definitive Design
Layouts of plans and elevations, selection of essential control dimensions and modules, description of basic materials and products in generic terms, outline specs, etc.
Construction details appropriate to design goals and choices for materials and products.
Dimensions and planning modules for the selected materials, systems, and products.
General design data required for final selection and the development of the details and specifications.
Final Design/Contract Documents
Complete construction details, complete specifications with precise descriptions of quality, performance requirements, etc.
Verification of latest codes, products, utilization limits, compatibility, etc.
Latest data from post-occupancy evaluation studies, latest research, recent building failures.
Verification of costs and design goals.
Construction
Requested substitutions, changes, alterations for mistakes and unanticipated conditions.
Required inspections and tests.
Determination of essential criteria for design selections, acceptable range for performance of alternatives; actual options for specified products.
Established procedures for inspections, tests, certifications, etc.

Figure 3.3 Use of information about building construction in the design process.

Sources: Recent local construction; national cost reports, with regional adjustments.

3. What building forms are possible, in view of the spatial requirements, anticipated building size, and most likely types of structure and general construction? *Sources:* Published recent examples (from books and periodicals); recent local construction (with some post occupancy evaluation); industry data for various products and systems (especially any priority items, such as the steel open web

joist and joist girder system or a curtain wall system from a particular manufacturer).

4. What modular planning dimensions derive from choices of construction materials and systems?
   Sources: Data from industry, in terms of standard production units, range of size variations, usage limits (spans of structural decks, etc.); general data summaries and system comparisons in books.

In general, the type of data needed for preliminary design stages tends to be of a broad, comprehensive basis. General references, such as *Architectural Graphic Standards*, tend to be useful at this stage. Anything that presents general summaries or comparative evaluations to support the scanning of alternatives will help to simplify the task of considering all the options.

Some specialized books, dealing with a specific construction (steel, space frames, high rise, etc.) or a specific building type (school, residence, office, etc.) may also be useful at preliminary stages of design. Books of general usage or ones that deal with materials related to a common design problem may be worth owning for this purpose, unless access to a library is possible.

As design work proceeds through developmental and final stages, required information becomes increasingly specific and detailed and needs generally to be more reliable and more up-to-date. For final design some major typical questions are the following:

1. What specific material or product is to be used for a given item of the construction (roofing, door, ceiling, etc.)?
   Sources: For most up-to-date information, industry-supplied data (from *Sweets Catalog Files* or directly from manufacturers, suppliers, or industry organizations); for comparative evaluations of competing products, various books with basically unbiased views on product use.
2. What specific data must be recorded (on drawings or in the specifications) to completely identify a chosen material or product for design purposes?
   Sources: Master reference specifications; industry-recommended specifications (once you have decided to use their products); construction drawings and specifications for recently built construction (updated, if necessary, for code changes, etc.).
3. What are appropriate details for the construction of various elements of the building?
   Sources: Any that help—previous work, industry recommendations for proper use of their products, books with construction details. (Beware here of regional differences, timeliness, product obsolescence, and so on.)

If the building designers are really concerned about costs, technical feasibility, appropriateness and adequacy, and the code acceptability of the final design work regarding construction materials and details, these inquiries and their supporting data must be pursued with some vigor. Designers must establish some organized working process and set specific goals and objectives that reflect their personal design philosophies and values. If the designers leave these matters for others to worry about, the design will usually suffer a lot of distortion between conception and constructed reality.

3.4 SOURCE ANALYSIS

The following is a brief analysis of the various sources for information about building construction. Special attention is given to sources that are specifically useful for design data and the development of construction planning and detailing.

3.4.1 Construction Documents

Construction is mainly done with materials, methods, and details that have been tested in use. Construction drawings for completed buildings are a major reference.

A major concern in using previous work as a reference is its timeliness. Materials, products, construction methods, codes, and general usage change with time. Changes may be minor or may consist of the total elimination of certain items (asbestos fiber products, lead-based paints, steel rivets, etc.). Anything that is borrowed from previous work must be tested or reviewed for timeliness.

If at all possible, the building itself should be inspected. Two concerns are for the actual compliance with the drawings and specifications (no major changes made during the construction) and the effects of time, weather, and usage. The latter is especially critical for finish materials.

While construction drawings provide the most directly useful materials for designers, it is advisable to review the specificatons for various points of information not shown on the drawings. For example, the forms of details may be clearly shown for a building with exposed concrete construction, but the ingredients and proportions of the concrete mix, the type of forming materials, any special curing or refinishing of the cast concrete, and other data critical to the quality of the finished work are most likely not on the drawings.

3.4.2 Standard References

Certain references carry a degree of authority that makes them qualify as widely used standards. This may be true because of some sanction by a real authority, such as a building code, an industry-wide organization (AISC, ACI, etc.), a professional group (AIA, ASCE, etc.), or a standards-making body (ASTM, ANSI, etc.). Or, a reference may simply be so widely accepted and used that it is recognized as a major source of data in a given field.

A few references approach the status of design standard as a result of their wide use. Examples are *Architectural Graphic Standards* and the multiple volume *Time-Saver Standards.* However, this status is more often reserved for items that do not come from general publishers. Model bulding codes, such as the *Uniform Building Code* and *BOAC Code,* are in this category, together with publications of the ASTM, ANSI, and other relatively independent organizations. Some industry-sponsored publications also earn some stature from their general acceptance, examples being the *AISC Manual and Specification* and the *ACI Code.*

Designers generally need some access on a continuing basis to those standard references that deal with their primary areas of concern. A minimal reference library for building designers is discussed in the next section.

One approach in identifying important references is to see what the books or writers that you respect use as references. This can be an inbred group, but repeated citing of the same references by many writers should indicate some respect for the sources.

3.4.3 Special References

Every major type of building material or product has one or more organizations to represent its interests, as do all the major divisions of design professionals, including journals, newsletters, manuals, handbooks, recommended standards, and so on. A collection of all of these, spanning the whole range of the building design and construction businesses, would be overwhelming and highly redundant. Still, a few are truly indispensable, and individual designers must establish their own critical list.

3.4.4 Detail Manuals

Of special interest for the materials developed in this book are publications that provide details for building construction. Detail reference books can be quite helpful to the experienced designer; the usual cautions should be recognized with regard to timeliness and regional acceptability. Details can be frozen in form for only the most ordinary forms of construction and for materials and products of enduring usage. As with any references, local usage and codes, emergence of new materials and products, effects of recent research or use-tested results, and other effects that change what we do over time must be acknowledged.

3.4.5 Case Studies

Documentation of noteworthy buildings in books or periodicals sometimes include information about their construction. These may be of some use, but presentations are usually brief and quite short in terms of comprehensive development. Treatment in depth in a book may be of some value, but few examples exist. Presentations more often feature photographs and presentation-type drawings, rather than messy details of the building construction—no matter how informative the latter may be.

Construction details are more likely to be given in books or articles that deal with some construction issue (curtain walls, precast concrete, etc.). However, narratives of design or construction problems can also be informative. The latter are more apt to be featured in engineering books or periodicals. You can look at the pretty pictures in *Progressive Architecture* and then see if there are any stories in *ENR* magazine or other construction-oriented periodicals that tell about what didn't go well during construction.

3.4.6 Periodicals

These abound in great number, if all branches and groups in the business of design or construction are included. Add the areas of manufacturing, marketing, investing, financing, insuring, inspecting, and maintenance, and the flow of information is enormous. One should be quite judicious and preferably pick a few notable ones to subscribe to and then really read them.

Magazines intended for wide readership, like *Progressive Architecture* and *ENR*, have considerable advertising, which can be a source of much information about construction materials and products. Ads may also be a means of access to additional information.

One notable advantage of periodicals is the relative timeliness of the materials presented. Monthly magazines are usually prepared only a few months in advance, although news stories may be more up-to-the-minute. And, of course, publications that are weekly or daily are even more timely in this regard.

3.4.7 Catalogs and Brochures

Individual manufacturers and suppliers and many industry-wide organizations spend vast amounts on advertising in order to attract the attention of buyers. For the building market, the real buyers may often be the purchasing agents of building owners or contractors, but a major access to sales is also through building designers and specifications writers.

Slick, handsome, expensively printed brochures, files, notebooks, and even multiple-volume sets of materials are produced and distributed (mostly free to the user) by the sellers of products and equipment. Major groupings of these advertising materials constitute the various sets of *Sweets Catalog Files,* which are reissued annually. Qualifying for a set of *Sweets Files* requires some evidence of being in business that involves a considerable volume of construction.

As discussed in Chapter 1, the Masterformat system tends to serve the purposes and tasks of the specifications writers and purchasing agents somewhat more fully than it does the work of the designers. It is nevertheless essential for all users of information in the building field to become familiar with the system. Both *Sweets Catalog Files* and *Architectural Graphic Standards* —two primary and indispensable sources for design information—are organized with this system.

3.4.8 Access

Obtaining information often involves overcoming some obstacles. Cost, time, and effort may be major concerns. Books are expensive, not so easy to find out about, and even less easy to be sure about their real worth. Maintaining a personal, up-to-date library is in general expensive and takes considerable time and effort.

Some materials are obtainable directly only by persons with some qualification (professional status, volume of business, etc.). Trade magazines and journals typically have restrictions on who is allowed to subscribe, usually in the interest of protecting some special status of the publishers or in assuring advertisers as to the nature of the subscriber.

Industry-supplied materials are often provided without cost, but usually only to persons with some qualifications. Obviously, materials intended primarily for promoting sales are prepared and distributed with the intent of placing them in the hands of people who affect sales (mostly purchasing agents, specification writers, and professional designers).

There are very few well-maintained technical libraries that are accessible by the public. Those developed for professional schools of architecture and civil engineering represent the largest group. If you are near such a school, you should find out what your opportunities are for using the library facilities. A major feature of these libraries is a typically quite extensive periodical section, where both recent issues and bound previous issues are accessible.

Whether you work alone or in a large organization; whether you are in or near a large urban center, or out in the boonies; whether you are broke or rolling in cash; whether you are broadly engaged in building design work, or highly concentrated in some specialty; whether you are experienced or just beginning; whether you have very limited responsibilities, or are the single, designated person doing what you do;

whether you are a specifier, or are a buyer and user—all relates to what specific type of information is important and what form of access is available and useful.

In most cases it is harder to find out that particular information exists and is really useful to your purposes than it is to get access to it. If you really need it for design work, and your design choices may affect somebody's purchases, industry-supplied information is readily available. If you may actually want to purchase a book, publishers and booksellers want desperately to know about you. Don't be afraid to ask them what they have that might help you.

3.5 THE MINIMAL CONSTRUCTION DESIGN LIBRARY

Every designer eventually acquires a personal reference library, partly consisting of materials that are actually used with some frequency. The need for such a library depends greatly on the working context of the designer in terms of type of work and where the work is done.

If needed materials are readily available from a school or office library, there may be no real need for maintaining a personal set of materials. If such is not the case, some plan should be established for acquiring a minimal library of materials.

The following list includes materials that are essential for a good starting library on building construction, oriented to the interests of the general building designer. The whole collection will cost about $500 (allow for inflation). Some materials are not for sale, so access must be established by other means.

New publications come forth as time goes by, of course. Also, price change—mostly steadily upward. Technical books often sell in quite small quantities, become rapidly obsolete, and are relatively costly to produce, so prices are usually high, except for a very few highly popular titles. Because there are several commercial publishing companies, alternatives often exist for a specific type of book, and it is hard to identify the best choice for everybody. Bear in mind, no two persons would compile the same list for this purpose, although there would certainly be many similarities.

1. *Sweets Catalog Files*

There are several Sweets files, but the one most generally useful for the building designer is subtitled "Products for General Building and Renovation." The files are basically not available for purchase, so you are out of luck unless you work in a medium to large office or have access to a library copy. Most of the individual entries in the files are available directly from the groups that supply them to Sweets, so you can probably obtain the ones you actually need directly from the source. To help with this, you can buy the first volume of the aforementioned set from Sweets; this contains lists of the contents of the file and names of suppliers, as well as other useful information to help with selections of materials and products. Files are also available in electronic form. Write to Sweets Division, McGraw-Hill Information Systems Company, 1221 Avenue of the Americas, New York, NY 10020.

2. *Architectural Graphic Standards*

This has been a must for every building designer for more than 50 years. The latest edition (eighth) costs about $160, but an abridged edition of the seventh edition is available in a Student Edition for about $60, and has much of the pertinent, relatively unchanging material; worth the price for a personal copy, if the latest, unabridged version is accessible somewhere. This book has grown in size over the years to over 800 pages in the 8th edition. Much of the material consists of data and planning criteria—of use for a full service design office. However, if your interest is really limited to details for building construction, an abridgement (shortened version) of the 8th edition titled *Construction Details for Buildings* generally contains only that; at about half the cost of the whole book.

3. *Building Code*

You should have the latest edition of one of the model codes (UBC, BOCA, or SBC), which cost around $60. However, if you do most of your work in one region, you should also have copies of the local city, county and state codes. There also may be special codes for housing, schools, energy use, access for the handicapped, and so on. Get only what you plan to use a lot and keep up-to-date. Some code publishers have update services to which you can subscribe. For further discussion of codes see Chapter 4.

4. *Basic Building Construction Text*

It is essential to have some unbiased, comprehensive reference on building construction materials and processes. Best are those which are prepared for use as texts, such as the following:

Fundamentals of Building Construction: Materials and Methods, 2nd edition, by Edward Allen, Wiley, 1990, $60.

Construction Materials and Processes, 3rd edition, by Don Watson, McGraw-Hill, 1986, $50.

Building Construction: Materials and Types of Construction, 6th edition, by D. Ellison, R. Michkadeit, and W. Huntington, Wiley, 1987, $35 (paper).

5. *Basic Dictionary of Construction*

The areas of architecture, building construction, and various technical fields related to buildings have a considerable mass of terminology that is frequently used in publications and for communicaton between elements of the industry. It is quite important to acquire some familiarity with frequently used terms and to have some reference for finding definitions of unfamiliar ones. Accuracy of usage is essential for communication and to avoid embarrassment. Some books have glossaries, but for the inexperienced it is good to have at least one really complete source. Some current general references are the following:

Architecture and Building Trades Dictionary, 3rd edition by R. Putnam and G. Carlson, Van Nostrand Reinhold, 1983, $20.

Dictionary of Architecture and Construction, by Cyril Harris, McGraw-Hill, 1988, $20 (paper).

Construction Glossary: AN Encyclopedia Reference and Manual, by J. Stein, Wiley, 1986, $40 (paper).

6. *Basic Structures Reference*

It is hard to find a single book that treats the subject of building structures comprehensively, including aspects of construction planning and detailing. One that does so reasonably and is generally readable by persons with little engineering background is:

Building Structures, by James Ambrose, Wiley, 1988, $70.

For the slightly more serious structural designer the following two books deal extensively with structural engineering, although somewhat less with architectural issues.

Building Structural Design Handbook, edited by C. Salmon, Wiley, 1987, $85.

Structural Engineering Handbook, 3rd edition, edited by E. Gaylord and C. Gaylord, McGraw-Hill, 1990, $70.

7. *Basic book on General Building Technology*

A single volume that covers most of the important topics of building technology, other than structures and basic construction is:

Mechanical and Electrical Equipment for Buildings, 7th edition by B. Stein, J. Reynolds, and W. McGuiness, Wiley, 1986, $60.

A book that covers building technology generally, including some treatment of structures and architectural planning, is:

Building Design and Construction Handbook, 4th edition, by F. Merritt, McGraw-Hill, 1982, $65.

A book that also deals quite broadly with architectural technology with a special emphasis on use of systems design (integrated design; optimization, value analysis, etc.) is:

Building Engineering and Systems Design, 2nd edition, by F. Merritt and J. Ambrose, Van Nostrand Reinhold, 1989, $55.

8. *References for Construction Details*

This is an area that is hard to fill up, as good references are in short supply. Be aware that detailing practices are in many cases quite personal, regional in character, and subject to rapid change as codes are revised and the technology advances in general. The following are some current useful references.

Construction Details for Commercial Buildings, by Glenn Wiggins, Whitney, 1988, $30 (paper).

The Professional Handbook of Architectural Detailing, by O. Wakita and R. Linde, Wiley, 1987, $55.

Wall Systems: Analysis by Detail, by Herman Sands, McGraw-Hill, 1986, $40 (for curtain walls).

Of course, *Architectural Graphic Standards* was originally developed primarily as a detailing manual and still contains a large collection of construction details.

3.6 FINDING INFORMATION SOURCES

Finding information is an individual task that relates to the working context and real needs of the seeker. Serious researchers and scholars can use library access facilities of considerable sophistication to pursue obscure bits of information. Working building designers in general operate at a much less sophisticated level in pursuing the information critical to design work. However, the age of the computer is steadily producing systems for sharing of information that are generally working out to an expanding audience of users. *Sweets Files* are now computer accessible, although the bound volumes are still the main vehicle for information sharing.

It is hard to be certain whether there are more people looking for information or more seeking to give it out. In any event, there is a mountain of available information on just about anything, if you want to make a serious effort to go after it. There are also a lot of people in the business of assisting both those who are looking for information and those who are trying to give it out. Publishers, booksellers, compilers of bibliographies, librarians, and operators of computer data banks are all working to bring the seekers and the givers of information together. Let them help you.

4

Building Codes and Industry Standards

What it is that constitutes "common practice" in building design and construction at any given time is documented in various published codes and standards. These publications may carry legal status, as in the case of enforceable building codes, or may simply prevail because of their recognition and general acceptance. This chapter briefly discusses the issue of reference standards.

4.1 GENERAL SOURCES OF STANDARDS

Documentation of what may or may not—should or should not—be done with regard to construction of buildings is forthcoming from many sources. A first issue in evaluating the importance of any standard is the nature of its source. Two primary concerns in this regard are the actual binding nature of the standard (as legally enforceable, for example, as in the case of building codes) and the reliability of the source with regard to its authoritative competence and its general neutrality or isolation from special interest.

Standards are widely used and vital to the practical accomplishment of design work, but designers should be very clear about the character of the standards as influenced by their originators and publishers. Many standards used for design work are developed by special groups with vested interests in the particular issues with which the standards deal. The highly respected and widely used ACI Code (*Building Code Requirements for Reinforced Concrete*) is accepted by most building codes and forms the basis for the most design work for concrete structures. Nevertheless, it is published by the American Concrete Institute, which exists frankly primarily as a promoter of the interests of the concrete industry; meaning mostly manufacturers of cement and accessory products for concrete construction and contractors who specialize in concrete work.

Building codes, which are generally subject to a lot of review in the public interest, tend to promote standards which are on the edge of marginal acceptability, by defining the least acceptable quality of construction. This least (minimal, barely adequate) level of quality quickly becomes the upper, rather than lower, limit to which design work is aimed. Satisfying the code requirements for construction that is resistive to fire or earthquakes does not result in optimally fire or earthquake-resistive buildings. Pushing the limits for span length of floor joists or roof rafters virtually guarantees bouncy floors and sagging roof surfaces.

Designers must carefully assess the real nature of any standards that are used for quantified design data. The nature of any standards (marginally adequate versus, conservatively excessive) should be understood in relation to the particular design goals.

4.2 USE OF STANDARDS FOR DESIGN

Buildings are complex and their design involves dealing with a virtual mountain of minute detail. "Standard" details and specifications are used for much of the final design documentation. "Custom" design consists more often of the customized packaging or arranging of standard parts. Windows, doors, plumbing fixtures, roofing, flooring, and ceilings are selected from available product lines. Use of standards in this situation occurs mostly before building designers are involved. Designers' inputs may be limited to ascertaining what standards (if any) were used and to exercising some judgment about the appropriate standards.

The single major standard for most building design work is the prevailing building code, which tends to deal more specifically with matters for which decisions are influenced by designers. These may relate to building form and size, to selection of basic types of construction, and to many fine points regarding planning and detailing of the construction. Selections and details for the basic building components are also tightly regulated. Add the matter of the code's legal status, and its prestige becomes considerably enhanced.

In the end, hardly any single designer can expect to fully understand and evaluate all the standards that affect the design work. Only considerable design experience can show which design guides or standards are reliable and useful, and deserve some critical attention to assure the quality of the design work.

4.3 CODES

It is hardly possible to build *anything* in the United States without being subject to some regulatory building ordinance. Enforceable building codes are mostly enacted ordinances by some political entity (city, county or state). However, they may also be adopted and implemented by some agency, such as a Board of Health, a Coastal Commission, a State Board of Education, or a state or federal housing authority.

In any event, code criteria are seldom developed entirely by the publishers of the legally-adopted code. Most of what codes consist of come from a vast collection of government and industry standards and a basic form developed as a model code. Model codes are produced by groups with collective interest in the regulation of building construction and use.

By deliberate study or by steady friction, building designers eventually become aware of the many ways in which codes affect their work. Codes are helpful in some ways, but are largely viewed as antagonists by designers. Pushing against or artfully dodging the code limits is a continuing game with many designers. In a more constructive view, the ability of designers to misuse codes feeds back to the development of more clear code directives and data.

Of specific concern for any individual design project is the code that directly affects the work. This generally means the code for the site where the building is to be built. The specific implementation of this code of jurisdiction is in the form of the permits and inspections that are required to assure acceptability of the construction. This process is described in the next section.

For particular regions of the United States, one of the model building codes (UBC, BOCA, or SSBC) generally forms the basis for most local codes. For smaller municipalities or counties, the model code may be adopted virtually intact. For more populous areas, there is usually a specially-written code; but its basic content is typically still largely the same as the general model.

In fact, *all* building codes have more in common than they have differences. They mostly all deal with the same basic issues and use the same authoritative sources for standards. Learning any single code in detail provides a basis for a quick pickup of familiarity with a different code. Nevertheless, doing design work in many locations means learning a lot of different codes with their individual pecularities.

As legal ordinances, codes are subject to ongoing revision by acts of their publishers. The model codes are also frequently revised. In the end, any building is subject to the code as it exists at the time of issuing of the permit for the building construction. For large projects, this can mean some considerable time after the beginning of design work, and some code changes are thus inevitable after initial design decisions are made. There are no general rules for maintaining the flexibility to deal with this situation, but the possibility should be anticipated and code conformance should be verified at stages in large projects.

4.4 BUILDING PERMITS AND INSPECTIONS

Enforcement of building codes takes the form of permits and inspections by agencies empowered by the publishers of the codes. A permit to build is initially issued upon the satisfaction by the agency that the work as designed and specified is in conformance with the code, is competently described in the contract documents, that the designated builders are properly licensed and bonded, and that a legal contract for the work exists. For the building designers, the review for issuance of a permit is equivalent to a final exam or a final juried crit on their work; a traumatic event for the inexperienced and an ongoing nuisance for old timers.

While getting a permit in the first place is a major task, keeping it is subject to review as well. Periodic inspection of the work is usualy done to assure conformance with the design as initially presented. In the end, a final inspection will precede a final permit for occupancy of the building; a privilege, not a foregone conclusion for the owner!

Getting a permit involves mostly demonstrating that the design directives (drawings and specifications) are complete and in conformance with code requirements and acceptable professional design standards. It also usually involves submission of some supporting materials, such as computations for structural designs and energy consumption, an environmental impact study, a geological analysis of the site, and other items appropriate to the project and its location.

The issue there is that buildings—especially large ones—are of interest to the community at large, as well as to their owners and users. How they are built and used is of some concern to the neighbors and to the whole community. The laws exist to force owners, builders, and the designers they employ to recognize those concerns, whether they wish to or not.

4.5 STANDARD SPECIFICATIONS

This book deals with construction mostly in terms of the physical form and detail of the elements of the buildings. However, the precise nature of the materials and components of the construction are also strongly controlled by the design specifications (called the *specs* for short). While the details—essentially presented graphically—exercise major control on form and appearance, the written specifications largely control the *quality* of the construction. The degree of this control is not always appreciated by designers, who tend more often to be graphically and visually-oriented, and frequently have little to do with the actual preparation of specifications.

Just as the drawings for a particular building must be somewhat uniquely developed, so must the specifications. No two buildings are *exactly* the same, and the package of drawings and specs is tailored to various circumstances, as well as design differences. Nevertheless, both the drawings and specs typically consist mostly of standard items; lifted from standard references and sometimes slightly modified for the situations at hand. The unique package for a single building thus consists mostly of the specific collection and arrangement of the standard parts; maybe with some light trimming.

Most large design organizations have their own collection of "standard" details and specifications. Increasingly, however, use is made of major sources for standard specs, developed within a system that is coordinated with other aspects of the building design and construction processes. A major system of this kind is that developed by the CSI (Construction Specifications Institute), an organization that most working specification writers belong to. However, other organizations and individuals also publish packages of standard specifications.

This book is not about how to write specifications—a major effort, now recognized as a principal sub-specialty and profession in its own right. However, designers who do not develop a serious awareness of specifications and their influences on design work are ripe for shocks with regard to their inability to control the finished quality of their work.

Sources For Further Study

Architect's Handbook of Construction Detailing, by David Kent Ballast, Prentice Hall, 1990. An excellent source for standard construction details with accompanying data keyed to various industry standards and to the CSI *Masterformat* indexing system.

The Architect's Studio Companion: Technical Guidelines for Preliminary Design, by Edward Allen and Joseph Iano, Wiley, 1989. Contains an excellent digest of building code issues and data affecting architectural planning.

5

Architectural Components

Constructing an object that sits on the ground, houses some form of human activity, gets rained on, is in place through many seasons and years, and must be built by persons engaged in the construction business using presently available materials and products produces some very common results over time. Buildings have typical parts, and building design involves dealing with basic factors relating to the functions of those parts and to the problems of making them.

This chapter presents material relating to the most ordinary and generally indispensable parts that occur in the majority of buildings and with some of the design concerns that arise in producing them. While designers are necessarily concerned with all of the building's parts, the concentration here is on those parts most directly influential in achieving the construction of the building, which is the central topic of this book.

Design of any building part must be done in the context of the problems of the whole building. Dealing with the individual elements of building construction requires some attention to their eventual coordination in the whole constructed building. Putting the parts together is the theme of the work in Chapter 10 of this book, where the whole construction of several buildings is presented.

The objective in this chapter is to deal generally with the various concerns for the basic building component parts. For real situations, consideration must be given to the means for generating the building parts in terms of available materials and products; these issues are discussed in Chapters 6 through 8.

5.1 INDIVIDUAL COMPONENTS

For the purpose of discussion, we will consider a two-level set of basic architectural components. On the first level are those major elements that principally determine the structure and basic character of the construction of the building. As shown in Figure 5.1., these major elements are the roof, floors, exterior walls, interior walls, foundation elements, and elements for vertical circulation (stairs, elevators, etc.).

On the second level are elements which, although necessary, have generally less influence on determination of the basic form of the structure or the choice of the principal materials and methods of construction. The second level includes windows, doors, ceilings, and various items involved in the building service systems (for power, lighting, HVAC, etc.). The second level may also contain site features outside the building.

In any given situation one or more of the second-level elements may take a dominant role in the set of values for design. For example, in a high-rise office

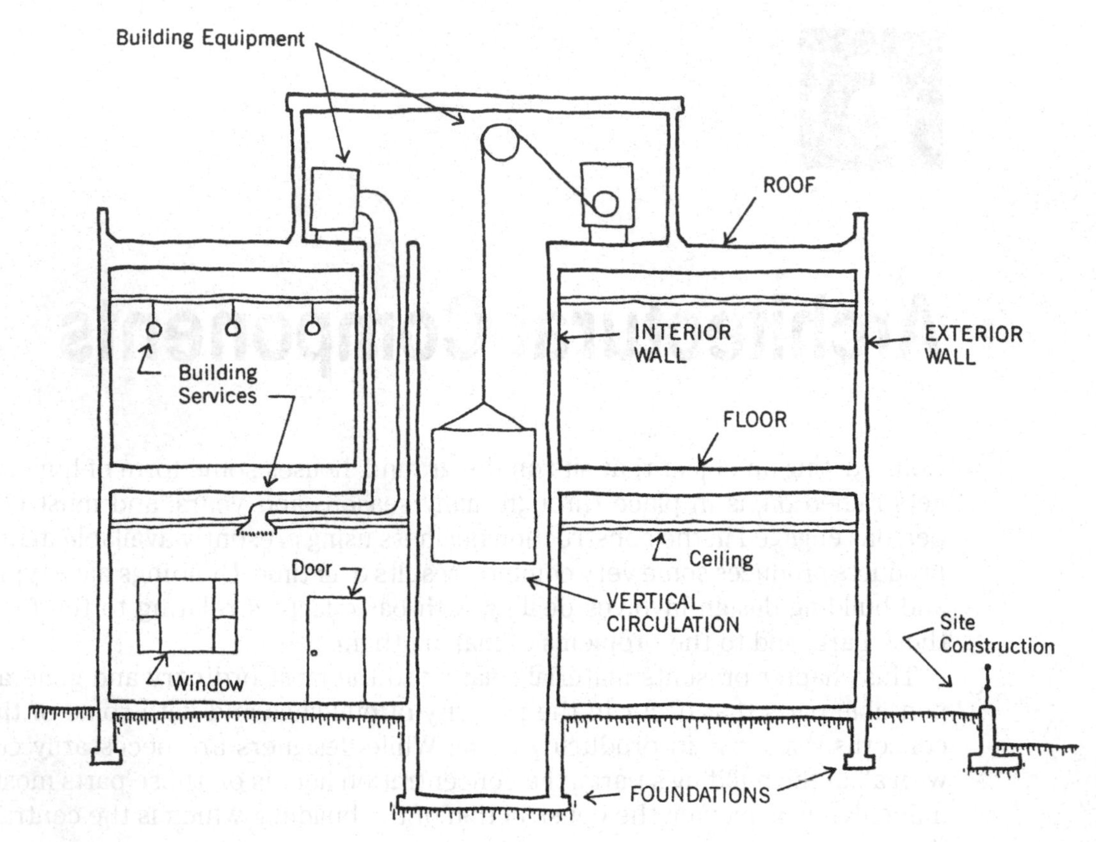

Figure 5.1 Architectural components with construction identity.

building, windows may well be of major design concern, relating to choice of structure, planning modules, and general development of the exterior walls.

A primary set of elements is the group that forms the enclosing shell of the building; the entity that creates the basic shelter function by separating indoors from outdoors. This is of major concern to the building occupants for obvious reasons. However, it is also of major concern to the architect and to the community because of its high visibility. A major portion of the effort in architectural design is devoted to studies and development of details for this enclosing shell.

The following sections consider individual elements of the building construction, beginning with the major components of the enclosing shell.

5.2 EXTERIOR WALLS

For most buildings the exterior walls are the most prominent architectural features of the building construction. Most of our experience with the viewing of buildings consists of seeing them from the outside, as pedestrians or as passengers in vehicles. Most of us have seen hundreds or thousands of buildings from the outside for every one we have seen from the inside. Thus buildings have their greatest impression on people as objects viewed from the outside.

Design of the exterior walls is a major concern for architects. Quite often, major elements of the building structure occur at the building perimeter and must be integrated with other features of the construction. Exterior surfacing materials must be carefully selected for both appearance and durability in the exposed situation.

The building's exterior walls may be visualized as constituting a controlled barrier between the building's interior environment and the outdoors. The exterior wall is a major element in the control of the building's internal conditions, functioning as both a barrier and a selective filter, as shown in Figure 5.2. These functions extend to the whole building enclosure, but have their major impact on the design of the exterior walls.

Many of the specific design criteria for physical performance and details for exterior walls derive from the needs relating to defined filter functions. For any given design situation, a list of specific desired filter functions can be developed, and criteria can be stated in both general terms and some actual performance data. Items such as the required amount of thermal insulation will derive from such criteria.

Basic concerns that must be addressed in the design of exterior walls include the following:

1. Exterior finish: The exterior surface finish may be that of the raw, untreated structure, or materials may be conceived of as primarily decorative, but they must also be weather resistant and often afford protection for the materials they cover.
2. Structure: Walls must be adequate for their own support in any case, but must often also serve other structural functions, such as support of roofs and floors or bracing for wind or seismic effects on the building (as shear walls). Even when walls need only support themselves, other elements of the building structure must often be integrated into the construction of the building exterior—most notably, columns and spandrel beams.
3. Interior finish: Exterior walls have inside surfaces that are usually part of some occupied space. For complex buildings, the same general exterior wall, possibly with a uniform exterior surface treatment, may need to provide for many different interior conditions.
4. Fenestration: This is the term to describe the general development of the building windows. Windows are major architectural design features: however, they ordinarily have specific purposes that relate to interior spaces. They must be made to integrate into the general exterior wall construction, serve the

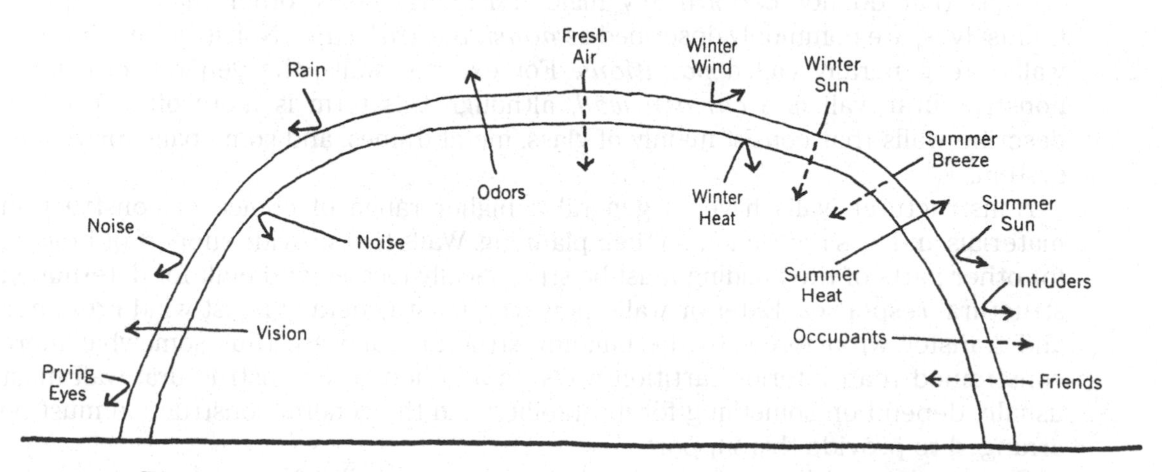

Figure 5.2 Visualization of the building enclosure as a barrier with selective filter functions.

required interior functions for light, vision, and ventilation, and still look great on the building exterior. They must also relate to the barrier and filter functions of the walls of which they are parts and not detract from the walls' required structural functions.

5. Doors: Exterior doors, like windows must be integrated into the exterior wall construction, while also performing the specific tasks required for some controlled entry and exit of people and other items.
6. Climate control: For most buildings it is necessary to maintain some control of interior conditions while exterior climate varies. This involves the blocking of entry of wind, rain, and snow, and maintaining some limited range of the interior temperature, humidity, and general air quality. The exterior walls, together with the doors and windows in them, are major components of the system for control of the interior environment.
7. Sound control: Unwanted exterior noises, as well as some interior sounds that may disturb neighbors, may be blocked by the exterior walls.
8. Fire: Walls serve as barriers to fires preventing exterior fires from penetrating to the building interior and interior fires from spreading to nearby buildings. If a wall is a major structural element for the building, its loss during a fire is even more serious. Critical concerns may include those for flame spread on exterior or interior surfaces, combustibility of the construction materials, spread of fire inside hollow spaces in the walls, and the function of wall materials as protection for other parts of the construction, such as steel columns.
9. Durability: Walls must maintain their integrity through the life of the building. Surface finishes are most critical in this regard. However, the actions of thermal expansion and contraction, freezing, wetting and drying, wind or seismic movements, and general aging can conspire to cause deterioration of many elements of the exterior wall construction.

Exterior walls must sustain the direct effects of wind, consisting of both inward and outward pressures on the wall surfaces, as well as surface shear from wind blowing parallel to the wall plane. This applies to all exterior walls, regardless of any other structural requirements for the walls. For walls that serve some other structural purpose, the wind pressure effects must be combined with other loading conditions and structural actions.

Walls that do not perform any major structural tasks, other than to support themselves, are commonly described as *nonstructural walls.* Nonstructural interior walls are generally called *partitions.* For exterior walls, the general term for a nonstructural wall is a *curtain wall,* although this term is more often used to describe walls that consist mainly of glass, metal frames, and some panel or veneer system.

Nonstructural walls have in general a higher range of choice of construction materials and less restriction on their planning. Walls that provide support or bracing for other parts of the building must be strategically located and built for determined structural responses. Exterior walls, however, must generally resist wind pressures and transfer wind forces to the building structure, and are thus somewhat more constrained than interior partition walls. In addition, any nonstructural wall must usually depend on something for its stability, and the general construction must be arranged to provide this support.

Commonly used forms of construction for exterior walls often relate quite directly to local climate, local building traditions, local building codes and zoning, and local availability of building materials and building crafts. Thus, concerns for heat loss in very cold climates, for heat gain in hot climates, for effects of locally heavy rains or snows, for freezing, for salt air, and for windblown sand can produce strong constraints on wall design. The logical size, type, and locations of windows, the need for various types and amounts of insulation, and the requirements for sealing of construction joints will be affected by these concerns.

Forms and styles of architecture have often derived from responses to local conditions. Transplantation of architectural styles from one region to another has often resulted in some inappropriate buildings.

Walls are often used for hanging or displaying items, such as pictures, shelves, display cabinets, and signs. For walls that are not flat or vertical, or to which direct attachment is otherwise difficult, fulfilling these functions may require some special attention.

Walls are frequently the locations for power outlets, phone connections, lighting switches, thermostats, and air registers or exhaust grills. In addition, they must incorporate the necessary wiring, piping, or ducting to service these items. Some forms of wall construction can accommodate these items with ease, while others cannot do so, requiring services to be surface mounted in some cases.

Most forms of wall construction are semiporous, permitting some flow of heat, sound, and moisture. Where control of these is important, it is often necessary to enhance the basic construction by the addition of some materials. Thermal insulation is a common addition of this type.

Most walls are assembled from many parts. A major factor in controlling flow through the wall thus becomes the achieving of a tightness or sealing of the joints. Joints in the basic wall construction may be of concern, but more often the critical joints are those that occur around window and door units. Airtight and watertight seals at these locations are critical for reduction of airflow (and the accompanying heat loss in cold weather or heat gain in hot weather) and for keeping water from entering the building or from soaking into the construction materials inside the wall.

5.3 ROOFS

For many low-rise buildings the roof constitutes a major portion of the building exterior. For this reason, the various barrier functions that were described for exterior walls may be of major concern for the roof as well. (See Figure 5.2.) For the problem of solar heat gain in hot weather, the roof may be particularly critical, as the sun shines on this surface (the top of the building) for the longest part of the day.

Needs for thermal insulation, for a water-repellant surface, for a barrier to airborne moisture flow, and for weather seals in general are critical concerns for most roofs. The general major need, however, is simply that for a water-repellant surface and the drainage of water off the roof. The type of roof surfacing used and the performance requirements for it are based on the basic requirements for watertightness and drainage.

Roofs are broadly distinguished as slow draining or fast draining, with the speed of draining related essentially to the angle of slope of the roof surface. Using old carpenter's lingo, slopes are often quoted in terms of numbers of inches of rise per foot of horizontal run. Using the foot as 12 in., a slope is said to be 4 in 12 indicating a vertical rise of 4 in. for a horizontal run of 12 in., or an angle of about 18.5 degrees. For low angles the slope is sometimes described in percent; thus the 4 in 12 slope would be described as 33 in 100, or a 33% slope.

In most cases the type or roof surfacing (called the *roofing*) is related to the slope of the roof. Shingles and tiles can be used only on relatively steep slopes (fast draining), and generally the steeper the better. Typically, each type of shingle or tile has some minimum required slope and some recommended range of slopes for best performance.

For roofs described as flat or low slope, a real waterproof membrane roofing is usually required—one that tightly seals the roof surface. The traditional solution for the flat roof is a built-up roofing, with multiple, overlapping layers of felt, coated with hot asphalt or coal tar. This is still a common solution, although various other forms of membrane roofing are also possible.

A special type of roofing is that produced with jointed metal sheets of copper, aluminum, steel, or some special alloys. These can be used on fast-draining surfaces, or—with careful attention to drainage patterns—on near-flat surfaces.

Making the general roof surface water repellent is not usually the greatest problem in keeping water out of the building. Most roof leaks occur at points of discontinuity of the surface: at edges, ridges, valleys, and at points where vents, chimneys, or skylights penetrate the roof. Development of proper detailing of the construction and the flashing and seals used at these locations is one of the most important problems in construction design.

Another critical concern for roofs is that of thermal expansion and contraction. Roofing materials must maintain some resilience and resist cracking or separation at joints. A major problem is sometimes that of the differential movements between the various elements of the construction, typically consisting of different materials—often with significantly different rates of thermal expansion. For very long buildings, it is usually necessary to provide expansion joints in the construction at intervals, to relieve the stresses in the materials due to thermally induced movements.

Movements can also occur due to structural deformations, induced by live load deflections, by wind, or by seismic activity. As with thermally induced movements, these may be partly controlled by special jointing, involving concerns for both the structural effects and the integrity of the roofing materials.

Most roofing materials do not function well as paving, so special provisions must be made if regular traffic must be accommodated. If the traffic is localized, catwalks or strips of raised decking may be provided, but if the traffic is generally dispersed on the roof surface, a terrace surfacing must be developed and watertightness must generally be developed as it is for underground buildings. (See discussion of underground buildings in Section 5.14.)

The complete design of a roof must include some considerations for where water goes as it drains off the roof surface. Shaping of the form of the roof surface can be a means for directing the water flow to certain points; however, it is seldom acceptable to simply dump the water off the roof.

Disposal of roof runoff water must be developed in conjunction with the general planning of the building and the site. Heavy streams of water can wash out plantings, build up water against basement walls, flood walks and driveways, or intrude on other properties.

Various devices are used to control water flow on a roof or to function in various ways to get it off the roof and away from the building. As shown in Figure 5.3., some common devices are the following:

Saddles and dams: These are built-up elements on the roof for localized control of flow—to divert water away from or towards some point. Typical common problem locations are the backs of chimneys in sloping roofs and the nonguttered, down-

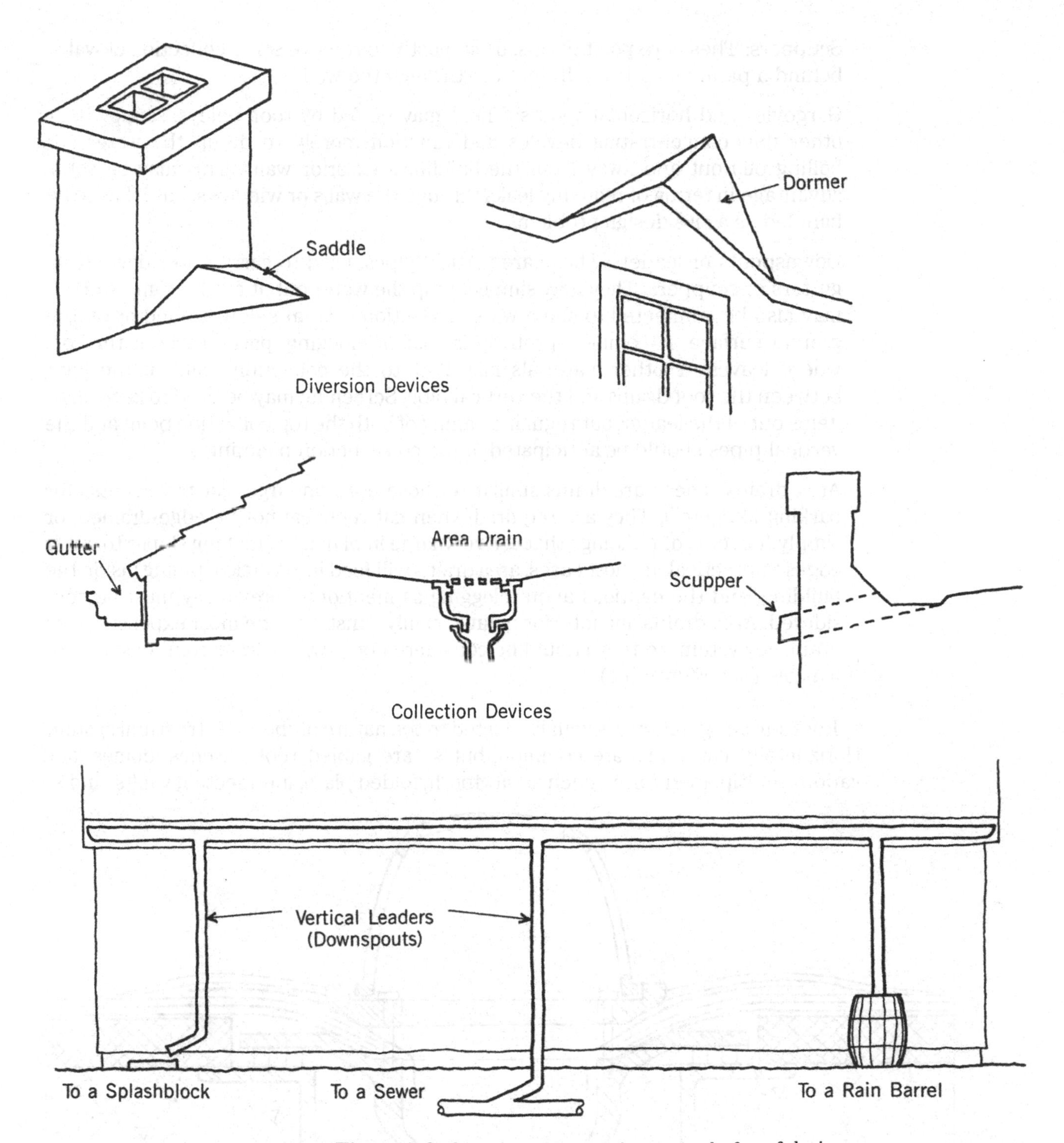

Figure 5.3 Common elements for control of roof drainage.

slope edge of a roof over an entrance door. Good planning may avoid these situations, but long experience with bad planning has produced a collection of tricks for expedient solutions.

Gutters: These are horizontal troughs, generally placed to catch water running off the lower edge of a roof. They can be built into the roof edge construction, but are more often simply hung from the face of the edge. Attention must be paid to the fact that they need to slope significantly along their lengths for good drainage.

Scuppers: These are point drains, used mostly to remove some collection of water behind a parapet wall and drain it out through the wall.

Gargoyles and horizontal spouts: These may be fed by roof valleys, scuppers, or other flow-concentrating devices and function merely to dump the water (or boiling oil) out and away from the building's exterior walls. This may be some advantage in terms of reducing leaks through the walls or windows, but it has to be handled as a site design problem.

Downspouts or leaders: These are vertical pipes, used to carry water down from gutters or scuppers. They may simply dump the water out at the bottom, but they may also be connected to some water collection system below ground or on the ground surface. A common problem is that of clogging, particularly at the top, where leaves or other materials may flow to the collection point at the joint between the roof drains and the vertical pipe. Screening may be used to keep large items out of the leader, but regular cleaning of both the top collection point and the vertical pipes should be anticipated in the construction planning.

Area drains: These are drains similar to those used on large, flat pavements (for parking lots, etc.). They are required when flat roofs cannot be edge-drained, or simply for roofs of buildings that are very large in plan area, making slopes to outer edges impractical. In most cases, area drains will feed into vertical piping inside the building, and the cautions about clogging as mentioned previously must be considered. Area drains and interior piping usually constitute the most expensive roof drainage system, so they should be considered only when other solutions are not feasible (see Figure 5.4).

Roof surface geometry is usually related to the nature of the roof structural system. Horizontally flat spans are common, but so are gabled roofs, arches, domes, and various multiplaned forms, such as sawtooth, folded plate, intersecting vaults, and so

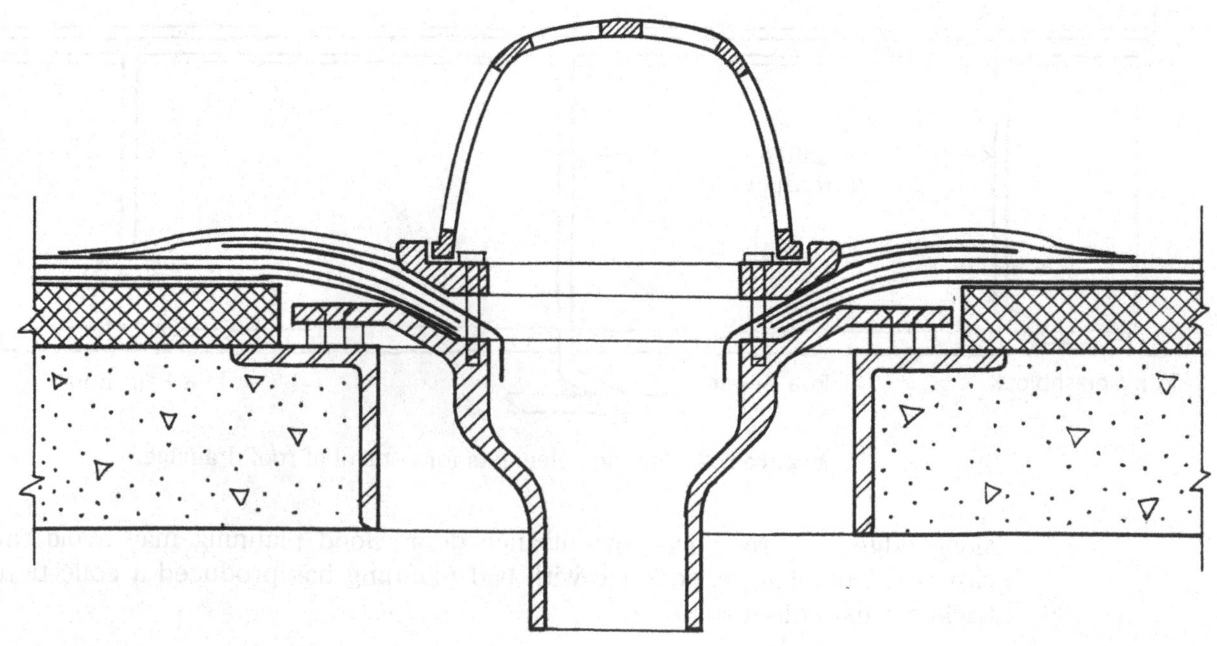

Figure 5.4 Typical form of roof area drain in a flat roof with built-up roofing.

Figure 5.5 Building with an architecturally-prominent roof; making the roof form and materials of considerable concern for appearance.

Figure 5.6 Interior with architecturally-prominent form reflecting the roof structure.

on. In fact, exotic and complex structures are most often used for roofs, where concerns for vertically flat walls or horizontally flat floors are not the issue.

Roof surfacing often becomes a major architectural design concern for nonflat roofs, as the surfaces are often considerably in view. For low-rise buildings they may well be very dominant parts of the architecture of the building exterior (see Figure 5.5). In many cases the choice of slate shingles, clay tiles, copper sheet, or other quite expensive roofing may be justified for the impact of the finished surfaces on the general building appearance.

Roof geometry can also be dominant to interior spaces when the roof form is directly reflected on the building interior (see Figure 5.6). This is even more the case when the structure is directly exposed, so that both its general form and details of its parts are in view.

5.4 WINDOWS

Windows typically begin life as holes in walls. Some initial concerns therefore relate to the various problems of creating the hole in the structure of the wall. The more structural tasks that the wall must perform and the larger the window opening, the greater the structural concerns for the window.

In most situations some consideration must be given to the spanning of the wall construction across the window opening. Again, the size (span length in horizontal distance) of the opening and the load on the wall will determine the magnitude of this problem. Various methods for spanning can be employed, depending on the form and materials of the wall construction. While the opening disturbs the wall, the wall and its structure may also disturb the window. For large openings, it is necessary to develop the construction details so that any significant structural movements of the wall (most notably, vertical deflection of the structure spanning the opening) do not distort the window, possibly resulting in cracked glass or impaired movement of sash in operable windows.

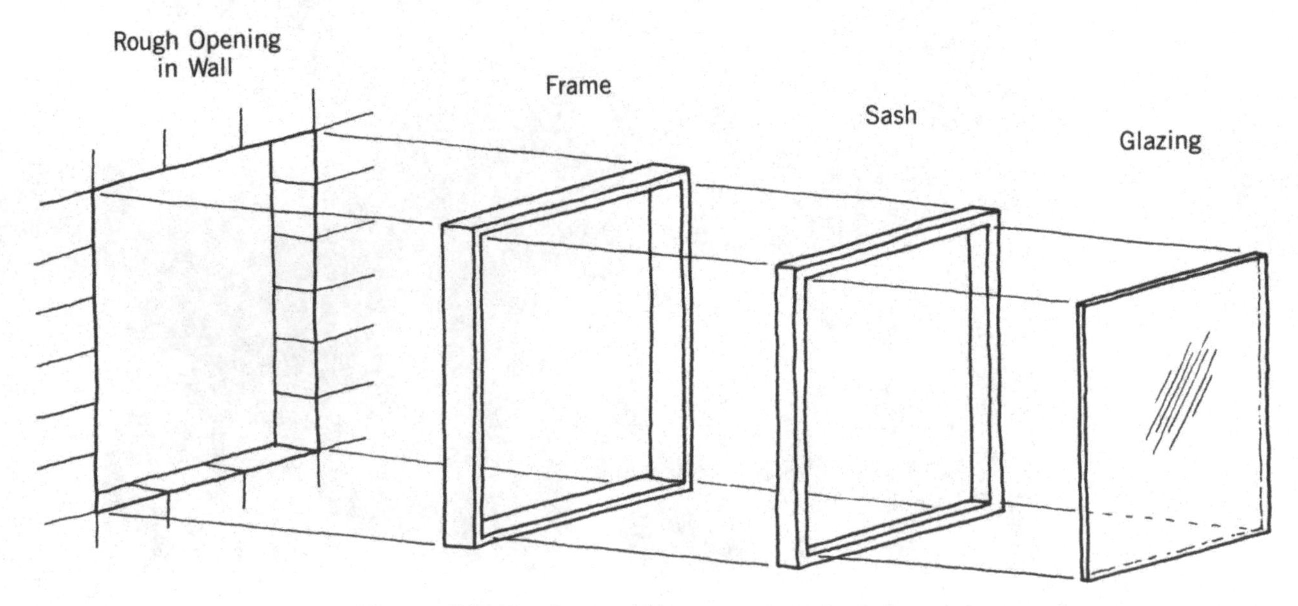

Figure 5.7 Fundamental components of windows.

The typical window consists of the basic elements shown in Figure 5.7. The initial definition of the window is made by the hole cut in the wall; this is called the *rough opening.* The term *rough* is used advisedly here, as the feasible limits of construction accuracy require some overcutting to assure that the window itself can be accurately placed.

The window opening is typically more finely developed by the insertion of a *window frame* in the rough opening. This establishes the basic window-to-wall interface, and the matters of accuracy of placement of the window and management of the movements of the wall are handled here. Any trimming up of the joints are also developed to assure a neat, weather-tight relationship.

If the window is required to be capable of being opened (the technical term is *operable*), there will typically be a second frame—called the *sash* —that slides or swings with respect to the window frame. The *glazing,* which permits light transmission and possibly vision through the window, is installed in the sash.

If there is no requirement for opening of the window after it is installed, there is no basic need for the sash. In this case, the glazing may be installed directly in the frame. This is the typical situation for most high-rise buildings with full interior air management which does not require ventilation from the windows. However, the development of the construction details and on-site installation sequences may result in the use of a sash for practical purposes. Many variations of this process are possible, as discussed in Sec. 7.9 and for Buildings 6 and 7 in Chapter 10.

With the operable window there are three different joints, all of which must be made weather-tight. These include the joints between the frame and the wall, between the sash and the frame, and between the glazing and the sash. The most challenging of these is the one between the sash and the frame, as this must allow for movement of the sash, while achieving some weather resistance when the window is closed.

The planning and detailing of windows is a major architectural design task. Windows have major impact on the appearance of the building exterior and simultaneously serve specific required functions (for light, vision, ventilation) for interior spaces. Windows are dual-facing, and choices for the window materials, forms, and sizes and of the placement of the windows in the walls must satisfy both the exterior and interior concerns; a situation not always easy to resolve with success in both areas of concern.

Window size and placement, choice of glazing, the weather resistive character of joints, and the control of operable windows can all affect the management of air quality, temperatures, and sound conditions inside the building. Other concerns may include those for security (breaking glazing or opening sash to enter), ability to locate furniture or partitions, maintenance of visual or acoustic privacy, and various relations to fire safety.

All in all, there is a lot to consider in designing windows, and the immense range of possible solutions in terms of materials, standard products, and construction details makes the design choices all the more difficult.

A final issue to consider is that windows quite frequently get some treatment beyond the basic construction. Use of exterior sun-shading devices (awnings, louvers, etc.) and interior shades, curtains, drapes, or blinds, can alter appearances as well as some of the filter barrier functions. Planning for these is sometimes complicated by the uncertainty of occupant modifications. Still, some provisions for these enhancements are frequently made. Various basic purposes for enhancement of the building's enclosing shell in general are discussed in Chapter 8.

5.5 EXTERIOR DOORS

There are many similarities between windows and exterior doors, and much of what was mentioned in the preceding section applies to doors as well. As with a window, most doors begin as holes in walls, have the hole trimmed with a frame, and have a movable unit (sash/door) that is contained by the frame, but must be "operated."

As shown in Figure 5.8, the basic construction elements typically consist of a rough opening in a wall, a door frame, and the operable door. In some situations, where traffic control is not required, the door itself may be omitted. In this case, all that is required is the hole in the wall (possibly with a frame to trim it), which is then described simply as a *doorway*. Even when a door is added, the hole plus the frame is frequently referred to as the doorway for the door.

Doors serve a primary function as traffic control devices. Many specific requirements for a particular door derive from this function. For exterior doors, a fundamental need for security is typical. In most situations, exterior doors are also required fire exits, bringing a number of code requirements, including those for the size, means of unlatching, direction of swing, location in the wall, and the fire-resistive character of the construction.

As with windows again, exterior doors are subject to the general concerns for barrier functions of the building exterior enclosure. The door itself, plus the various joints between door, frame, and wall, must satisfy some criteria for control of passage of air, heat, and sound.

Doors typically have a number of hardware accessories that must be carefully chosen to satisfy all of the functional requirements for the door. This makes it necessary in the construction design process to do a very thorough analysis of the door requirements and to specifically identify each door in the building drawings. This is normally done by giving each door an identifying symbol or number, which is

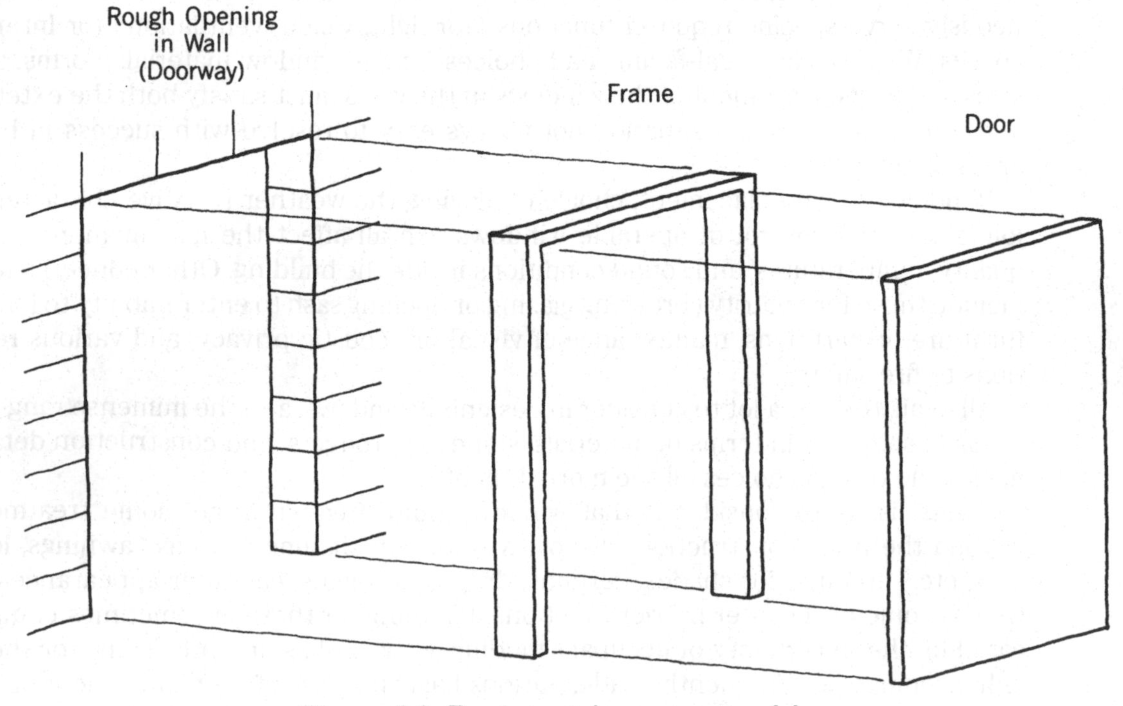

Figure 5.8 Fundamental components of doors.

referred to a table (called the door schedule) and/or the specifications, where the door itself, all the hardware and other accessories, plus the details for the frame, will be precisely described.

Not a small concern for exterior doors—particularly those at major entrance points to the building—is their appearance and general welcoming (or not) character. This applies not only to the door itself, but to the general setting, which may include side windows; a canopy; a recessed, sheltered area; special lighting; and other features. Functional needs cannot be ignored, but the entrance door is where people meet the building first, and it is usually desired that the meeting be friendly.

Doors are potential major barriers for persons with limited facilities, such as those who are wheelchair-bound. As for fire exits, codes may require certain dimensions and special features for door operation for a truly barrier-free environment. Facilities designed specially for the aged, the very young, or persons with particular handicaps will have particular concerns. However, most buildings for public use or those where persons are employed must now be designed to some degree as barrier-free.

5.6. FLOORS

Floors are working platforms and must respond to the needs of the building users. Ordinarily, this means the desire for a dead flat surface and an exposed finish that responds to practical concerns for the type of traffic, anticipated maintenance, budget, and various functional demands deriving from concerns for fire, sound control, and so on.

Floor finishes endure considerable wear and soiling and are often replaced or reconditioned with some frequency. For this reason, finishes are often dealt with as applied surfacing that can be removed and replaced or restored without damage to the supporting structure. The supporting structure must facilitate the installation—as well as the replacement—of the applied finish, and should usually be able to facilitate some changes in future types of finishes.

There are two major categories of supporting floor structures: those placed directly on the ground, and those that span over open space beneath them (called framed floors). Floors on the ground are most often achieved with concrete pave-

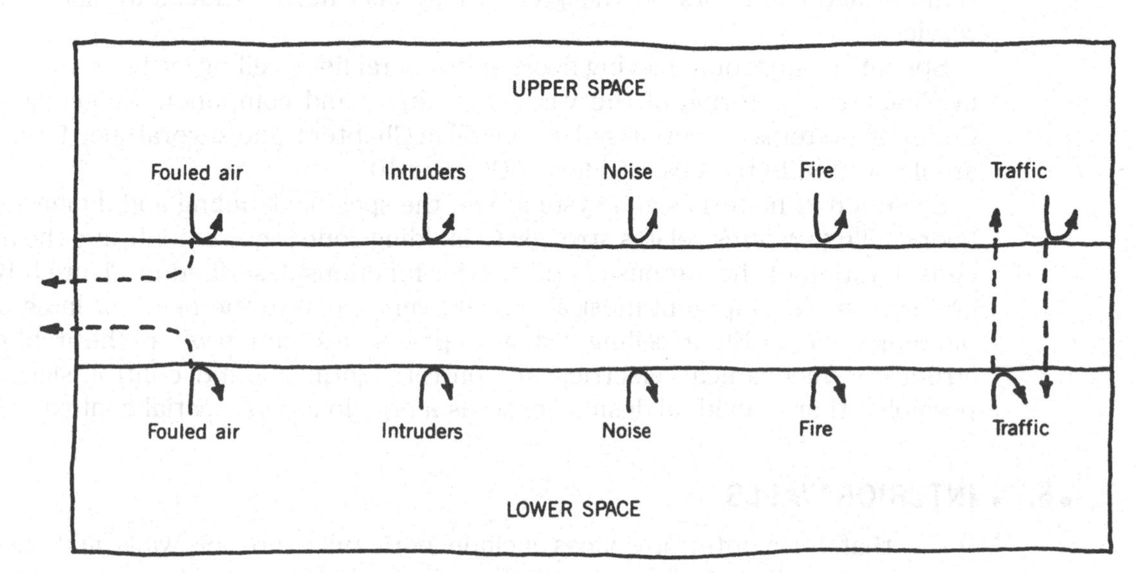

Figure 5.9 Barrier functions of floor/ceiling systems.

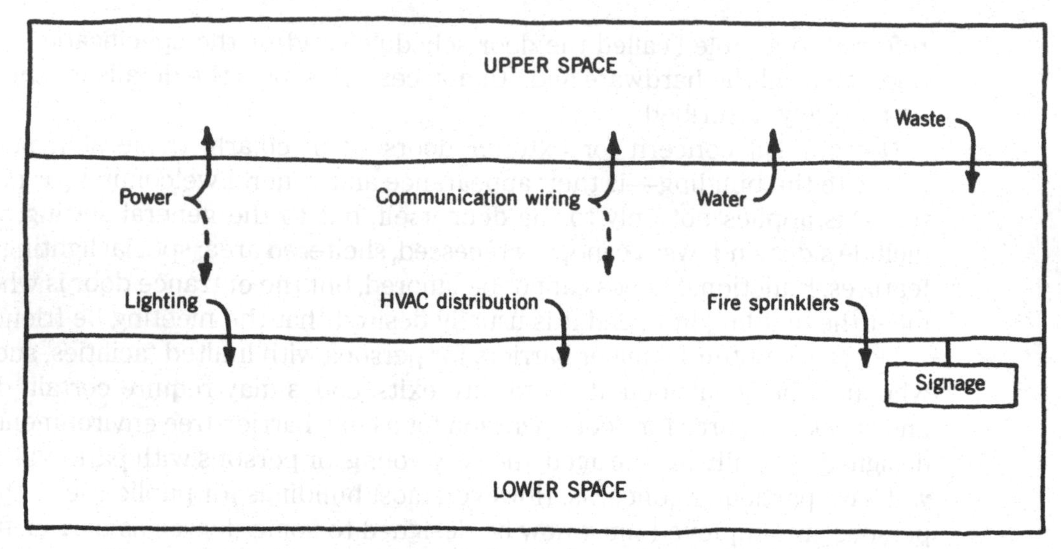

Figure 5.10 Accommodation of building services in floor/ceiling systems.

ments. Framed floors can be achieved with a wide variety of structures, depending on the length of spans, fire requirements, and the materials of the general construction of the building.

When floors are framed, they become separating elements between spaces. This brings three major relationships into consideration. The first of these relationships involves the needs of the underneath space as regards the underside surface of the floor structure. This surface is called a ceiling, and is discussed in Sec. 5.8.

The second relationship derives from the relations between the two separate spaces—above and below the floor structure. The floor/ceiling system represents a barrier between these two spaces, and must deal with the various concerns illustrated in Figure 5.9.

The third relationship involves the various building service elements that must typically be incorporated into the floor/ceiling system. This usually relates more to the lower space, but may also relate to the upper space, as shown in Figure 5.10. In some situations, floors on the ground may also need to facilitate some of these services.

Specific means for achieving floors and general floor/ceiling systems are discussed in Chapter 7 in terms of the various products and components commonly used. Common systems are discussed in general in Chapter 9 and several specific examples are illustrated in the case studies in Chapter 10.

Selection of materials and systems and the specific detailing and dimensioning of floor/ceiling systems relates strongly to building code requirements and the need for consideration of the various spatial barrier functions described in Figure 5.10. General system development must also relate very much to the need for integration of building systems. Floor/ceiling systems represent a situation where the architectural, structural, mechanical, electric, fire control, lighting, sound control, signage, and possibly other individual design concerns merge in a tight spatial context.

5.7 INTERIOR WALLS

Walls that form interior spaces include both fully interior walls and the inside surfaces of enclosing, exterior walls. Exterior walls are major elements in the control of the building's interior environment, functioning as barriers between the indoors and outdoors, and functioning as selective filters, as shown in Figure 5.2. The needs for these barrier and filter functions provide major criteria for the general design of the exterior walls.

Interior walls—that is, those that separate only individual interior spaces—also usually have some barrier or selective filter functions, although their general situation is different from that of the exterior walls. What is generally not critical here is the need to maintain thermal differences between the spaces on the two sides of the wall; a *major* issue with exterior walls. Likewise, the need for air tightness and general weather sealing is absent. While possibly exposed to interior view, the interior walls do not constitute major elements of the building's exterior appearance. Finally, interior walls seldom contain windows—a principal feature of most exterior walls.

On the other hand, walls are often made with the same basic structures and materials, regardless of their location. All walls may need to accommodate doors, with the various requirements described in Sec. 5.5. And, in general, there are many typical requirements that are the same for all walls.

Walls must be adequate for their own support in any case, but must often also serve other structural functions, such as support of roofs or floors or bracing for wind or seismic effects on the building (as shear walls). Even when walls need only support themselves, other elements of the building structure must often be integrated into the wall construction, so as not to otherwise intrude on interior spaces. Walls that do not perform any major structural tasks, other than to support themselves, are commonly described as *nonstructural walls.* Nonstructural interior walls are generally called *partitions,* although the term *partitioning* refers in general to the defining of individual interior spaces.

Two major concerns for interior walls are their ability to function in the control of fire and sound. These properties may also be of concern for exterior walls, but do not hold quite the same relative level of importance in design. Major considerations for these and other special functions for walls are discussed in Chapter 8. Concerns for fire are the motivation for a major portion of building code requirements.

5.8 CEILINGS

The overhead boundary of an interior space is called its ceiling. From a construction point of view, a ceiling may be a separate element, or simply be the underside of an overhead structure. Ceilings are highly visible, and are usually of major concern in interior design.

Ceilings are often the location of many building service elements, such as lighting, HAVC registers, fire sprinklers, smoke detectors, and signs. The separate design of each of these elements or subsystems, as well as their total integration with the general ceiling construction, is a major design activity.

In construction work interior spaces are sometimes said to have "no ceiling," meaning that nothing special is done to create a ceiling—it is simply the unfinished, underside of the overhead structure (see Figure 5.11). More often, however, something *is* done, if nothing more than some careful cleaning or painting of the underside of the structure.

In many situations, a separate ceiling construction is developed, either attached directly to, or suspended from, the overhead structure. Thus a floor/ceiling or roof/ceiling sandwich of sorts is created, and often encloses an interstitial space that is used to contain various building service elements, such as wiring, piping, ducting, and minor equipment. The full design of this system and all of the contained elements is a major integrated design problem.

(a)

(b)

Figure 5.11 Ceilings developed as the directly expressed underside of the overhead construction: (a) Sitecast concrete "waffle" construction. (b) Wood rib and plank deck units. (c) Exposed steel framing system.

(c) **Figure 5.11** *(continued)*

In some ways, ceiling finishes represent a major opportunity for freedom of choice of materials. There is a general lack of concern here for effects of contact wear; ceilings are not walked on (except in the movies), and are generally out of reach. Fragile materials can be used, although there is nothing wrong with hard, durable ones either.

5.9 INTERIOR DOORS

Most of the fundamental concerns for doors as construction elements are discussed in Sec. 5.5 with regard to exterior doors. Many of the basic features of interior doors are not distinct from those for exterior doors, except for the necessary considerations for the building enclosing shell functions.

Doors serve primary functions as traffic control devices, and many specific functions derive from this requirement. Doors are usually placed in walls and must also relate to the functional requirements for the wall in terms of barrier and filter requirements. A soundproof wall with a leaky door is useless.

As shown in Figure 5.8, the basic construction elements typically consist of a rough opening in a wall, a door frame, and the operable door. In some situations, where traffic control or barrier functions are not required, the door itself may be omitted. In this case, all that is required is the hole in the wall (possibly with a frame to trim it), which is then described simply as a *doorway*. Even when a door is added, the hole plus the frame is frequently referred to as the doorway for the door.

A major distinction is made for doors that occur in the exit pathway for occupants in case of fire. Codes require many features for these doors, including considerations for size, operating hardware, direction of swing, and so on.

Doors typically have a number of hardware accessories that must be carefully chosen to satisfy all of the functional requirements for the door. This makes it necessary in the construction design process to do a very thorough analysis of the door requirements and to specifically identify each door in the building drawings. This is normally done by giving each door an identifying symbol or number, which is referred to as a table (called the door schedule) and/or the design specifications, where the door itself, all the hardware and other accessories, plus the details for the frame, will be precisely described.

5.10 STAIRS

Buildings with floors at more than one level ordinarily require stairs. Other devices for intra-level movements may also be provided—such as ramps, elevators, or escalators—but the stair is the most primary device.

Movement in tall buildings will occur mostly with elevators, but stairs are still necessary for fire exits. In fact, the code requirements for fire stairs are usually primary sources of criteria for the design of most stairs. This includes the design of the stair itself, the construction surrounding it, required lighting and the access doors to the stairs.

Planning of stairs is a major design problem (see Figure 5.12). The stairs themselves take up considerable plan space which is generally not usable for the functional activities of the building occupants. Thus tight planning is usually desirable. Placing of stairs in the general building plan is also critical to their accessibility and safe use as fire exits. While the rest of the floor plan may be essentially a two-dimensional planning problem, stairs are vertically continuous and must be planned three-dimensionally.

When a stairs is designated as a fire exit, its design is highly constrainted by code requirements. These deal with the stair dimensions (width, riser and tread), construction, stairway details, and door details. Special stairs, not designated as fire exits, may be free of some of these constraints, but must still be safe in general and subject to some limitations.

5.11 ELEVATORS

Elevators are now quite common in all buildings with more than one level, other than single-family houses. They may well account for the majority of intra-level movement, but do not qualify as fire exits—thus stairs must usually also be used. The stairs are also options when the elevators are not operable for any reason.

Elevators are expensive and are generally designed to minimum requirements for economy. Major concerns are for speed and the total number of users. For commercial buildings it is common to have more than a single elevator, and, of course, for very large and very tall buildings, many separate elevators.

As with stairs, in addition to their basic cost, elevators take up considerable space that is not usable for other building functions—notably those deriving income for the building owners (see Figure 5.13). Besides the plan area occupied by the elevators

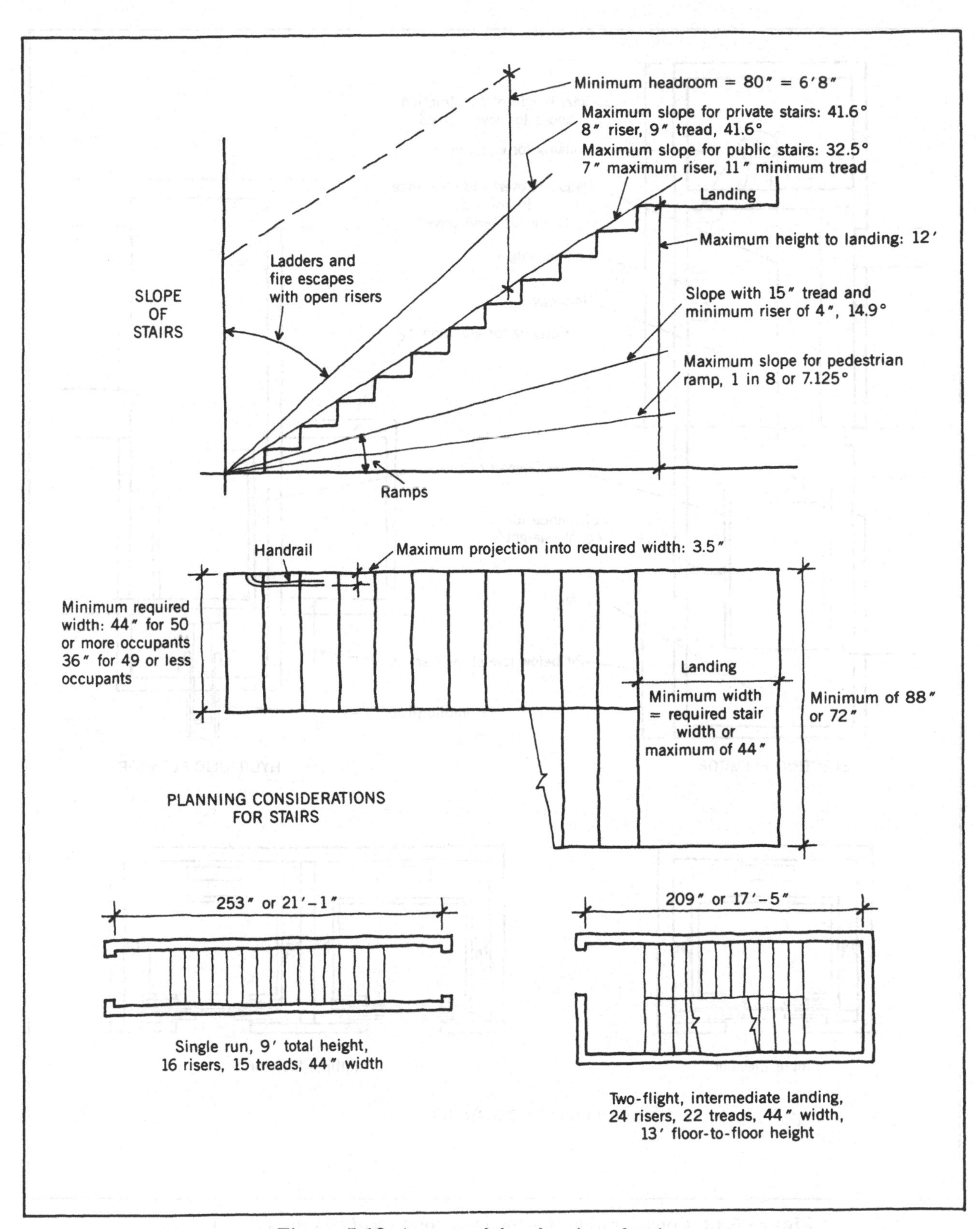

Figure 5.12 Aspects of the planning of stairs.

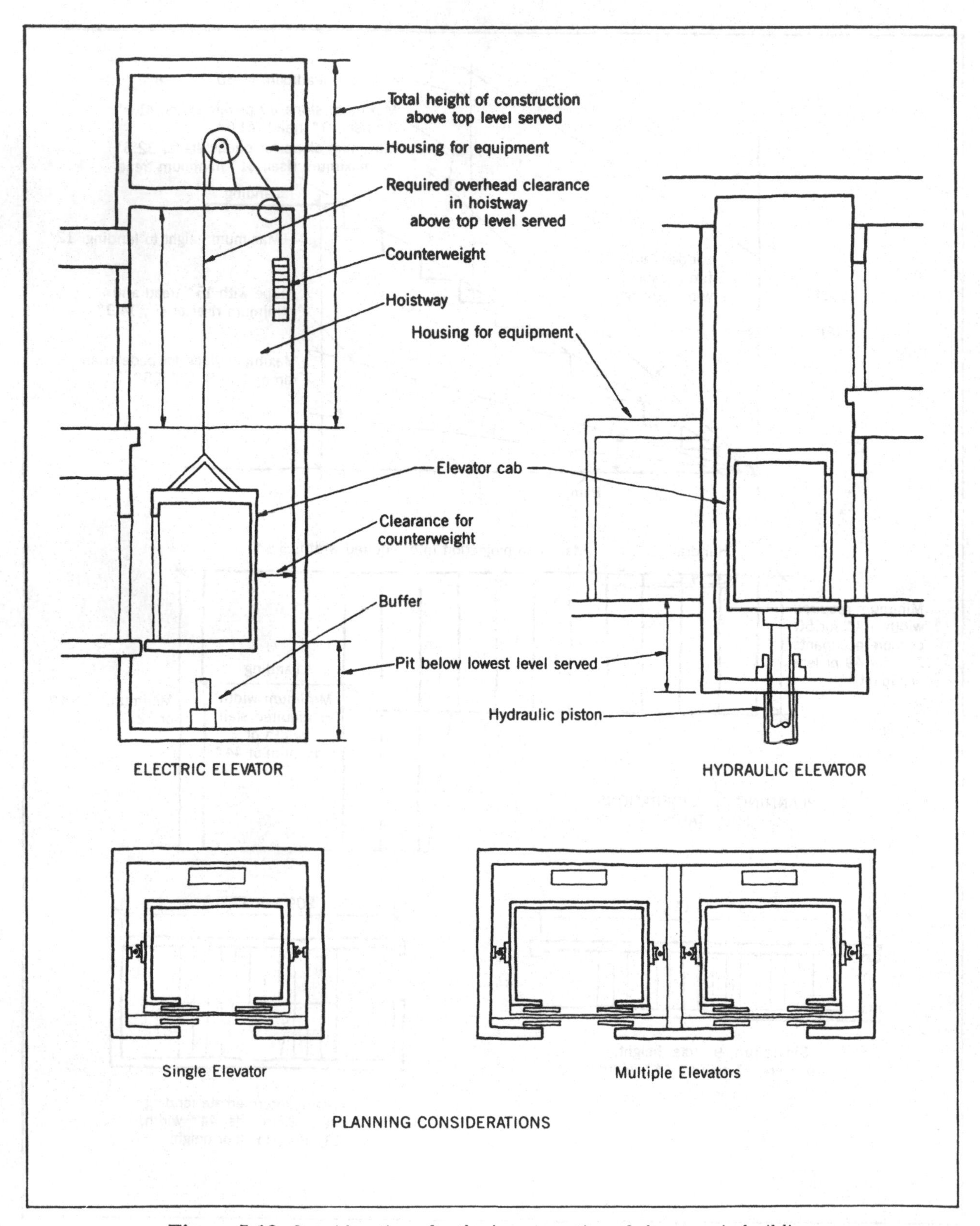

Figure 5.13 Considerations for the incorporation of elevators in buildings.

and the construction surrounding them, space must be provided above the highest level served and below the lowest for construction and equipment. It behooves designers with responsibility for building planning to become familiar with the ordinary space demands for elevator systems.

REFERENCES

Architectural Graphic Standards, 8th edition, Ramsey and Sleeper, Wiley, 1988.

Mechanical and Electrical Equipment for Buildings, 7th edition, Stein, Reynolds, and McGuinness, Wiley, 1986. Chapter 23.

Sweet's Catalog Files: Products for General Building and Renovation, annually published, McGraw-Hill. Division 14.

■ 5.12 ■ BUILDING FOUNDATIONS

Foundations essentially consist of transitional elements between the building's above-ground superstructure and the ground. Basic problems involve the resistance of vertical gravity loads, resistance to lateral load effects (wind and earthquakes), and generally providing a base for construction of the building.

The three basic types of foundation elements are those shown in Figure 5.14. The simplest and most common foundations consist of so-called shallow bearing foundations; also called footings. These typically consist of bearing pads of reinforced concrete, of sufficient plan size to resolve vertical forces into a low value of bearing pressure on the soil.

When soils of sufficient bearing capacity do not exist at the location of the bottom of the building—or when the loads are simply too great for bearing pressure resolution—special foundations, called deep foundations, must be used. One form is the pile, a shaft-like element which is driven into the ground like a large nail. Timber poles (tree trunks) were used for this in ancient times and still are. More often, however, other forms are now used, as discussed in Section 4.5. Accuracy of placement of piles is quite limited, so piles are ordinarily used in groups.

The other form of deep foundation consists of an excavated shaft which is advanced down some distance and then filled with concrete to form a stilt-like column under the building. These are called piers or caissons.

Both piles and piers may be seated in soil or advanced to bear on rock. When the excavated shaft bears on soil, it is common to splay or spread its bottom, forming a bell-shaped, truncated cone at the bottom of the round shaft, as shown in Figure 5.14.

These basic elements—footings, piles, and piers—are typically used with various other elements to form a complete building subgrade base. This may include grade beams, foundation walls, ties for isolated foundations, pile caps, and so on. Exact requirements will depend on the building size and form, the type of construction, the general type of foundation system, and various possible soil conditions. Several different situations are illustrated in the case studies in Chapter 10.

■ 5.13 ■ BASEMENTS

Many buildings consist essentially only of above-grade construction and spaces, with a minimal ground penetration made basically to achieve the foundations and a base for construction. However, if interior spaces are extended below the ground level, the below-grade portion is called a basement.

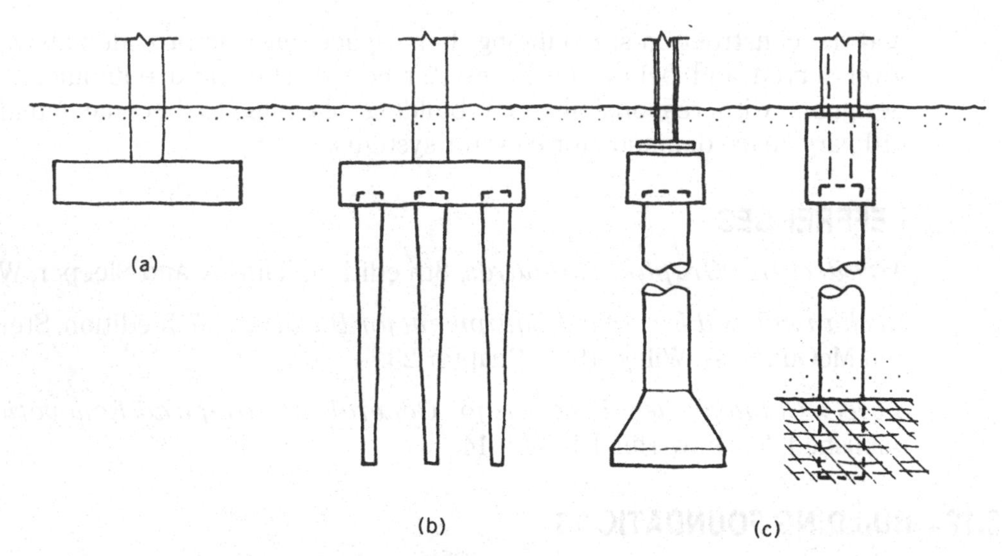

Figure 5.14 Basic types of building foundations: (a) Shallow bearing footings. (b) Driven piles. (c) Excavated piers.

Basements that exist only beneath the overhead building consist primarily of enclosing walls and floors. Essential problems, as shown in Figure 5.15, consist of retaining the soil outside the walls, preventing intrusion of water, and handling various factors for comfort of occupants when the spaces are used other than simply for equipment or storage.

When basement exist, they are usually points of entry and exit for various building services that are delivered underground—typically water, gas, and sewers, but also possibly underground power and telephone lines.

Basement walls also frequently serve as bearing walls or spanning grade beams in conjunction with the building foundation system. Architectural planning and general detailing of the construction must deal with all of these relationships.

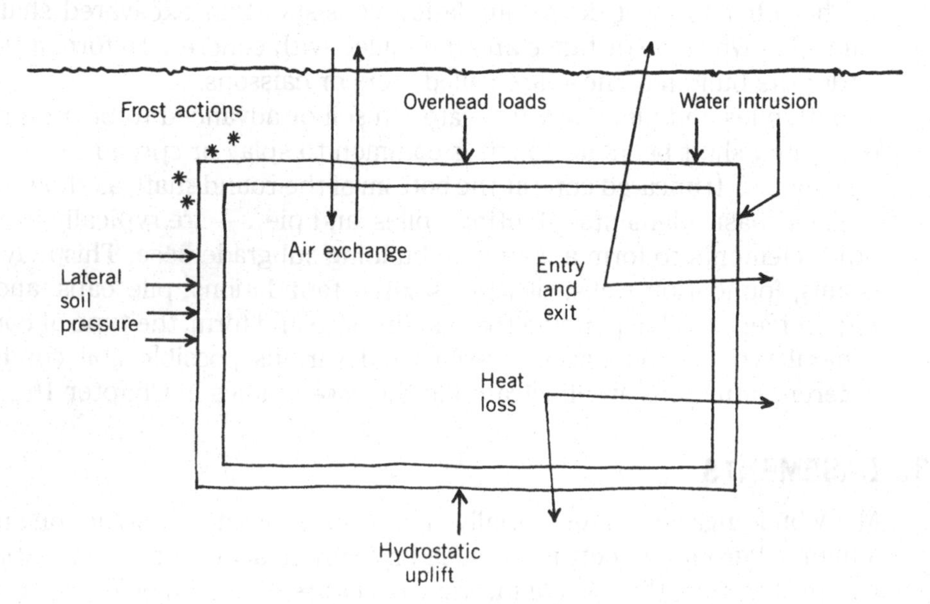

Figure 5.15 General concerns for below grade spaces.

When basements are deep, extending considerably below finished ground level, a major consideration is often the achieving of the excavation for their construction. This is especially difficult on tight urban sites, where the basement walls may be extended to the property lines.

5.14 UNDERGROUND BUILDINGS

Buildings are occasionally built entirely underground. However, the most common occurrence of underground buildings is with basements that extend horizontally beyond the perimeter of the building above ground. In either case, the element added to the basement structure is that of a roof that forms a terrace or supports soil.

Placing a terrace or soil on a roof presents major structural and waterproofing problems. These are discussed extensively in Sec. 7.15. There is actually little comparison to ordinary roof construction, except for the waterproofing problem. Truly isolated underground buildings have many additional problems regarding planning and construction. Entry and ventilation must be achieved, offering some special additional considerations for both the structure and general construction detailing. Otherwise, the underground building has all the general problems encountered with basements.

5.15 UTILITIES AND BUILDING SERVICES BELOW GRADE

Most buildings are served by connections to electric power, phone lines, and piped water and sewers. Many must also relate to gas piping, trash collection, and delivery services. Both building and site planning must relate in various ways to some orientation and connection to these services.

Water, sewer and gas piping are generally underground and related to mains that are off the site in a particular direction and at a specific elevation. Elevation is particularly critical for sewers, which must work by gravity flow. Frequently, in developed areas, there are separate sanitary sewers (building waste water) and storm sewers (roof and surface drainage runoff).

Placement of the building on the site and design of the building subgrade base and foundations must anticipate the facilitation of attachment to underground services. Where extensive regrading and/or site construction is planned, it must also be carefully coordinated with services, as well as existing site edges at streets and adjacent properties.

5.16 MISCELLANEOUS BELOW GRADE CONSTRUCTION

In addition to the usual elements of foundations, basements, and underground services, various other items may be built into the ground on a site. Some such items are the following:

Vaults and Manholes. For service to a site or in conjunction with general area development, providers of utilities may need to install some underground structures on the site. These take up underground space and also need access through surface level covers.

Tunnels. For large projects, underground services may be placed in tunnels, permitting continuous access for alteration or service without disruption of the site surface. Disruption of drives, parking areas, or extensive terracing or landscaping can thus be largely avoided. Tunnels may also accommodate pedestrian or vehicular traffic, connecting underground building spaces to streets or subways, connecting separate buildings on the site, and so on. This is being increasingly done for building complexes in very cold climates.

Existing Services. By previous arrangements, underground service mains, subways, tunneled drainage channels, or other elements may already be in place on an otherwise undeveloped site. This is most likely to occur in highly developed areas, such as urban centers.

5.17 CONSTRUCTION FOR SITE SURFACE DEVELOPMENT

The design and construction of a building must be done cooperatively with the necessary work for the full development of the building site. The following are some principal concerns for site development.

Site Structures

Development of almost any building site involves some site surface construction—that is, some elements built outside the building. Sidewalks, drives, curbs, planters, and fences are most common. For sloping sites, earth berms, retaining structures, terraced slopes, and other elements for slope control and elevation changes are usually necessary. This may be modest for the average small project with the building occupying a major portion of the site. It can also be quite extensive for large, park-like sites.

This book is basically about *building* construction, so the treatment of site construction is dealt with minimally—involving only the most common elements. However, the full management of the site as a resource and complete, well-coordinated design of the whole building and site development should incorporate all of the construction work.

Site construction of various forms is discussed in Sec. 7.17 and many typical elements are illustrated in the building cases in Chapter 10.

Easements

As previously mentioned for underground elements, such as subways, prior arrangements may be in place to give certain rights of access or use to a property. One such arrangement is an easement, which reserves certain rights for someone else regarding a given property. Air rights for overhead power lines are one such item. Others include rights to extend roads, sewers, water mains, and so on. This does not always block the possibility of some site surface development, but it must be done with the risk of disruption in mind.

Plantings

Development of plantings on sites may consist of preserving of existing materials, of insertion of new plantings to replace or augment the existing site, or of wholly new plantings. Whenever new materials are to be developed, there is usually some construction associated with the plantings. At the least, this consists of the preparation of the soils into which the plantings are placed for growth. In many instances this means the importing or moving of soils to provide the desired type of environment for the planting.

This book will not treat the general design of landscaping, but will consider some of the construction problems associated with plantings.

Soil Structures

Undoubtedly, the major "element" in site construction is the site itself, as constituted by the soils that define its surface and immediate subsurface layers. Making of a site as a constructed object means working extensively with the site soils. This is a significant aspect of foundation design, pavement design, landscape design, and so on.

Besides providing support or encasement for various objects, soils are also used frequently for some forms of direct construction. Although topsoil, plantings, paving, or various ground covers may be used to develop surfaces, the general surface is usually developed by the underlying soils. Achieving this general site "construction" means working with soils as construction materials. It is necessary to have this view in order to appreciate the need to understand something about the structural character—and limitations—of soils.

Use of soils in various situations is discussed throughout this book. A general discussion of soils and their properties is provided in the Appendix.

5.19 ELEMENTS OF BUILDING SERVICES

Major concerns for building services must be given in the development of the building construction. A first concern is simply for the physical incorporation of the elements of the systems: piping, wiring, ducts, equipment, and various built-in fixtures. However, the building construction itself sometimes becomes an element of these systems; as when the passive insulative properties of the building shell affect the design of the HVAC system. The following discussions treat some of the issues for the most common service systems as they relate to concerns for the development of the building construction.

HVAC Systems

Heating, ventilating (air freshening), cooling, and the general treatment of the building's interior thermal and atmospheric condition is a multi-dimensioned task. This is often lumped under the general activity described by "HVAC," which stands for heating, ventilating, and air conditioning.

By itself, the term *air conditioning* is often taken to mean essentially the cooling of interior air. However, even in this limited context, it typically also involves the removal of moisture from the air. As referred to in the HVAC term, air conditioning is meant to stand for the complete conditioning of the air, which may include the filtering out of dust and odors, freshening with outdoor air, adjustment of the temperature, and adjustment (up or down) of the relative humidity.

The general purpose of having an HVAC system is to maintain some level of healthful and comfortable conditions for people inside a building (see Figure 5.16). This problem has many variables, including the following major ones:

Local Climate. The *need* to heat or cool, or to add or remove moisture in the air, depends on the local climate, as well as the season of the year.

Building Occupancy. Who is occupying the building (athletes, hospital patients, etc.) and what they are expected to do in the building (sleep, perform aerobics, etc.) makes a difference for what is an acceptable interior condition.

Building Size, Shape, Construction Type, etc. This affects what form of system may be used and how the operating equipment may interact with the general building construction.

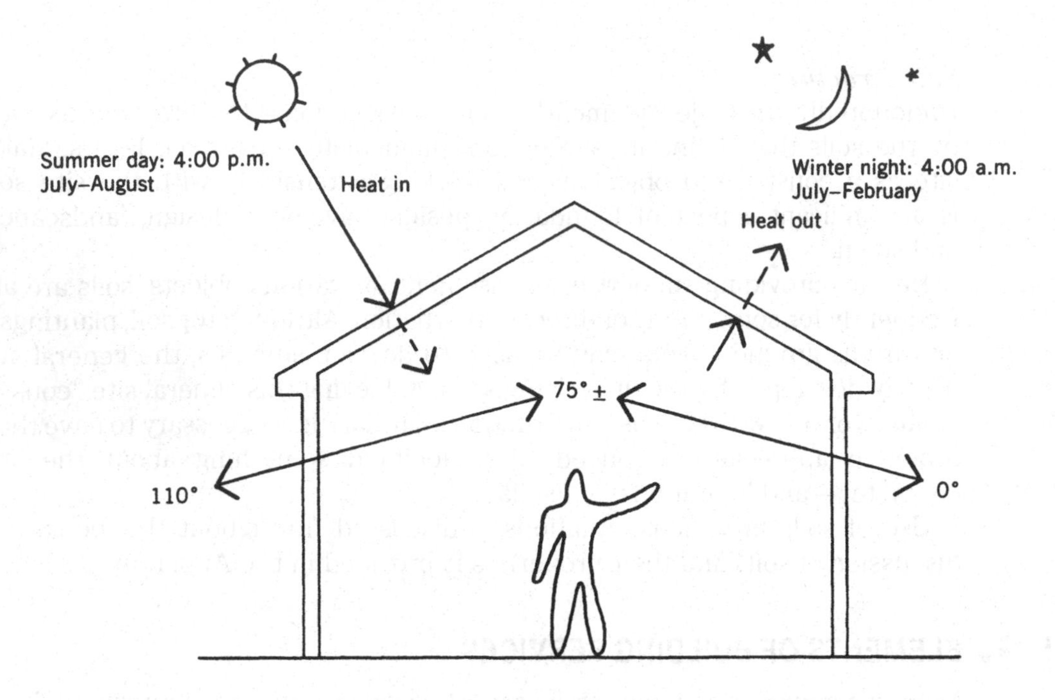

Figure 5.16 Fundamental concerns for human comfort in enclosed buildings.

HVAC systems generally include the following basic components:

A heat generating system (furnace, fireplace, etc.).

A cooling system (refrigeration unit, evaporative cooler, etc.).

An air handling system (circulating, exhausting, refreshing, etc.).

A control system for hand adjusting and/or automatic monitoring of the system operation.

Functions can be handled individually, or combined in various ways. In a so-called "all air" system the basic heating or cooling may be achieved as part of a whole process of conditioning the air. Cooling is most often achieved this way, although other heating methods are sometimes more effective.

Major operating concerns are for the control of the system and the amount of energy consumed. Cooling in steamy climates and heating in frigid ones are major sources for energy consumption and typically receive a lot of attention in terms of achieving optimal designs for energy savings.

From an economic point of view, the building construction represents primarily a major, one time cost investment. Initially installed HVAC systems are similar, but the more critical concern may be for their operating costs over some period of time. If modifications of the building construction can reduce these long term costs, they may justify some higher initial investments.

Other factors that affect the relations between HVAC systems and the general building design include the following:

Space for HVAC Equipment. Space must be allotted for equipment, and it must often be strategically located for its feasible operation.

Noise and Vibration. These are critical factors for cooling equipment and large fans. Equipment must also be accessible for maintenance and replacement.

Space for Air Duct Systems. These must serve the building's interior spaces and link up with the operating equipment. Air intake and exhaust must connect to outdoor air at appropriate locations.

Properties of the Building Enclosure. Thermal responses and air leakage of roofs and exterior walls will affect the task required of the HVAC system.

Building Planning. Various factors of the building form and interior arrangement will establish the nature of the HVAC problems. Compass orientation for sun exposure, locations and numbers of windows, number of discrete interior spaces, and many other factors will affect the complexity and efficiency of the HVAC system and the choices for basic types of systems.

For various purposes, it is advisable to closely integrate the designs of the HVAC and lighting systems for a building. Various aspects of this are discussed in Chapter 9 with regard to the development of integrated systems.

Lighting Systems

Lighting (actually illumination) is usually achieved in buildings by some combination of daylight and electrical lighting fixtures (see Figure 5.17). During daylight hours the building site and general exterior, as well as some parts of the interior, are typically illuminated by daylight. In these energy-conscious times, it is usually a design goal to make *maximum* use of daylight for energy and cost savings.

At night, as well as deep inside the building at any time, some other form of lighting must be provided; typically from electrically-powered lighting systems. This involves the installation and maintenance of the light fixtures, installation and occasional maintenance or rewiring of the electrical power distribution system, and some means of control of the system (switches, dimmers, etc.).

For large commercial and institutional buildings—especially those having long hours of usage—the powered lighting system may be a major user of energy. In these cases, striving for an efficient lighting system may become a major design goal for the building. While this effects the design of the lighting system itself, it also quite often relates to the general planning of the building to reduce the demands for lighting.

In reducing the need for powered lighting, the least flexible item is the need of the building user. This may involve the necessity to perform some specific act (walk down a stair, read a book, etc.), usually defined as the visual task. Or, it may relate to some goal, such as the proper viewing of a displayed sales item. These are the things that need doing, and the whole lighting system must respond. Proper selection of lighting equipment and proper installation in terms of positioning units may permit the powered system to do an optimal job of providing the necessary illumination.

The planning and detailing of the building form and its construction have a considerable effect on defining the illumination problems. Selection of internal finish materials in particular can have a major effect on the operation of the internal illumination systems; as can ceiling heights, room height-to-width ratios, and the number, height, size and placement of windows.

In warmer climates, where solar heating is a major thermal comfort problem, the design for use of daylight must be very carefully integrated with that for solar shading from heat gains. And now, in any climate, it is generally desirable to have close coordination in the general designs for HVAC concerns and lighting, as part of an overall energy use reduction effort.

Figure 5.17 Interaction of natural and artificial illumination for buildings during daylight hours.

Lighting presents many technical problems, requiring good engineering analysis and design. However, illumination is also a major architectural design problem: if the building can't be seen, how can it be appreciated architecturally? Therefore, for purely *visual* reasons, the architect must be sensitive to the issues and solutions for lighting problems. Equally important, the implications of choices of building form and detail with respect to influence on lighting problems must be appreciated.

Power Supply Systems

The major power source for buildings is electricity; typically generated somewhere off the building site and delivered in wired form to the building (see Figure 5.18). This *power* source is actually an *energy* source with a potential for many uses; some of which can be done by other sources, such as gas, oil, steam, hydraulic force, or the sun. But to operate the artificial illumination system, run the computers, make the fans blow, and operate numerous appliances, we need electricity.

The electrical supply system is permanently attached (hard-wired) to many elements, such as the HVAC equipment, lighting fixtures, elevators, and so on. However, it must also supply an open plug-in system for the more-or-less random attachment of portable items, such as TVs, desk lamps, power tools, and kitchen hand appliances.

Installation of the electrical power system usually has minor influence on architectural design or construction development issues. There are many special requirements for the details of the electrical systems themselves, but only a few that affect the general construction; some of which are discussed in the case studies in Chapter 10.

A major concern for electricity is the potential danger for fire or for severe electrical shock. Codes have many requirements for the systems that originate as concerns for safety. In some situations the building construction is relied on to protect occupants from the electrical system, where elements are buried inside the

Figure 5.18 Electric power, delivered from an off-site location; a less than gracious element of the visual environment.

construction. Today, however, the systems themselves are mostly made as user-safe as possible on their own.

For large buildings there are many pieces of equipment that must be housed. This involves both a space commitment and some attention to the security of the equipment—providing protection against misuse and protecting occupants from accidental dangerous shocks.

With our ever-expanding technology, demands for use of electricity often increase after initial construction of a building, or user needs change. Access to the electrical equipment and wiring must be reasonable to permit alterations of the system.

Signal and Communication Systems

Just about every residence these days has a doorbell, a telephone system, and a thermostatic control for its HVAC system (see Figure 5.19). Each of these systems is typically operated with a separately-wired, low voltage, direct current electrical system. In many buildings there are also cable TV systems, computer networks, public address or intercom systems, and fire alarm systems; all using their own wired systems.

Many of these wired systems may be built into the construction, in the same general manner as the basic electrical power system; although safety requirements for low voltage systems are considerably eased. Depending on the degree of permanence of the systems, however, they may be somewhat less *hard wired;* that is, they

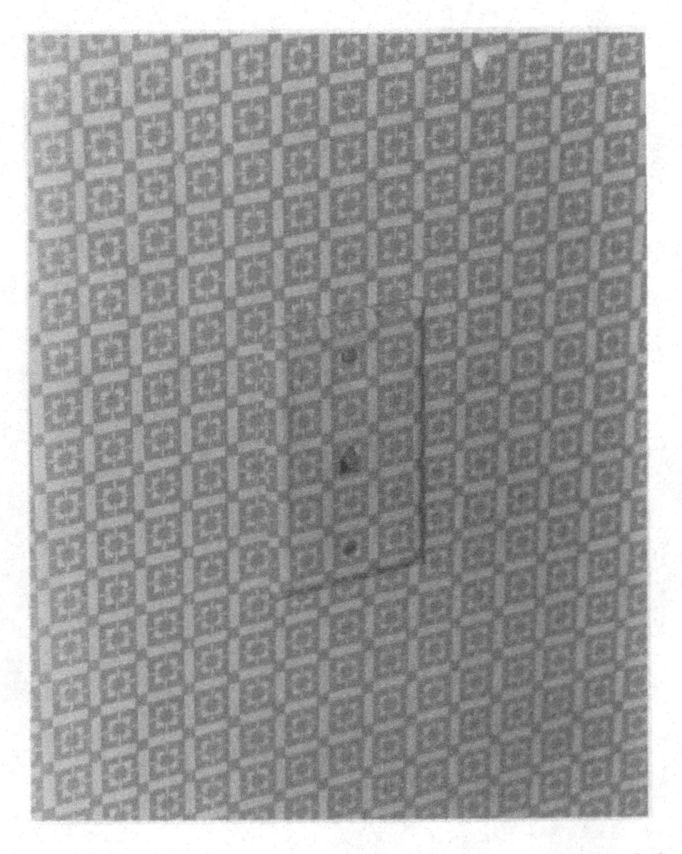

Figure 5.19 Ubiquitous elements of modern interiors: the phone jack and the thermostat. Can be camouflaged, decorated, or disguised, but not gotten rid of.

may edge over to the class of systems more similar to those typically installed by building users for their personal hi-fi systems or small scale computer networks.

Some issues regarding these systems are discussed in the building case studies in Chapter 10.

Piped Systems

The most common piped systems are those installed for the water supply, waste drainage, and natural gas supply systems (see Figure 5.20). Each of these has special concerns; for example, pressurization in the water supply and gas supply systems, and gravity flow for drainage in the waste systems. These systems also typically connect to off-site mains, presenting problems for the site design and the general building base and foundation, as well as details for incorporation into the building construction.

Piping may also be required for fire sprinklers or other fire-fighting purposes, for irrigation, for internal roof drainage (usually separate from the system for sanitary waste from toilets, etc.), and sometimes for pressurized air or other special liquids or gasses in hospitals or laboratories.

Piped systems usually have three components: a supply point (delivered from the mains, for example), the piped distribution or collection system, and terminals at some form of fixture (toilet, sink, shower, etc.). The building entry from the mains, the piping network, and the fixtures must all be accommodated in the construction. Concerns for the various common systems are discussed for the building case studies in Chapter 10.

Sound Control Elements

Control of sound in buildings is not so much done with a "system" as it is with special attention to the form and details of the construction. Design for sound control begins with an analysis of the potential problems (maintaining privacy, containing noisy sound sources, enhancement of acoustic conditions in single spaces, etc.). Solutions for these may typically involve combinations of efforts: planning for separation, choice of surfacing materials, shaping of interior forms, and modifications of the construction of separating elements (walls, roofs, floors, etc.).

Sound may be carried by the air or be transmitted through the solid elements of the building construction. Both transmission paths need consideration and involve different properties and details of the construction.

Special concerns for sound in some particular situations are illustrated in the case studies in Chapter 10.

Fire Control Elements

Fires in buildings are best "controlled" by eliminating the hazard for their ignition and spread in the first place. This can only be partly controlled by the building designers; building users can create hazardous conditions by moving dangerous or simply highly flammable materials into any building, by overloading electrical equipment or circuits, by careless use of matches or smoking materials, and so on.

It must therefore be assumed that a fire can occur in *any* building, and the scenario for fighting the fire and for safely removing the building's occupants must be played out. A major portion of building code requirements is devoted to these concerns, and no other single concern has a greater impact on the restriction of use of the materials, form, and details of building construction.

In addition to the determination of the basic construction, there are often special

Figure 5.20 Building hookups to off-site sources for water and gas.

elements and systems required for the total fire control, detection, alarm, and suppression. This may involve lighted exit signs, automatic door closers, water sprinklers, fire hose cabinets, special water supply for fire fighters, smoke alarms, and so on. The total fire control "system" thus virtually involves the *whole* building.

Some general concerns for the general fire resistance of the building construction are discussed in Sec. 8.1. Some special construction details relating to fire control are illustrated in the case studies in Chapter 10.

6

Materials

Materials for construction of buildings include traditional ones long in use and new ones emerging as recent technological discoveries and developments. The following discussion treats primarily the most common and widely used materials for current building construction.

6.1 WOOD AND WOOD FIBER PRODUCTS

The enduring large stands of forests in North America constitute a resource for wood for many purposes (see Figure 6.1). For buildings, major uses are the following:

Solid wood, sawn and shaped: This includes structural lumber plus elements used for trim, shingles, flooring, window sash, and door frames.

Glued laminated elements: These include plywood, timbers made from multiple lumber pieces, and various other products.

Fiber products: These are produced with wood fiber as the primary bulk ingredient and include paper, cardboard, compressed fiber panels, cemented shredded fiber units, and cast products with wood fiber as the principal aggregate.

Wood is organic and is subject to decay over time (by simple aging or by decomposition), to rot, and to consumption by insects. A critical aging process occurs in the early weeks and months after a tree is cut down. For solid sawn pieces, a major property is the degree of retained moisture, which generally indicates the extent to which the wood has cured from the fresh-cut (green) condition. The moisture itself is not of concern, but the extent to which the natural shrinkage of the material has occurred is, as this is mostly what causes warping and splitting as well as other flaws, shape changes, and dimensional changes.

Most wood must be finished or protected in some manner if exposed to weather. A few species—such as those used for shingles, siding, and poles—can weather naturally. Coatings and cladding may be essentially for protection, or may be primarily decorative.

While wood comes from renewable sources, the quickest renewal is achieved mostly with trees that are functional primarily for fiber sources and not for solid sawn products. Large old trees, of the types used for the best solid wood products, are a dwindling source. For this, and other reasons fiber products and various forms of laminated products are steadily replacing solid wood in many applications. Similarly,

Figure 6.1 Common building elements of wood and wood fiber.

fiber products are also displacing plywood from many uses which involve low structural demand.

Wood is still the material of choice in the United States for most ordinary construction and is mostly displaced only when there is a heightened concern for fire or the need for some special property that is beyond the potential of the material. Even if it eventually recedes significantly as a solid sawn product, wood will endure as the basic source for paper and other fiber products.

6.2 METALS

Steel is the most widely used metal for construction, including a range of size and usage forms that is greater than that for any other material (see Figure 6.2). Hardly any form of construction can be achieved without some steel products. Nails and other connectors for wood, reinforcement for concrete and masonry, and innumerable hardware items are required in every building.

The most impressive uses of steel, however, are for the large rolled beams and columns for high-rise building frames and the stranded cables used for tension structures. Even for structural applications, however, extensive uses are for modest elements, such as formed sheet steel decking.

When exposed to air and moisture, steel rusts—a progressive condition that eventually destroys the material completely. Large, thick pieces may take a long time to rust away significantly, therefore they endure for a reasonable time. The thinner the piece, or the more crucial it is to maintain its full cross section (for structural use, for example), the more serious it becomes to mount some rust prevention method. The general means is by coating the steel surface with other metals or a rust-inhibiting paint. Steel reinforcing bars may be coated with plastic where corrosion inside of concrete is a concern.

Steel structures are noncombustible and are used to replace wood where fire is a concern. The light wood frame (2 × 4s et al.) can be duplicated in light-gage steel (formed sheet) elements. However, steel elements heat rapidly and soften quickly, so some protection is required for steel structures where any significant fire resistance is required.

Aluminum is used only sparingly for structures, as its cost is prohibitive. However, it has considerable corrosion resistance in some applications, weighs about one-third as much as steel, and can be extruded into complex, sharp-cornered forms—giving it an edge for usage in various situations (see Figure 6.3). Intricate framing elements for windows, doors, and curtain walls are such a case.

Other metals are used for special purposes, where their properties are significant. They may also be used for coatings, such as the ultrathin silver and gold films used to develop reflective glazing.

6.3 CONCRETE

Concrete is a material consisting of loose, inert particles that are bound together by some infilling binder (see Figure 6.4). That definition could be extended to include many mixtures, but the principal ones covered are the following:

Structural concrete: This is the familiar material produced for sidewalks and streets as well as foundations and building frames. Aggregate is typically locally available gravel, crushed stone, sand, clam shells, and so on—selected in a range of

a

b

Figure 6.2 Utilization of steel for building construction.

c

Figure 6.2 *(continued)*

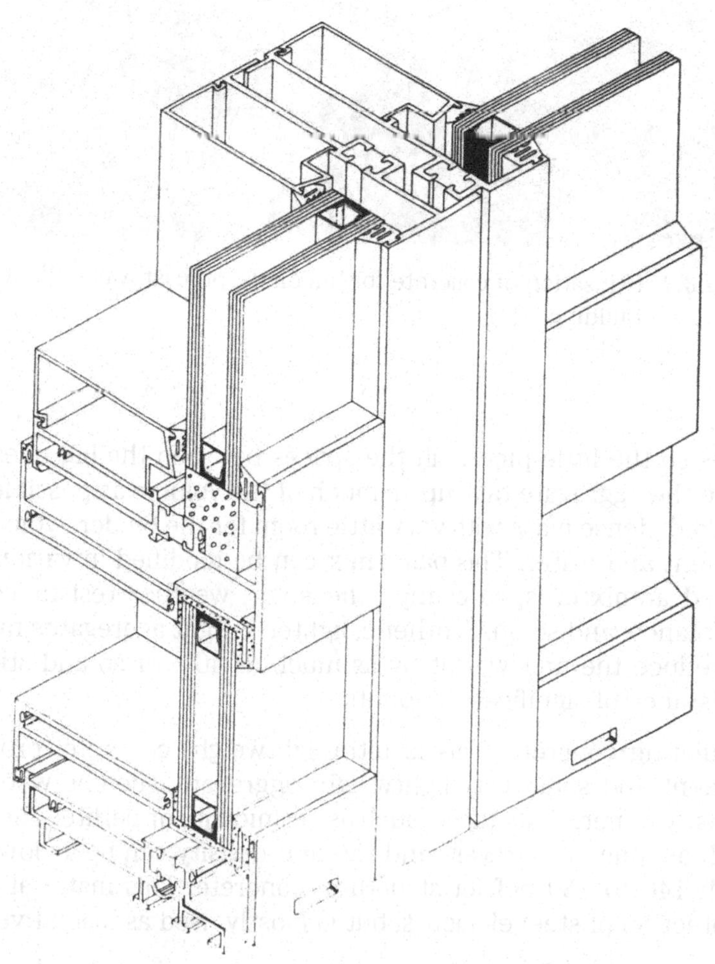

Figure 6.3 Utilization of aluminum for buildings: complex extruded elements form a priority window wall framing system. 8900 Series by Custom Window, Englewood, CA.

Figure 6.4 Utilization of concrete for buildings: precast wall units form a two-story building.

sizes so the little pieces fill the spaces between the big ones. The objective is to have the aggregate take up as much of the volume as possible, producing a tightly packed, dense mass with very little room for the binder, which consists of portland cement and water. This basic mix can be modified in various ways by additions, called admixtures, to change the color, weather resistance, water penetration resistance, and so on. Synthetic, lighter-weight aggregates may replace the gravel to reduce the unit weight by as much as 40% or so and still produce structural resistance of significant amounts.

Insulating concrete: This is ultra-lightweight concrete, produced with portland cement and some very lightweight aggregate, such as wood fiber or some low-density mineral substance, such as vermiculite or perlite. The mix is lightly foamed with as much as 20% air, and the unit density can be as low as 30 pcf, compared with 140 to 150 pcf for structural concrete. This material may be used for fire protection of steel elements, but is mostly used as insulative fill on roof decks.

Fiber concrete: Research and experimentation have resulted in the use of various forms of concrete with fiber materials in the mix. The general purpose of the fibers is to add some additional tensile capacity to the concrete, reducing or eliminating the need for the usual steel reinforcement used to alter the tension-weak character of the concrete. Roofing tiles, siding, and pavements are some of the current applications for this form of concrete.

Gypsum concrete: Gypsum can be substituted for portland cement in plaster or concrete. A lightweight roof deck material is produced with gypsum and an aggregate of wood chips. This is an insulative material with significant strength and stiffness for roof structural applications in controlled situations. Priority systems consist of combinations of gypsum concrete cast over modular metal supports and forming units, the forming units developing ceilings in some applications.

Bituminous concrete: Substitute a hot bitumen (coal tar, etc.) for the portland cement and water in ordinary concrete and you have asphalt pavement. Not much used for buildings, but a significant material for site development.

In most applications, structural or not, some compensation or modification must be made to overcome the tension-weak character of concrete. Steel reinforcement in the form of rods or welded wire mesh is the usual solution. The concrete encasement serves to protect the steel to prevent rusting or to insulate it from fire. Other means for improving tension resistance include the use of fiber materials in the mix (as mentioned previously) and *prestressing,* which consists of inducing compression in concrete structural members before loadings occur that produce tension. Prestressing is usually induced by high-strength steel strands stretched inside the concrete and anchored or bonded to the concrete.

Concrete must be cast in forms and cured over some period of time after it is mixed and placed in the forms. Hardening is produced by a chemical reaction in water-mixed concrete, and takes some time to occur in ordinary situations. The cast concrete must be kept moist and at a reasonable temperature during the curing period.

Mixing, transporting, casting, finishing, curing, and reinforcing or prestressing of concrete represent major cost factors. Reducing the cost of these items is often more significant than attempts to make the mix itself cheaper. Better to go for the best mix and find ways to reduce other costs—particularly the time-consuming and labor-intensive ones.

Concrete is generally the most durable material for ground-contacting construction, and is thus extensively used for foundations, basement walls and floors, pavements, and various site constructions. It is also quite weather resistive and can function reasonably in an exposed condition as a raw, unprotected material. Finally, it is considerably fire resistive. All of this frees it from the need for the many concerns for protection required with wood and steel.

6.4 MASONRY

The term *masonry* covers a range of construction that overlaps the areas of tiling, cladding, and precast concrete (see Figure 6.5). Traditional forms of brick and stone masonry can still be obtained, but other processes and materials have largely displaced them. Most structural masonry for bearing walls, shear walls, or foundation walls is now produced with units of precast concrete (concrete blocks—called CMU for concrete masonry unit). What appears from the outside to be a building with brick walls these days is often a frame structure with thin "brick" tiles adhesively bonded to a backup material.

Figure 6.5 Utilization of masonry in buildings: structural wall with precast concrete masonry units (concrete blocks), reinforced with steel rods in concrete-filled voids.

Stone is mostly used decoratively these days. Rough stone is developed as either veneer or facing on other structures. Cut stone (granite, marble, etc.) is cut in increasingly thin slices to form facings for what is usually a framed curtain wall system.

In the face of these trends, the crafts of stonemasonry and bricklaying slowly decline and become ever less obtainable. The enduring popularity of the form of construction (or at least its appearance) keeps it alive—but increasingly as illusion rather than fact.

Classic masonry consists of some inert units bonded into a contiguous mass with a binder between the units. The binder is usually mortar—similar in form but quite different in its ingredients and properties from that used in earlier times. Units may be stone—cut or unprocessed—bricks of cast and fired clay, precast concrete (mostly as hollow block), or other elements, such as hollow tiles of gypsum or fired clay.

It is desirable to have masonry consist mostly of the units, with a minimum of mortar in thin joints between the units. It is best to rely on the mortar only for compressive resistance, although it is desirable to have it adhere to the units to bond them together. Strength and stability of the construction depend on the integrity of both the units and the mortar joints, and to a considerable extent on the craftsmanship of the workers who prepare the units, mix the mortar, lay up the units, and protect the construction while the mortar cures.

Units must be closely fit together, so the size and shape of the units, the patterns of their arrangement, and general form of the construction they are forming, and the required edges, corners, openings, intersections, supports, and other form-related concerns must be carefully developed. This is especially critical with CMU construction; bricks and stones can be cut to trimmed size, but concrete blocks must be used in whole-piece form. Special sizes and shapes of blocks can be provided, but some dimensional and modular control is necessary.

Like concrete, masonry is tension-weak, which needs consideration in structural applications. The strongest masonry structures are usually those developed as *reinforced masonry,* which uses steel rods in a manner similar to that for reinforced concrete. This is of greatly heightened concern for structures required to resist major earthquakes.

6.5 PLASTICS

Plastics are used extensively in buildings as solid formed elements for trim and hardware, as film for moisture barriers or in laminates, as foamed insulation, and for many forms of coatings. As in its early days, plastic is often used to imitate other materials. Thus wood paneling is often a photograph embedded or laminated into plastic and mounted on a compressed wood fiber panel; wood strip siding is duplicated in form and appearance with formed vinyl sheets; metal trim or surfacing is actually thin metal film on plastic; brick, Spanish clay tile, mosaic tile, marble, and just about anything can be imitated with plastic in the Disneyland style.

Chemistry, forming methods, and combinations of plastic with other materials can be varied extensively to respond to many demands. Potential problems include concerns for fire resistance, low strength or stiffness, wear resistance, stability over time, and possible toxic effects.

Many paints, varnishes, and other coatings have a plastic base. Multiple coatings may produce combined effects, with each coating adding something additional to the total effect.

6.6 GLAZING

Glazing for windows and doors is usually of glass. Various forms are used, depending on the size of glazing unit required and various special considerations (see Figure 6.6).

Single-Thickness Glazing

This mostly consists of a single piece of glass—float glass for small panes, polished plate or other forms with truer surfaces for larger, thicker pieces. Transparent glazing is developed to be colorless, although some modification of transmitted light always occurs. For privacy, sun control, or heat transmission alteration, the glass may be coated, tinted integrally, or have one or both surfaces altered by grinding or other surface texturing.

Multiple-Thickness Glazing

To reduce heat flow through the glazing it is possible to use more than a single thickness. In the old days, this usually meant a second window, called a storm window. A newer technique is to bond two panes of glass together with a small sealed space between them, producing a thermos bottle effect. For additional insulating effect, a third pane may be added, with a second trapped air space in the sandwich.

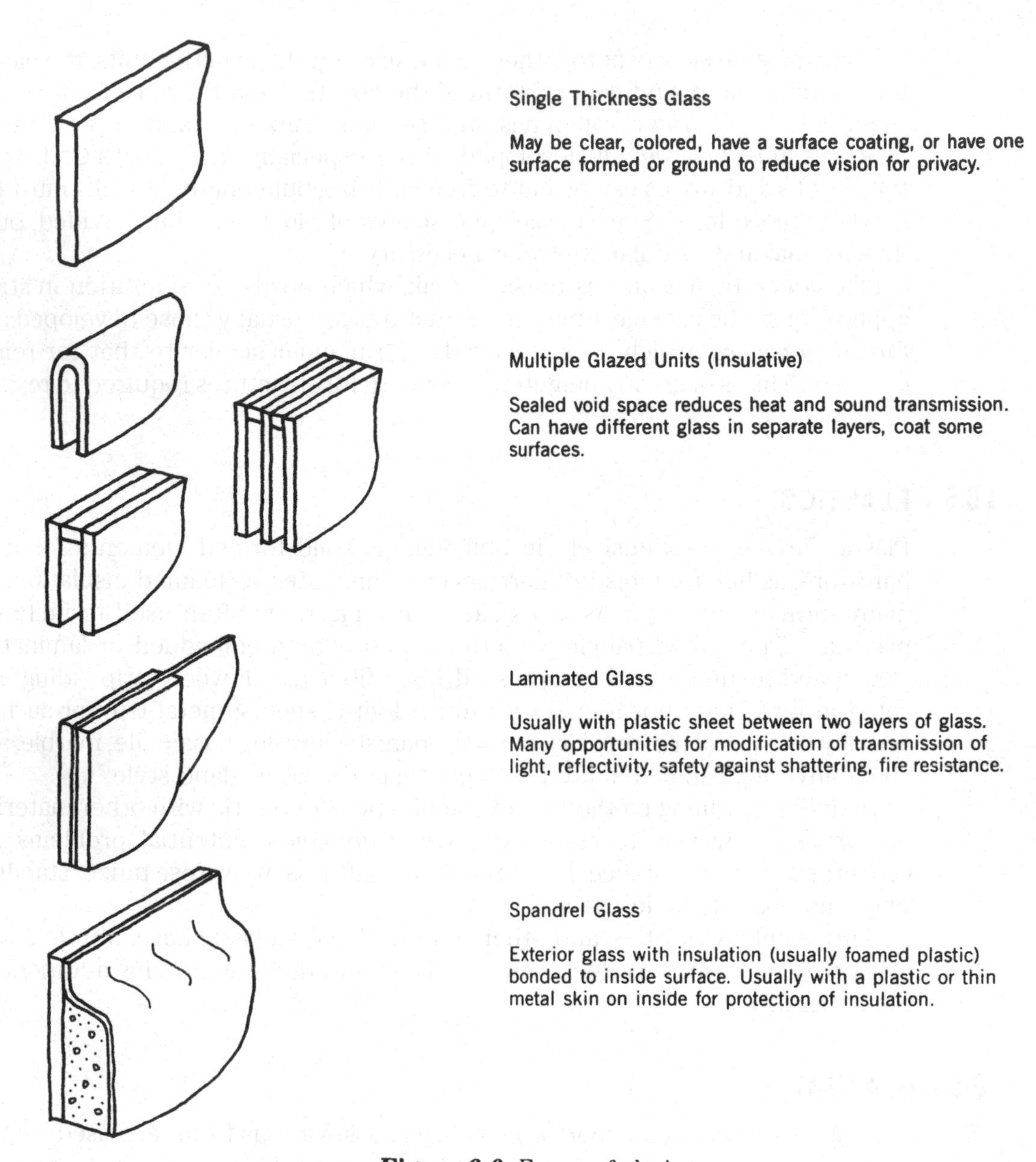

Single Thickness Glass
May be clear, colored, have a surface coating, or have one surface formed or ground to reduce vision for privacy.
Multiple Glazed Units (Insulative)
Sealed void space reduces heat and sound transmission. Can have different glass in separate layers, coat some surfaces.
Laminated Glass
Usually with plastic sheet between two layers of glass. Many opportunities for modification of transmission of light, reflectivity, safety against shattering, fire resistance.
Spandrel Glass
Exterior glass with insulation (usually foamed plastic) bonded to inside surface. Usually with a plastic or thin metal skin on inside for protection of insulation.

Figure 6.6 Forms of glazing.

Laminated Glazing

Laminates can be formed with various materials for different purposes. So-called safety glass usually consists of two sheets of glass with a sheet of plastic in the center; the adhesion and toughness of the plastic prevents the usual form of breakage of the unlaminated glass. An open-mesh steel fabric may be inserted for additional breakage resistance, producing a form of glazing used for some fire doors and skylights. There are now many forms of laminated glazing, permitting many special modifications for safety, color change, reflectivity, and so on. Safety concerns for falling broken glass from curtain walls in urban areas are now resulting in the use of laminated glazing for many such buildings.

Spandrel Glass
This is not actually glazing, but consists usually of using similar units to form the opaque portions of curtain walls that occur at spandrel beams, columns, or portions of walls not containing windows (below the sill, when floor-to-ceiling glazing is not used). Although glass is used, and installation may be similar to that for the windows, special products are usually used, often in the form of a sandwich with some insulation materials bonded to the inside surface of the glass.

Plastic Glazing
Hard plastic can be used for glazing in some situations, although plastic is used more often in laminates. Major problems are the scratch resistance (as compared to glass) and the fact that some forms of commercial products for cleaning can dissolve or etch the plastic surface. Plastic glazing is not much used for ordinary windows, but is used for skylights, sky windows, and some large curtain wall or storefront systems. Special sandwich panels with plastic facings bonded to metal frames may be used in some situations. The single-piece, domed plastic "bubble" is one type of element used for skylights.

6.7 WATERPROOFING MATERIALS

Water control is a major problem in building construction. Various materials are used in the forms of membranes, coatings, sealants, flashing, and miscellaneous joint-sealing devices (closers, gaskets, water-stops, mouldings, etc.). The range of materials is considerable, including metals, rubber, plastics, and various bituminous compounds.

This subject is best addressed in terms of various specific types of problems and particular construction elements. It is treated in various sections in Chapter 7 and in several of the building case studies in Chapter 10.

A distinction is made in construction between *waterproof* and merely *moisture-resistive;* the first implying a total form of protection or resistance and the other some degree of enhancement. For example, roofing for a flat, horizontal surface must be quite waterproof, as also must be the seals in curtain walls and the liner for an indoor swimming pool in an upper floor. On the other hand, the outsides of basement walls in ordinary conditions need only be moisture resistive to a degree, as also must be the wall surfaces in bathrooms and floors in kitchens.

6.8 SOILS

A major material used for site development is soil. Building in and on the ground literally means using soils as construction materials. The general nature of soils—including their structural properties—must be reasonably understood by any designer involved in site, foundation, or below grade construction.

For landscaping development it is also necessary to understand the nature of soils related to establishment and nurturing of plantings. Where extensive planting is to be developed, an early issue to be settled is the potential for using existing site materials and the means necessary to preserve them for final landscaping work.

This book deals essentially with construction, so we will not attempt to fully treat the subject of planting design, except for some aspects related to construction problems. The general engineering concerns for soils, however, are treated in the discussions in the Appendix. Various construction problems with soils are discussed in Sections 7.11 through 7.15 and illustrated in the case studies in Chapter 10.

Information Source
Simplified Design of Building Foundations, 2nd ed., James Ambrose, Wiley, New York, 1988.

6.9 PAVING MATERIALS

Surfacing of the ground can be achieved with raw soil, with general plantings (grass, etc.), or with a variety of pavings. The desire or need for paving usually arises in anticipation of some form of traffic, and a primary concern must be for the support of the type of traffic. Fine gravel or pulverized bark may surface a lightly trodden garden pathway, but something serious is required for heavy trucks on a daily basis. Pavings for various purposes and with various materials are discussed in Sec. 7.14 and illustrated in Chapter 10.

6.10 MISCELLANEOUS MATERIALS

Many materials can be used for various purposes for building construction. Some are used in unique applications; some are used as substitutes for more traditional materials—for cost savings or for some enhanced property that is superior to the material it is replacing.

Paper
Paper is used with various elements of building construction. Two common uses are the facings for gypsum drywall units (a sandwich with a gypsum plaster core) and the backup material for stucco. Paper can be coated—usually with plastic, but also with aluminum foil, bitumen, or wax—or can be laminated. For use as a moisture barrier a laminate produced uses two layers of paper with a bituminous material in the center. As with other multilayered elements, reinforcement may be added, often in the form of strands or mesh of fiberglass or strong plastic threads.

Cardboard in solid form is more or less thick paper, intermediate between ordinary paper and various forms of compressed fiber panels. Many variations, including ones with coatings or lamination, are possible. Concrete forming, such as that for round columns, is sometimes accomplished with heavy cardboard elements.

Corrugated cardboard is actually a sandwich of alternating layers of flat and corrugated paper, produced essentially to obtain a material with strength and stiffness but light weight. As with other products, coatings and reinforcements are possible.

Composites
Increasing use is being made in construction of materials that are classified as *composites.* These are materials produced as blends or mixtures of more than one basic material. So called fiber concrete, for example, is made by blending mineral or other fibers into the concrete mix. However, the term composite is also used to describe elements assembled with multiple materials, as in the case of laminated glazing, which consists of alternating layers of glass and plastic film.

A composite material or element may consist essentially of a modification of some basic material to add or improve some characteristic. Thus steel bars cast into concrete or glass fibers encased in plastic result in major enhancements of the basic material; adding tensile resistance to the concrete and considerable tear resistance to plastics.

However, a new composite may also result in a truly new product, with a whole array of new properties. Good old wood, steel, aluminum, concrete, and masonry will surely endure, but the future really belongs to the new and improved materials that are emerging from materials science and experimentation.

General References for Materials

Construction Materials: Types, Uses and Applications, 2nd edition, by Caleb Hornbostel, Wiley, 1991.

The Science and Technology of Building Materials, by Henry J. Cowan and Peter R. Smith, Van Nostrand Reinhold, 1988.

7

Construction Elements

This chapter deals with the component parts from which buildings are created. Taken in its entirety, this is an immense topic—one that could easily fill a very large book or a multivolume set. There are indeed many books that treat the general topic, as well as the many fractional parts of it. Several sources for more detailed information are discussed in Chapter 3. The purpose here is to give a concise overview with a concentration that deals only with major issues and elements of building construction. In addition, the viewpoint emphasized is that of considerations most significant to design—of the building in general, and of the whole construction.

7.1 WALL STRUCTURE

There are many ways to build walls, but most methods fall into one of three basic categories: frame plus surfacing, monolithic, or panel. Many factors determine the choice of methods, including the choice of materials, desired appearance, and various functional requirements for the wall.

Frame Plus Surfacing

The light wood frame with 2 × 4 studs (see Figure 7.1) is the most common structure of this type, but many walls are also produced with steel frames, using many of the same surfacing options as those for the wood frame. A variation of this type is the curtain wall, in which surfacing units may be installed between framing members, as well as on the surface as covering on the frame.

For structural walls, vertical frame members usually function as vertical columns, and their size and spacing may need to relate to the specific tasks for load support. Surfacings on framed walls do not usually function to assist in support of vertical loads, but they may be used for bracing of the frame in the direction parallel to the wall surfaces, even serving to develop shear walls for the lateral bracing of the building.

Besides basic stud framing, various other framing members are required—to form edges, to surround openings, and sometimes simply to provide for fastening of the surface materials at joints in the surfacing.

Surfacing must first of all provide some form of rigid structure across the space between framing members, as a roof or floor deck must. Loads are usually less demanding here, but some structural spanning strength is desired.

Beyond its structural tasks are the need for considerations of the contributions of facings to appearance, fire resistance, thermal flow, acoustic separation, security, and any other needs for the wall in general. The problems of both exterior and interior

Figure 7.1 Development of building walls as surfaces applied to a structural frame: plywood and stucco on light wood frame.

facings are considered in the next two sections. Enhancement of walls for various performances is considered in Chapter 8.

Framed walls ordinarily produce void spaces inside the wall. These are handy for the placing of wiring, piping, and various other elements of the building services to maintain an uncluttered interior space. However, fires can spread in these spaces out of view of occupants and fire fighters, so some deliberate blocking of the continuous form of the enclosed voids is required.

Monolithic Walls

Masonry and concrete construction usually produce walls of a solid, monolithic character, although voids in blocks or cavities in multiwythe walls may create some hollow spaces. The basic idea here is that structure and surface are produced at the same time (see Figure 7.2). There is no basic need for surfacing, although various ones are often applied.

These walls do not provide for the ease of installation of wiring and piping or other contained elements, or for their easy modification in the future. This calls for some higher degree of integration and coordination of the design work, as deliberate provisions must be made in the detailing of the construction.

Many of the same surfacing materials used with framed walls may be applied here. If applied directly to the structural surface, they need not be structural in nature to the degree required when they are placed on a frame. However, surfaces are sometimes applied on framing strips attached to the wall (called furring), and the surfacing is thus in a situation similar to that with a framed wall.

Monolithic walls of masonry or concrete are usually quite thick in comparison to framed walls. Framed walls may be deliberately thickened for various reasons, but masonry and concrete construction usually requires considerable thickness. If surfacing materials are added, thicknesses are increased further. One possible exception is the precast concrete wall, not cast in vertical forms.

Panelized Walls

Various forms of construction can be used to create walls in preformed, panelized units (see Figure 7.3). Individually framed units, sandwich panels, precast concrete units, and various curtain wall systems are of this type.

Panels may be developed strictly for the development of the exterior walls, or they may be part of a general modular, prefabricated system for the entire building construction. In either case, the panels may develop only the wall structure (as in the case of curtain wall systems) or may serve structural purposes for the rest of the building.

Figure 7.2 Monolithic wall construction: thick walls produced with masonry construction using precast concrete units (hollow concrete blocks); may be left with surfaces unfinished, producing an entire wall with a single unit (structure, exterior finish, and interior finish in one piece). This is the closest modern equivalent of ancient construction processes, such as that with adobe bricks.

Figure 7.3 Walls prefabricated with large panel units: precast concrete walls cast flat on the site and lifted into place (called tilt-up construction).

Prefabricated panels are often factory produced, where material use and type of work possible may permit some different forms of construction than normally associated with site construction work. On the other hand, site joining of units should be as simple as possible to permit quick, easy assemblage to capitalize on the time savings potential with such systems.

Custom design of a panelized system may be justified if there is sufficient amount of the construction. However, custom designs for factory production and the transporting, handling, site assemblage, and site finishing work require a great amount of design effort and considerable knowledge of all the processes involved.

Many systems of this type are priority products of manufacturers with experience in the whole range of activities, from initial design for specific requirements to final installation. Possible direct use or adaptation of such systems, available in the region of a proposed building, should be investigated before proceeding with a complete, original design effort.

Special consideration must be given to interfacing prefabricated elements with the rest of the building. This applies to foundations and other supports, connections to other construction components, and installation of wiring, piping, and other built-in elements for building services.

▪ 7.2 ▪ EXTERIOR WALL SURFACES

Surfacing of exterior walls must usually be done in response to two major considerations. These are:

Appearance—as a major architectural design consideration; the degree of concern depending on the character and use of the building, as well as the relative visibility of the wall (street side, alley side, etc.).

Weather resistance—depending on climate conditions.

Surfaces may be self-generating, as the natural result of some form of construction—such as masonry. Many surfaces, however, are developed by adding separate elements, as cover on a frame or as a covering strictly to develop the surface itself.

Natural Surfaces

Some natural surfaces consist of the raw, untreated construction materials. Concrete, brick masonry, stone veneers, glass, and some very weather-resistive woods such as cedar present this as a possibility (see Figure 7.4). Often, however, the "natural" finish may be enhanced, either to simply protect the surface or to slightly modify its appearance.

Clear sealers may be used on wood and masonry to provide protection and help preserve the materials. These may let the material's natural character remain basically unaltered, or they may provide some alteration such as a slight color change.

Some materials that are left essentially untreated may produce a natural surface over time that is significantly modified from that of the original surface. Copper will develop an oxide coating that presents the characteristic mottled green coloration. Redwood and cedar will weather to shades of green, gray, and black. Steel will rust—normally an undesirable effect, except with grades of steel deliberately produced to form a self-arresting rust coating.

Weathered appearances may also be speeded up or artificially induced by special processes. Wood may be stained to the weathered look at construction time; metals may be treated to speed up the corrosion processes that may take some time if left to nature. Materials that may tend to weather to a nonuniform texture or color may be helped a bit to develop more uniformity of appearance.

Natural appearances that develop over a long time may be reasonably predictable for common materials used in familiar circumstances. For new materials, this experience may be yet to be gained, and may bring some surprises.

Natural appearances gained over time may amount to buildups of dirt or corrosion, and it may be desirable to clean the surfaces with some frequency to preserve a "natural" look that is somewhat fresher.

Frame Covers

There is an interactive relationship between frame covering and the framework being covered. Overall size of the frame, spacing of frame members, material and form of frame members, and feasible means for attachment to the frame must be considered in establishing the range of alternatives for the frame covering (see Figure 7.5).

This works in two directions, of course, so that spacing of frame members may be determined by the form of covering chosen. In the end, the whole construction must be compatible and work together to perform the various tasks required for the wall, including structural and nonstructural ones. Typically, the frame will function to respond to some tasks (support of vertical loads) and the covering to others (resistance to thermal flow), and the two will combine for some (development of shear wall action).

A primary function for covering is the need to span the distance between framing members. For exterior walls, this involves the resistance to wind pressure as a

a

b

Figure 7.4 Exterior finishes expressing the natural character of the construction materials: (a) Board-formed, sitecast concrete with no applied finishing materials. (b) Glass panels and metal retaining frames in their "natural" condition.

a

b

Figure 7.5 Single piece, surface-developing materials on frames, without separate applied finishes: (a) Plywood sheathing panels with finished surfaces. (b) Stucco developed directly on the wood stud frame. Both wood siding and stucco are also frequently developed as applied surfaces over a separate structural surface.

uniformly distributed load. Both the covering and its attachments to the frame must resist this form of loading.

A secondary primary consideration is the exterior appearance of the building. Single-element coverings may achieve both structural infill and finish, but this is not always the case. For appearance or for functional reasons, such as weather protection or fire resistance, a finish material may be added. This may simply be a coat of paint or other sealing, or some additional covering.

The possibilities for development of the covering of the exterior of buildings are quite extensive, running the whole range of available materials—wood, metal, masonry, concrete, plastics, and innumerable composites and mixtures. At any given time, in a specific location—for a particular size, shape, and type of building—product marketing, code requirements, and prevailing design styles will tend to shorten the list considerably.

Traditional finish materials sometimes endure in popularity, even beyond the time of their continued feasibility as basic construction elements. These days what appears to be wood strip, stone, or brick masonry, or clay tile roofing is quite often some other material imitating the original—and possibly actually excelling it in some ways, such as cost, durability, speed of installation, insulative value, and so on.

Selection of the exterior covering, as an architectural design decision, must be related to the complete appearance of the exterior wall, which includes the type, materials, form, and arrangement of windows and doors, the treatment of the bottom, top, and edges of the walls, and possibly—if it is in view—the roof covering. In addition, the exterior covering is only a part of the wall, and the whole wall construction must be considered, especially for some of its major physical properties, such as thermal flow, acoustic transmission, and so on.

Applied Surfacing

Applied surfaces are those that are mostly used to cover structural surfaces for one reason or another (see Figure 7.6). There may be real functional needs for this, such as weather protection, but for whatever reason it exists, the applied finish is what is seen and its appearance is a critical concern.

Stucco (exterior plaster), for example, was originally developed as a protective covering for adobe or other highly vulnerable masonry, or simply as a crack filler to make walls airtight and water repellent. But, as with many basically functional necessities (such as waterspouts, column capitals, cornerstones, etc.), it soon became a medium for expression by designers. Now it is often imitated with an acrylic plastic finish (which is considerably more resistive to weather, and probably would have been chosen by the ancient builders if it had been available).

Whatever the underlying wall structure may be, applied coverings may be used to develop an immense range of exterior appearances. If you ask them, most manufacturers or distributors of a particular form of covering will show you how to apply their product to just about any supporting structure you can imagine.

Of course, you may feel that the applied covering should not be used essentially as a *cover-up,* but should be compatible with and express the rest of the construction. This is a philosophical issue that will not be resolved here.

Designers should be aware of the full range of possibilities for the exterior finishing of buildings. They should also be aware of the feasible means for achieveing them in today's construction: as natural or applied surfaces; as real or fake. They should also have access to the necessary information to evaluate their relative cost and various properties that may have significance in any particular situation.

Moisture Content

Wood shrinks and swells with changes in moisture content. To minimize dimensional change after installation, install siding at a moisture content which matches the local climate as closely as possible. If climate in a particular region causes wood to maintain 8% to 13% moisture content annually, then the most ideal siding would be installed at a moisture content within that range, and the material would be stored, stickered and protected for a week to ten days prior to application (acclimatization).

Extra precautions must be taken if unseasoned materials are used: 1) allow material to acclimatize (as described above) before installation or application of finish; 2) use patterns which allow for some shrinkage (bevel, channel rustic with an adequate tongue, board & batten, etc.); 3) use as narrow a width as possible.

Refer to "Guidelines for Installing and Finishing Wood Siding over Rigid Foam Sheathing" (WWPA A-31) for additional information.

Priming

Often material which has been properly seasoned, stored and handled, will pick up moisture after installation and prior to painting. Later, when the siding releases its moisture, joints may open up or buckling may occur.

Priming or prefinishing all sides, edges and ends ***after*** it has reached climatic balance and ***before*** it is installed, provides extra protection. Prefinishing can also minimize objectionable unfinished lines where joints open up due to face width shinkage.

Apply Siding over Building Paper

Nails and Nailing

Correct nails and nailing practices are essential in the proper application of wood siding. Nail locations are included with individual patterns; however, the following information will help in selecting and using the right nail for the right use.

Siding and Nailing Patterns

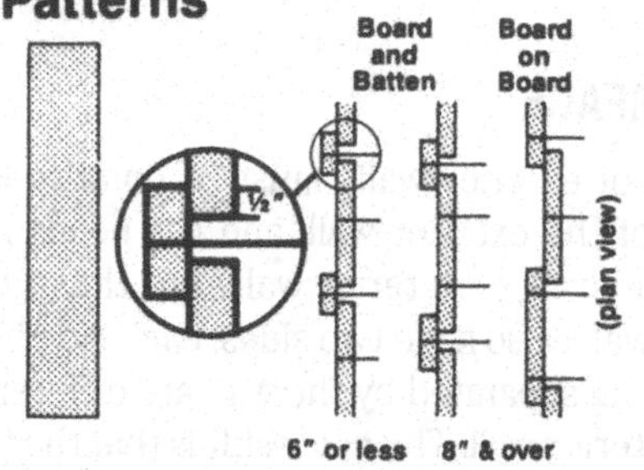

Board is available surfaced or saw textured (see "thicknesses" and "widths" under "sizes," on back).

Widths 6" or less require a ½" overlap, for widths 8" and over, increase overlap proportionately and use 2 nails 3-4" apart. Vertical application only.

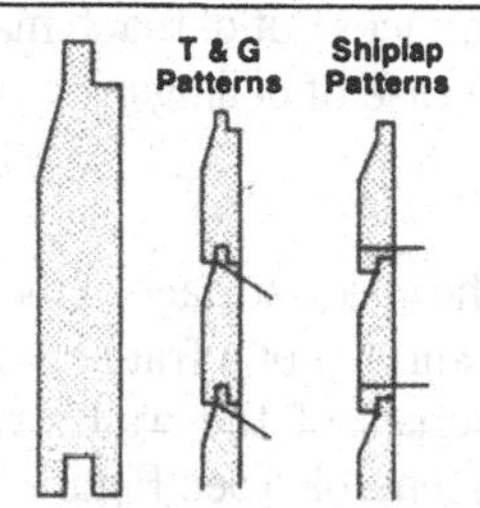

Drop is available in 13 different patterns. Widths 8" and over use 2 nails 3-4" apart, as shown in Channel Rustic. Some T & G (as shown), others shiplapped. Refer to WWPA "Standard Patterns" (Form G-16). Horizontal application only.

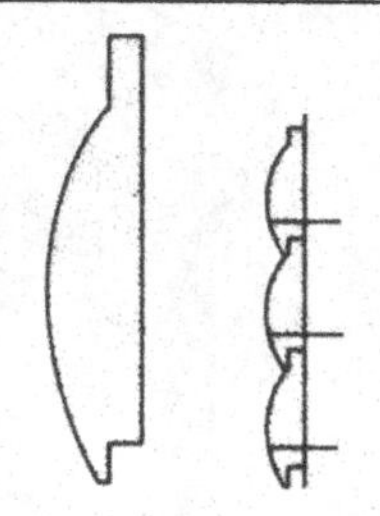

Log Cabin is 1½" at thickest point.

Nail 1½" up from lower edge of piece.

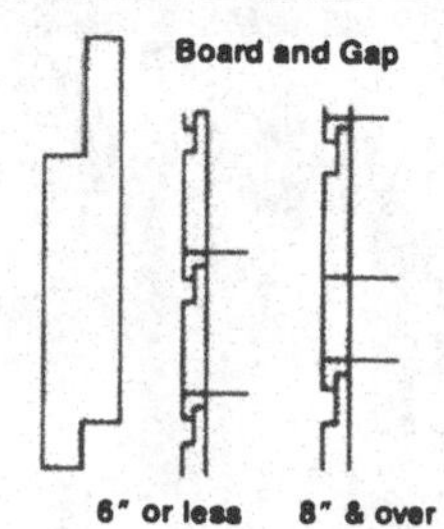

Channel Rustic has ½" overlap and 1" to 1¼" channel when installed.

Widths 8" and over use 2 nails 3-4" apart.

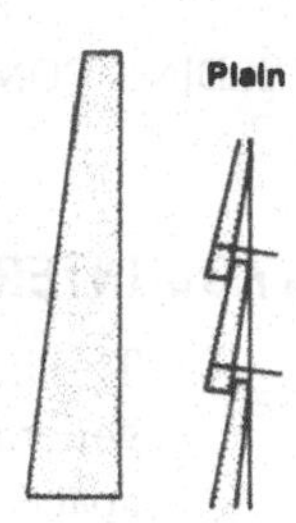

Bungalow is thicker and wider than Bevel Siding. Plain Bungalow or "Colonial" may be used with smooth or sawn face exposed. Horizontal application only.

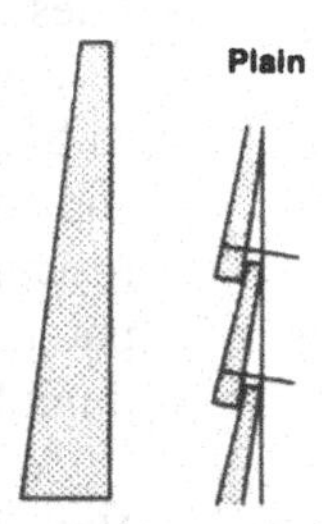

Plain Bevel may be used with smooth face exposed or sawn face exposed for textured effect. Recommend a maximum width of 10".

Recommend 1" overlap on plain bevel siding. Horizontal application only.

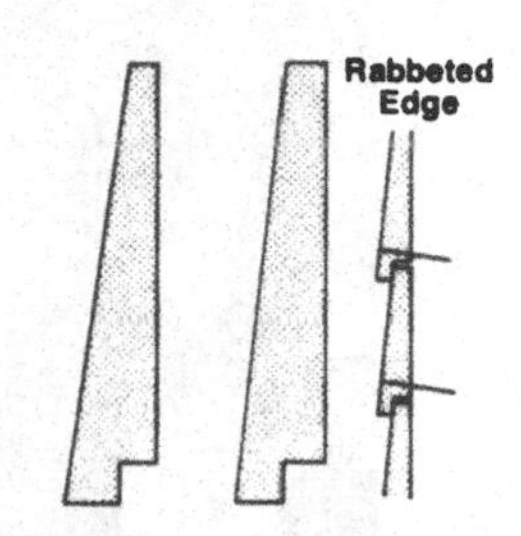

Dolly Varden is thicker than Bevel Siding and has a rabbeted edge. Horizontal application only.

Widths 8" and over use 2 nails 3-4" apart, as shown in Channel Rustic.

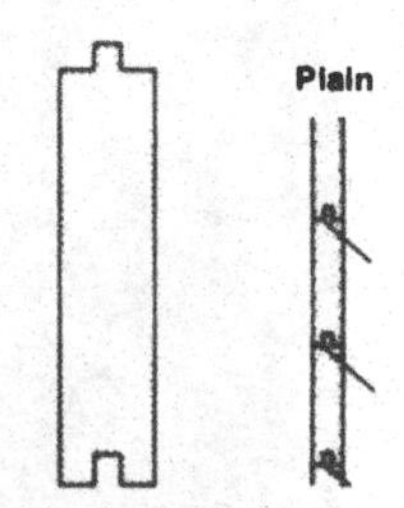

Tongue & Groove is available in smooth surface or rough surface.

Use single finish nail as shown for 6" widths or less. Wider widths, face nail twice per bearing.

Figure 7.6 Installation of wood siding. Reproduced from *Wood Frame Design*, with permission of the publisher, Western Woods Products Association.

7.3 INTERIOR WALL SURFACE

The interior surfacing of exterior walls must respond to two design concerns: the general development of the exterior wall, and the needs generated by the interior space for which it is a surface. Exterior walls are thus typically unsymmetrical in construction. Interior walls also have two sides; each side being related to a separate space. If the spaces being separated by the wall are different, there may be different finishes—as for the exterior wall. The net result is that the finishing of the surfaces of each wall—whether exterior or interior—must respond to the conditions on both its sides.

As with the exterior surfacing, finishes for interiors may be achieved naturally or with applied materials; both structure and appearance must be given consideration; and the range of possibilities is enormous. While there is no concern for weather resistance, the interior surfaces within arm's reach are generally subject to more potential wear and soiling. Surfaces out of reach may potentially be covered with less durable materials—as in the case of ceilings.

Natural Surfaces

Natural finish may occur as the inside surface of monolithic construction (masonry or concrete) or the unfinished surface of a frame covering. Depending on the interior space and the specific materials of the wall structure, the generally unadorned structural surface may be acceptable (see Figure 7.7).

Figure 7.7 A common "natural" interior finish: the inside surface of a glass and metal window wall.

Even if left exposed, however, it may be possible to achieve a finish of some higher appearance quality than that normally obtained. Structures that are built with the intention of being covered are sometimes left in relatively sloppy condition. This may produce a raw, brutal appearance that is desired for some effect, but some minor additional care or cleaning might be able to produce a much neater appearance. This is a place where careful specification writing is required and workers should be alerted to the differences from the usual job situation.

Some natural surfaces that need protection on the exterior may be left exposed or untreated on the interior, where weathering is not a problem. On the other hand, other issues become of concern for interiors; for example:

Flame spread on the wall surface or combustibility of the surface materials.

Resistance to abrasion, puncture, or other effects of wear or damage from contact.

Reflective or absorptive properties of the surface for light and sound.

Appearance is also conditioned by the usually more intimate condition on the interior. Tactile response may also be an issue for surfaces within reach of viewers.

For various reasons, there are many materials that are only feasible for use on either the exterior or interior, and some that can be used in either location. Plaster, for example, can be used in either location, although a somewhat different form will be used for the two. Gypsum drywall, on the other hand, is primarily limited to interior application if exposed and only painted. These are applied surfacings, however, and most "natural" surfaces—such as wood siding, masonry, and concrete —can do service in either location.

Frame Covers

These materials are subject to many of the considerations discussed for exterior frame covering in Sec. 7.2. Prime concerns are their structural requirements related to the frame and their appearance (see Figure 7.8). Also of concern are all of the issues discussed in the preceding discussion regarding the special needs for interior surfaces.

One-piece coverings are more frequently used on the interior, the most popular being gypsum drywall, a sandwich of two paper faces over a gypsum plaster core. Other one-piece paneling can be used, but most finishes are developed as applied coatings or veneers over a plastered or drywall substructure.

However, if an applied finish is planned for, various materials can be used for the structural sheathing of the frame, providing greater structural strength, accommodation of stronger anchorage, better insulation, better fire resistance, higher reduction of sound transmission, or whatever may be desired for the wall.

What is required for design is a full definition of both the tasks required for the interior space and the exterior, enclosing wall. One issue that may be of concern is what occupies the void inside the framed wall. The initial installation—and possibly access for future changes or repairs—of items in this space may require consideration of both the materials used and the construction sequence for the interior cover. Typically, exterior walls are "closed up" by completion of the exterior covering before the interior covering, and cutting the wall open in the future is most likely to be done from the inside.

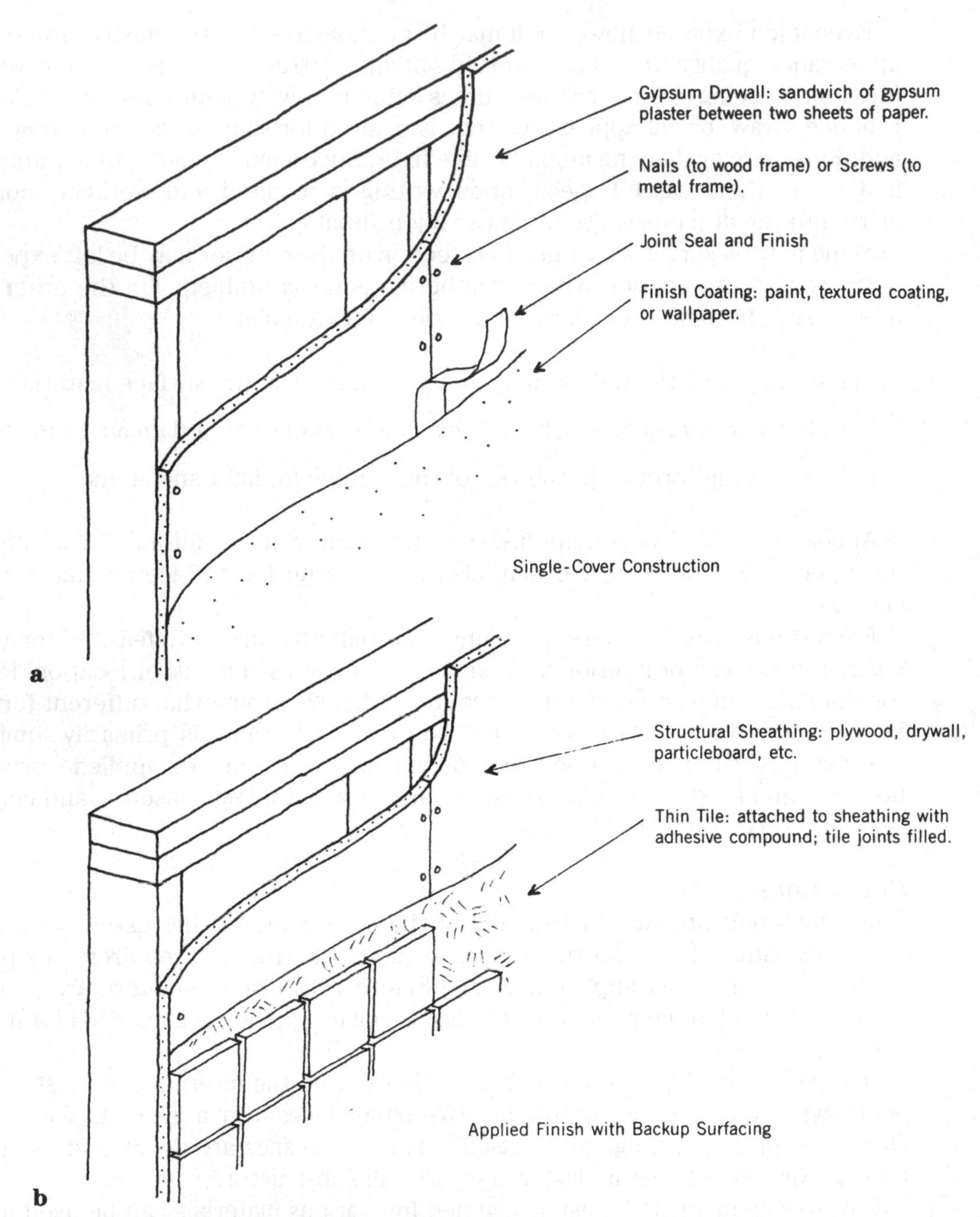

Figure 7.8 Development of the interior surface of exterior framed walls: (a) Single-piece covering with developed finished surface. (b) Rough-surfaced, structural sheathing with separate applied surface.

Applied Surfacing

Applied finishes for interior walls include a really vast array of possibilities. All the options for exterior applied finishes may be considered, and indeed have probably been used somewhere. With the lack of concern for weather, however, the barriers are down and just about anything goes. Paper, fabrics, and carpeting can be used; the last often for its sound absorbency.

Refinishing without the need to disturb the structural wall is often a necessity, especially in commercial rental occupancies (stores and offices). Finishes that can be easily removed or covered up are preferred for these situations. A lot of natural finishes get covered over when occupants get tired of them or others move in.

Creation of some finishes may involve some additional construction, as when furring strips are used to permit attachment of paneling to a concrete or masonry wall. Interior tile, brick, or stone surfaces can be achieved with the real materials, requiring considerable preparation and adding significant additional thickness; but the more frequently used solution is to use very thin elements attached by adhesives —a much faster, thinner, and more economical solution.

Wood paneling, if one with real wood at all, is usually also in the form of a thin layer of the finish wood. This may be done by using actual solid strips of the wood on a structural plywood; or by using a face ply on a thinner back-up of plywood or compressed wood fiber which is then attached with adhesives to the wall structure.

7.4 ROOF STRUCTURE

Designing a roof requires dealing with two major functional problems: the generation of the general roof construction, form, and structure, and the development of the means for making the roof-water-resistant. As an interior/exterior barrier and basic enclosure element for the building, a roof usually has many additional requirements, but the fundamental requirements for functional construction and lack of water penetration are primary concerns.

This section deals with the basic development of the roof form and construction (see Figure 7.9). Section 7.5 deals with the general problems of resisting penetration of water.

Roof Form and Structure

For various reasons, roofs allow for the greatest freedom in selection of types of structural systems. Some reasons for this are as follows:

Roof structures do not present the need for a flat top surface or a nondeflecting, nonbouncing, generally "solid" feel that is required for floors.

Restraints on the overall depth (vertical thickness) of the structure are less critical than for floors, especially floors in multilevel buildings.

Drainage actually requires a nonflat top surface, making arches, domes, shells, and gabled trusses actually useful for their geometry.

Selection of the basic structure and general roof construction must respond to various issues, including the following:

The materials and basic construction of the rest of the building.

The dimensions of required horizontal, clear spans.

Special needs or desires for a particular geometric form for the roof surface or for the ceiling of the space beneath the roof.

The form and layout of supports for the roof structure: bearing walls, columns, widely spaced piers, and so on.

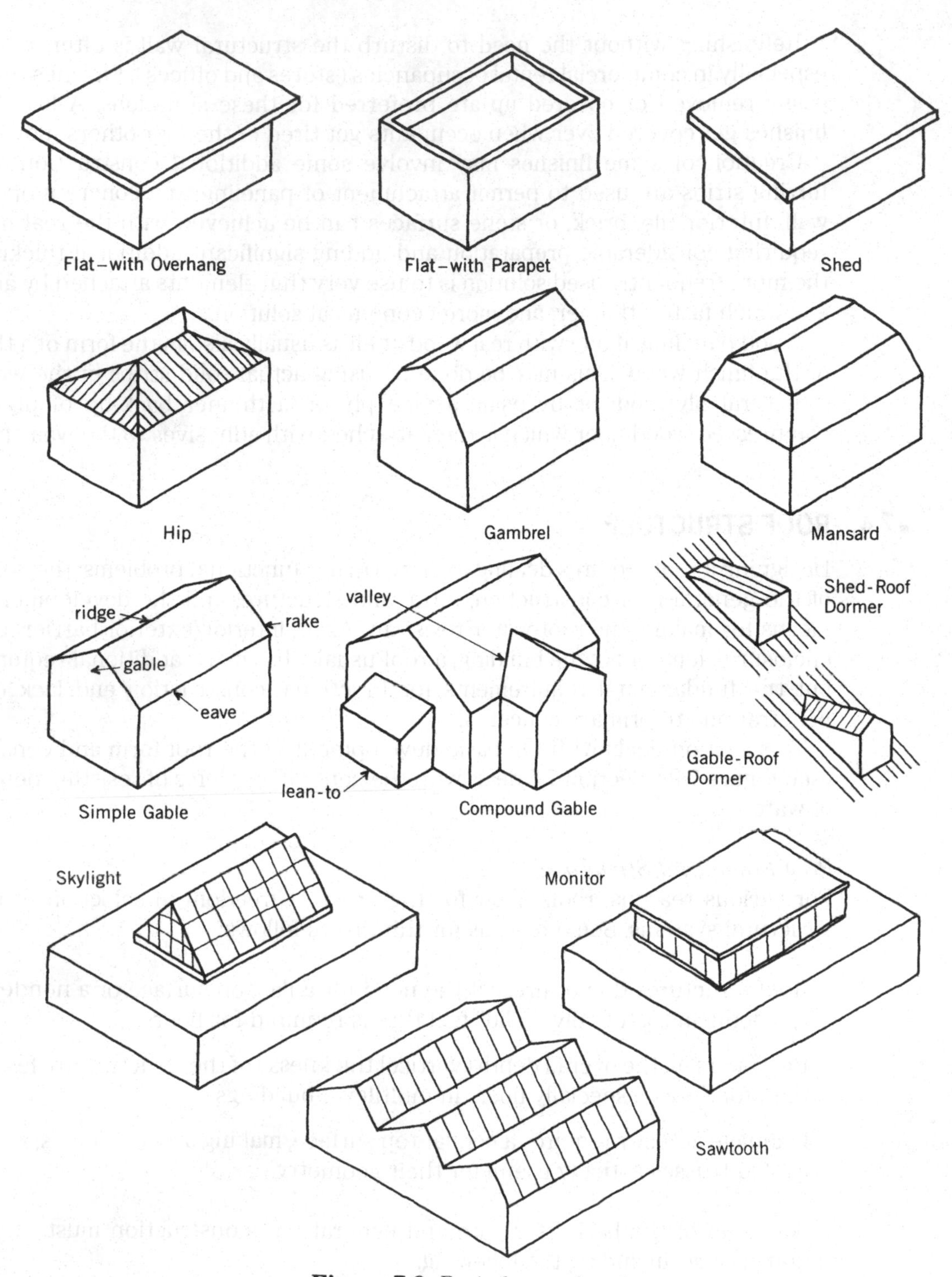

Figure 7.9 Basic forms of roofs.

Penetrations of the roof for vents, elevators, skylights, and so on.

Need for the placement of equipment or provision for traffic of people or vehicles on the roof.

Acceptable range for deformations (deflection, etc.) of the roof structure for visual effects, or for relations to construction beneath the roof, or for assurance of planned patterns of drainage.

General architectural impact of the roof in terms of its form, the visibility of the top surface, and the possible exposure or expression of the underside of the structure as the ceiling of interior space.

The horizontal span range of feasibility for various structural elements and systems must be recognized. Flat plywood sheets, for example, can span only a few feet, while thick timber deck, formed sheet steel, or precast concrete slab units can achieve considerable spans. Flat-spanning beam systems are generally quite limited for very long spans, while trusses, arches, shells, and cable systems can achieve enormous spans.

While span dimension is a major concern, the form of the roof is also strongly related to the type of structure. Some structural systems, such as the truss, can be adapted to a wide range of different forms, while others, such as concrete shells or catenary cables, have highly determined forms that cannot be arbitrarily modified, as their form is strongly related to their structural nature and behavior.

In general, the selection of the structure and basic construction system for a roof cannot be done as an isolated design exercise, but must be done as an integrated consideration in relation to the whole building design and the overall development of the construction.

Frame Systems

Frame systems of various forms are used for most roof structures (see Figure 7.10). The complete structure consists of a structural surfacing (roof deck or sheathing) in combination with the frame. Many combinations of frame materials and deck materials are possible.

Framing for flat surfaces may be developed with ordinary beams, joists (called rafters, if sloped), or light trusses. Members that support decks must be spaced a distance that does not exceed the spanning capability of the deck. In general, the longer the span, the larger the required spanning elements; thus spacing must also increase for appropriate utilization of the increasingly heavy members. For very long spans, there will usually be a hierarchy of spanning elements, with the largest members at one end and the smallest—deck supporting—members at the other. (See Figure 7.11). Framing may also be used with arches, dome ribs, or other nonflat geometries of the roof surface.

Roof framing may often be quite light in weight, as loads are usually low in magnitude, except for very heavy snow packs or high wind pressures where local conditions make these situations critical. In addition, the deflection character of roofs (bounciness, excessive deformation) is seldom a major concern, whereas it is usually of concern for floors. Thus, although systems used for floor structures may generally also be used for roofs, other—lighter—systems are also possible for roofs. The really ultra-lightweight systems (trussed space frames, cable nets, pneumatically sustained fabrics, etc.) may usually be used only for roofs.

Decks

An essential function of decks is the support of roofing. This may occur directly, as when shingles are nailed on top of wood decks. However, there are various considerations that must often be made in the development of a complete roof system that may affect the choice or detailing of the deck, such as the following:

Nature of the deck surface: Of note are the formed steel decks, which require some filler or cap material to provide a smooth surface for roofing—notably membrane roofing.

a

b

Figure 7.10 Roofs developed as framed structures: (a) Wood girders, purlins, rafters, and plywood deck. (b) Exposed light steel trusses with formed sheet steel deck. (c) Arched roof form produced with "bowstring" trusses of bolted lumber, with wood rafters and board deck.

Figure 7.10 Continued.

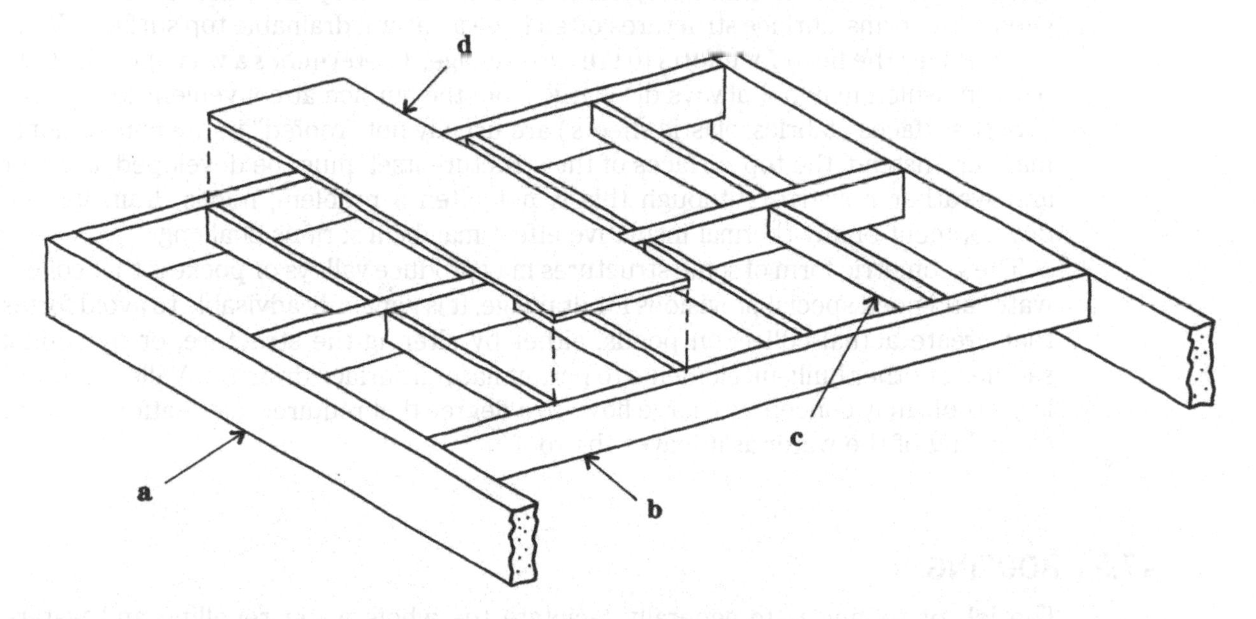

Figure 7.11 Fundamental form of a framing system for medium to long span roofs: (a) Major spanning elements—large girders, trusses, etc. (b) Moderate span elements to achieve span between widely spaced major elements. (c) Short span elements, spaced to limit of decking material. (d) Roofing surface-producing decking, usually of modest span capability and chosen to accommodate attachment of roofing materials. Same type of system can be used for floors, but floors typically have shorter spans.

Development of thermal resistance: Some decks may make significant contributions as insulators, but most roofs need some degree of enhancement in this regard (see Chapter 8).

Development of a moisture (vapor) barrier: Metal decks, if fully sealed at joints, may provide this effect; others generally do not.

Provision for drainage: If the roof framing is essentially flat, the deck—or something on it—may need to provide for the low-slope drainage of a membrane-roofed surface.

Support of ceilings: Because of widely spaced deck supports or other reasons, suspended ceilings are sometimes hung from decks. The attachments and the concentrated loads must be accommodated by the deck.

Attachment of roofing: Most roofing is held onto the roof surface by mechanical means—often nailing. Decks—or something—must be capable of being nailed to and be of sufficient thickness and strength to develop withdrawal resistance of nails.

Surface Structures

Various forms of *surface structures* are described in Sections 9.7.7 and 9.7.8. Hard surfaces (concrete, wood, metal) present the same general potentialities and problems as the decks of framed systems (see Figure 7.12). Because of their usual geometric forms, surface structures often have a natural drainable top surface. While this reduces the need for efforts to create drainage, it determines a very specific drain pattern, which may not always discharge from the surface at convenient locations.

Soft surfaces (fabrics, plastic sheets) are usually not "roofed" in the conventional manner. Instead, the top surfaces of the structure itself must be developed as water and weather resistive. Although this is not often a problem, nor is drainage, the development of any thermal insulative effect may be a serious challenge.

The geometric form of some structures may produce valleys or pockets that collect water and need special provisions for drainage. It is generally advisable to avoid forms that create actual collection ponds, either by altering the structure, or by adding saddles or other built-up elements to permit natural surface drainage. Valleys on very large roofs may concentrate large flows to a degree that requires major effort to avoid cascading of the water as it leaves the roof.

7.5 ROOFING

The job of roofing is to generally facilitate the whole water-repelling and water-removing functions required at the top of a building. The term *roofing* is often applied only to the elements developed to make the general roof surface water repellent, but the full job includes development of the various edge flashings, roof penetration flashings, drainage, collection, and disposal elements as required. The idea is to keep the water from getting into the building or inside the construction, and to get it off and away from the building.

Roof surfacing materials (henceforth called "roofing") are broadly divided between those that are functional only for fast-draining surfaces (high slope angle) and those that can develop the so-called flat roof (see Figure 7.13). Roofs are hardly ever totally dead flat (construction accuracy pretty much guarantees it) and usually slope

a

b

Figure 7.12 Roofs developed with surface structures: (a) Hard surface, as a concrete shell-hyperbolic paraboloid (saddle-shaped) surface with large edge girders and abutment supports. (b) Soft surface, as a fabric sustained in tension by air pressure.

somewhat, not only to create minimal drainage, but to direct it where it needs to go to get off the roof.

In truth, there are more than two classifications of slopes, with some that are intermediate between the almost-flat and the really steep. Most shingles, for example, drop off at a reasonably high minimum slope (15 degrees or so), but other materials may be used at this slope and slightly shallower ones before a really watertight membrane is required.

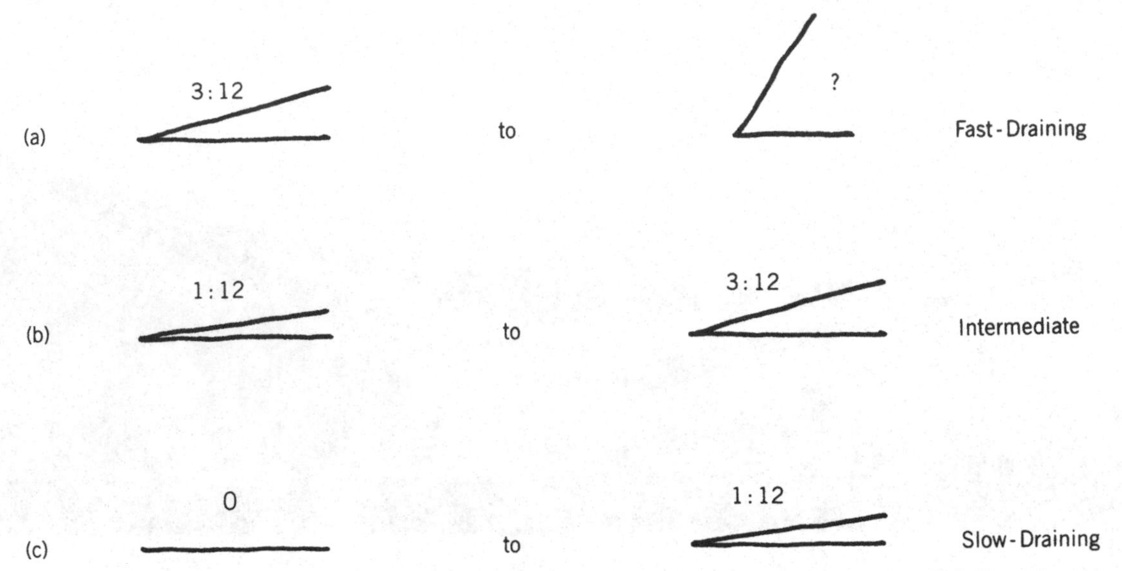

Figure 7.13 General classification of roof surfaces by angle of slope: (a) Fast-draining. (b) Intermediate slope. (c) Slow-draining.

Fast-Draining Surfaces

As stated previously, fast-draining surfaces are usually visualized as those that permit shingles. Shingles (and other overlapping units, such as tiles) do not develop a watertight seal of the surface, and require a fast, single-direction flow of the drained surface (see Figure 7.14). In most cases some form of liner is used between the shingles and the deck, but this is also not of a fully water-sealed, membrane character.

Shingles and tiles can be used on slopes from about 3 in 12 (an old English carpenter's lingo: 3 in. up for 12 in. across, or about 15 degrees) to a full vertical position in some cases. Indeed, some shingle materials require a greater minimum slope, established by the building code or the manufacturer's warranty requirements. Shingles may consist of a variety of materials, including wood, stone, precast concrete, impregnated felts, impregnated glass fiber mats, plastic, and metal.

As with all roofing, it is necessary to deal with the general surfacing in terms of shingle orientation, pattern or arrangement, overlap required, attachment, underlayment (water-repellent liner), and the general requirements for handling the specific shingle materials. In addition, to fully utilize the roofing—it is necessary to provide properly for the downslope edges (called eaves), building end edges (at gables, etc.), ridges, valleys, and the joints at various roof-penetrating elements, such as vents, chimneys, roof hatches, and skylights.

A major design concern for shingles is their appearance, as the higher slopes usually put them in view. For a one-story building with large overhangs, for example, the roof surface may be a major part of the viewed building. The whole roof form, surfacing, and general detailing of edges, ridges, and any rooftop elements can be a major design element.

Selection of shingle materials must be coordinated with the selection of other exterior finish materials. It must also be properly related to the slope conditions, structural support for attachment, local weather conditions, construction budget, and other practical concerns.

Some materials used for the near-flat roof surface can also be used for low slopes that are essentially fast draining. The only significant limitation comes from topping materials that may not stay in place if the slope is considerable. Membrane materials, however, are usually quite expensive compared to other options for roofs of reasonable slope.

There is a slope classification between the minimum for shingles and one so low that a membrane is the only option. Membranes can be used here, but so can some special forms, such as roll roofing (impregnated felts with shingle-like toppings) and

a

b

Figure 7.14 Roofing for fast-draining surfaces: (a) Wood shingles, cedar shake. (b) Sheet-form, asphalt shingles. (c) Mission clay tile, the real thing. (d) Clay tile, developed with fiber-reinforced concrete.

c

d

Figure 7.14 Continued.

various metal roofing. Roll roofing is generally quite inexpensive (and looks it!), while metal roofing of any kind is generally expensive, although some very handsome surfaces can be developed with expressions of the exaggerated rib systems of the sheet metal roofing.

A special roofing problem is presented by forms such as that of a dome, where the surface changes slope from flat to steep, all on a continuous surface. Very few materials can fully accommodate that whole range, except some of the single-ply coatings, as described in the next section.

Membranes

The traditional membrane roof is one described as *built-up*, having multiple, overlapping layers of impregnated felts, mopped for adhesion and sealing, and topped with hot bitumen of asphalt or coal tar pitch (see Figure 7.15). The degree of overlap results in from three to five layers of the felts; thus the system is said to be three-ply, four-ply, or five-ply.

The built-up roof must have the top felt layer protected by a glaze coat of bitumen, by a layer of gravel or ceramic particles embedded in a flood coat of bitumen, or by a cap sheet material of mineral felt or glass fiber mesh. Gravel is generally used on very low slopes. Cap sheets may be used on slopes for 1/2 : 12 to 6 : 12.

Membranes may also be applied as single-thickness sheets, called *single-ply membranes*. These may be continuously adhered, mechanically attached, or simply held down by materials placed on top (called ballast)—such as several inches of gravel or thick, preformed paving units. Joints between sheets are sealed by methods relating to the composition of the sheet material and the general form of attachment.

A special form of membrane can be formed by liquid application, usually in multiple coats with separate coats of a different color to assure coverage.

There are many forms of membrane roofing systems. Many of these systems are produced as priority items, with all the required materials and accessories supplied by a single manufacturer, who may also deal with installation through franchises of suppliers or contractors.

Selection of a membrane may be quite simple, because of the nature of the building project or the limited options available on a regional basis. Typically, however, there are many competing systems and contractors, each claiming to have the best system. With all the factors involved, selection is difficult. However, a specialty area of consultation—consulting roofing designers—is rapidly developing.

Leaking roofs generally represent one of the most nagging problems for designers and builders, probably producing more civil suits than any other situation. Membrane roofs lead the pack—not necessarily because of the membranes themselves, but because of the generally greater and potentially more critical problems of flashing and sealing at all the roof edges, intersections, penetrations, and general discontinuities.

Metal Roofing

Roofing with sheets of metal is a very old method, predating the present forms of most membranes and shingling with felts and other currently used materials. Copper, lead, and other special metals were used in ancient times. Coated steel, aluminum, and some special alloys are used today (see Figure 7.16).

Methods and details of installation used today retain much of the appearance of ancient work, although some newer technology is used in both the forming and sealing of joints and the effecting of the attachments to supports.

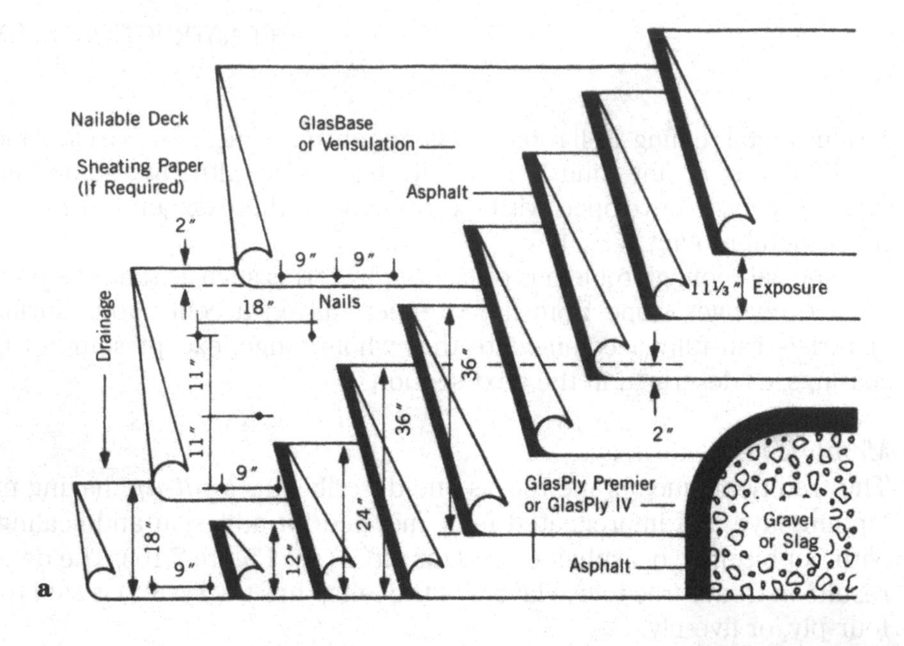

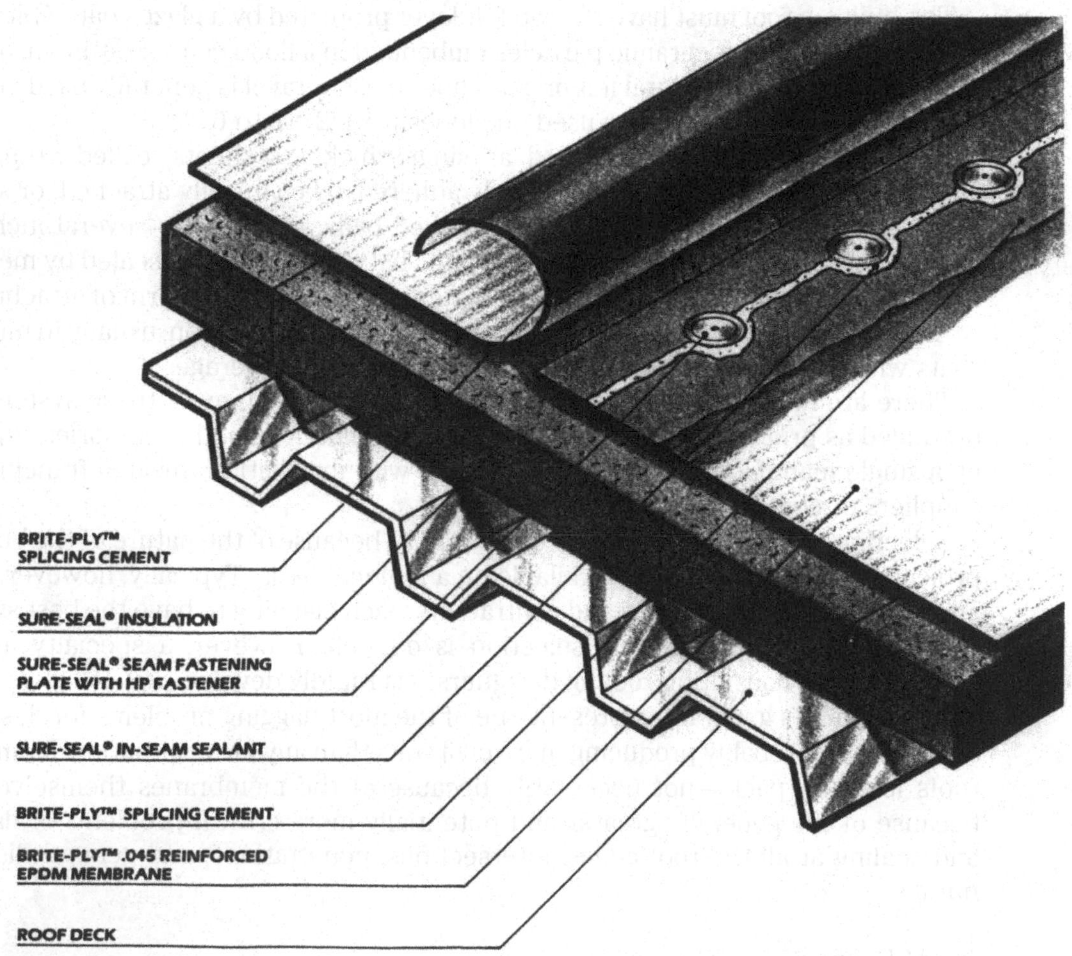

Figure 7.15 Membrane roofing for slow-draining roofs: (a) Basic form of built-up roofing, consisting of multiple plies, hot-mop sealed with bituminous material (coal tar, asphalt, etc.), courtesy of Manville Roofing Systems Div., Denver, CO. (b) Single-ply, preformed membrane, courtesy of Carlisle Syntec Systems, Carlisle, PA. (c) Liquid-applied surfacing for a continuous covering of a complex roof surface.

c

Figure 7.15 Continued.

The general form used is that of long, narrow sheets, running in single pieces downslope and joined at their edges to each other. Edge joints are usually achieved with seams that protrude upward to form ridges (called *standing ribs*), resulting in a strongly expressed striped texture.

Metal roofing can generally cover the range of slopes from very low (1/2 : 12 or so) up to vertical. The basic roof materials and details can be used for siding with the joint/ribs running vertically.

As with priority membrane systems, metal roofing—together with the various accessories required for a complete installation—is often supplied in a system "kit" by manufacturers.

Flashing

Poor flashing—in design, choice of materials, bad detailing, or sloppy installation—is usually responsible for most of the leaks that occur with roofs and many walls. Flashing occurs at some type of joint in many cases, and must do its job without interfering with other joint functions (structural connection, fire-stopping, expansion control, seismic separation, etc.); this is not so easy in some cases. Joint actions may cause damage to the flashing and a good flashing installation is lost.

Basic situations requiring flashing and many of the typical forms of flashing have been around a long time (see Figure 7.17). Flashing materials have changed considerably, however, and now include plastics and special fabrics as well as traditional metals such as copper.

The broader array of sealant materials now available has simplified some flashing details, and even eliminated some traditional needs for flashing. Water-repellent coatings, liquid-form membranes, and other modern materials and methods have also reduced or eliminated the need for some flashing. Still, the traditional roof edges, joints, penetrations, and intersections continue to require very careful detailing for adequate flashing.

Flashing is needed with any type of roofing, from the simplest, cheapest shingles to the most elaborate, expensive membranes. The amount of flashing is generally

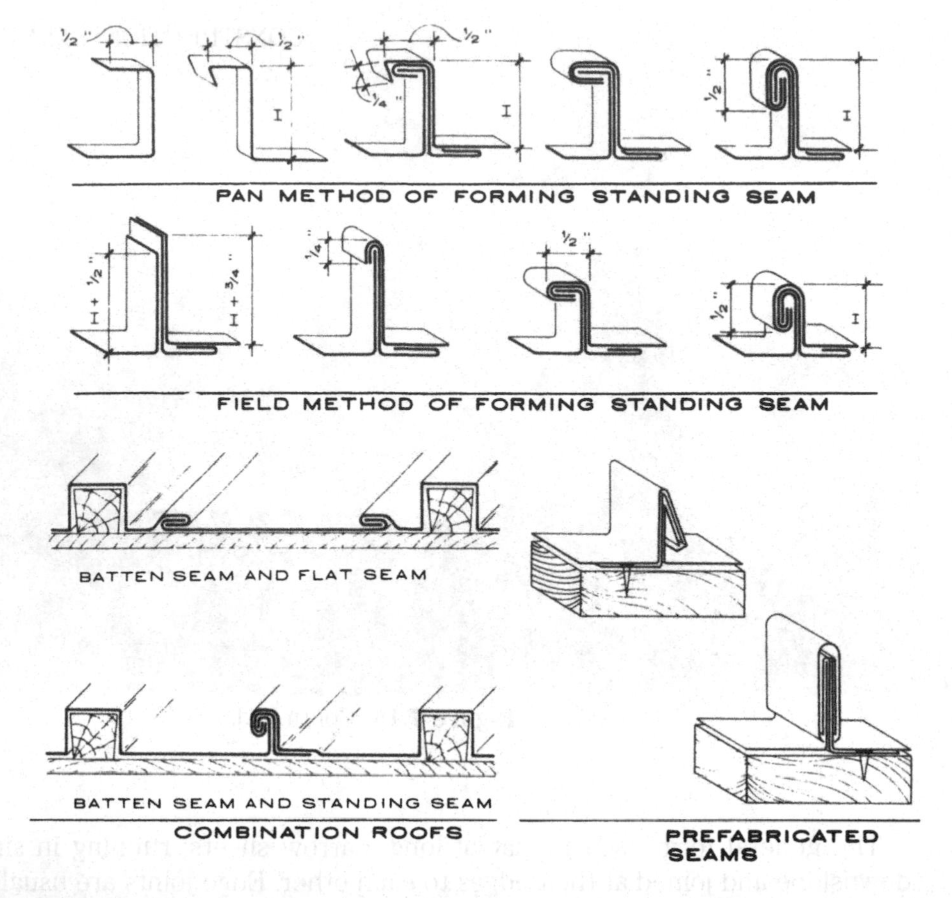

a

b

Figure 7.16 Roofing with sheet metal elements with upturned, exposed seam joints between single pieces: (a) basic formation of seams (reprinted from *Architectural Graphic Standards,* Student edition, with permission of the publisher, John Wiley & Sons. (b) Batten-seamed surfaces of complex form. (c) Simple, intermediate slope surface.

c

Figure 7.16 Continued.

related to the complexity of the roof form and construction. Roofs with parapets, multiple planes or levels, or many penetrations will require a lot of flashing, regardless of the basic type or cost of the roofing materials. Extensive flashing increases the odds in favor of leaks and adds to the construction cost—a double liability. Lesson: Use simple roof forms and avoid penetrations if you are really concerned about leaks and cost.

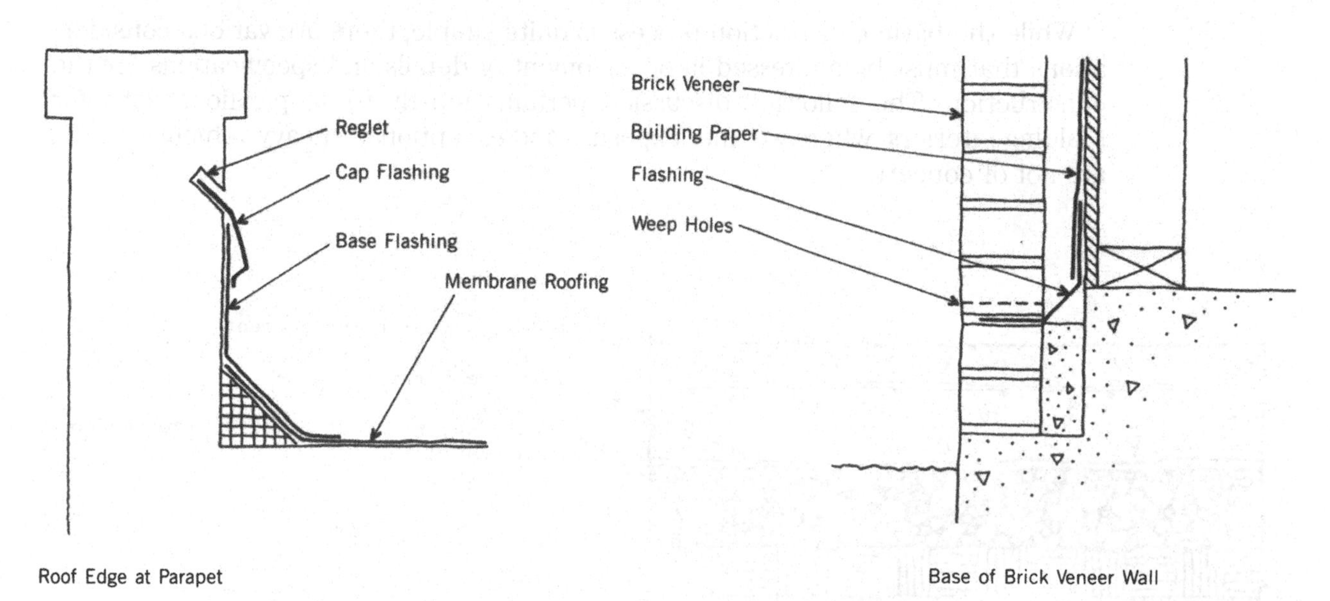

Figure 7.17 Development of flashing to ensure water resistance at vulnerable construction joints. Details respond to the nature of water flow, to potential movements in the joints, and to ordinary construction procedures.

Roof Drainage
Draining a roof has the primary purpose of keeping the water from entering the building or penetrating the construction to cause damage. However, the water has to go somewhere. This presents the problem of disposal of the water as a waste (unwanted material) item. In some situations it may be acceptable simply to dump it off the edges of the roof, but more typically it must be controlled in some manner.

7.6. FLOOR STRUCTURE

General problems of floors and floor/ceiling systems are discussed in Sections 5.6 and 5.8. The following material presents information on the two basic types of floor structures—pavement slabs and framed systems with decks.

7.6.1 Pavement Slabs

The usual means of achieving a working surface for sidewalks, driveways, basement floors and floors for buildings without basements is to cast concrete directly over some prepared base on top of the soil (called *slab on grade* construction). As shown in Figure 7.18, the typical construction for this consists of the following components:

1. A prepared soil surface, graded to the desired level; compacted, if necessary to avoid subsidence.
2. A coarse-grained pavement base, usually predominantly fine gravel and coarse sand; also compacted to some degree, if more than a few inches thick. Three or four inches is common for floor slabs.
3. A membrane of reinforced, water-resistant paper or thick plastic film (6 mil minimum), used where moisture intrusion is especially severe.
4. The concrete slab.
5. Steel reinforcement; often of heavy gage wire mesh.

While the basic construction process is quite simple, there are various considerations that must be addressed in development of details and specifications for the construction. The following discussion pertains mostly to simple floor slabs for building interiors, where weather exposure and exceptionally heavy vehicular traffic are not of concern.

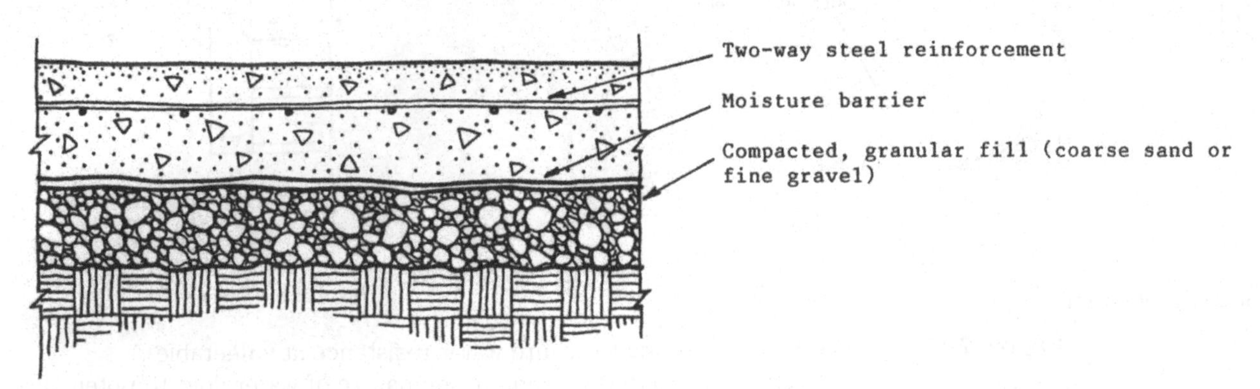

Figure 7.18 Common form of concrete pavement (slab on grade) for a building floor.

Thickness of Pavements

For residences and other situations with light traffic, a common thickness is a nominal 4 in slab; actually 3.5 in., if standard lumber 2 by 4s are used as edge forms. With proper reinforcement and good concrete, this is an adequate slab for most purposes.

Where some vehicular traffic is anticipated, or where other heavy concentrated loads—such as those from tall, heavy partitions—are expected, it may be desirable to jump to the next logical thickness of 5.5 in., based on using 2 by 6s for edge forms. At this thickness, it may be possible to avoid providing individual wall footings for non-structural partitions, a simplification in construction probably justifying the cost of the additional concrete for the slabs.

For very heavy trucks, for storage warehouses, or other situations involving heavy concentrated loads, thicker pavements may be necessary. However, the nominal 4 in. and 6 in. slabs account for most building floors.

Reinforcing

For 4 in. slabs reinforcing is usually accomplished with welded wire mesh. For thicker slabs—or as an alternative for 4 in. slabs—small diameter rebars may be used with somewhat wider spacings than those in the wire mesh. Even when mesh is used, some extra rebars may be provided at edges, around openings, or at other critical locations.

The objective for the reinforcement is to reduce cracking of the concrete, due primarily to shrinkage during curing, differential volume changes due to temperature changes, and some unequal settlement of the pavement base. Cracking is especially undesirable in the top (exposed surface) of the slabs, so the reinforcement should be held up during casting to be relatively close to the top surface.

Another means for reinforcing—or basically altering—the concrete is to add fiber materials to the mix; a growing practice for all slabs exposed to severe weather. In cases where considerable movement of the base is expected, pavements may need to be prestressed or developed as framed systems on grade.

Joints

It is generally desirable to pour paving slabs in small units; primarily to control shrinkage cracking. This results in construction joints, which may not relate well to development of floor finish materials. Where permanent wall construction is established, joints should be located at these points, rather than in the middle of floor spaces.

Larger pours can be made if control joints are formed or cut in the slab. Movements at these joints may be small for interior floors, but they should still be located at points that will not cause problems with floor finishes if possible.

Surface Treatment

If the top of the paving slab is to be exposed to wear, it should be specially formed for this purpose during casting of the concrete. A hard, smooth surface is generally desired; typically developed with fine trowelling with a steel trowel, and possibly the addition of some hardening materials to the finished surface.

However, a smooth surface is not always desirable. For a slip-resistance surface, intended to reduce accidents when the floor is wet, a grit material may be added to the surface. Smoothness is also generally not desirable when other finish materials are to be used, especially ones that must be attached with adhesives. In the latter case, the steel trowelling may be omitted, and a rough leveling of the surface acceptable.

7.6.2 Framed Construction on Grade

As previously described, it may sometimes be necessary to provide an actual framed construction on top of the ground. Two situations that often make this a possible necessity are the following:

1. When finished grade must be achieved with a considerable amount of fill, making it likely that major settlement of the pavement base will occur.
2. When deep foundations are used to support columns and walls with very little settlement. Even though the paving base may be good, some relative settlement is likely.

Framed systems used in this case are usually variations on the typical slab-and-beam systems, described in the next section. For modest spans, beams may be created by simple trenching, as shown in Figure 7.19(a). For longer spans, beams of considerable depth may be required, and a separate forming of beams required, as shown in Figure 7.19(b).

A special framed structure that is developed as a system cast directly on the ground is that used for a basement floor that must resist considerable upward ground

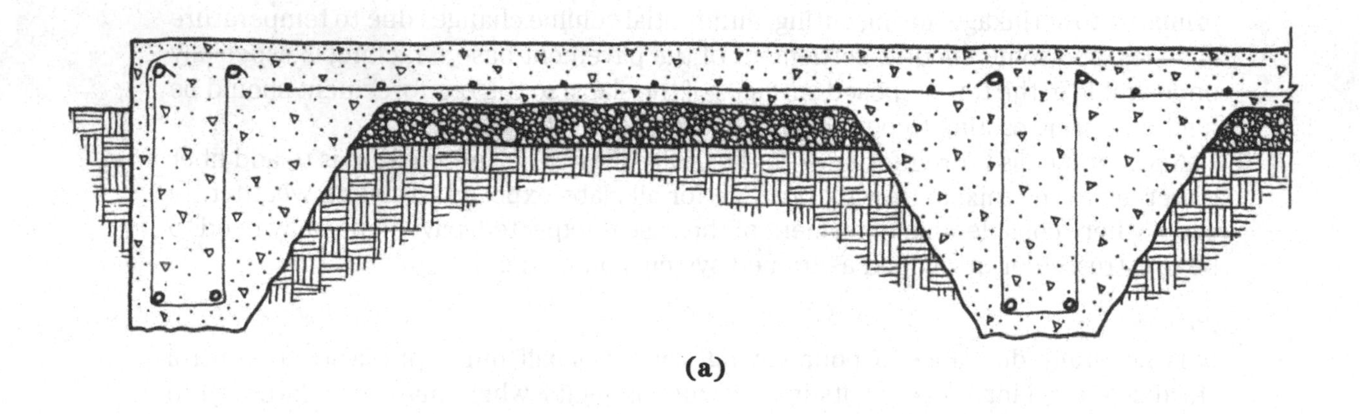

(a)

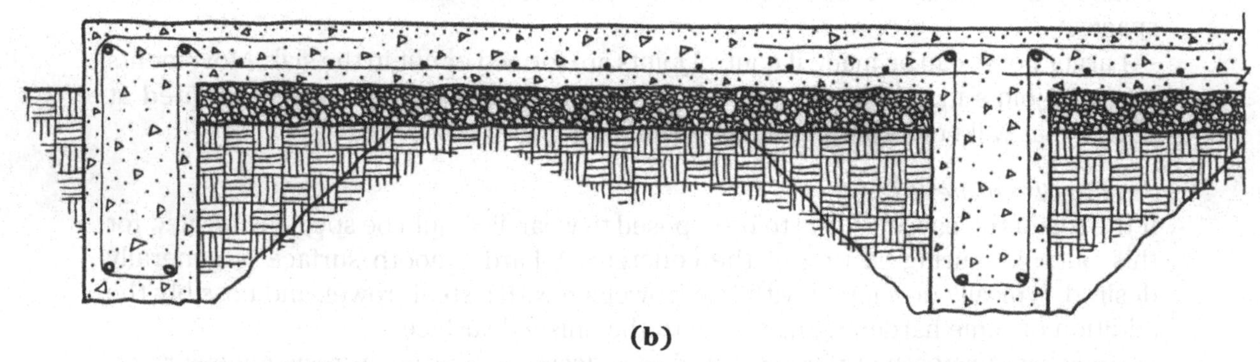

(b)

Figure 7.19 Common forms for framed concrete floors cast directly on the ground: (a) Beams formed by trenching and cast continuously with slabs. (b) Beams separately formed and cast, with slabs cast over backfill. Both systems are based on having the beams supported by other construction, so that settlement of the soil from under both the slabs and beams is not a concern for movement of the floor surface.

water pressure (called *hydrostatic uplift*). This is required when the basement extends below the ground water level. An example of such a structure is illustrated for Building 7 in Chapter 10.

7.6.3 Framed Floors

When another space exists below a floor, a framed structure must be used to span some clear distance to achieve a platform for the floor (see Figure 7.20). Various materials and systems can be used to achieve the spanning structure, with choices relating to many considerations, including the following:

1. The materials and basic construction of the rest of the building.
2. The dimension of required horizontal, clear spans.
3. The form and layout of supports of the floor structure: bearing walls, columns, widely spaced piers, and so on.
4. Penetrations of the basic floor structure for stairs, elevators, duct shafts, or other elements.
5. Any special requirements for loads: exceptionally heavy loads (such as in warehouses or libraries), concentrated loads (wheel loads, equipment, etc.), or hanging loads.
6. Special considerations for floor finishes or supported ceiling construction.

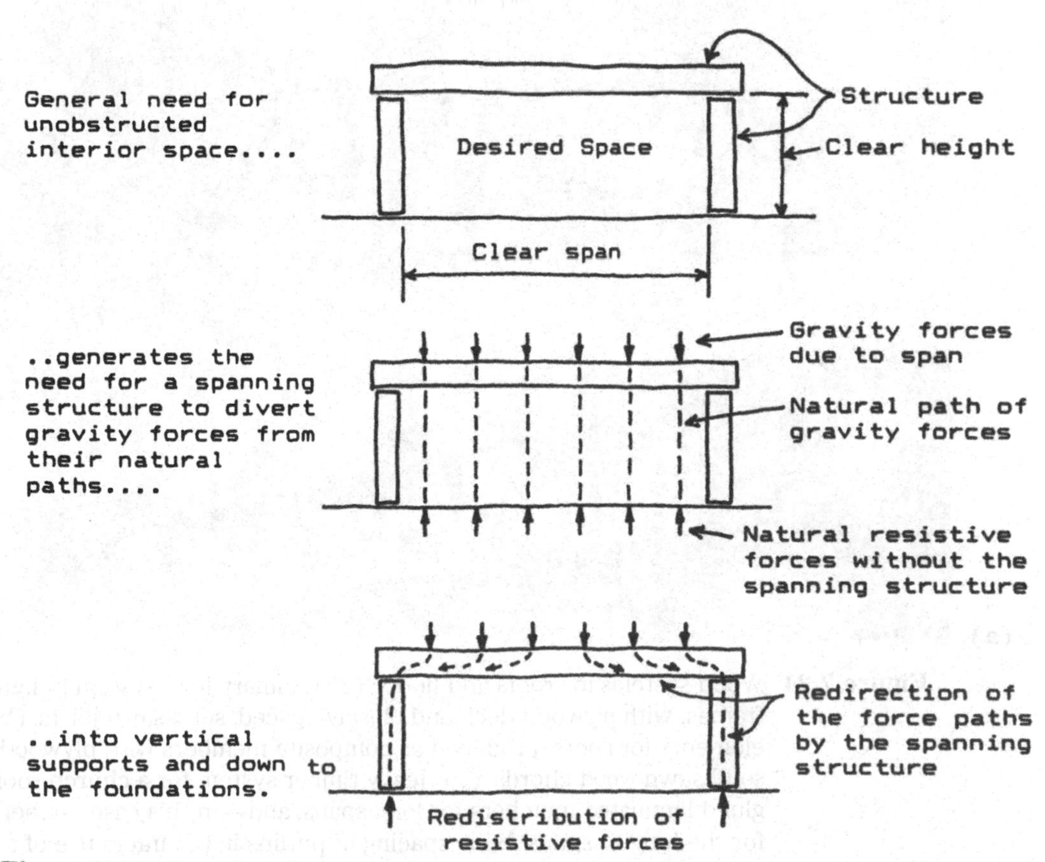

Figure 7.20 Considerations for framed roof and floor structures developed to create open interior spaces.

Many special systems can be used, but the most common systems are variations on simple deck and beam systems, with a deck of some flat material supported by regularly spaced beams. Most floor structures have relatively modest spans permitting a wide range of spanning elements; typically of wood, steel, or reinforced concrete.

Figure 7.21(a) shows a common system in light wood frame construction, consisting of a structural plywood deck and very closely spaced wood joists. This is reasonable for spans up to about 20 ft. Beyond this span, other elements may be substituted for the solid lumber joists; one example being the composite plywood and timber I-beams shown in Figures 7.21(b) and (c).

Common in the past, but less widely used today, is the so called *heavy timber* system that uses a nominal 2 in. thick plank decking and larger timber beams at wider spacing. (Ordinary wood joists are typically 12, 16, or 24 in. on center.) The exposed heavy timber structure offers the advantage of some fire resistance, while the light wood frame must be covered to obtain any rating.

Although the exposed, heavy timber frame was common in the past for both utilitarian (barns, etc.) and higher quality buildings (churches, etc.), the construction is now usually premium priced and the craft for its execution not so easy to obtain.

As is frequently mentioned here, the light wood frame is the definitive choice for the structure of buildings in the United States, except when circumstances prevent

(a)

Figure 7.21 Wood systems for roofs and floors: (a) Ordinary joist system in light wood frames, with plywood deck and closely-spaced, solid sawn joists. (b) Joist elements for floors, produced as composite members with plywood webs and solid sawn wood chords. (c) Heavy timber system for a church roof, using glued laminated members for long spans, and—in this case—steel girders for the longest spans. Wide spacing of purlins in (c) infers use of a longer-spanning deck, probably of wood plank deck units for their exposed appearance.

(b)

(c)

Figure 7.21 Continued.

its use. Other choices must be compared with the light wood frame and justified on the basis of some extenuating circumstances; a major one sometimes being code requirements for fire ratings.

The general form of the light wood frame deck and joist system can be developed with steel elements, as shown in Figure 7.22(a); in this case using steel open web joists. Analogous to the heavy timber system is the system shown in Figure 7.22(b), using a slightly heavier deck and heavier beams (rolled W sections here).

While wood framing is used almost exclusively to support plywood, wood plank, or wood fiber decks, steel framing can be used to support almost any deck system, including plywood, sitecast concrete, or precast concrete, as well as a wide range of steel decking. For wood and steel decks it is common to use a concrete fill on top of the deck for many reasons, such as:

1. It adds a fire barrier/insulation on top of the floor system.
2. It creates a smooth surface, especially on formed steel decks.
3. It adds stiffness and a feeling of solidity to the deck.
4. It improves the acoustic separation character of the floor system.

In the past, it was common to use the existence of the concrete fill as an opportunity to bury wiring systems for power and communication. This is still done, although other options exist for many situations; most of which allow more ease of modification as needs rapidly change in many commercial occupancies.

Concrete floor systems may be sitecast or precast. As mentioned previously, concrete decks may be used with steel framing. In fact, concrete will ordinarily be used in one way or another with any steel floor system (see Figure 7.23). This may occur in any of the following ways:

1. As inert concrete fill on top of a structural deck of formed sheet steel; the steel deck carries the concrete, as well as other loads.
2. As a composite deck system, with the concrete fill and the formed steel deck interlocked to work together as a single structural unit.
3. As a structural concrete deck, using a formed steel deck strictly for the casting of the concrete.
4. As a structural sitecast concrete deck poured on removable forming.
5. As a precast concrete deck, with units attached to the steel framing.

Each of these choices has some circumstances that make it a more feasible alternative; however, they are also somewhat competitive for similar situations, with choices made as much on the basis of personal preferences as any other reason.

Wholly sitecast concrete systems include the common options shown in Figure 7.24. The wood or steel deck and beam system can be emulated in cast concrete (Figure 7.24(a)), and is indeed probably the most common system, due to its great adaptibility. Other systems may be justified or preferred in various situations. The flat slab and flat plate systems (Figures 7.24(b) and (c)) offer a clean underside form, which may be functionally useful in obtaining overhead clearances in the space below the structure, or simply be cleaner in appearance where this is a factor.

Narrow ribs and thin decks can be used to produce relatively light, longer-spanning systems in sitecast concrete. With ribs in one direction only (Figure 7.24(d)) this is called concrete joist construction. A unique possibility in the cast concrete material is to create a two-way, intersecting rib system (Figure 7.24(e)), called a waffle system,

(a)

(b)

Figure 7.22 Steel systems for roofs and floors: (a) System with closely-spaced, light steel trusses (open web joists) and formed sheet steel deck. (See also Figure 7.10a.) (b) System with widely-spaced, rolled steel beams (W shapes) and longer span steel deck units. Many variations are possible, accommodating a range of spans for both the deck and the deck-supporting elements.

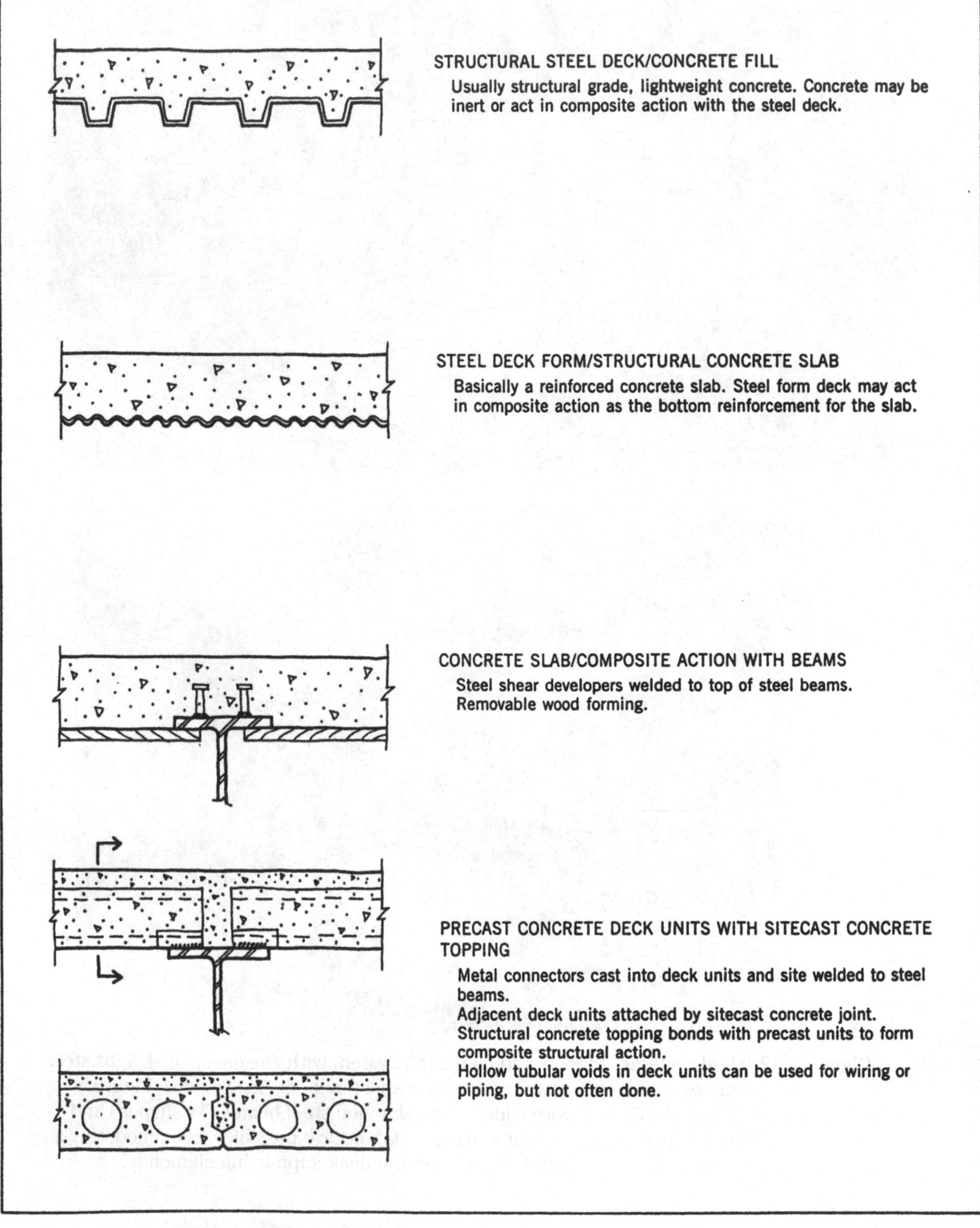

STRUCTURAL STEEL DECK/CONCRETE FILL
Usually structural grade, lightweight concrete. Concrete may be inert or act in composite action with the steel deck.
STEEL DECK FORM/STRUCTURAL CONCRETE SLAB
Basically a reinforced concrete slab. Steel form deck may act in composite action as the bottom reinforcement for the slab.
CONCRETE SLAB/COMPOSITE ACTION WITH BEAMS
Steel shear developers welded to top of steel beams.
Removable wood forming.
PRECAST CONCRETE DECK UNITS WITH SITECAST CONCRETE TOPPING
Metal connectors cast into deck units and site welded to steel beams.
Adjacent deck units attached by sitecast concrete joint.
Structural concrete topping bonds with precast units to form composite structural action.
Hollow tubular voids in deck units can be used for wiring or piping, but not often done.

Figure 7.23 Concrete floor decks used with steel framed structures.

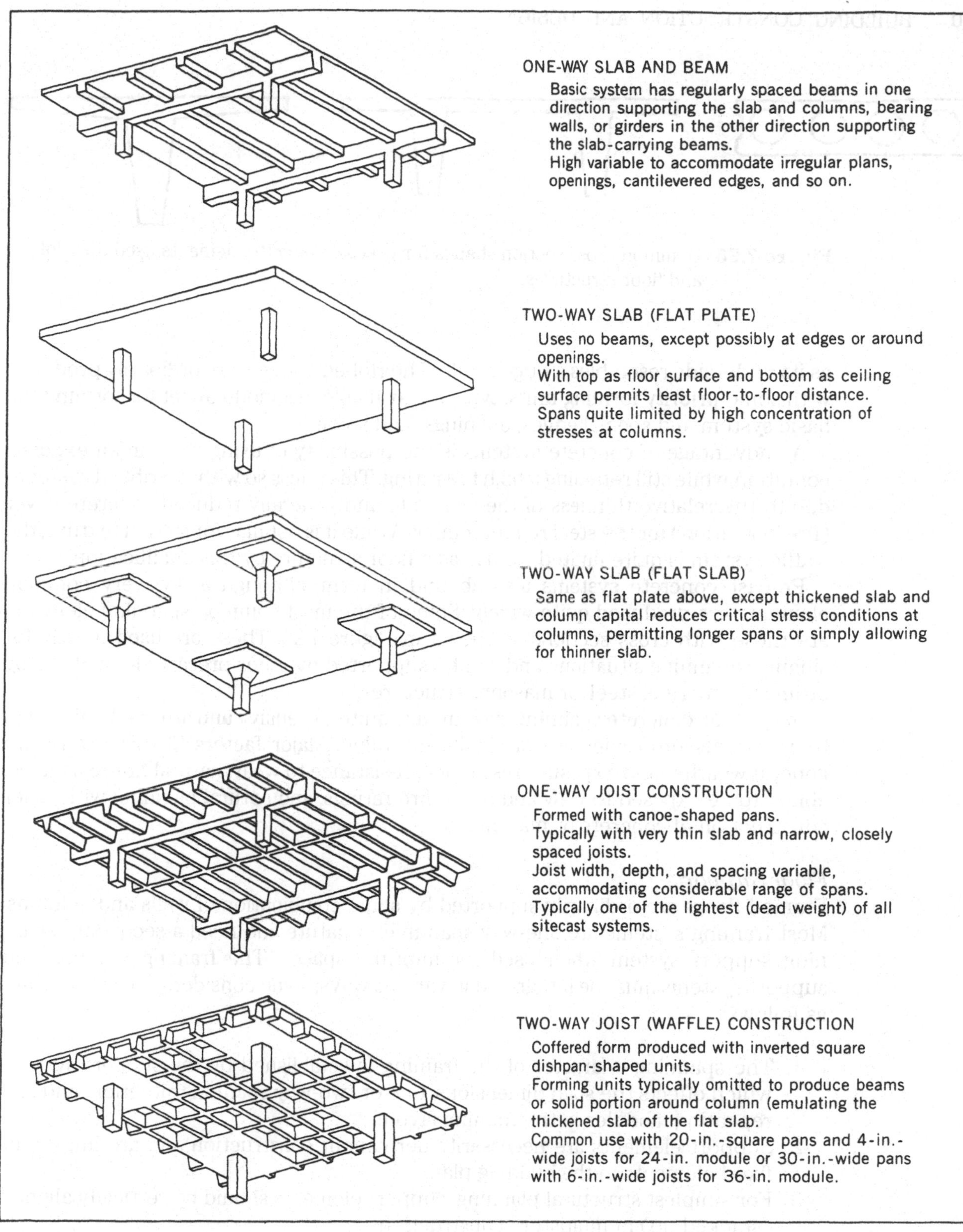

ONE-WAY SLAB AND BEAM
Basic system has regularly spaced beams in one direction supporting the slab and columns, bearing walls, or girders in the other direction supporting the slab-carrying beams.
High variable to accommodate irregular plans, openings, cantilevered edges, and so on.
TWO-WAY SLAB (FLAT PLATE)
Uses no beams, except possibly at edges or around openings.
With top as floor surface and bottom as ceiling surface permits least floor-to-floor distance.
Spans quite limited by high concentration of stresses at columns.
TWO-WAY SLAB (FLAT SLAB)
Same as flat plate above, except thickened slab and column capital reduces critical stress conditions at columns, permitting longer spans or simply allowing for thinner slab.
ONE-WAY JOIST CONSTRUCTION
Formed with canoe-shaped pans.
Typically with very thin slab and narrow, closely spaced joists.
Joist width, depth, and spacing variable, accommodating considerable range of spans.
Typically one of the lightest (dead weight) of all sitecast systems.
TWO-WAY JOIST (WAFFLE) CONSTRUCTION
Coffered form produced with inverted square dishpan-shaped units.
Forming units typically omitted to produce beams or solid portion around column (emulating the thickened slab of the flat slab.
Common use with 20-in.-square pans and 4-in.-wide joists for 24-in. module or 30-in.-wide pans with 6-in.-wide joists for 36-in. module.

Figure 7.24 Common forms of sitecast concrete floor systems.

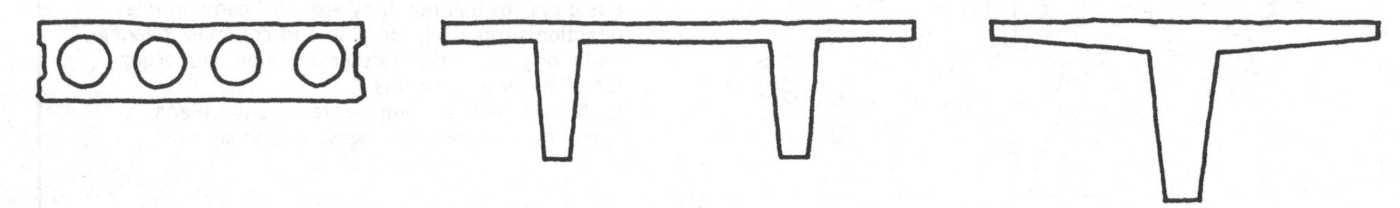

Figure 7.25 Common cross section shapes for precast concrete elements used for roof and floor structures.

as its underside resembles a huge waffle. The ribbed systems are ordinarily produced with some priority forming units, typically available in a modular set for forming the basic system and special edges, openings, and so on.

An advantage of concrete systems is the possibility of using them in an exposed condition while still retaining a high fire rating. This is less so with the ribbed systems, due to the relative thinness of the elements and generally reduced concrete cover (fire insulation) for the steel reinforcement. While it is popular for its rich texture, the waffle system is quite limited for use as a floor structure by this qualification.

Precast concrete systems also abound in form, although a few very common elements are produced quite widely. Some of the most common spanning elements are those with cross sections as shown in Figure 7.25. These are used mostly for singular spanning situations and can be supported by other precast elements or by ordinary concrete, steel, or masonry structures.

In general, concrete spanning systems are quite expensive and are used only when their various properties are of significant value. Major factors in this regard are general weather and exposure resistance, resistance to wear, natural fire resistance, ability to be exposed to view and retain fire ratings, and natural blending with other construction of concrete or masonry.

Floor Supports

Framed floors are ordinarily supported by some combination of walls and columns. Most framing systems are one-way spanning in nature and need a secondary spanning support system when used for multiple spaces. The framing systems and support systems must be integrated in various ways; some considerations for this are as follows:

1. The spanning character of the framing is established by the support system, which effects the span dimensions, one or two-way spanning situation, and any repeating modules in the framing layout.
2. Support elements are necessarily permanent construction and are important fixed elements in the building plan.
3. For simplest structural planning, support elements should be vertically aligned (stacked up) in multistory construction.
4. Lateral bracing is usually achieved with vertical elements at the location of supports for the framing system; therefore the framing support and lateral bracing functions must be planned for simultaneously.

Bearing walls can be achieved with various forms of construction, the most common being the following (see Figure 7.26):

Figure 7.26 Common forms for structural bearing walls: (a) Closely-spaced wood 2 by 4s form a wall in a light wood frame. (b) Structural masonry, developed with units of precast concrete (concrete blocks).

1. Stud framing—mostly in light wood (2 × 4, etc.) and mostly for support of light wood, joist or rafter framing.
2. Structural masonry—mostly with CMUs (concrete blocks); used to support just about any kind of framing: wood, steel, or concrete.
3. Sitecast concrete—used mostly with other concrete construction, but can support any framing. Foundation and basement walls represent the most frequent use.
4. Precast concrete—used mostly for exterior walls, mostly with tilt-up construction.

Bearing walls represent highly fixed, generally solid plan elements which require serious architectural planning. Some openings are possible, but the basic form must be that of solid wall construction. If the need exists for considerable open plan space, or for future rearrangement of walls, the use of bearing walls is probably questionable.

Columns may be arranged as an independent system, or may be developed in conjunction with a wall system. Small columns can be buried inside stud walls to eliminate any plan intrusion. (Developed as multiple studs or as 4 × 4, 4 × 6, etc., inside a 2 × 4 stud wall.) Pilasters can be used with masonry or concrete walls, representing lumps in the wall form, but at least not a whole freestanding presence in a plan.

Columns are locations of concentrated loads and the framing that delivers the loads must be acknowledged in planning. Vertical wiring, piping, ducting, and any openings in the floor system must be coordinated so as not to conflict with locations of major framing elements.

Columns also need major foundations, which must be considered in planning elevator pits, under-floor drains, and other elements. A column may fit into the plan at a floor level, but the whole vertical structure should be considered. In tall buildings, column sizes will probably change from top to bottom of the structure, and planning at all levels should be considered. For example, an elevator shaft must be constant in dimensions throughout its height, even though bearing walls or columns at its edges change in successive levels.

The pattern of supports establishes the nature of spanning required by the framing system supported. One way span systems need support only on two sides; continuous framing needs repetitive, reasonably equally spaced supports; two-way spanning systems need special supports that permit the appropriate two-way action.

Planning of support systems must be done as a coordinated effort with general architectural planning development. Architectural planning may lead structural planning and the logical framing system may emerge. On the other hand, if an early decision is made to use a particular type of structure—for example, when exposed construction is used—the needs and limitations of the structure must be dealt with in developing architectural plans.

7.7 FLOOR FINISHES

Most floor structures receive some form of added finish as a wearing surface. This is more or less functionally required, depending on the materials of the construction of the structural deck. Concrete slabs can be left exposed and finished to a high quality surface condition, if so desired. On the other hand, structural plywood decking is usually quite rough, and gets rougher as nailing of edges is achieved.

Finishes abound, but a few common ones account for most floors. The following discussions deal with common situations, with primary attention given to construction concerns.

Concrete

Concrete top surfaces, particularly of pavements or structural decks, can be brought to a finished surface condition in a variety of ways. Of primary concern usually is the simple truing up or smoothing out of the surface and the development of as flat a surface as possible. It is also usually essential to achieve a good, dense concrete surface before the concrete attains its initial hardening—called the set.

If an applied surface, or simply additional construction, is to be placed on top of a concrete surface, the truing up may be done with the accomplishing of a relatively rough and slightly porous surface. In many situations, a thin additional layer of lean concrete fill is used to achieve a finer, flatter, generally truer finished surface or to receive thin surfacing materials, such as linoleum or vinyl tiles.

When fill is placed on top of a structural concrete slab, it may be used in a manner similar to that with fill on a steel or wood deck; for example, to contain buried wiring systems. This is usually a simpler way to contain wiring, rather than within the structural concrete. It also allows for a total rewiring in the future without disturbing the basic structure.

Concrete surfaces left exposed can be finished in various ways, depending on usage and maintenance. The concrete itself may be made as smooth as possible, or may be deliberately roughened to reduce slipperiness when wet or oily. Grit of some form may also be embedded to achieve the non-slip surface. On the other hand, paint, sealers, or wax may be used to enhance the appearance or smoothness of the surface.

For construction detailing and building dimensions, it is important to anticipate the desired finishes of floors, particularly when the surface of the structural concrete must be depressed to permit additional thickness for construction for finishes.

Information Source

Portland Cement Association

Wood

Wood flooring is still widely used, although other floor finish materials have displaced it in many situations. Hardwood flooring used to be common for residential occupancies, but is now rare, since carpeting is now so widely used.

Wood flooring is attractive, and is now mostly reserved for feature usage—as with marble, ceramic tile, and so on—except for some specific uses, such as gymnasiums. Wood flooring is now mostly installed in prefinished blocks or strip panels and is often attached with adhesive materials, rather than nailed down.

Information Source

National Oak Flooring Manufacturers Association

Carpet

Wall-to-wall carpet has emerged as a universal floor covering. Tough, stain-resistant synthetic fibers and mass production, as well as simplified installation, have made this a diverse, easily installed, and easily replaced covering material. Reduction in the number of smokers has also made it more feasible; burns being the nemesis of synthetic fibers.

Carpet with reasonable padding is also fatigue-reducing for occupants who must walk or stand on the floor for many hours. In general, it is usually a pleasant and highly popular floor surfacing for building users.

Carpet must be applied over a reasonably smooth surface, or else it will eventually tend to reflect the surface flaws beneath it. Smoothing elements (called *underlayment*) are thus often used; such as concrete fill or a layer of wood fiber panels. This was not a problem when floors were routinely finished with hardwood flooring or tile, but floors are now often left with rough, structural finishes, anticipating a fully-developed finishing of some kind.

There is a great range of quality and type of carpeting, permitting considerable variety within a simple basic construction provision.

Information Source
Carpet and Rug Institute

Thin Tile
Thin tile—literally dimensionally thin, although installation may require additional materials—may be of various types, including the following common ones:

1. Linoleum or square units of vinyl or other flexible materials.
2. Thin-set ceramic tiles, attached with adhesive materials.
3. Thin wood blocks or strips, attached with adhesives.
4. Relatively thin units of stone, fiber reinforced concrete, and other priority materials, adhered to emulate classic stone flooring.

As with carpet, the supporting surface must be quite smooth and flat, as bumps and cracks will eventually be reflected. Thus, while the finish materials themselves are thin, necessary fillers may add to the total construction that must be allowed for by the supporting structure. Linoleum placed directly over plywood will eventually reflect not only the plywood joints and nails, but even the wood grain of the face plies of the plywood.

Most hard floor finishes—of wood, stone, and ceramic tile—can now be produced with less dimensional allowance in the construction. Not quite as thin as a coat of paint, but mostly a lot less than was required in the old days.

Information Source
Tile Council of America

Thick Tile
Some forms of finish, essentially executed in the classic manner, may still require considerable allowance for the thickness of the complete work. Field stone, thick ceramic tiles, precast concrete tiles, or other elements may be placed in a mortar bed in the classic method. Depending on the size of the units, this may require several inches or as little as 1.5 to 2 in. for development on top of a structural concrete surface.

Terrazzo is essentially—in its present form—a concrete finish material; developed by a special concrete mixed with coloring agents and typically marble or other stone chips. When hardened, the surface is ground and polished to a very smooth finish. Brass strips are commonly used to separate zones of different colors to produce large floor patterns.

Terrazzo is a very functional, durable finish, although quite expensive when installed in the classic manner, with a build-up of as much as 2.5 in. of total thickness, including the final, thin layer of finish material. Many alternative techniques can also be used to simplify an installation, including use of thin-set, precast units. The good stuff, however, still takes a lot of time, a lot of money, and considerable thickness for construction.

Information Source

Tile Council of America

7.8 CEILINGS

Ceilings are the upper bounding surfaces of interior, enclosed spaces. Structures for ceilings may be developed in a number of ways, the simplest being that of no structure at all; that is, the ceiling is simply the raw underside of the overhead roof or floor structure. However, when the exposed structure is not desired, some form of ceiling construction must be developed. The following discussions treat the common means of achieving ceiling construction.

Directly Attached Ceilings

Ceilings may be developed by direct attachment of surfacing materials (drywall, etc.) to the overhead structure (see Figure 7.27(a)). This is possible when the form, spacing, and materials of the structure provide adequately for such attachment. In residential construction, drywall is ordinarily attached directly to the underside of the closely-spaced, light wood rafters or floor joists. This framing is ordinarily closely spaced to accommodate thin roof or floor decks of plywood, and the dual role of ceiling framing is easily accomplished.

When framing is more widely spaced—as when heavy timbers or steel W sections are used—it is usually necessary to use some intermediate framing attached to the major structure and spaced in response to the needs of the ceiling surfacing materials (see Figure 7.27(b)). Intermediate framing (furring strips, etc.) may also be necessary if the form or the materials of the major structure do not permit easy attachment of the ceiling materials.

Direct attachment will yield the minimum total thickness of the ceiling plus the overhead structure, which is generally more desirable with floor construction in multistory buildings, as total floor-to-floor height will be the least. However, it will also provide the least *interstitial space* (between the ceiling below and the floor surface above) and may present problems where considerable space is required for building service elements, such as ducts and recessed lighting fixtures.

Direct attachment of ceilings will also provide a direct reflection of the general form of the overhead structure; not so much a problem with floors, but possibly one with non-flat roof structures, For this, or other reasons, it may be desired to produce an essentially independent ceiling structure, as discussed in the following sections.

Suspended Ceilings

A separate ceiling structure may be hung from the overhead structure, deriving support from it, but not necessarily reflecting its profile or detail (see Figure 7.27(c)). This is often done to create needed space for equipment and services, but also to create a different form or simply a lower ceiling level.

Suspension must be achieved with materials and details that relate to both the

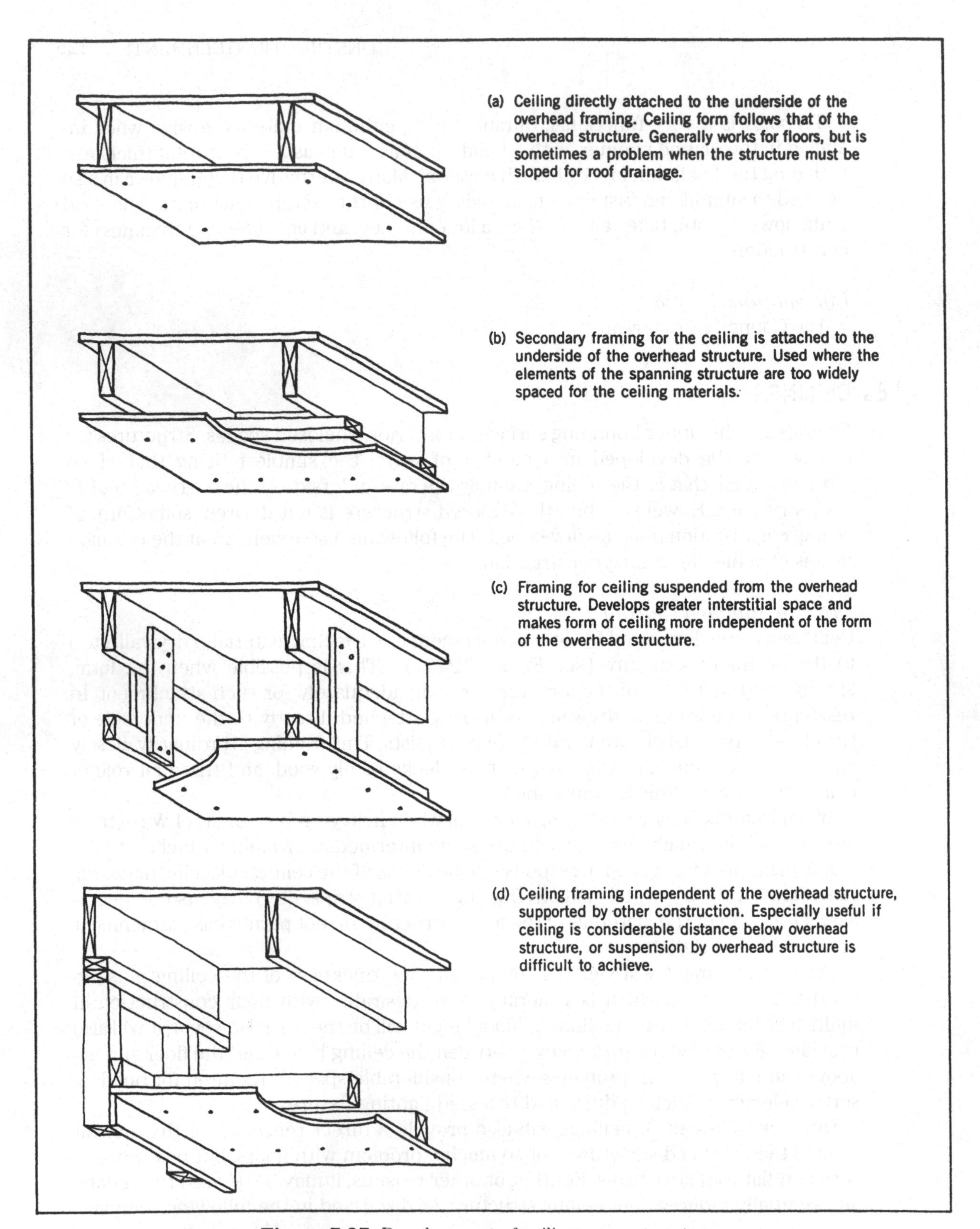

(a) Ceiling directly attached to the underside of the overhead framing. Ceiling form follows that of the overhead structure. Generally works for floors, but is sometimes a problem when the structure must be sloped for roof drainage.
(b) Secondary framing for the ceiling is attached to the underside of the overhead structure. Used where the elements of the spanning structure are too widely spaced for the ceiling materials.
(c) Framing for ceiling suspended from the overhead structure. Develops greater interstitial space and makes form of ceiling more independent of the form of the overhead structure.
(d) Ceiling framing independent of the overhead structure, supported by other construction. Especially useful if ceiling is considerable distance below overhead structure, or suspension by overhead structure is difficult to achieve.

Figure 7.27 Development of ceiling support systems.

overhead structure and the form of the ceiling construction. Typical methods and materials used relate to common associations of structures and ceiling constructions.

Many ceiling surfacing materials can be used in this situation, as well as with direct attachment to the overhead structure. Intermediate framing may be somewhat different if it is suspended instead of being directly attached. Suspension elements can be made independent of the spacing of the major structure; as when hangers are attached to the decking, instead of the framing—a common practice with decks of steel or concrete.

Many priority modular ceiling systems are developed primarily for use in suspended construction (Figure 7.28). These ordinarily have modular surfacing panels and a network of framing, and may even include special hangers. They are often also coordinated for use with modular lighting and HVAC units in a fully integrated system.

Because they exist often to provide for service systems overhead, suspended ceilings may need to provide for access to the elements of the enclosed services. This sometimes favors the use of the priority, modular unit systems, whose individual surface units are usually quite small and readily removable.

Separately Developed Ceiling Structures

In some situations, ceilings may be developed with construction that is totally independent of that overhead (see Figure 7.27(d)). This may simply be the practical means for achieving the ceiling, or it may be necessary where the ceiling is a great distance below the structure above it.

Where room sizes are small, and wall construction permits it, a ceiling may be developed with separate framing supported by the room walls. This framing may be quite modest, if it only needs to support the ceiling and does not provide for a floor for the space above it. Design live loads for ceiling spaces with limited access are usually only 10 psf. Span limits may be derived from critical concerns for deflection (visible sag) rather than from actual load-carrying capacity.

For some types of occupancies—notably speculative, commercial ones, such as offices and stores—use of independent interior wall and ceiling construction allows an increased level of freedom and ease for modifications of interior spaces.

Ceilings are considerably in view, and deserve serious attention where the quality of appearance of the building interior is of concern. This attention must be given to the ceiling construction in general, but mostly to the finishing of exposed surfaces.

Exposed Construction

The simplest ceiling finish is no finish at all, which may occur if the overhead structure is left exposed. However, as discussed for natural finishes on walls in Sec. 7.2, the unadorned and untreated ordinary structure may not be very appealing. Cleaning, a coat of paint, or other minor dressing up may be in order. More seriously, there may be some need to specify some greater care in the construction, in the selection of materials, in careful arrangements of parts that are not ordinarily treated with much precision of alignment, and so on.

The decision to expose the structure may be a practical one, as is done for garages, industrial buildings, chicken coops, and other not so glamorous interiors. It may also be a very conscious design decision for a relatively high class interior (Figure 7.29) where a very dramatic structure is deliberately used and exposed in a high-tech style. In the latter case, there may be a desire to be very honest to the ordinary details of the

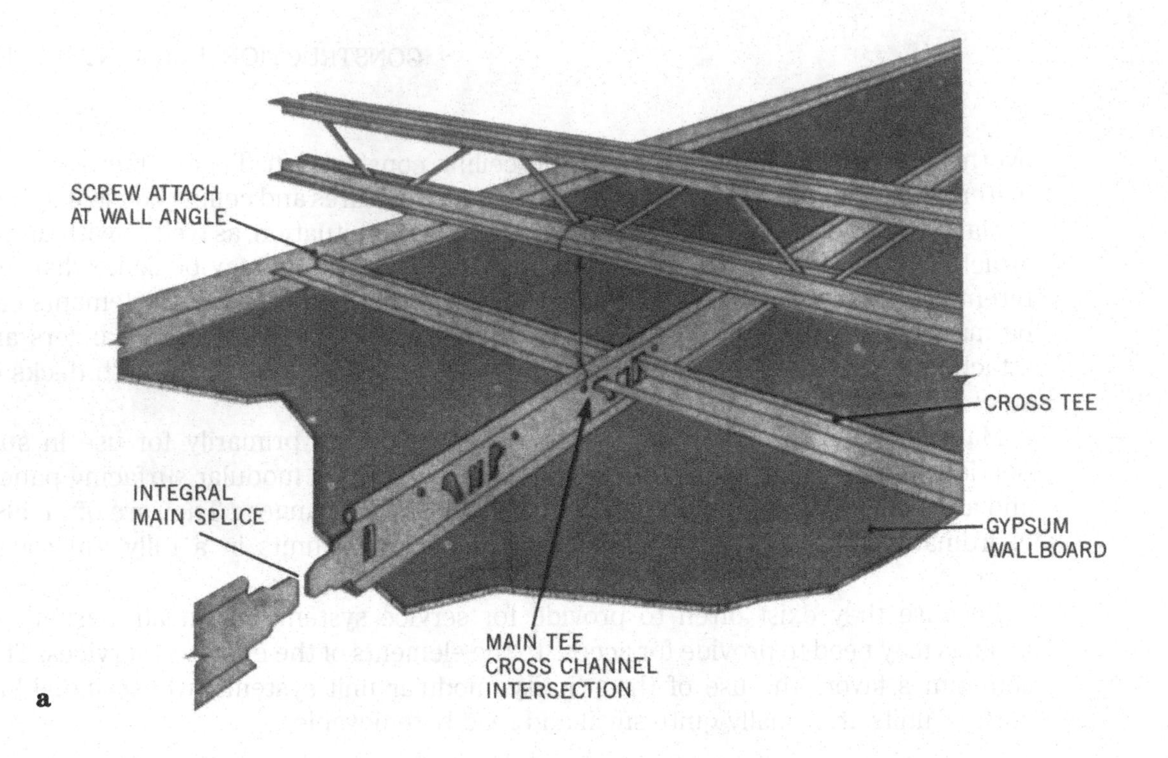

Figure 7.28 Suspended ceiling construction: (a) Gypsum drywall attached to a priority metal frame hung from the overhead structure with wires (courtesy of Gold Bond Building Products, Charlotte, NC). (b) Modular priority system with metal framing grid and lay-in units.

Figure 7.29 Ceiling developed essentially as the exposed overhead construction; may be cleaned or generally made neater, but is basically untreated for appearance.

proper construction, or some considerable design effort may be expended in developing a much more elaborate or generally spiffy looking structure.

Some structures may not endure exposure well and need some real protection, especially for fire resistance. The ability to leave structures exposed—especially wood and steel ones—may be qualified by various conditions pertaining to the building occupancy, size of the building, height of the structure above the floor, and so on.

Reasonably raw exposed structures on interiors are parts of some very dramatic and exciting historic architecture, such as the Gothic cathedrals, the great train sheds of the early industrial period, Paxton's Crystal Palace, the richly-formed concrete roof structures of Nervi, Candella, and Torroja, and the convoluted tension structures of Frei Otto. One cannot imagine having a suspended ceiling beneath such awe-inspiring displays of structure.

Drywall

The ubiquitous drywall panel accounts for the vast majority of ceiling surfaces in present construction. This is, in general, the all-purpose cover-up, forming both vast, uncluttered flat ceilings, and boxing around various complex, coffered, coved, curved, and multi-faceted forms.

There are indeed many types of drywall panels and many different special details for their installation. Modifications may be made for increased fire resistance, greater strength, better sound-response, or other desired properties. Special panels may be used for exterior sheathing, plaster bases, reflective insulation, and other purposes.

Fastening of panels to supporting framing and treatment of joints are major detail concerns, also having many variations. Long experience has resulted in some very standard forms of construction, but the technology continues to evolve.

As with any material or form of construction, drywall has its limits leaving some situations where other construction is clearly indicated. However, after its long, hard-fought battles with trade unions, code writers, and insurers, and its emergence from the stigma of being a cheap imitation of real plastered (wet-wall) construction, drywall is now pretty much the definitive wall and ceiling surfacing material, to be supplanted only for cause in special situations.

Plaster

Plaster is a very ancient material, originally developed primarily as a coating and protective covering for soft, vulnerable structures, such as adobe and soft-mortared brick or stone masonry. In many cultures, the smooth plaster surface became a canvas for painters, producing some very rich interior adornments. Plaster also became a sculptor's medium, producing rich forms, such as those in the Baroque period.

As with many historic materials, plastered construction is now often imitated, rather than real. Even stucco (exterior plaster), although still extensively used, is now also frequently imitated in plastic-based finishes on multi-layered, exterior insulation systems with fiber glass reinforcement and foam plastic backup.

However, real plaster can still be installed, pretty much in the traditional, "wet" method. A very high quality, durable finish can still be produced with a multi-coat plaster finish, ending with a very fine-textured, hard surface. It may be expensive and time-consuming, but will have properties that are hard to compete with.

For practical purposes, plaster is now sometimes installed over perforated gypsum wall panels, becoming transitional as an independent wall material and marginal between old style plaster and merely a very thick coat of paint. Some form of backup is generally needed, and is now likely to be either a formed steel product or some composite of paper, moisture-resistive coating, and steel reinforcement.

Gypsum plaster—the same material used in drywall—can be installed wet, but is too soft in general to be used for surfaces within reach of occupants or otherwise subject to contact or wear. It is, however, very low in density (lightweight) and may be a good choice for ceiling surfaces.

Really hard plaster is basically portland cement-based and forms a very dense, hard material. Stucco, applied in a reasonable thickness and reinforced with steel wire, literally forms a thin reinforced concrete slab unit with major contribution to lateral stiffening of a building, particularly one with a light wood frame.

Sculpted while wet, plaster surfaces can be given various textured forms besides dead flat and smooth. Other materials can also be worked into the surface just before hardening, such as paint (fresco), small tile, grit, crushed glass, and so on. Decorative plasterers were once prima donnas of the building trades, but the craft is now largely non-existent.

Modular Systems

Modular systems may consist of units adhered directly to the underside of a concrete slab or to an untreated gypsum drywall surface. More often, however, modular systems occur as a priority system using separately supported frameworks and small lay-in panel units (Figure 7.28).

Modular systems typically use some dimensions that relate to other ceiling items, such as light fixtures and grilles or registers for HVAC systems. Some priority systems are fully developed to incorporate these service elements in a totally integrated manner. Many commercial interiors are developed with these systems.

The dimensions and repeating modules of a ceiling system may be essentially divorced from those of the structure from which they are suspended. However, when interior partitioning must intersect the suspended ceiling, coordinated wall/ceiling systems may be used. There are many alternatives for this situation, including ones emanating from ceiling system manufacturers and wall system manufacturers. However, custom designs are not all that difficult either. Most modular ceiling systems can be trimmed to fit at their edges to accommodate some non-modular wall construction or other special construction situations.

If a modular ceiling is to be used, however, it will come off much better if everything related to it is kept neatly in the modular order, including lighting, HVAC elements, fire sprinklers, audio systems, signs and any decorations.

Miscellaneous Ceiling Finishes
Being typically out of reach, ceiling surfaces may be developed with just about any material—and probably have been. Wood, metal, plastic, paper, and fabrics are common. Egg cartons, growing plants, and glass are not unknown.

Fire codes and reasonable safety must be considered, but the highly visible ceiling surface is a blank canvas for the interior designer, and fair game. Conventional finish materials, used for interior or exterior walls, and even for floors, may be used. This is probably the safest place for designers to cut loose, as the fewest functional constraints exist. As with any interior surface, various physical properties must be considered however.

7.9 WINDOWS

The basic forms of windows and the basic concerns for their design are discussed in Sec. 5.4. This section presents some of the issues relating to choice of window materials and the problems of their installation in wall construction.

Types of Windows
As described in Sec. 4.4, windows are either fixed (not able to be opened) or can be operated (opened and closed). Operating windows (called *operable*) take various forms, mostly either sliding or swinging to move to an opened position (see Figure 7.30). The selection of the form or type of window depends on several considerations, such as:

Need for operation: Besides their basic purposes of providing for vision and admitting light, windows may function to permit fresh, outdoor air to enter the building for ventilation of interior spaces. This form of ventilation may be required —or in some cases may be undesirable because of its effect on the ability to control air distribution with a mechanical HVAC system.

Architectural design concerns: Shape, size, position, materials, and details of windows affect their appearance on the building exterior. Use of similar windows, matched sizes, vertically or horizontally aligned rows of windows, and other concerns may be added to pragmatic concerns for the window uses.

Typical concerns for effects on interior environmental conditions: These may involve thermal comfort, use of daylight, or prevention of direct sun rays on contents or occupants. Weather sealing of operable windows is typically more

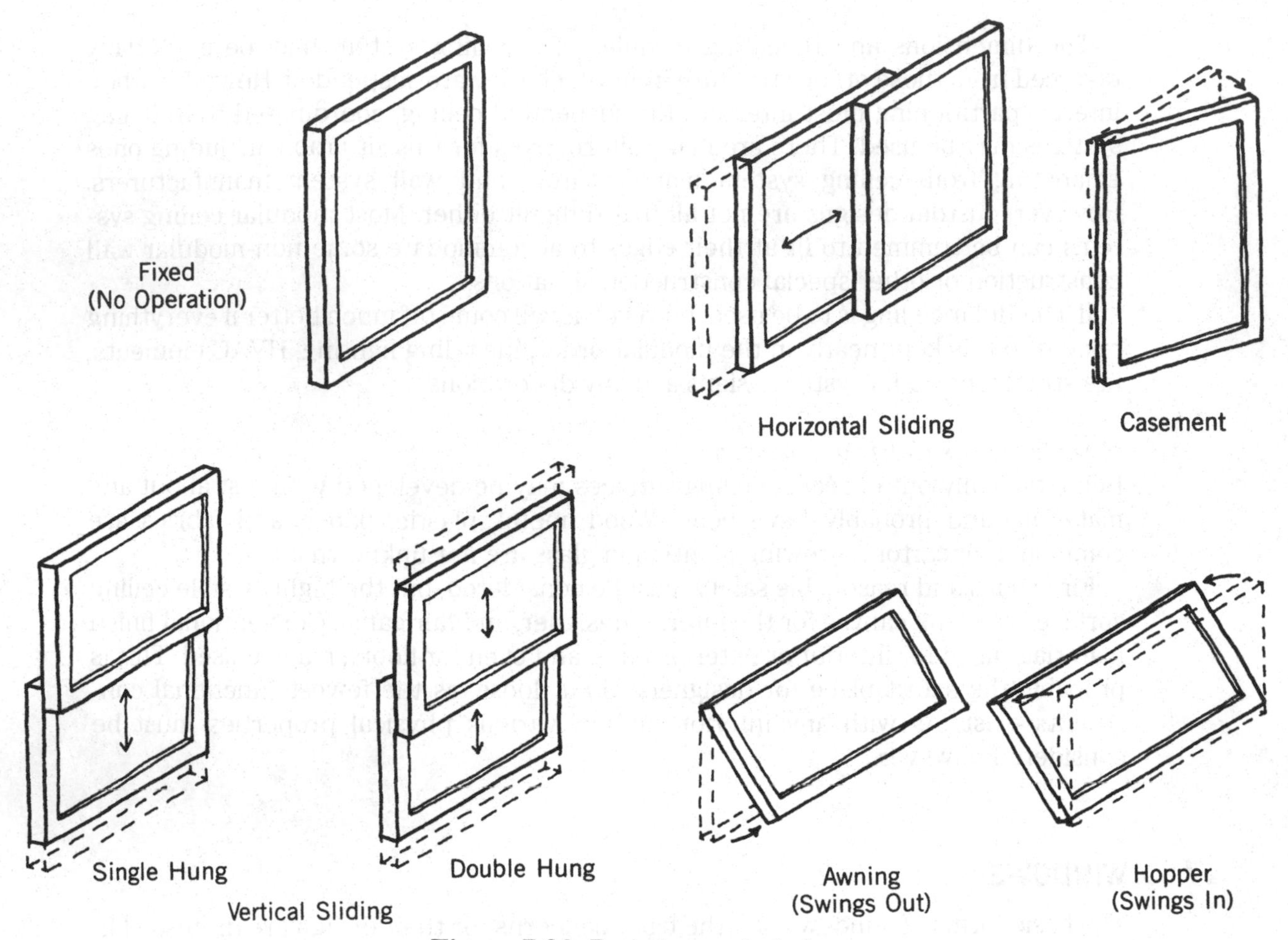

Figure 7.30 Basic forms of windows.

effective with some types of windows; generally good with single-hung windows and not so good with sliding or casement windows.

Use of any necessary auxiliary devices: Window operation may be a concern when there are screens, awnings, drapes, slatted blinds, automatic closers, alarms, and so on.

Access and operation by occupants: The type of occupants (small children, elderly, etc.) or the location of windows (high on a wall, for example) may present special problems for window operation. Auxiliary opening and closing devices may be required in some situations—working better with some types of windows.

Selection of window type may also relate to needs for security. The fixed window is obviously most secure, the single-hung second best, and others less so. Sliding windows can sometimes be relatively easily derailed and swinging ones easily pried open.

Window Components

For ordinary windows (see Figure 7.31), there are usually a set of components that include the following:

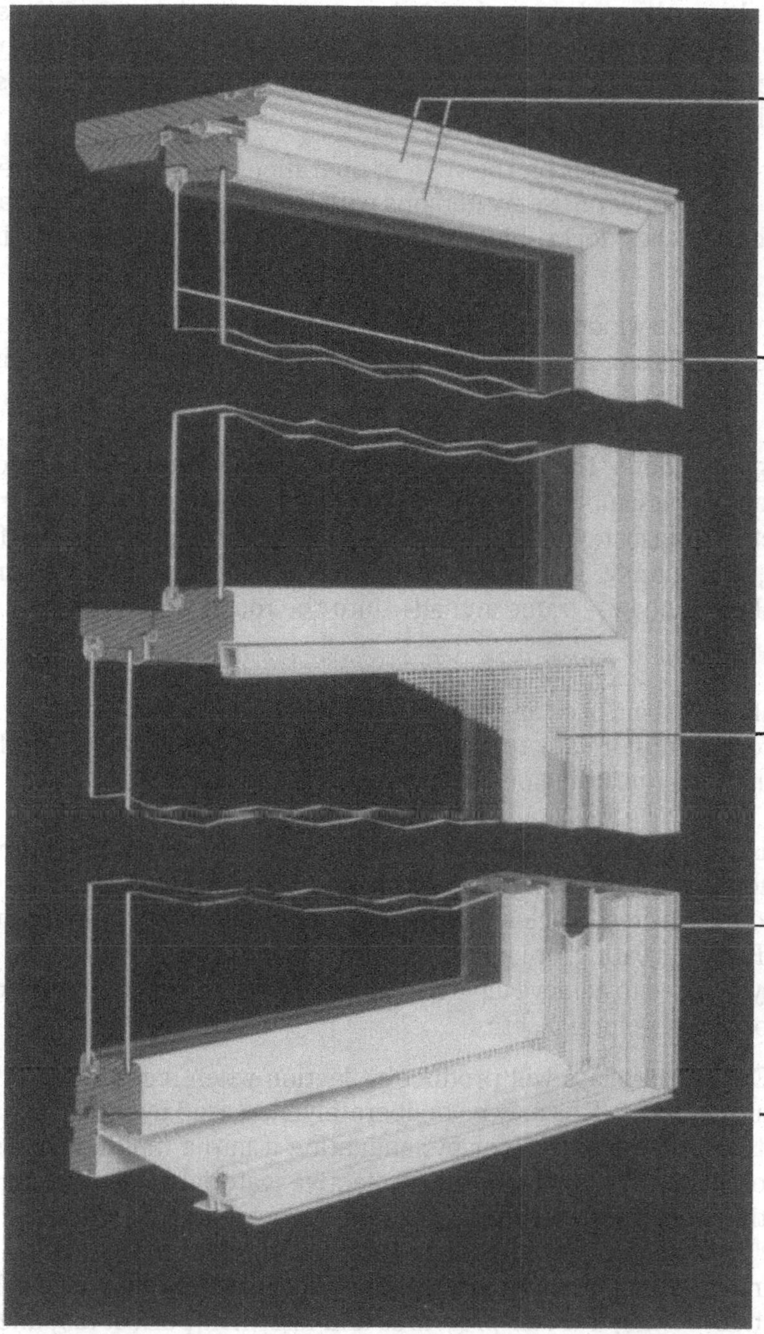

Inside a Pella® Clad Double-Hung — shown from exterior. Wood double-hung is similar except for exterior finish.
LOW-MAINTENANCE EXTERIOR of clad units is aluminum with baked enamel finish, mechanically bonded to wood parts. Sash joints are lapped and sealed.
Wood double-hung windows are as shown except for all-wood exterior. Wood units are factory-primed on all exterior surfaces to protect them prior to final finishing.
DOUBLE GLAZING PANELS create 13/16" air space between exterior glazing and the removable panels to provide maximum insulation and condensation control. Other options in the Pella glazing system to enhance energy efficiency: Sunblock™ and Heatlock glass and Slimshade® blinds. Total-unit U value with the closed Type E Slimshade is .23, R value 4.35 — equal to or better than equivalent "low-E" films. Heatlock glass has similar heat-reflecting characteristics; U and R value performance equals or exceeds that of triple glazing.
EXTERIOR SCREEN is fiberglass; aluminum frame is white (wood units) or matches exterior finish (clad). Half screen is shown. Full height screen is optional on traditional units only, not available on clad contemporary double-hung windows.
JAMB LINERS of tough, resilient vinyl conceal spiral spring-steel balances and insulate against heat and cold. Concealed aluminum continuous springs hold jamb liners snugly against sash and weatherstripping for tight seal and allow sash to pivot for washing from inside.
WEATHERSTRIPPING is vinyl-wrapped foam at all sealing points — head, jamb liners, sill and checkrail. With static air pressure of 1.57 psf (equivalent 25 mph), *maximum* infiltration in independent tests is 0.15 cfm/ft of crack.

Figure 7.31 Typical prefabricated window unit with frame, sash, glazing and hardware in a single package, ready to be popped into a rough wall opening and trimmed up. Pella Casement Window courtesy of Rolscreen Company.

Frame: This is what gets installed in the wall. It must fit into the rough opening in the wall structure, be firmly supported, and be trimmed and sealed into the finished wall construction.

Sash: This fits into the frame, and it must have some slightly smaller dimension than the frame opening if the window is to operate (open and close). Weather sealing and also permitting operation is a challenge. Repeated operation and effects of shrinkage, thermal expansion and contraction, or the warping or sagging of the frame or the sash may impair the perfect fit between the sash and frame over time (the typical case), and repair may be required to maintain both smooth operation and good weather seals.

Sash hardware: This depends on the type of operation and special requirements, such as those for security or use by persons with limited capacities. Typical items include handles, latches, locks, hinges, sliding tracks, and swing-restricting elements.

Glazing: What makes a window a window, and not just a hole in the wall. With some fixed windows, the glazing is set directly into the frame; otherwise it is set into the sash.

For most single windows, the frame, sash, and basic hardware are composed as prefabricated, factory-assembled units. Windows can be custom-built, but are more frequently selected from manufacturer's catalogs, within the range of options for size, details, finishes, and glazing. Installation typically consists of setting the prefabricated, fully finished window—frame and all—into the rough wall opening.

Glazing

Glazing materials and their uses are described in Sec. 6.6. The selection of glazing materials and forms must relate to appearance concerns, but also typically to a number of other issues, including one or more of the following:

Light transmission: There is really no such thing as completely transparent glass; any glazing modifies the light that passes through it in some manner. If essentially no modification is desired, a "clear" glass may be specified or no mention of the issue raised. Tinting, coating, or laminating of the glazing, or use of special glazing materials, may be used to achieve a particular modification, affecting color change, glare reduction, heat reduction, etc.

Reflectivity: Even clear glass will produce reflection when it is dark on the side opposite from the viewer. This may be deliberately increased by use of a coating of reflective material on the surface or by laminating it into a multilayered glazing. Color can also be modified by the type of reflective material used. Another reflection effect is that consisting of distortions of the reflected images when the glass is not perfectly flat. Use of heavy, polished plate glass may reduce this, but on building exteriors, reflections developed by multiple glazing units cannot be very true, due to their bowing from air pressures and less-than-perfect alignment on a single plane.

Energy concerns: Use of double and triple glazing, transmission modifying glazing, or other techniques generally increases the window cost, and its cost-effectiveness must be studied together with the whole design of the energy-using systems for the building HVAC and lighting.

Breakage: Replacement of broken glazing must be anticipated; preferably with some ease and with the least disturbance of the occupants.

Glazing is still mostly achieved with ordinary, clear glass, However, use of laminated glazing is becoming more prevalent, especially for the vast surfaces of curtain walls in highrise buildings. The laminating process is expensive, but allows for great variety of manipulation of various properties and some significant improvement of one major one—breakage; which is a serious concern in urban areas where falling glass is a critical safety issue.

Special Windows

Windows may occur in special circumstances or may serve some special purposes, such as the following:

Door glazing: Doors may double as glazed elements, performing some window functions. One function is often that of permitting users on opposite sides to see each other and not cause injury by the opening of the door. For stair doors, or other fire-separating doors, special glazing is required; otherwise appearance and any other functional requirements may determine the selection of the glazing materials.

Interior windows: Windows may occur inside buildings to develop partitioning but facilitate vision or light distribution. Interior corridors or other totally interior spaces may have walls with some glazing, permitting the borrowing of light from adjacent rooms with windows in exterior walls (see Figure 7.32).

Privacy windows: Some windows, such as those in toilets or bathrooms, may be used for light and ventilation, but restrict vision. This is ordinarily achieved with

Figure 7.32 An interior window wall; achieving some wall functions (privacy, security, traffic management, etc.) but permitting vision and borrowing of light.

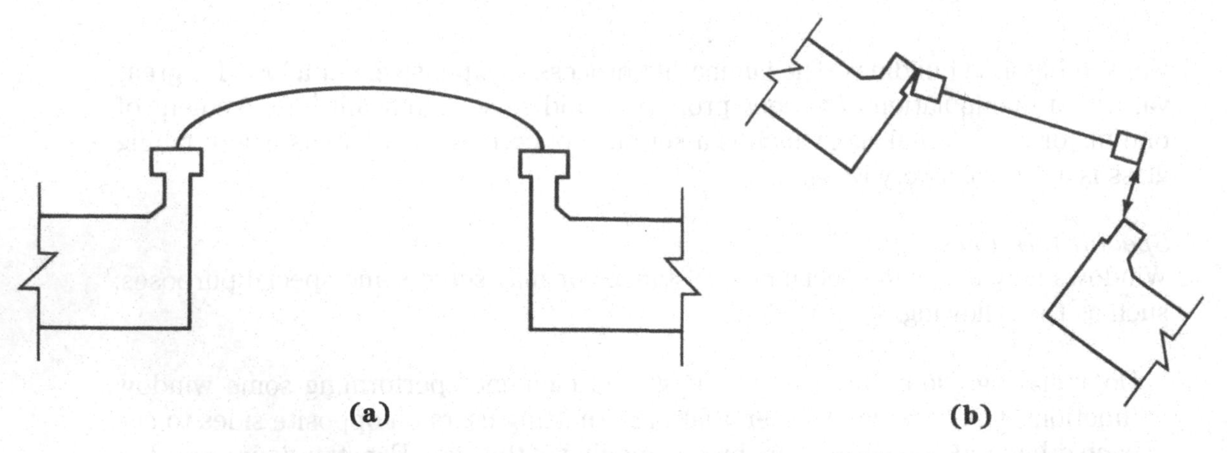

Figure 7.33 Basic forms of windows in the roof: (a) Skylight with fixed glazing. (b) Sky window, operable for ventilation.

glazing that is not clear, which can be achieved with white "milk" glass or by using a formed surface on one side to scatter light.

Skylights and sky windows: Windows sometimes occur in roofs; called *skylights* when they have fixed sash and *sky windows* when they are operable (see Figure 7.33). In addition to the usual requirements for windows, there are usually increased concerns for breakage of glazing and for control of direct sunlight.

7.10 DOORS

Basic uses and components of exterior doors are described in Sections 5.5 and 5.9. In this section, some considerations for selection of doors are presented.

As with single windows, doors are ordinarily selected as catalog items, using the options offered to "customize" a particular door. Doors are intended for repeated usage, and their support, hardware, and their own structural strength must be generally somewhat sturdier than those normally required for windows. Operation is also typically somewhat more complex, possibly involving such special items as hold-open devices, automatic closers, kick-plates, two-way swinging action, and panic hardware for fire exit use (see Figure 8.1(c)).

Types of Doors

Exterior doors provide entrance and exit from the building. This traffic may be limited to people, but must also frequently deal with movement of vehicles and the passage of various items. Furniture, supplies, waste materials, and various items must pass through doors on the way in or out.

The type of door selected, as well as its size, shape, materials, and general operation must be considered for the specific forms of traffic and general usage.

Most small doors swing horizontally, with hinges on one vertical edge and a latch on the opposite edge. However, many other forms are possible, as shown in Figure 7.34.

When doors move, they must move to some place other than their normal positions. This presents various problems for planning or other concerns. Swinging doors may swing into people on the opposite side, into furniture, or into a position blocking some traffic or another door. Door swings are ordinarily shown on architectural plans to illustrate their actions and show the space they need for operation. Swings may be restricted in some cases by bumpers or other devices.

For various situations there are minimum dimensions for the door opening provided when the door is moved out of its closed position. Use as a fire exit or need for access by handicapped persons may provide specific minimum dimensions. Judgment is required for some situations where no real criteria exist, such as for closet doors. Rules of common practice and the actual size range provided routinely by manufacturers establish some reference.

Building entrances are often places which designers develop as friendly, pleasant, inviting—even dramatic and exciting ones. This is the welcoming place, the first impression. The front door or main entrance may need to develop this character, and selection may be principally considered in these terms while still meeting functional requirements. If a custom-made door is desired, this is the place for it, although manufacturers have many grand alternatives.

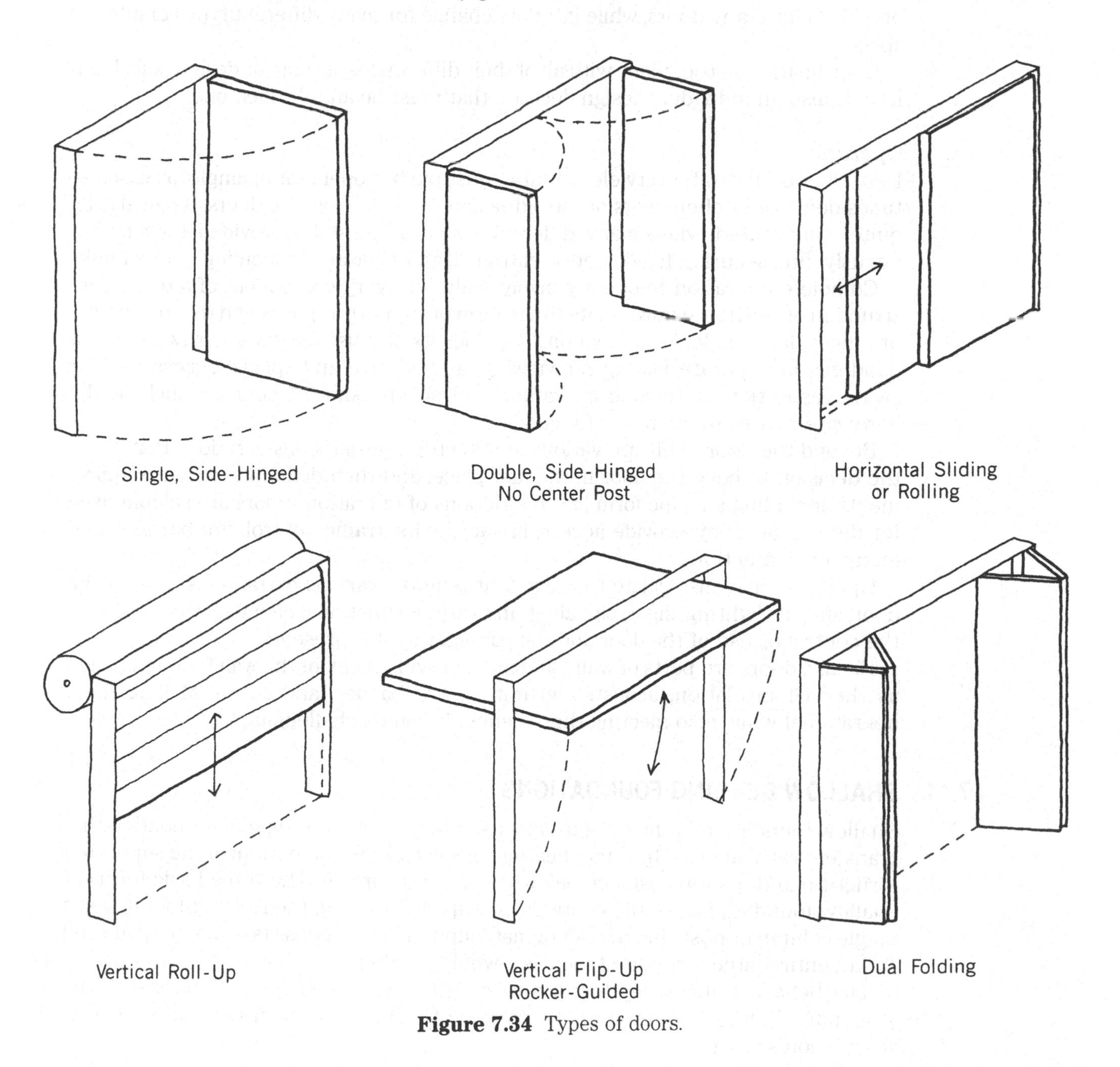

Figure 7.34 Types of doors.

Door Components

As described in Sec 5.5, doors ordinarily have basic units that include a frame, the door, and the various accessories that are used to facilitate its installation and operation (see Figure 5.8). For the door taken from a catalog, the kit may include almost everything required, or it may be limited to the door itself. Complete packages are somewhat less frequent here than with windows. Part of this has to do with the greater variability of doors, particularly as regards their hardware and other accessories.

A given building may use a number of doors, with many the same for design coordination of appearance. Yet practically every door may have *something* different: direction of swing, restriction of swing, type of latch or lock, glazing, hold-open, closer, and so on. Doors also have two sides, and one side may be the same (the outside?) for many doors, while interiors change for many different types of interior spaces.

Coordination of the doors, with all of their differences, is a major design task. Each item is also an individual design decision that must be an informed one.

Operation

Door operation can be very elementary—just push it open, for example, as is sometimes done for kitchen doors or the old-fashioned saloon or cafe doors. It can also be quite complicated when many different concerns must be provided for simultaneously: fire, security, handicapped entrance, and pleasant welcoming, for example.

Complete operation frequently involves almost every component of a door. Construction of the frame must relate to the form of support required and the installation of hinges, latches, locks, and so on. The door itself must facilitate hardware, resist handling, incorporate glazing for viewing, and receive any special accessories for swing restriction, automatic operation, and so on. And, of course, much of the hardware relates to operation functions.

Beyond the door itself, are various architectural design considerations that affect the decision to have the door in the first place, and then define its various requirements, including specific form and restrictions of operation. Doors are seldom used for decoration; they provide access, privacy, exits, traffic control, fire barriers, and many other functions.

Operation may also relate to usage that requires various extra items besides the door, such as lighting, signs, signaling, and various functional elements that deal with the control or use of the door and the passageway it represents.

Finally, doors are parts of walls, and must provide many of the wall functions, such as thermal insulation, acoustic separation, fire barrier, and so on. Making them operational while also meeting these demands can be challenging.

7.11 SHALLOW BEARING FOUNDATIONS

Shallow foundation is the term usually used to describe the type of foundation that transfers vertical loads by direct bearing on soil close to the bottom of the supported structure and a short distance below the ground surface. The three basic forms of shallow foundations are the continuous strip wall footing, the isolated footing for a single column or post, and the raft or mat foundation that consists of one big pad used for an entire large structure (such as a whole building).

The behavior of bearing footings relating to soil properties is discussed in the Appendix. Figure 7.35 illustrates a variety of elements ordinarily used in bearing foundation systems.

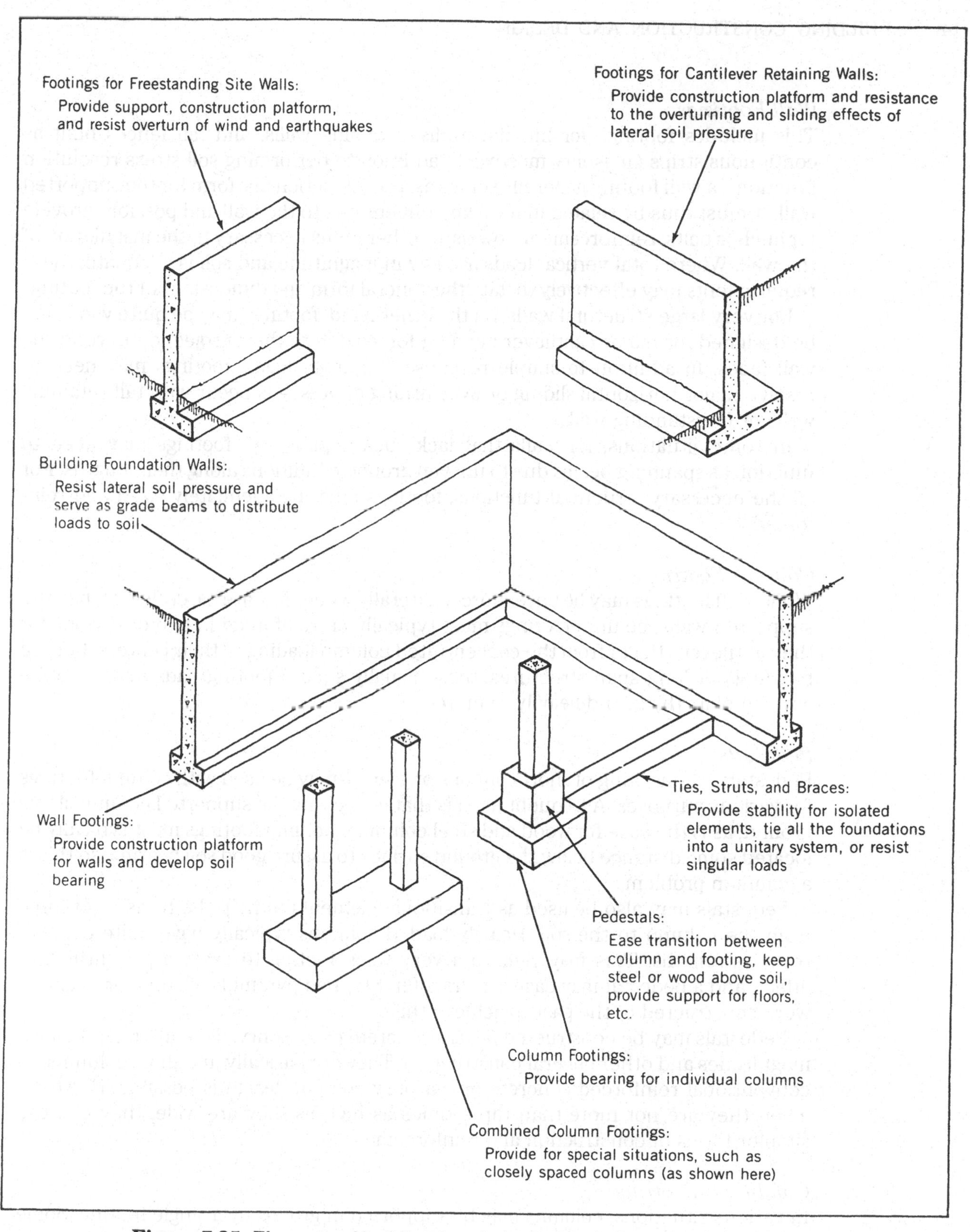

Footings for Freestanding Site Walls:
Provide support, construction platform, and resist overturn of wind and earthquakes
Footings for Cantilever Retaining Walls:
Provide construction platform and resistance to the overturning and sliding effects of lateral soil pressure
Building Foundation Walls:
Resist lateral soil pressure and serve as grade beams to distribute loads to soil
Ties, Struts, and Braces:
Provide stability for isolated elements, tie all the foundations into a unitary system, or resist singular loads
Wall Footings:
Provide construction platform for walls and develop soil bearing
Pedestals:
Ease transition between column and footing, keep steel or wood above soil, provide support for floors, etc.
Column Footings:
Provide bearing for individual columns
Combined Column Footings:
Provide for special situations, such as closely spaced columns (as shown here)

Figure 7.35 Elements of building foundations with shallow bearing footings.

Wall Footings
This includes footings for building walls, retaining walls, and tall fences built as continuous strips (masonry mostly). In addition to performing soil stress resolution functions, a wall footing generally serves as a construction platform for the supported wall. It must thus be related in form and dimensions to the wall and possibly provide for anchor bolts, reinforcement dowels, or other items necessary to the installation of the wall. Where total vertical loads are low in magnitude and soil is adequate, these requirements may effectively dictate the general form and dimensions of the footing.

For very large structural walls, on the other hand, footings may be quite wide, and be designed for major cantilever bending forces due to their projection beyond the wall faces. In addition to simple response to gravity loads, footings may need to resolve major horizontal sliding or overturning effects—as occur with tall retaining walls or freestanding walls.

In some situations, for walls that lack such capabilities, footings may need to function as spanning beams due to uneven ground conditions along their lengths. For all the necessary structural functions, footings must be adequately sized and reinforced.

Column Footings
While wall footings may be unreinforced laterally when they are barely wider than the supported walls, column footings must typically be reinforced in two directions for the cantilever effects from the concentrated column loading at their centers. For tall buildings or long span structures, these usually square footing pads may be quite large in plan, thick, and heavily reinforced.

Pedestals
Pedestals—consisting of short columns or piers—may be used with column footings for various purposes. A frequent need is that of keeping the supported column above ground, as is the case for wood and steel columns. Column footings must typically be located some distance below the ground surface to assure good soil bearing, so this is a common problem.

Pedestals may also be used as transitional elements to help the transfer of force from the column to the soil. Heavily loaded columns typically have quite concentrated strengths. This may require a very thick footing to avoid a punchthrough effect, and a pedestal may ease this transfer. Stepped pyramids of stone or masonry were constructed in the past to achieve this.

Pedestals may be constructed of cast concrete or masonry, depending on the load magnitudes and other general construction. They are basically just short columns, so conventional reinforced concrete or masonry column design is possible. However, when they are not more than three times as high as they are wide, they can use simpler forms of construction and reinforcement.

Combined Footings
In various situations, columns may be supported in groups by a single footing. Some reasons for this are the following (see Figure 7.35):

Closely-Spaced Columns. If required footings are large in plan dimensions, it may be difficult to use separate footings for closely-spaced columns. One solution may be to use oblong plan-shape footings, rather than the usual square. This is usually reserved for situations where columns are close to some other construction, such as an elevator pit. Two columns close together are more typically supported by a single footing, which is designed like a beam between them.

Cantilever Footings. Cantilever or strap footings are used where an exterior building column occurs close to the property line and the usual column footing at the building edge would project off the property. This may also be the case where a new building is built virtually up against an adjacent one, even though on the same property.

Sensitive Supported Structure. Occasionally a structure or some supported equipment has great sensitivity to movements and differences in settlement of separate footings may be critical; in which case, a single footing may be effective.

Pole Foundations

A special foundation is that of one in which a timber pole is placed in an excavated hole. This is most often used for signs, overhead transmission lines, and fence posts, but also sometimes for above-ground decks or even buildings. The pole bottom essentially bears directly on the soil, although a compacted gravel or poured concrete pad may be placed at the bottom of the hole.

Figure 7.36 shows some possible forms for pole foundations. Details of the supported structure and general development of the site will affect choices for treatment at the ground surface around the pole. Effects of lateral forces may also influence some concerns for the depth of the hole, the general anchorage of the pole in the hole, and the treatment at the ground surface. A confining concrete pavement around the pole will drastically change its lateral stability.

Information Source

Simplified Design of Building Foundations, 2nd ed., James Ambrose, Wiley, New York, 1988.

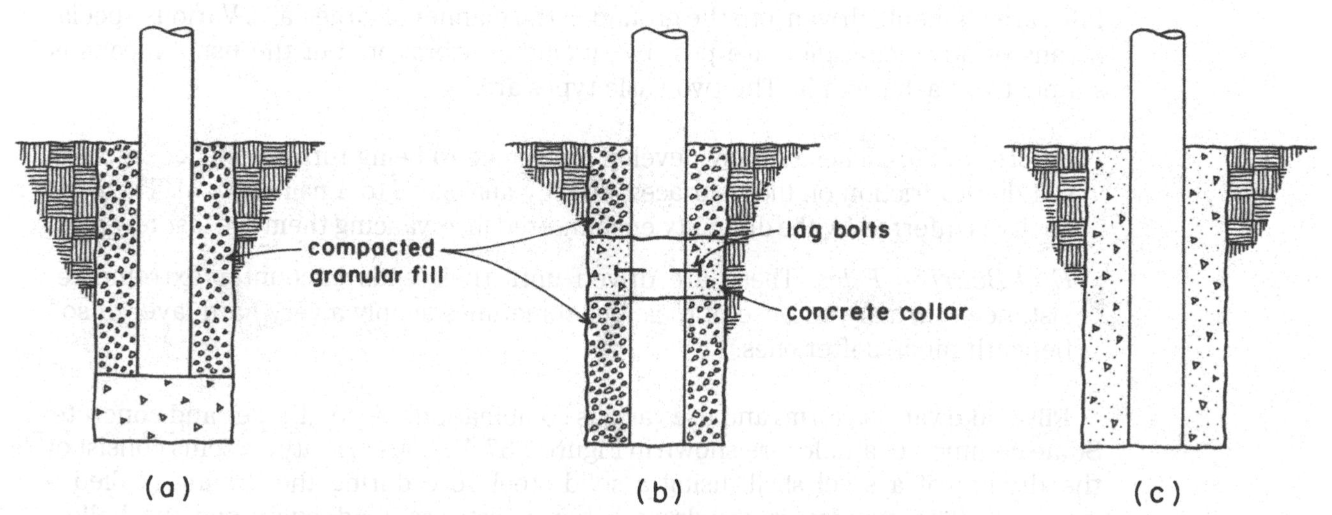

Figure 7.36 Methods of achieving pole foundations: (a) Concrete footing with compacted gravel fill. (b) Concrete collar for more positive lateral (horizontal) bracing. (c) Fully concrete-filled excavation for maximum bracing.

7.12 DEEP FOUNDATIONS

Far and away, the cheapest, simplest, most common foundation is the shallow bearing footing. However, the use of bearing pressure on yielding soil has some limitations which must be recognized; these include the following:

Soil Strength. Load capacity of soils—even the strongest ones—is limited. Good structural wood, for example, typically has a compression strength of around 2000 lb/sq in.—weak in comparison to steel or super-strength concrete. But a soil is considered super-strength if it can sustain as much as 10,000 lb/sq ft, or about 70 lb/sq in. Excessive loading, from heavy construction, tall buildings, or long span structures, can require footings with massive plan areas to keep soil pressure low.

Settlement. Soil is compressible; the greater the pressure, the more the deformation, which accumulates as downward vertical movement (settlement). With large deposits of clays, settlements may be progressive over time, accumulating in *feet,* not inches.

Instability. Soils—with discrete particles not generally tightly bonded or cemented—are subject to erosion, consolidation, organic or general chemical decomposition, and massive shrinkage or expansion due to major changes in water content.

For these, or other reasons, use of deep foundations may be favored. Actually, the simple reason for choosing what is almost always a more expensive foundation system is usually that the bearing capacity needed is not available at the location where bearing footings would need to be located.

Many basic forms of deep foundations are possible and many special ones are available as priority systems. Individual systems become popular regionally due to a combination of availability, appropriateness to local problems, and marketing concentration by companies.

Piles

Piles are elements driven into the ground in the manner of large nails. Various special means of advancing piles are possible, including vibration, but the usual means is simply to smash them in. The two basic types are:

Friction Piles. These simply develop resistance to being further pushed into the soil due to friction on their surfaces; directly analogous to a nail in wood. Their capacity is inferred by the difficulty encountered in advancing them the last few feet.

End-Bearing Piles. These are driven until their ends encounter extreme resistance—usually because of rock, but sometimes simply a very hard layer of soil beneath much softer ones.

Piles take various forms and use various combinations of wood, steel, and concrete. Some common examples are shown in Figure 7.37. Various priority systems consist of the driving of a steel shell, using a solid steel core during the driving (called a mandrel). The mandrel is withdrawn when resistance is adequate and the hollow shell is filled with concrete. In some soils, the shell can also be withdrawn during the placing of the concrete. In other systems the shell itself is driven, usually for development of an end-bearing condition.

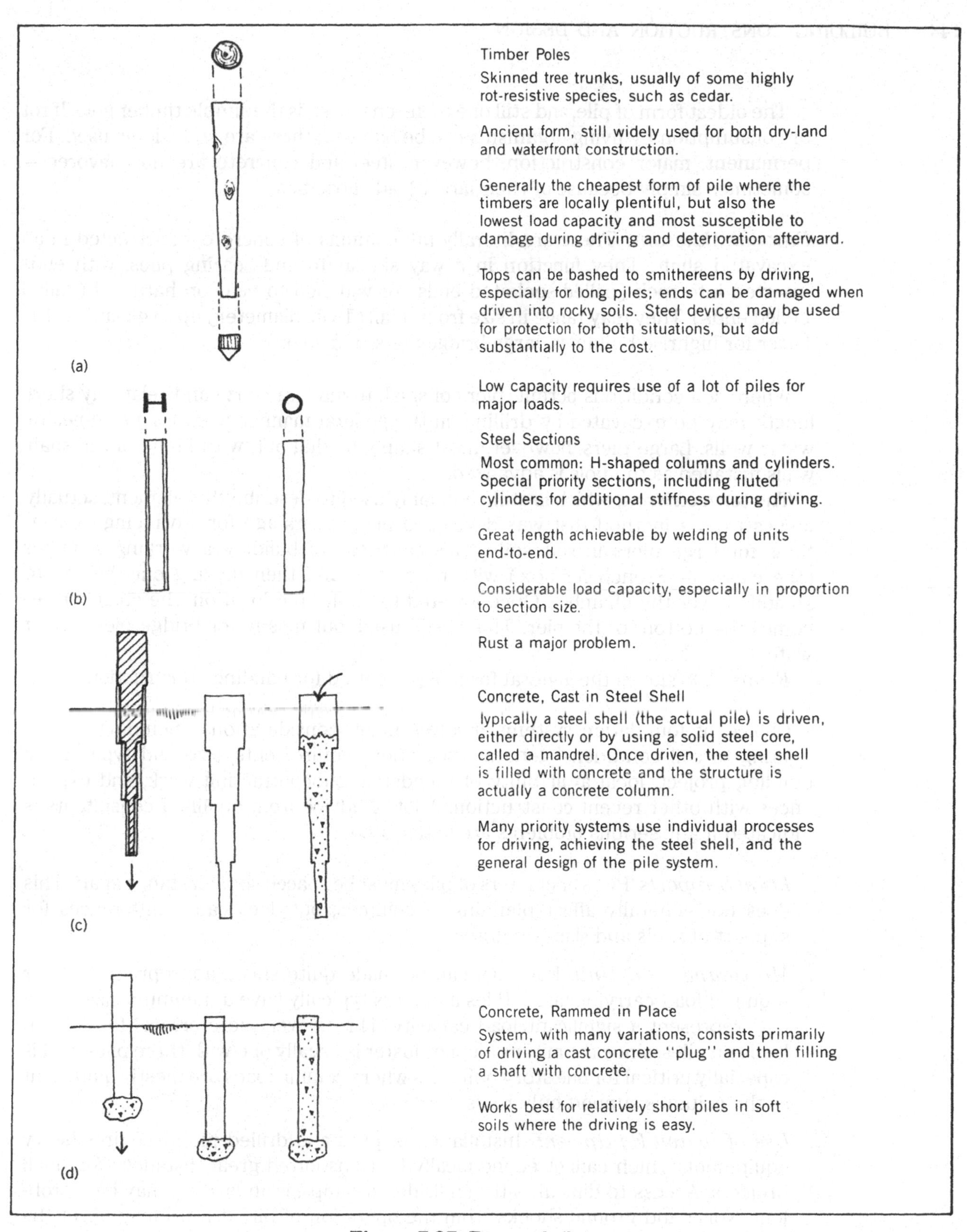

Timber Poles
Skinned tree trunks, usually of some highly rot-resistive species, such as cedar.
Ancient form, still widely used for both dry-land and waterfront construction.
Generally the cheapest form of pile where the timbers are locally plentiful, but also the lowest load capacity and most susceptible to damage during driving and deterioration afterward.
Tops can be bashed to smithereens by driving, especially for long piles; ends can be damaged when driven into rocky soils. Steel devices may be used for protection for both situations, but add substantially to the cost.
Low capacity requires use of a lot of piles for major loads.
(a)
H
O
Steel Sections
Most common: H-shaped columns and cylinders. Special priority sections, including fluted cylinders for additional stiffness during driving.
Great length achievable by welding of units end-to-end.
Considerable load capacity, especially in proportion to section size.
Rust a major problem.
(b)
Concrete, Cast in Steel Shell
Typically a steel shell (the actual pile) is driven, either directly or by using a solid steel core, called a mandrel. Once driven, the steel shell is filled with concrete and the structure is actually a concrete column.
Many priority systems use individual processes for driving, achieving the steel shell, and the general design of the pile system.
(c)
Concrete, Rammed in Place
System, with many variations, consists primarily of driving a cast concrete "plug" and then filling a shaft with concrete.
Works best for relatively short piles in soft soils where the driving is easy.
(d)

Figure 7.37 Forms of piles.

The oldest form of pile, and still one of the cheapest, is the simple timber pole. If rot or consumption by living organisms can be handled, these are still widely used. For permanent, major construction, however, steel and concrete are now favored—sometimes simply because of their larger load capacities.

Piers or Caissons. These are literally tall columns of concrete, constructed in an excavated shaft. They function in a way similar to end-bearing piles, with ends inserted into rock (called socketed ends) or widened to bear on hard soil (called belled ends). They may range in size from small (18 in. diameter) up to gigantic—the latter for highrise buildings, large bridge piers, and so on.

Where soil conditions permit, piers of small to moderate size and relatively short length may be excavated by drilling, in the general manner used for postholes or water wells. Large piers, however, must simply be dug out, with lining of the shaft walls installed as the hole is advanced.

The term *caisson*, which is still commonly used to describe this element, actually comes from a method that was developed many years ago for advancing excavations for large piers in soft soils. This consisted of building a working chamber (the caisson—French for box) with no bottom and then digging out the soil to steadily lower the chamber. Once lowered to its desired location, the chamber becomes the bottom of the pier. This is still used, but mostly for bridge piers under water.

Figure 7.38 shows the general form of piers used for building construction.

Choice of foundation systems involves many considerations, including factors relating to soil conditions, general excavation requirements, size and type of the building project, local availability of foundation subcontracting work, and experiences with other recent construction. A lot of advice from qualified consultants is indicated if any serious problems are anticipated.

Point Supports. Piers or clusters of piles must be spaced some distance apart. This does not generally affect planning of columns, but does cause differences for support of walls and slabs on grade.

Minimum Load Unit. Footings can be made quite small; appropriate to their required load-carrying tasks. Piles and piers typically have a minimum size which may represent a significant load capacity. This is even more critical for column loads on piles, since the minimum pile cluster is usually one with three piles, and is especially critical for one story buildings where column loads are mostly quite light as the columns support only roofs.

Use of Heavy Equipment. Installation of piles and drilled piers requires heavy equipment, which cannot economically be transported great distances for small projects. Access to difficult sites (hillsides, swamps, islands, etc.) may be a problem. Noise and ground shocks from the operation of pile drivers may upset the neighbors.

For any project, it is highly desirable to anticipate the form of foundation system very early in the planning stages for both buildings and general site development. This relates to both the economic feasibility of the construction and the intelligent planning of the work.

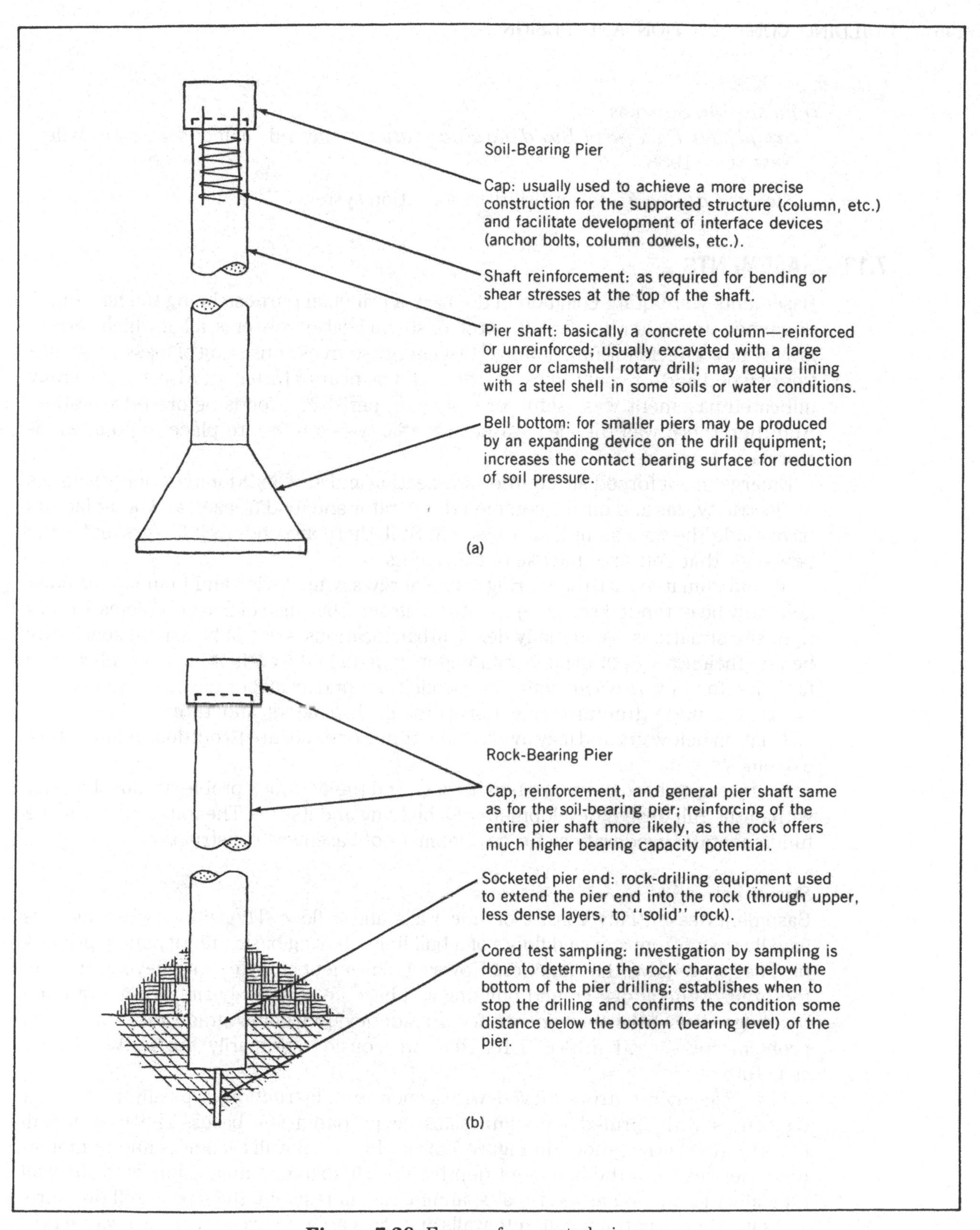

Soil-Bearing Pier
Cap: usually used to achieve a more precise construction for the supported structure (column, etc.) and facilitate development of interface devices (anchor bolts, column dowels, etc.).
Shaft reinforcement: as required for bending or shear stresses at the top of the shaft.
Pier shaft: basically a concrete column—reinforced or unreinforced; usually excavated with a large auger or clamshell rotary drill; may require lining with a steel shell in some soils or water conditions.
Bell bottom: for smaller piers may be produced by an expanding device on the drill equipment; increases the contact bearing surface for reduction of soil pressure.
(a)
Rock-Bearing Pier
Cap, reinforcement, and general pier shaft same as for the soil-bearing pier; reinforcing of the entire pier shaft more likely, as the rock offers much higher bearing capacity potential.
Socketed pier end: rock-drilling equipment used to extend the pier end into the rock (through upper, less dense layers, to ''solid'' rock).
Cored test sampling: investigation by sampling is done to determine the rock character below the bottom of the pier drilling; establishes when to stop the drilling and confirms the condition some distance below the bottom (bearing level) of the pier.
(b)

Figure 7.38 Forms of excavated piers.

Information Sources

Simplified Design of Building Foundations, 2nd ed., James Ambrose, Wiley, New York, 1988.

Sweets Catalog Files, for priority foundation systems.

7.13 BASEMENTS

Basements were quite common in the past; a principal purpose being the housing of equipment for either hot air, hot water, or steam heating systems, all of which needed gravity flow effects for functioning. Heat energy sources consisting of massive quantities of wood and/or coal had to be stored and be near to furnaces. Also, the generally unheated basement was useful for storage of perishable foods before refrigeration. And then there was the idea—if not the reality—of a secure place to go in windstorms.

Emergence of forced air, counterflow heating and air conditioning, energy sources of electricity, gas and oil, the common refrigerator and food freezer, and other factors have made the basement less necessary. Still, there are enduring factors and some new ones that continue the use of basements.

In cold climates earth sheltering is an energy-saving device and foundations must generally be extended some depth into the ground because of frost problems. In very tight site situations—especially dense urban locations—use of basement space may be an efficiency of planning. A building may be linked to others in a complex or to facilities for entry, waste removal, parking, ground transportation, or emergency shelter by underground connection through basements and tunnels. Pedestrian circulation below ground may avoid street traffic or exposure to outdoor temperature extremes.

In any event, basements are here to stay, and their various problems must be dealt with in the full design development of a building and its site. The following are some fundamental considerations for development of basement construction.

Structure

Basements essentially consist of side walls and a floor. The top of a basement is usually the bottom (ground floor) of a building—having basic identity and a primary purpose to achieve that function. However, basements may extend beyond the site footprint of the above ground building and become effectively simply underground buildings. This adds the concern for a grade level or below grade roof—a special problem considered in Sec. 7.16. Here we consider primarily the wall and floor structure.

Most basements are achieved with concrete construction, typically as sitecast structures with formed walls and slabs on prepared soil bases. The typical wall construction is that shown in Figure 7.39(a). Required wall thickness and reinforcement derives from the basement depth and wall-spanning dimension, with the wall typically functioning as a vertical-spanning slab in resisting the lateral soil pressure.

As in other situations, concrete walls and slabs are reinforced with two-way sets of reinforcing bars to resist temperature expansion and concrete shrinkage effects, as well as structural effects. If upward hydrostatic pressures from ground water exist, floor slabs may have major structural functioning as spanning elements; otherwise, they basically just lay on the ground beneath them.

Basement walls also frequently serve functions as general elements of the building base and foundation. Exterior building bearing walls will transfer their loads finally to

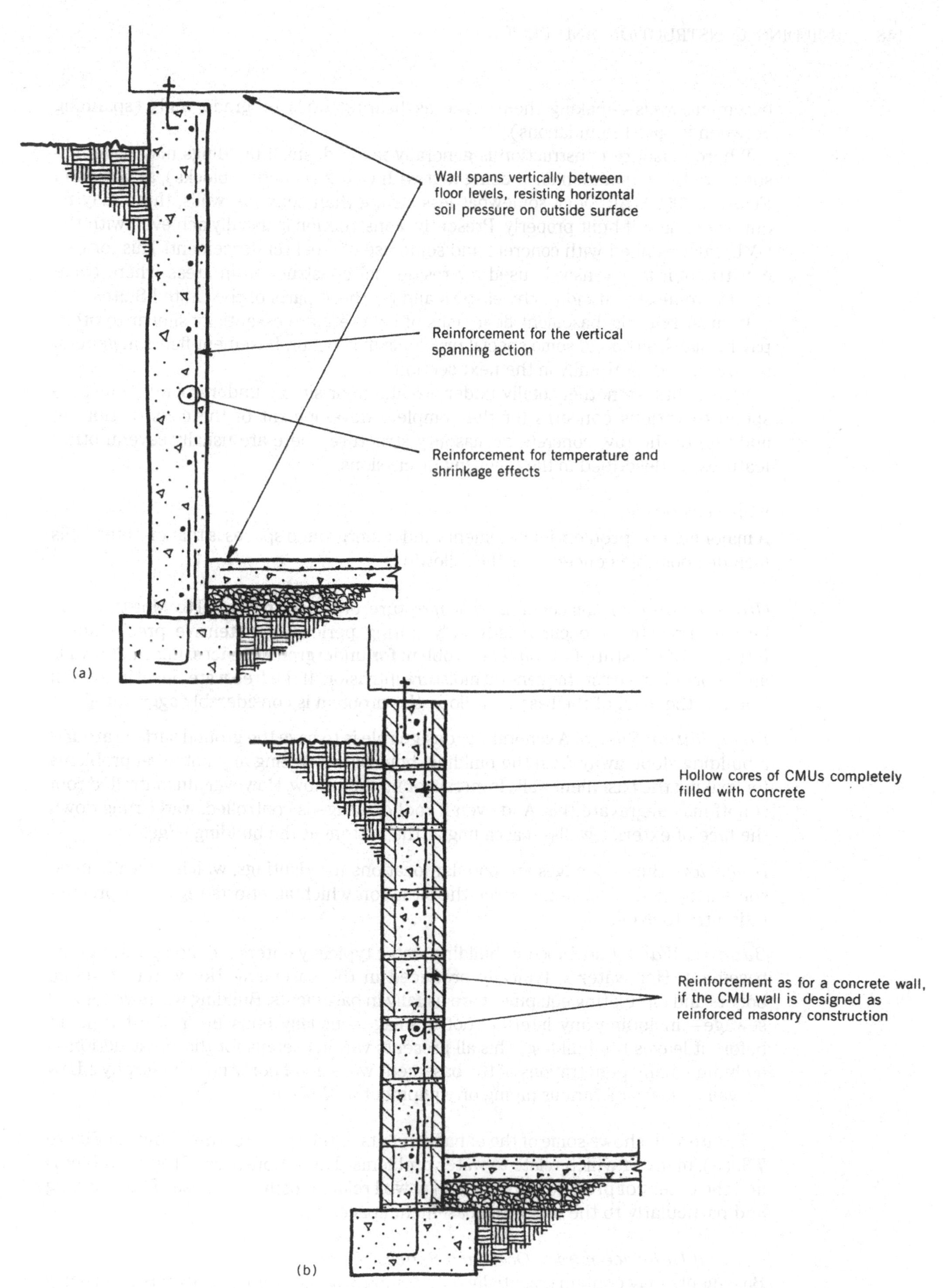

Figure 7.39 Basement wall structures: (a) Sitecast concrete. (b) Masonry with precast concrete units (CMUs or concrete blocks).

basement walls—making them serve as bearing walls or grade walls (spanning between isolated foundations).

Where masonry construction is generally favored, small buildings may have masonry walls—a common form being that with CMUs (concrete blocks), as shown in Figure 7.39(b). Although somewhat less strong than concrete walls, these may be quite adequate if built properly. Presently, construction is usually achieved with the CMU cavities filled with concrete and some use of steel reinforcement. This form of construction is extensively used for residential construction in areas where basements are most common—the eastern and northern parts of the United States.

In most regards, basement floor slabs of concrete are essentially similar to other paving slabs, although some special problems may occur. Basement floors in general are discussed more fully in the next section.

Basements—whether totally under a building or simply underground—must respond to various concerns for the complete development of the construction. In addition to the raw concrete or masonry structure, there are usually several other features, as described in the following discussions.

Water Control

A major general problem for basements and underground spaces is that of water. This includes possible concerns for the following:

Ground Water. All soil contains some moisture, or has the potential to. Moisture may be ever-present or occur mostly only during periods of extensive precipitation. Intrusion of moisture is a common problem for underground spaces. Basement walls and floors must resist this general moisture intrusion. If the free water level in the soil is above the level of the basement floor, the problem is considerably aggravated.

Precipitation Runoff. A general site design rule is to have the ground surface around a building slope away from the building; hopefully reducing any potential problems for soaking the basement walls from rain or melting snow. However, uncontrolled roof runoff may aggravate this. And even if roof drainage is controlled, water runs down the face of exterior walls—cascading at the bottom at the building edge.

Irrigation. Building edges are popular locations for plantings, which typically need some irrigation. This can produce the situation which all efforts regarding precipitation try to avoid.

Building Water Use. Piped in building water typically enters a building through its basement. Hot water is typically produced in the basement. Hot water or steam boilers and circulating equipment are usually in basements. Building waste water and sewage—including any interior roof drainage—usually ends up in the basement before it leaves the building. This all presents various details for the construction—including many penetrations of the basement walls and floor. And—if Murphy's Law prevails—leaks of various piping or equipment *will* occur.

Figure 7.40 shows some of the enhancements of the raw structure shown in Figure 7.39(a); many of which relate to water problems. The seriousness of these problems and the extent of provisions for them may well relate mostly to the use of the building and particularly to the use of the basement spaces.

General Enhancement of Occupied Spaces

Basements may contain essentially only equipment and storage spaces. Humans may enter, but do not *occupy* the space in the building code's meaning of the term. If

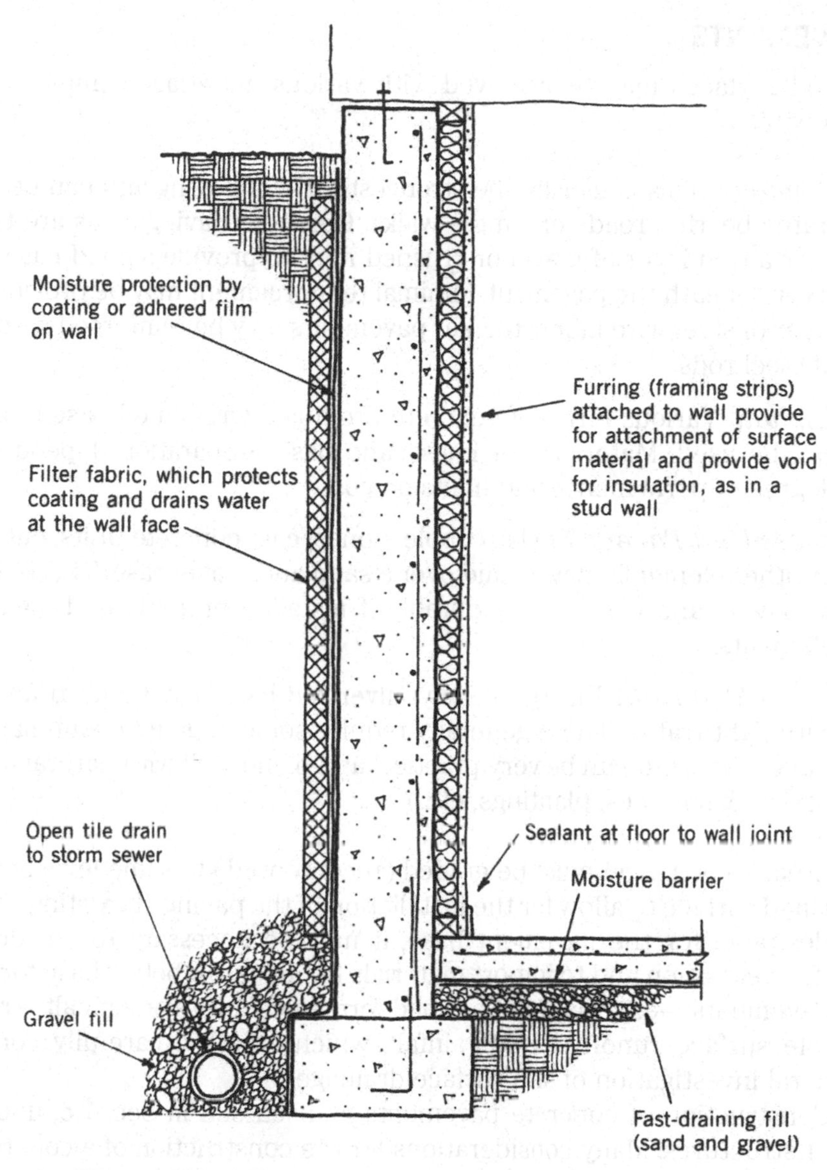

Figure 7.40 Basement wall construction—fully developed.

basement spaces are truly occupied, however, there will typically be heightened concern for water, thermal conditions, ventilation, and general finishing of interior surfaces.

Insulation of basement walls can be achieved in a variety of ways, depending on the primary effects to be obtained. General energy conservation is one concern; thermal comfort of occupants another. Insulation on the outside generally retards the effects of ground frost and cold air conditions at the ground surface. Insulation on the inside further reduces the cold, radiant effects of wall and floor surfaces.

Finishes of all the usual types can be achieved for floors and walls, if proper attention is given to installation details and choices of materials. Attachment to concrete or masonry surfaces is a first item to be dealt with. Enduring potential water problems is another.

Second to roofs, water problems in basements is probably one of the nastiest problems in building construction. This area deserves a lot of attention for successful development of building construction.

7.14 PAVEMENTS

Paved surfaces may be achieved with various materials, common ones being the following:

Concrete. This is usually the strongest form of paving and can be used for heavy traffic-bearing roads or simple walks. Concrete paving slabs are typically placed over a thin layer of gravel or crushed rock to provide a good base and a draining layer beneath the pavement. Minimal reinforcement may be provided with a single layer of steel wire fabric; thicker pavements may be reinforced with two-way grids of steel rods.

Asphalt. Various forms of concrete produced with an oil-based binder (tar, etc.) can be used. Materials, thickness, and base preparation depend on the desired degree of permanence and limits on cost.

Loose Unit Pavers. Bricks, cobble stones, cast concrete units, cut wood sections, or other elements may be laid over a sand and gravel base. This is an ancient form of paving and can be very durable if installed properly and made with durable elements.

Loose Materials. Fine gravel or pulverized bark may be used for walks or areas with light traffic. These generally require some ongoing maintenance to preserve the surface, but can be very practical and blend well with natural features of a site (existing surfaces, plantings, etc.).

Areas to be paved must be graded (recontoured) to some level below the desired finished surface to allow for the installation of the paving. If existing site materials are undesirable for the pavement base, it may be necessary to cut down to a lower surface elevation and to import materials to build up a better base for the pavement.

Pavements—especially of the solid form of concrete or asphalt—result in considerable surface runoff during rainfall, which must be carefully considered in the general investigation of site surface drainage.

Construction of concrete pavements is discussed in Sec. 7.6, under the topic of floor structures. Many considerations for the construction of a concrete slab cast on the ground surface are the same, no matter what its purpose. When intended for pavements, however, the condition of exposure to weather and need to support some forms of traffic must be dealt with.

Concrete pavements are highly durable and effective for heavy traffic. They are, however, quite expensive if correctly installed, so other forms may be used for economy or because of a desire for a different form of surface. Of course, most of these can be developed on top of a concrete slab—as they might be when used on the roof of an underground structure. The following discussion, however, deals primarily with their use in place of a concrete slab. (See Figure 7.41.)

Asphalt

Asphalt paving is essentially a form of concrete (sand, gravel, and a binder); made with a bituminous binder instead of portland cement. However, it is typically built-up in layers, with a somewhat denser, smoother topping developed for the top surface. It is naturally less permeable (allowing water penetration) than ordinary concrete, although rapid water accumulation during heavy rains will flow off a sloping surface of either concrete or asphalt at about the same rate.

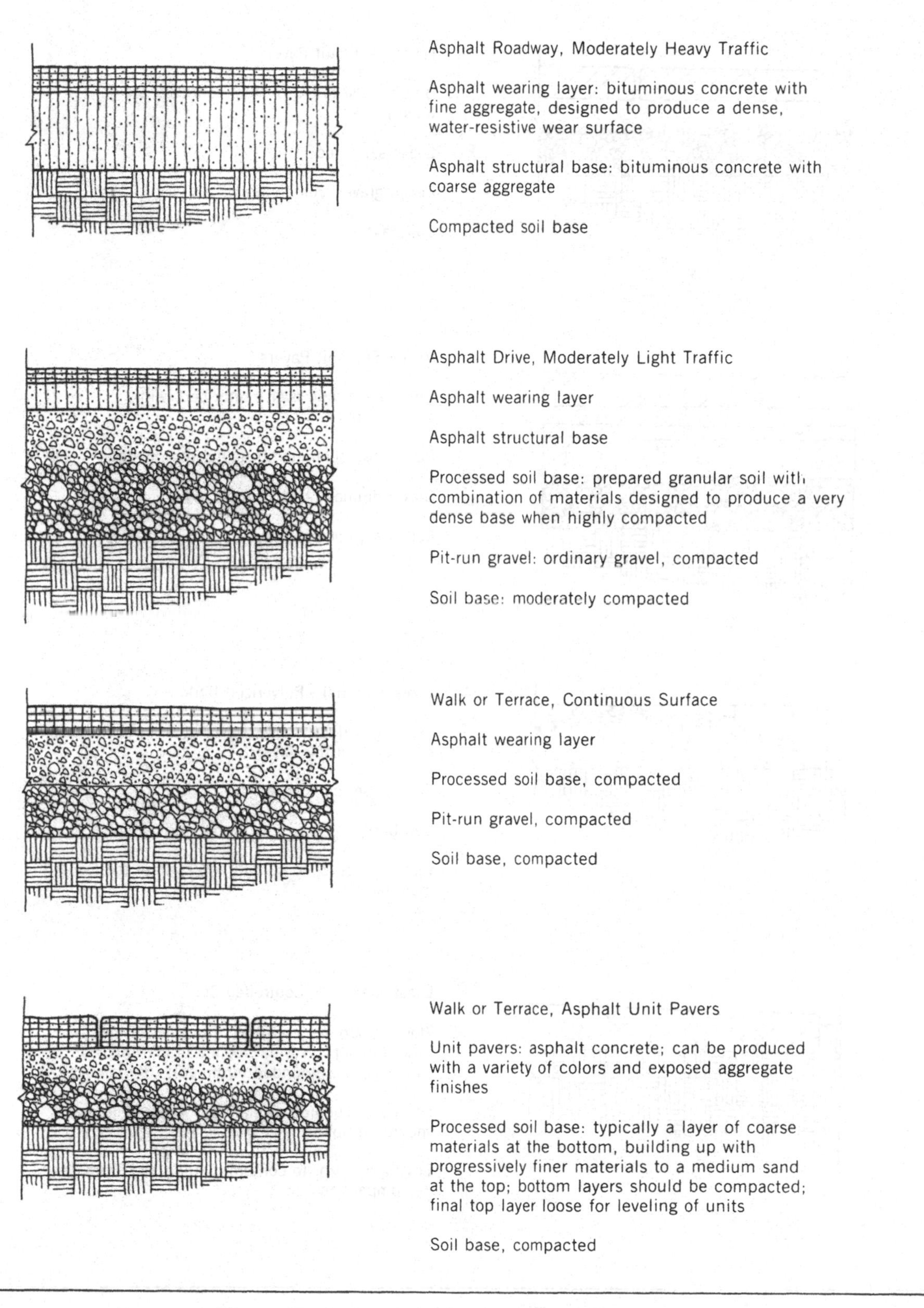
Asphalt Roadway, Moderately Heavy Traffic
Asphalt wearing layer: bituminous concrete with fine aggregate, designed to produce a dense, water-resistive wear surface
Asphalt structural base: bituminous concrete with coarse aggregate
Compacted soil base
Asphalt Drive, Moderately Light Traffic
Asphalt wearing layer
Asphalt structural base
Processed soil base: prepared granular soil with combination of materials designed to produce a very dense base when highly compacted
Pit-run gravel: ordinary gravel, compacted
Soil base: moderately compacted
Walk or Terrace, Continuous Surface
Asphalt wearing layer
Processed soil base, compacted
Pit-run gravel, compacted
Soil base, compacted
Walk or Terrace, Asphalt Unit Pavers
Unit pavers: asphalt concrete; can be produced with a variety of colors and exposed aggregate finishes
Processed soil base: typically a layer of coarse materials at the bottom, building up with progressively finer materials to a medium sand at the top; bottom layers should be compacted; final top layer loose for leveling of units
Soil base, compacted

Figure 7.41 Miscellaneous forms of paving.

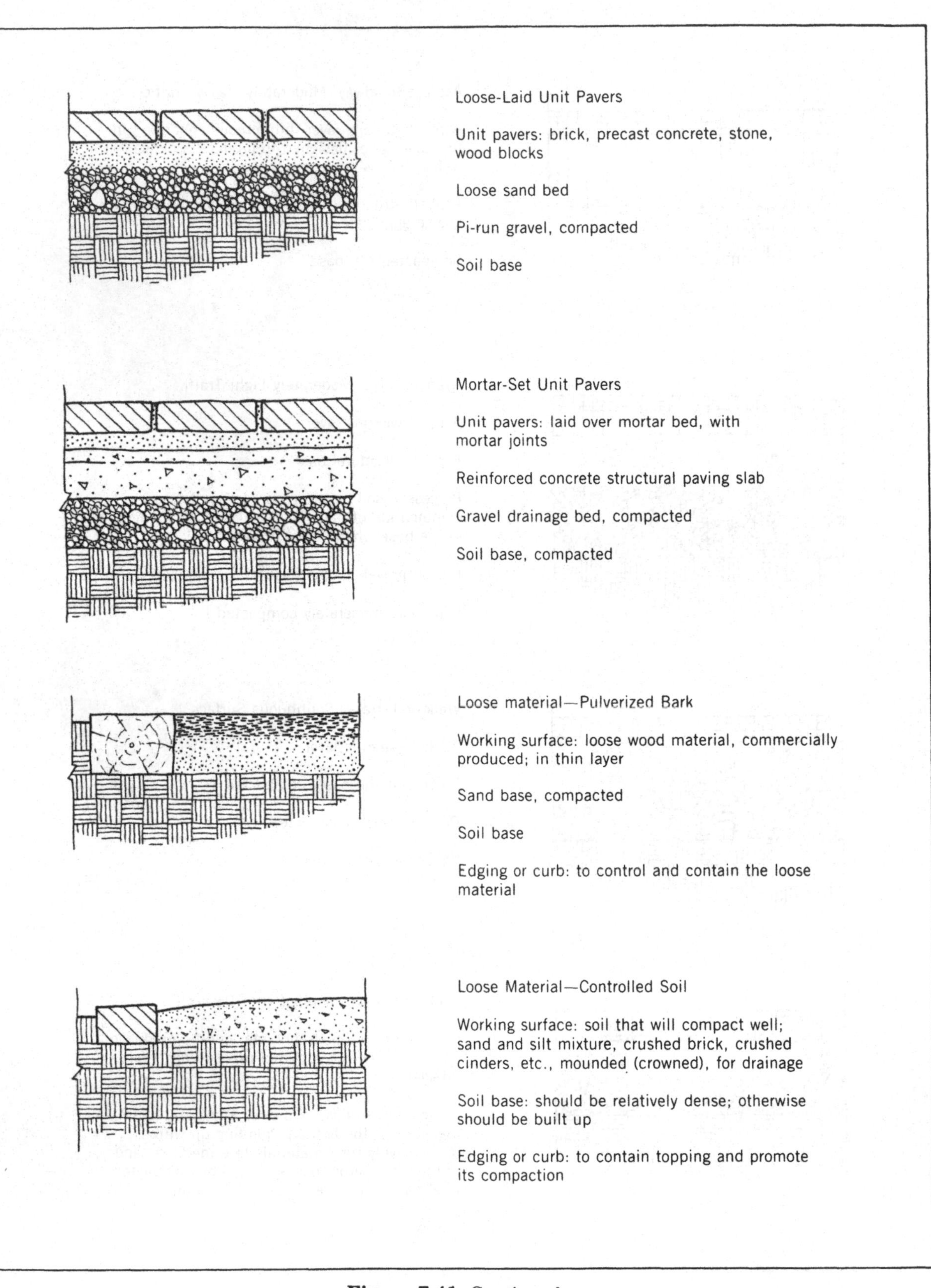
Loose-Laid Unit Pavers
Unit pavers: brick, precast concrete, stone, wood blocks
Loose sand bed
Pi-run gravel, compacted
Soil base
Mortar-Set Unit Pavers
Unit pavers: laid over mortar bed, with mortar joints
Reinforced concrete structural paving slab
Gravel drainage bed, compacted
Soil base, compacted
Loose material—Pulverized Bark
Working surface: loose wood material, commercially produced; in thin layer
Sand base, compacted
Soil base
Edging or curb: to control and contain the loose material
Loose Material—Controlled Soil
Working surface: soil that will compact well; sand and silt mixture, crushed brick, crushed cinders, etc., mounded (crowned), for drainage
Soil base: should be relatively dense; otherwise should be built up
Edging or curb: to contain topping and promote its compaction

Figure 7.41 Continued.

As with a concrete slab, a base should be developed for a good asphalt pavement. However, for lightly-trod walks, a very thin coating of the ground may suffice. Asphalt surfaces are somewhat less durable than hardened concrete, and constant traffic will wear them down. However, they can be repaired or restored by additional coatings on top of the old paving.

Unit Pavers

Ancient pavings were developed with stones laid to produce a relatively smooth surface. Very fine surfaces were developed with cut stones, and the floors over soil in many great works of architecture were developed as loose stone paving over a crushed gravel base.

Later, bricks and fired clay tiles were used in the same manner as cut stones. Matched sets of smooth field stones were used to produce cobblestone paving; basically considered as peasant-quality stuff—hence the word "cobble" which meant roughly formed.

As with any paving of loose units, the application may be made essentially directly over soil or developed more permanently. Bricks or tile, for example, may be set in mortar over a concrete slab for a really permanent, stable pavement.

In addition to stone, brick, or tile, just about any hard element can be used for paving. Wood blocks, planks, timbers, or cut sections of logs can be used. So can precast concrete units—now most commonly used when installation is directly in the ground.

Complete details for unit paving will vary depending on the units themselves, but also on the traffic to be borne and what is underneath the paving. Selection of the units is usually a landscape design decision, but also responds to availability of materials and general economics for the project.

Loose Materials

For some forms of traffic, or for special surfaces such as racetracks, playgrounds, or playing fields, various loose materials can be used for a developed surface. Widely spaced unit pavers may be used with an infill of grass for a form of paved lawn. Other non-grassy areas can be developed with pulverized bark, sawdust, fine gravel, or various mixtures of soil materials.

Natural soil surfaces may form relatively hard, durable crusts in some cases, or be encouraged to do so with some help. Adding some fine silt or dry clay to an existing sand or gravel surface and use of watering or compaction may produce such a surface.

Grass alone, when carefully cultivated and maintained, can form a type of pavement. Besides sustaining some degree of traffic, it may serve effectively to prevent surface erosion, improve precipitation runoff, or otherwise do what is generally expected from other forms of paving.

7.15 SITE CONSTRUCTION

Construction on building sites—other than that for the buildings—includes soil structures and various other forms of structures; such as pavements, curbs, walks, drives, planters, retaining walls, and stairs. There is also often considerable construction for the installation of various building services for electrical power, phones, water supply, gas, and sewers.

Pavements are discussed in Sections 7.6 and 7.14. The discussion in this section considers some other forms of common site construction.

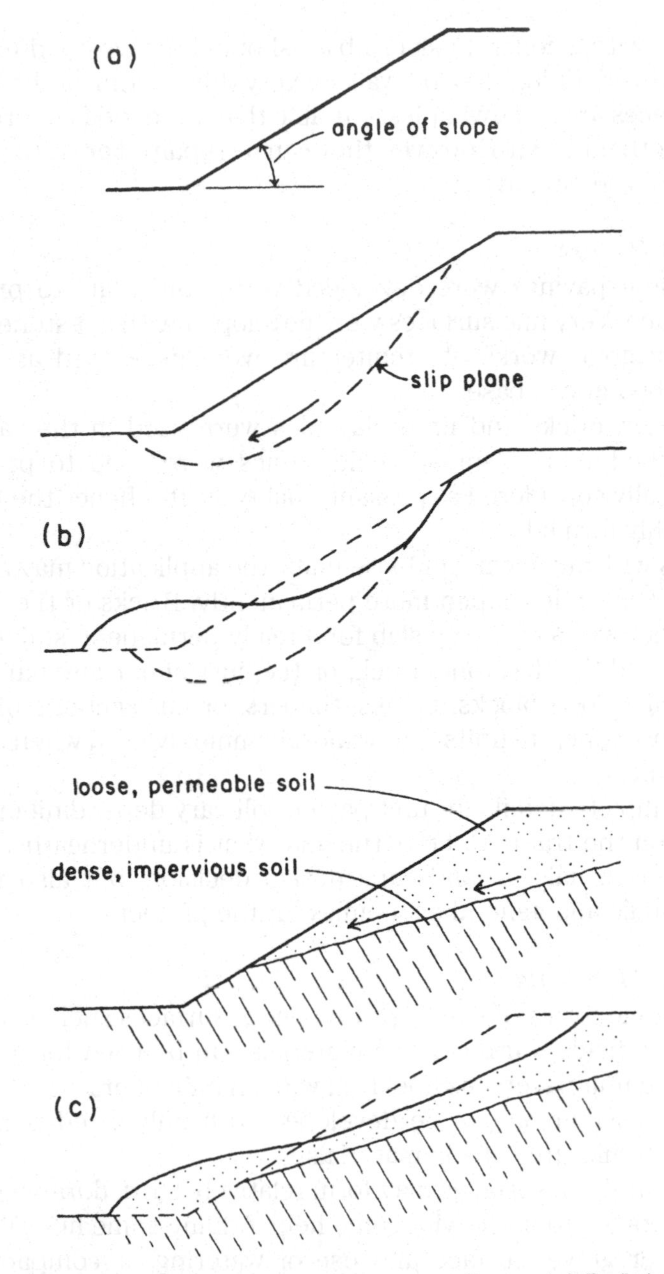

Figure 7.42 Considerations for failure of sloped soil surfaces: (a) Definition of the slope angle. (b) Rotational failure of loose material. (c) Slippage of soil layers in a stratified mass.

Soil Structures

Site development and building construction often requires the treatment of sloping ground surfaces. When this is necessary, a critical decision is that regarding the maximum feasible angle for the slope (see Figure 7.42(a)). The stability of a slope may be in question for a number of reasons. Two principal issues relating to the soil materials are the potential for erosion from excessive rainfall or irrigation, and the possibility for movement of the soil mass in a downhill direction.

A sloped surface may be generated in two different ways: by cutting back of existing soil, or by building up with fill. The relative stability of the slope will depend largely on the character of the soils at and near the surface of the slope.

For cut slopes the soil is largely existing and slope angle limits must be derived basically from the determination of existing soil conditions. For a relatively clean sand the angle of repose is quite easily derived, although erosion or shock may be critical; particularly for loose sands. For rock or some very stable cemented soils, a vertical cut face—or even a cave—may be possible.

For most ordinary surface soils of a mixed character, the proper slope angle must be derived from studies of potential loss of the slope face by various mechanisms of failure. Slope loss by either erosion or slippage is often due to a combination of water soaking of the surface soils and the presence of a relatively loose soil mass.

Erosion is a problem related to many considerations of the general site development. The slope itself or the water it must drain may be affected by plantings, irrigation, general site contouring, pavements, and both site and building construction. Slopes may be surfaced with plantings and various pavings to protect the surface and retard erosion.

Slippage of the soil mass on the slope may occur from a number of causes. A common one is simply the vertical movement of the soil mass due to gravity, as shown in Figure 7.42(b). This may occur as a semi-rotational effect along a slip plane, as shown, or by the sliding of one soil mass over another, as shown in Figure 7.42(c).

For many ordinary situations reasonably safe slope angle limits are established by experience and rules of thumb. Conservative limits may be established by building codes, although studies by a geotechnical expert or provision of some slope control elements may allow steeper angles.

Retaining Devices for Slopes

Slopes or abrupt changes in the ground surface elevation can be maintained or established by various means. Simple shaping of the soils may suffice if the mechanisms of slope failure are acknowledged, as discussed in the preceding section. However, various other elements can be used to help, as shown in Figure 7.43.

Plantings: Large plants with firmly-established roots can help to hold soil on a steep slope. If widely spaced, they may provide only partial restraint, so recommended slope angle limits should still not be significantly exceeded.

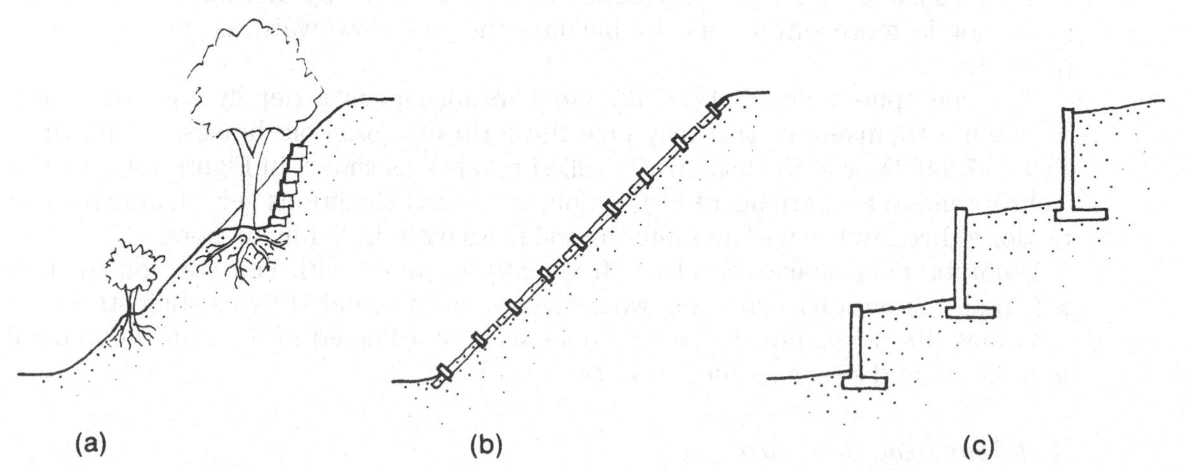

Figure 7.43 Devices for retention of sloped soil surfaces: (a) Plantings with established root growth. (b) Coverings with paving materials or specially-formed units. (c) Series of stepped retaining structures.

Surface treatments: These are basically means for keeping the surface itself in place. They do not generally affect deeper soil conditions, so cannot overcome slip plane failure or sliding, as illustrated in Figure 7.42(b) and (c). They are generally most effective in simply retarding erosion or the massive soaking of the soil from continued precipitation. Roots of various plantings can grab a significant layer of the surface once the plantings are well established. Coatings of asphalt, individual pavers, coarse gravel, or large rocks may be used. Use of geotextiles can also be one factor in a total slope stabilization system.

Stepped construction: Slopes may also be subdivided into stepped units. The "steps" may be quite large and individually maintained as close to flat—or at least a much shallower angle than the general slope. The "risers" of these steps can be any form of retaining structure, appropriate to the height of the riser. Individual steps may be occupied by plantings, walkways, driveways or roads or a series of buildings. Large developed hillsides are often stepped up with roads and rows of buildings in this manner.

Temporary Bracing

Various expedient methods may be used to maintain soil profiles during construction. Deep excavations for buildings or general construction on steep slopes may utilize these methods.

Shallow excavations can often be made with no provision for the bracing of the sides of the excavation. However, when the cut is quite deep, or the soil has little cohesive character (loose sand, for example), or the undercutting of adjacent construction or property is a concern, some form of bracing for the cut faces may be required.

One means for bracing consists of driving steel sheetpiling to form a wall which can then allow soil to be removed from one side, as shown in Figure 7.44(a). As with other temporary walls of this nature, the piling may stay in place, possibly used as the forming for one side of a cast concrete wall. However, if possible, the quite expensive piling units can also be withdrawn for reuse.

Another form of bracing wall is one constructed with vertical beams or posts (called soldiers) and horizontal boards or sheets (called lagging), as shown in Figures 7.44(c), (d), and (e). This construction may be achieved by driving the soldiers as piles, but is more often done by building the wall downward as the excavation proceeds.

Both sheetpile walls and those using soldiers and lagging generally need some form of bracing themselves. This may take the form of cross-hole braces, as shown in Figure 7.44(d) or individual struts (called rakers), as shown in Figure 7.44(c). For walls intended for permanent installation, where soil conditions permit, bracing may be done through the wall and into the soil mass with drilled-in anchors.

Major bracing of excavations is frequently required with construction in tight urban site situations, especially when the site is surrounded by existing streets or buildings. Its development must be done as a coordinated effort with the general foundation and site construction work.

Soil Retaining Structures

Site development frequently involves the use of various retaining structures. These help to achieve abrupt changes in the ground surface elevation. The form of construction often relates primarily to the height change achieved by the retaining structure.

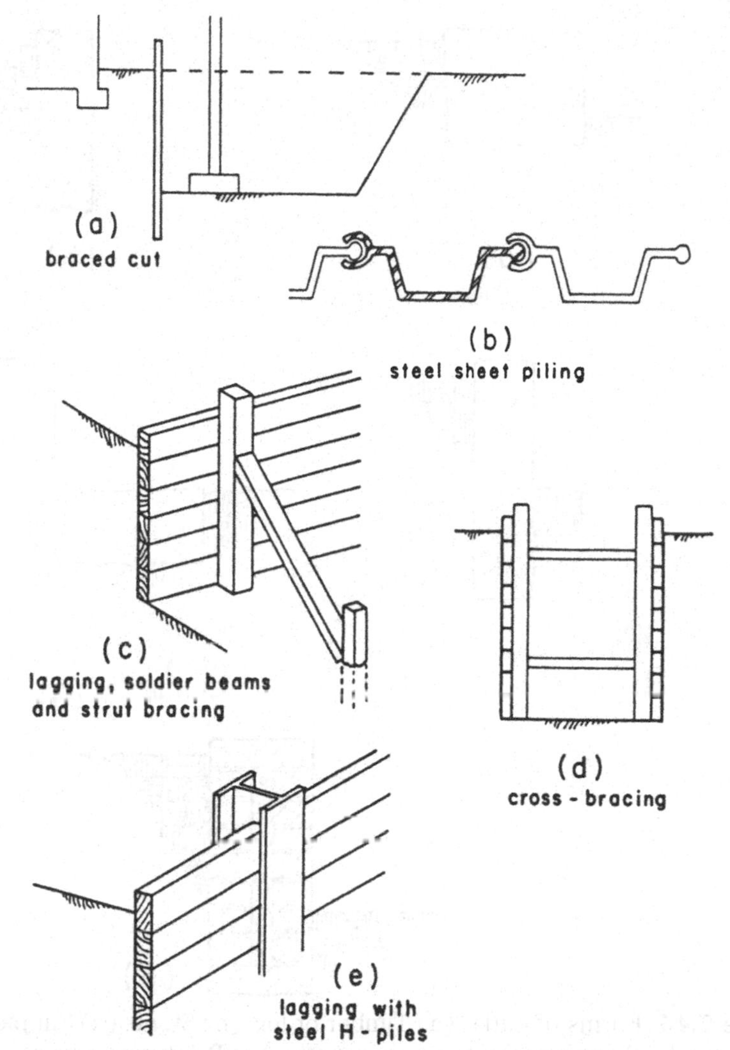

Figure 7.44 Elements used for bracing of vertical cuts: (a) Driven bracing of steel elements (called sheet piling). (b) Form of cross section of sheet pile units. (c) Soldier beams and lagging, shown with raker braces. (d) Narrow excavation with soldier beams, lagging, and cross-hole bracing (no rakers). (e) Lagging with steel column shapes used for soldiers; the soldiers may be driven or not.

Curbs

The smallest retaining structures are curbs, which may take various forms, as shown in Figure 7.45. Curbs are usually edging devices that define the boundary between different units of the site surface. They are frequently placed at the edges of pavements and thus often relate to the paving materials and forms. The form of drainage of the pavement may also affect the details of the curb.

Curbs are usually limited to achieving elevation changes of less than 18 in. or so. As the height of the retaining structure increases, a different general form of construction is required.

Loose-Laid Retaining Walls

For abrupt elevation changes of more than 18 in. or so, some form of wall construction is required. This may be achieved with loosed-laid stones (without mortar) or other

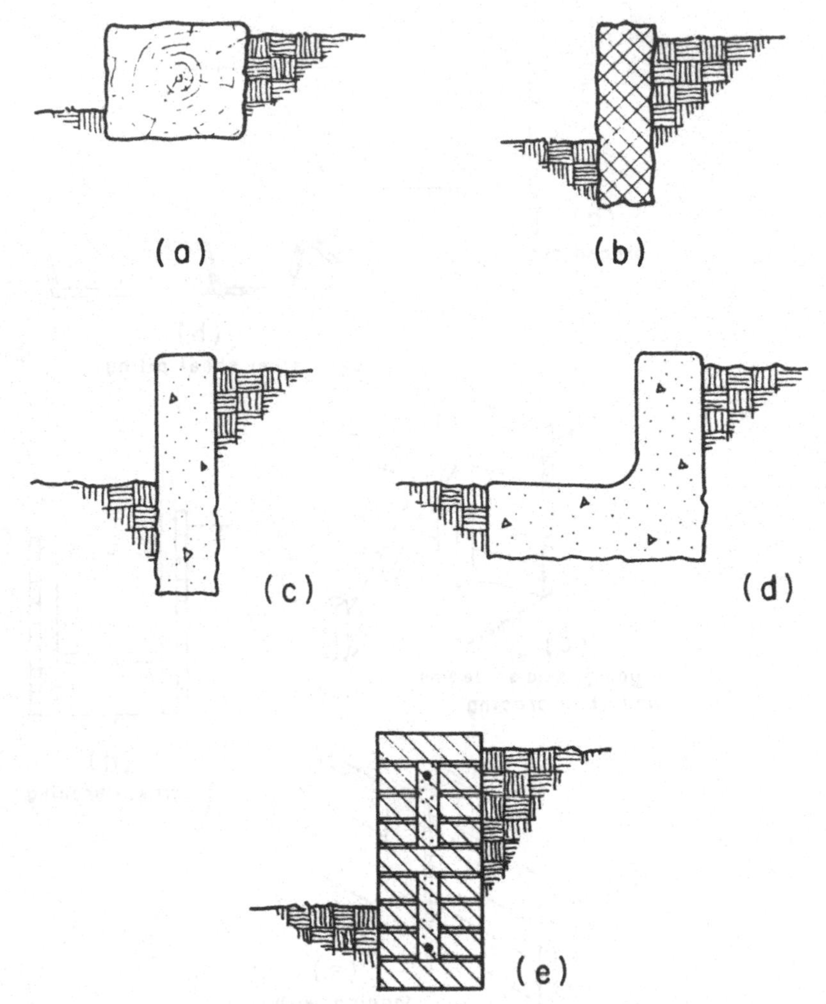

Figure 7.45 Forms of curbs: (a) Timber or log. (b) Stone. (c) Simple concrete. (d) L-shaped concrete to form gutter. (e) Brick masonry.

elements, as shown in Figure 7.46. Such construction must be banked, or leaned, into the cut to resist the horizontal force of the soil on the high side of the wall.

Walls of this type can be very effective and simple to construct. Executed with field stone or timber they may be quite attractive; especially when used with other largely natural materials, such as earthen surfaces or plantings.

A typical advantage of the loose-laid wall is its natural porosity, which allows ground water to seep through. This is especially critical when regular heavy irrigation of plantings on the high side occurs.

Cantilever Retaining Walls

The strongest retaining structures for achieving abrupt elevation changes are *cantilever retaining walls,* consisting of structural walls of masonry or concrete, anchored to a large footing. Some common forms are those shown in Figure 7.47.

A form used for walls up to around six feet high is that shown in Figure 7.47(a). (Height refers to the ground surface elevation difference on the two sides of the wall.) Critical structural concerns are for the rotational, overturning effect and the horizontal sliding effect; both caused by the horizontal soil pressure on the high side

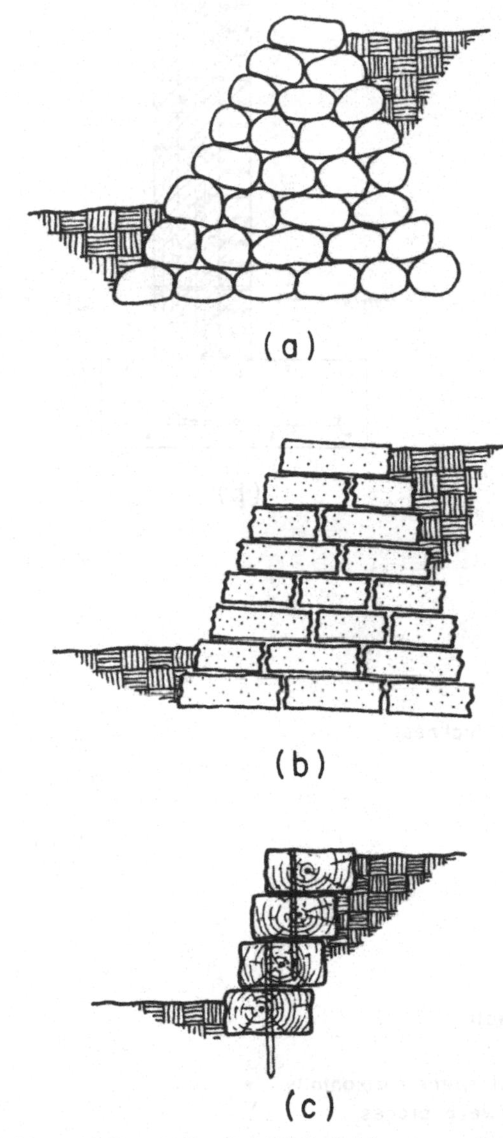

Figure 7.46 Loose-laid retaining walls: (a) With field stones. (b) With broken pieces of concrete slabs. (c) With timbers (railroad ties, etc.).

of the wall. The dropped portion of the footing (called a shear key) is commonly used to enhance the resistance to horizontal sliding.

Short walls may be achieved with masonry or concrete. If achieved with concrete, both the wall and the footing are usually reinforced as shown in Figure 7.47(a). Masonry walls may be constructed similarly—as shown in Figure 7.47(b)—or may be developed as gravity walls, relying strictly on their dead weight to resist the overturning effects, as shown in Figure 7.47(c).

As walls get taller, it is common to use a tapered wall form. Concrete walls are evenly tapered, as shown in Figure 7.47(d), while masonry walls are typically step-tapered, using regular units of the masonry, as shown in Figure 7.47(e).

Tall retaining walls may also be braced by buttresses or fin walls perpendicular to the retaining wall, as shown in Figure 7.47(f) and (g). If built on the back side, these do not affect the viewed side of the wall; although building them on the exposed, low

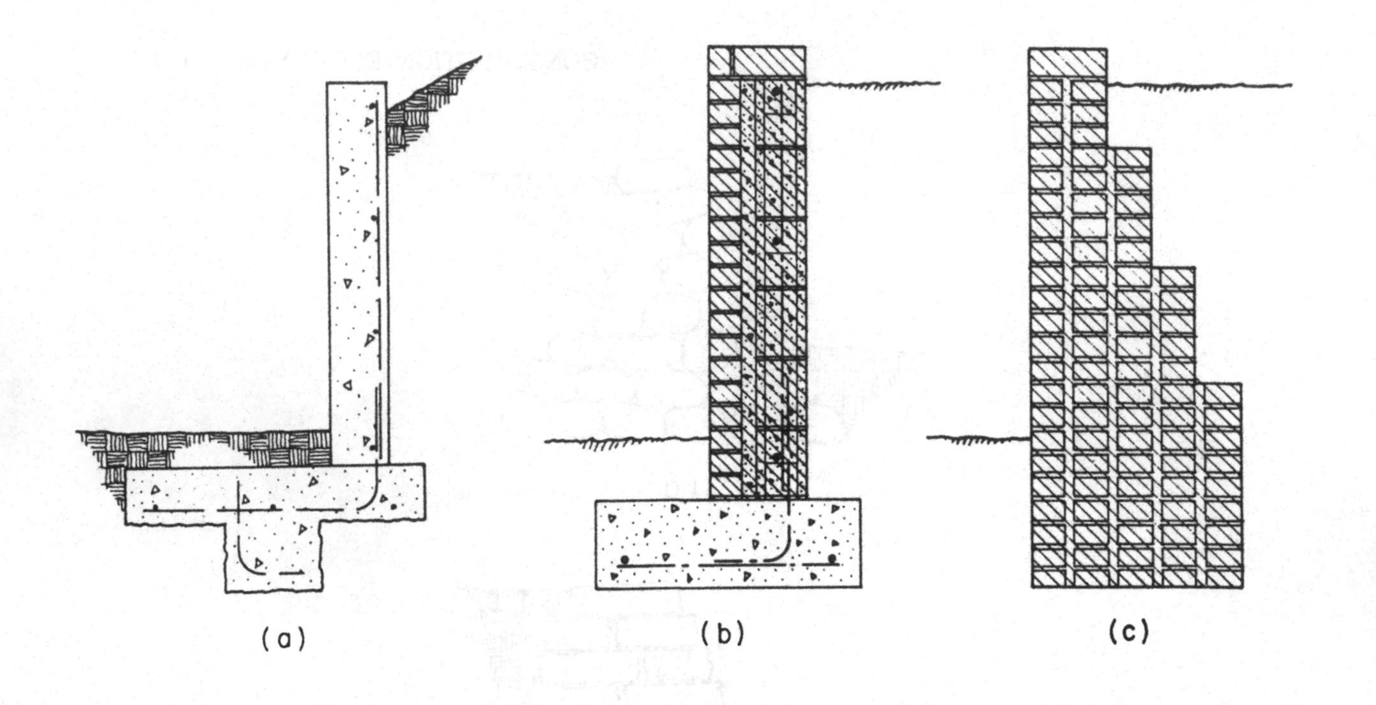

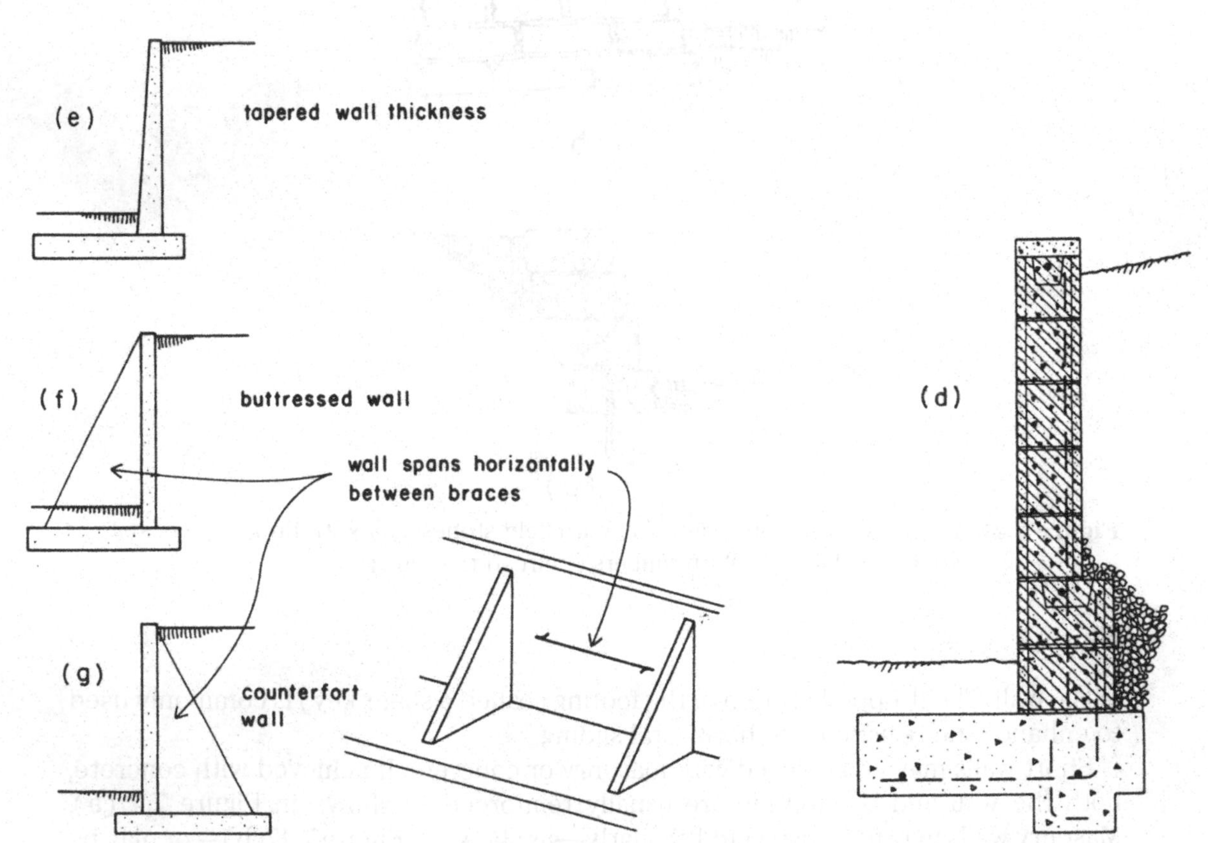

Figure 7.47 Forms of retaining walls: (a) Short reinforced concrete cantilever wall. (b) Reinforced concrete masonry wall (concrete blocks). (c) Brick construction, as a gravity wall relying on its mass for stability (an ancient method, now too costly). (d) Tall, tapered wall, achieved by stepped thickening of masonry units. (e) Tall, tapered wall, achieved by continuously tapered thickness of sitecast concrete. (f) Tall wall braced by exposed buttresses on the low side. (g) Tall wall braced by concealed counterforts on the high side.

side is usually easier and more economical, as it involves less excavation and backfill on the high side.

Retaining walls may also be developed as parts of building construction, becoming building walls with some bracing of the wall often provided by other elements of the building construction. The typical basement wall is generally a retaining structure, although not usually of the cantilever type, as it works by spanning between the basement and the first floor constructions which brace it laterally.

Information Source

Simplified Design of Building Foundations, 2nd ed., James Ambrose, Wiley, New York, 1988.

7.16 UNDERGROUND BUILDINGS

When a space exists truly underground, all of the problems described in preceding sections for basement walls and floors are present. A major additional concern, however, is that for an overhead structure that is *not* simply a floor of a building above. This structure—which we call the roof of the underground space—may support soil of some thickness or be paved as a terrace or plaza surface. With an extensive underground construction and major landscape development, all possible overhead conditions may exist.

Structure

Compared to ordinary roofs, those for underground spaces usually carry many times the total gravity loads. The construction of the structure itself is likely to be among the heaviest we use—sitecast concrete. No light trusses, space frames, or fabric structures here! Add to that either a heavy load of soil or the high live load usually required for a terraced roof area or a plaza (150 to 250 psf in most codes).

A further concern is often that of restricting the total depth of the structure in order to keep the total depth of the underground construction to a minimum. All of this conspires to generally keep spans short and try for some efficiency in the spanning systems if possible. The usual options for sitecast concrete floor systems are all possible, as shown in Figure 7.24. Among these the two way flat slab and two way waffle systems usually offer the smoothest underside and least overall depth and are favored—especially if the underside of the structure is left exposed.

Water Control

A roof is a roof, and the generally basically flat roof of an underground structure is not significantly different in many regards from any flat roof. A totally watertight membrane is required, together with all flashing and careful sealing of any penetrations. The membrane must be protected—especially here—and some insulation and a vapor barrier may be indicated.

Drainage is somewhat different here than for a conventional roof, but not any less indicated. The form of drainage will relate somewhat to what is above the roof—soil or paving—and probably to the development of a total drainage system for the whole underground construction. As in other situations, the roof must usually be drained essentially by gravity flow, involving considerations for the total slopes required and the vertical dimensions of the construction necessary to achieve drainage and removal of water.

As for walls and floors, concern for the watertight security of the underground space may be heightened if the space is occupied by people. However, there is hardly any middle ground for a watertight roof—it is or it isn't, so this is less a variable issue than it is for basement walls and floors in ordinary circumstances.

Earth-Covered Roofs

An especially critical water control problem is that of the roof supporting earth—and usually plantings. The earth may get wet from precipitation alone, but the plantings will require continuous moisture, so the water condition will be ever-present. But drainage must be effective to prevent a saturated, rot-inducing condition for the plant roots.

Another major consideration in this situation is the provision of sufficient depth of earth fill for sustaining of the planting. This may be minimal for grass, but of major proportions for trees. To save depth, large plants and trees may be placed in special planters integrated into the space of an otherwise not so deep underground construction.

Figure 7.48 shows some of the features of an earth-covered underground structure, sustaining only minimal planting in this case.

Paved Roofs

When close to the surface, underground spaces may have paved roofs—forming terraces or plazas for buildings, or simply parts of a general site development. Extensive underground parking garages are developed in many locations with parks,

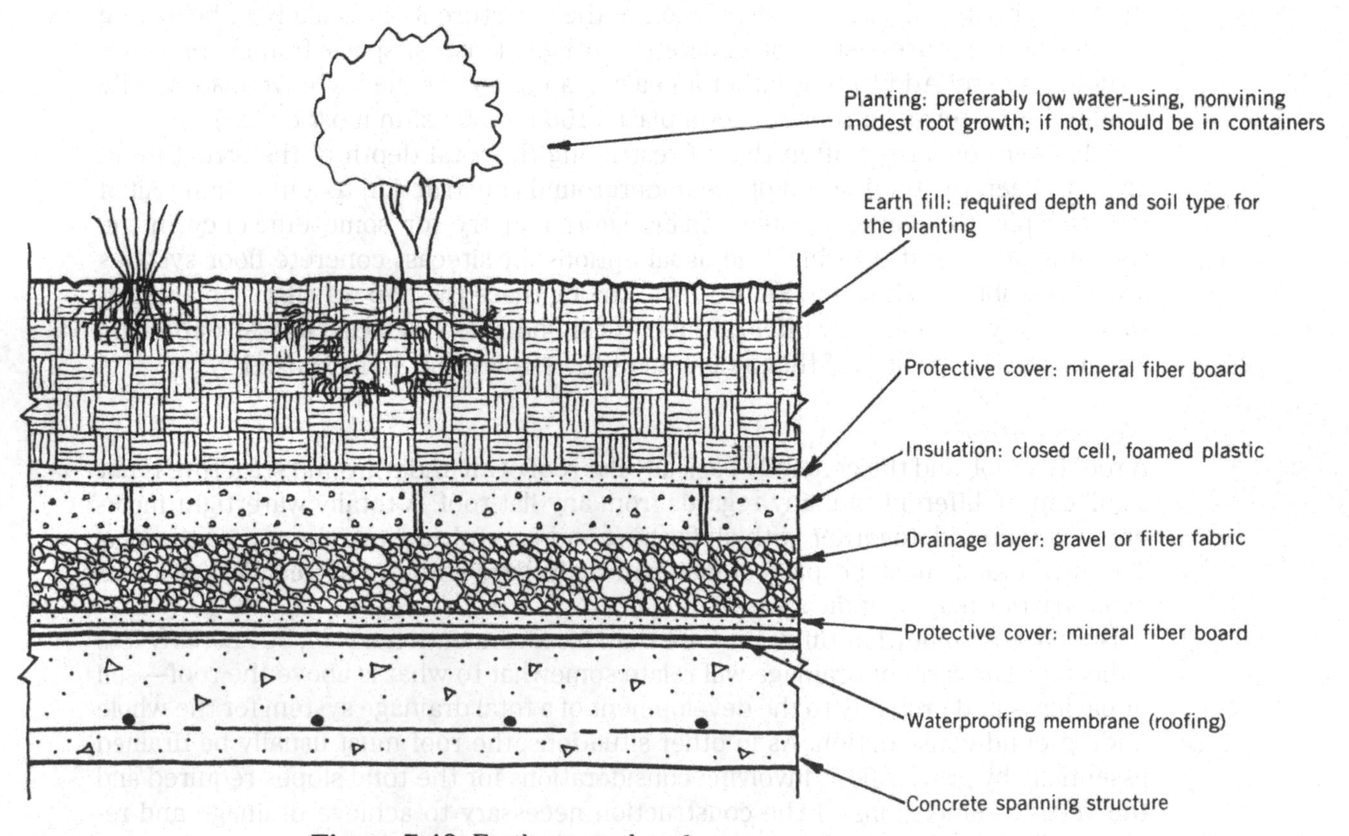

Figure 7.48 Earth-covered roof over an underground space.

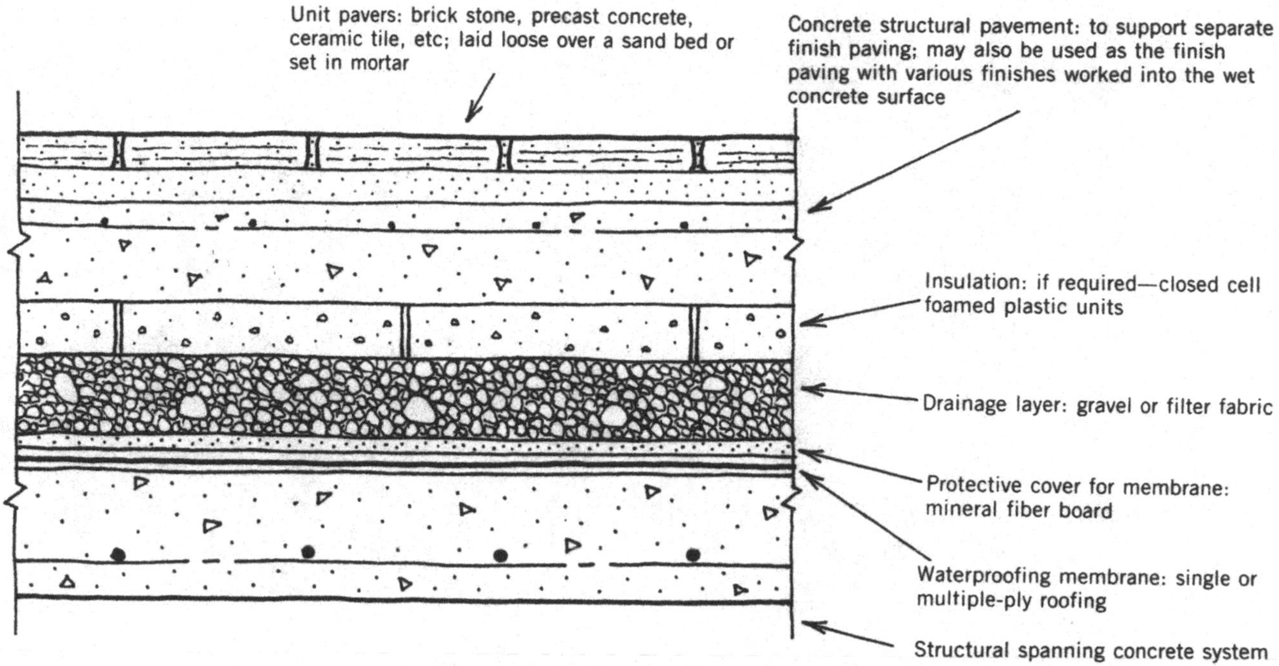

Figure 7.49 Paved roof over an underground space.

squares, or other open spaces on their roofs. Paving here is not essentially different from other outdoor paving, with all the usual options, a few of which are shown in Figure 7.49. The sub-base and general support may be somewhat different and must be generally integrated with the overall construction of the roof of the space below.

The same underground space may have both planting and earth for plantings on different areas of the roof. This calls for some coordination of the overall dimension for development of the two types of surfaces, if the supporting roof structure is constructed essentially flat.

Information Sources

Building Underground, Herb Wade, Rodale Press, Emmaus, Pennsylvania, 1983.

Earth Sheltered Homes: Plans and Designs, Underground Space Center, University of Minnesota, Van Nostrand, Reinhold New York, 1981.

Enhancement of the Construction

Basic forms of construction have particular properties with regard to various physical responses; each being more or less fire resistive, more or less resistant to sound transmission, and so on. Achieving better levels of response may be accomplished by simply choosing among common alternatives for the construction. However, it is also possible to effect various modifications to enhance the ordinary forms of construction.

In the typical situation, many different properties of the construction must be considered. The optimal solution for any single behavioral response is not likely to be optimal for all other desired responses. What is lowest in cost is not likely to be most durable, most attractive, most fire-resistant, and so on. Primary values must be given to important design goals and some other factors must be made less critical.

Some primary behavioral responses are considered in the discussions in this chapter.

8.1 FIRE RESISTANCE

With regard to the building construction, there are various issues that affect its design that derive from concerns for fire (see Figure 8.1). Major concerns, which in most cases are bases for the development of building code or zoning requirements, include the following:

Combustibility: Just about anything can be consumed or otherwise destroyed by fire. However, some materials and elements, such as thin pieces of wood or paper, can be quite easily burned. To the extent that parts of the building construction can add to the so-called fire load (total mass of combustible materials present), the construction represents a potential hazard on its own—regardless of the building contents or the activity of occupants.

Flamability: A major concern for surface materials for floors, walls, and ceilings is the degree to which fire may rapidly spread on the surface. This involves both the general flammability (combustibility) of the materials and the speed at which the surface burn may spread. Materials are given a *flame spread rating* for this evaluation.

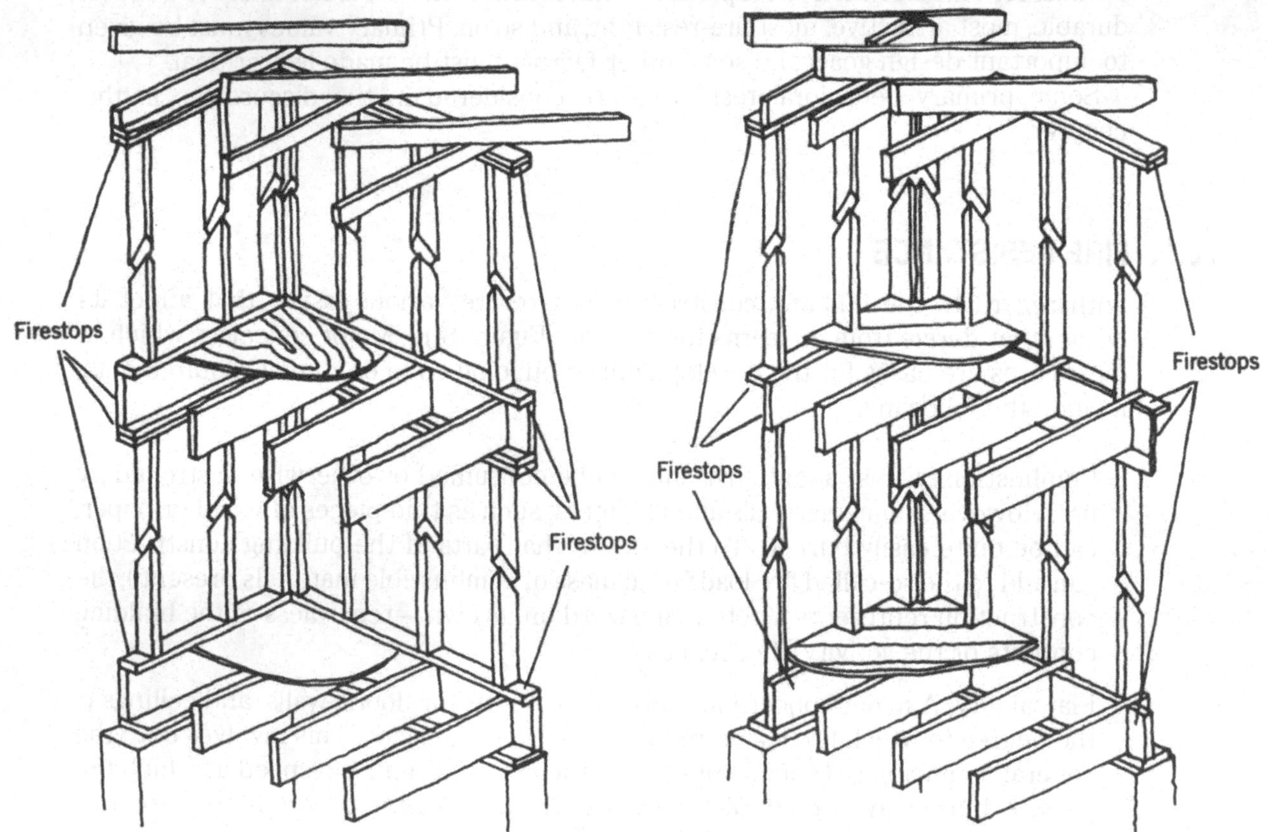

Figure 8.1 Development of the construction for fire safety: (a) Thermal insulation for vulnerable structural elements. (b) Blocking inserted in wall cavities to prevent invisible spread of fire inside the walls. (c) Door-opening hardware responsive to panic-stricken occupants.

c

Figure 8.1 Continued.

Vulnerability: To the extent that the construction is vulnerable or susceptible to fire, and likely to be easily and quickly made nonfunctional (for example, burned up), it is considered to lack resistance to fire. This is especially of concern to parts of the construction that provide structural support or that achieve spatial separation for portions of the building. Floors should neither quickly collapse nor quickly permit fire to spread by penetrating them.

Toxicity: Most materials release some toxic (deadly, life-threatening) materials when they burn. They also consume oxygen to produce a fire. Most loss of human life in building fires is due to either oxygen delpetion or inhaling of toxic materials (smoke, poisonous gas).

Form and Detail: Various aspects of the construction and building planning may affect fires or the safety of occupants and firefighters. A suspended ceiling is one of a fireman's worst enemies; fires can occur within their enclosed spaces and spread rapidly without being seen or otherwise detected, bursting suddenly out or causing collapse without warning. For doors, location, size, visibility, direction of swing, and operating hardware relate to the potential use of the doors as safe fire exits. Dead-end corridors (no way out, except back the way you came in) are firetraps; the longer, the more deadly.

Most building code regulations regarding fires were initially developed under sponsorship by insurance companies. Insurers often cover the building, its contents, and its occupants—all being of concern for claim settlements. Most building fires are relatively quickly contained and extinguished without loss of life or serious injury to occupants. Still, they cause damage to the building and its contents, and cost the insurers money. Insurers are not to be viewed as bad guys, but they are in business primarily to make money, not as humanitarian organizations.

Development of fire-safe buildings, with life safety as a prime concern and the building occupants and fire fighters as the main targets for protection, is in the process of emerging as a major research and development area. Effectiveness for life safety and quick, practical, and safe fire fighting is ascending over the concept of protecting the building and/or its inanimate contents.

For the building enclosure systems, some specific issues that relate to fire safety include the following:

Separation of buildings: Spread of fires from building to building is a community concern, so there is a desire to keep a fire in the building where it originates. This relates to roof and wall materials, to the size, location, number, and fire-resistive character of openings, and to the proximity of the building to its property lines or to adjacent buildings.

Surface spread: Fire can move rapidly along surfaces, depending on the finish materials. Most surfacing materials are rated for their flame spread character.

Accessibility to fire fighters: Site planning or building form may prevent fire fighters from getting themselves or their equipment to a fire.

Much of what is in building codes relates to fire safety. However, building code requirements in general are of a minimal, rather than an optimal, character. Making the most fire-resistant, life-safe building is seldom achieved by simply satisfying the minimum standards of building codes.

8.2 THERMAL RESISTANCE

Maintaining thermal comfort for building occupants generates various concerns for the enclosure construction. (See Figure 8.2.) Some primary considerations are the following:

Resistance to thermal flow, that is, the natural effect of the materials and the form of construction on direct air-to-air flow of heat into or out of the building.

Air leaks, resulting in the heat loss or gain represented by the volume of air change (generally called *infiltration*); cubic feet per minute flow translating to calories or Btus per hour losses or gains.

Miscellaneous effects, such as those of the interior surface on air movements, radiant losses to cold interior surfaces, and so on.

Ability of the construction to accommodate elements of the building's HVAC system.

Specific climate conditions, general form of the building HVAC system, and the needs of the occupants will determine the relative importance of various factors and the general effectiveness of various forms of enhancement of the enclosure construction.

Development of any enhancements of the building enclosure begins with an analysis of the properties of the basic construction. If the results of such an analysis

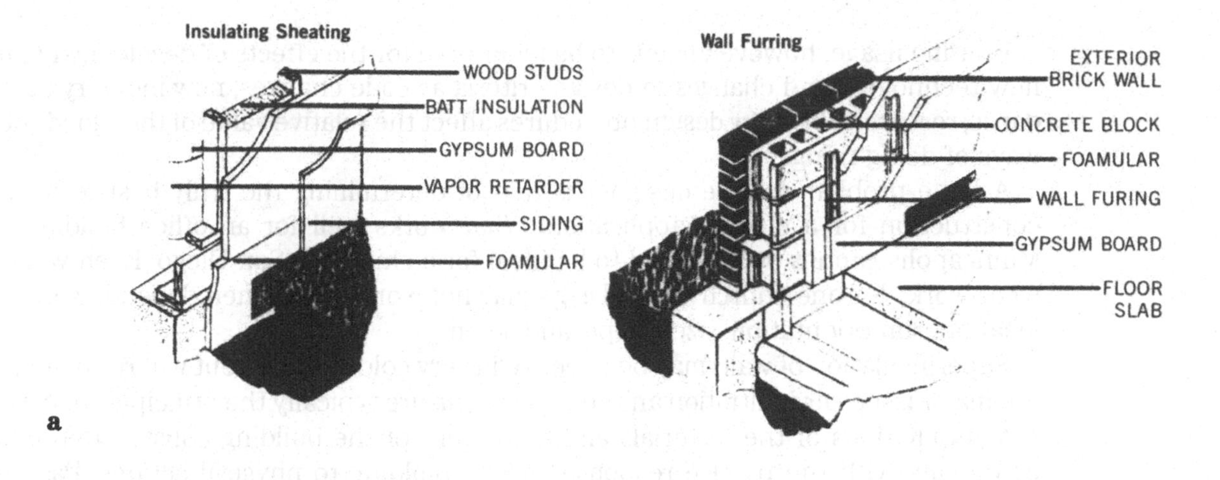

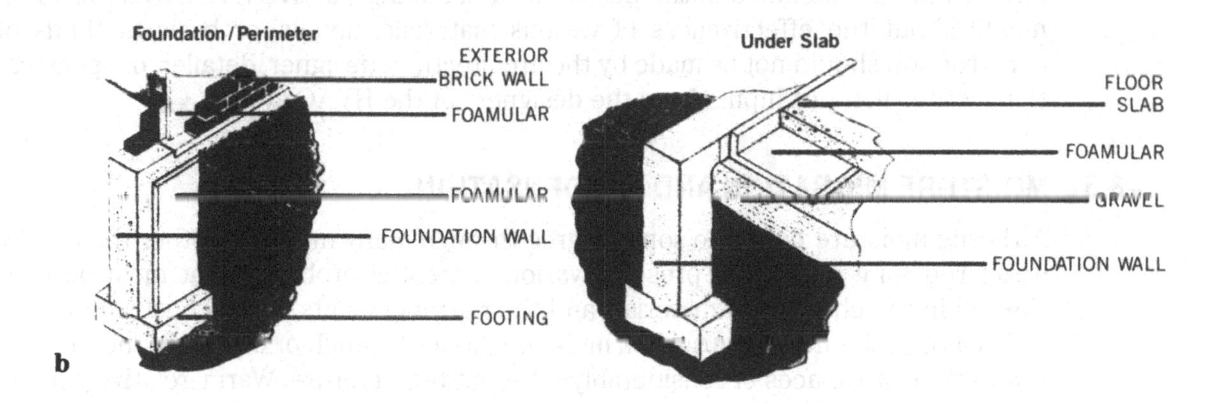

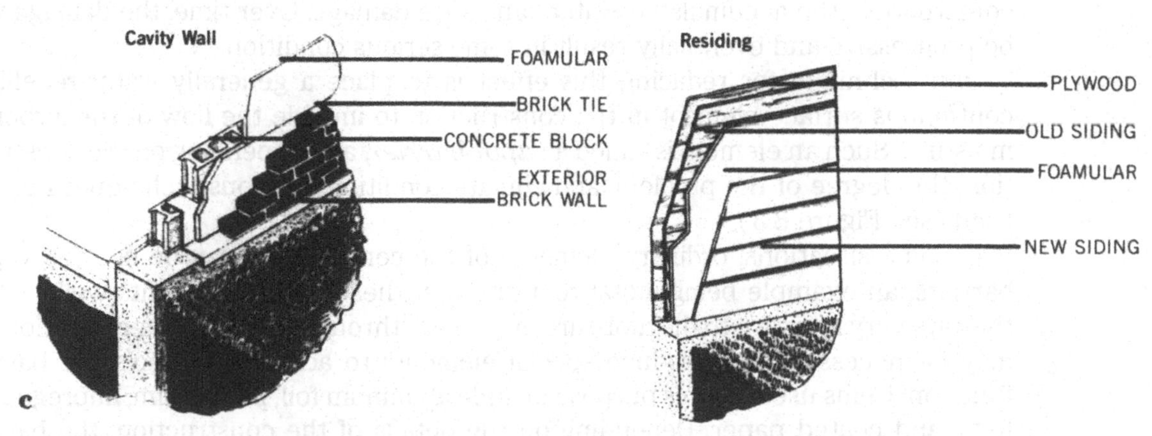

Figure 8.2 Use of thermal insulation for cold weather protection; applications of Foamular (foamed plastic) insulation elements, courtesy of UC Industries, Inc., Parsipanny, NJ.

produce values for behaviors that are less than satisfactory, then some modifications or additions are considered. Of course, smart designers will anticipate needs and make some considerations of these issues in their initial choices to avoid disruptive changes later. Routine, repetitive use of construction with demonstrated successful performance will give some assurance that additional modifications of a major nature will not be required later.

Routine usage, however, needs to be monitored for the effects of developments in new technology and changes in design criteria as code changes, new industry standards, research, and new design procedures affect the relative value of the "good" old ways of doing things.

A large problem for the designer is that of determining the truly best forms of construction for a specific application. What works well for an office building in Minneapolis cannot be expected to be ideal for a church in San Diego. Even worse, what works for one church in San Diego may not work for another, depending on its location, sun orientation, size, shape, and so on.

Superinsulation of walls may be effective in very cold climates, but where cooling is the major issue, air infiltration and solar heat gain are typically the principal concern.

Manipulations of the materials and the details of the building construction deal essentially with the passive responses of the building to physical actions. Passive effects must be included in the design of the building's active HVAC system. Judgments about the effectiveness of various materials, details, or basic methods of construction should not be made by the architectural designer, detailer, or specifications writer without inputs from the designers of the HVAC system.

8.3 MOISTURE MIGRATION AND CONDENSATION

Airborne moisture passes to some degree through many materials, other than really water-repelling ones. This presents various potential problems that must be considered in the choices for materials and their arrangements in the construction.

A major problem develops when moisture passes through or into construction that is between two spaces of considerably different temperature. Warm, relatively moist air moving to a colder location may reach a point where some of the water held in the air will condense due to the lowering of the temperature. If this occurs inside construction, the accumulated water can cause damage. Over time, the damage will be progressive and eventually result in some serious condition.

One technique for reducing this effect is to place a generally water-repelling, continuous surface element in the construction to impede the flow of the airborne moisture. Such an element is called a *vapor barrier*, and depending on the construction, the degree of the problem, and climate conditions, various techniques may be used (see Figure 8.3).

In some situations, ordinary elements of the construction can be used as vapor barriers, an example being metal roof decks. If these have their joints well sealed, they are very resistive to the moisture movement through them. In other situations, it may be necessary to introduce special elements to achieve the moisture barrier function. Items used for this purpose include aluminum foil, plastic film, impregnated felts, and coated paper. Depending on the details of the construction, the barrier element may be added as a separate piece, or it may be used to surface some other part of the construction, such as wall sheathing or an insulation unit.

A critical concern in using vapor barriers is to have them situated in the construction so as to impede airflow before a major temperature change occurs. If warm, moist air comes up against a cold moisture barrier, condensation will occur on the face of the barrier. If this is inside the construction, the water condensed can create a problem.

Condensation can also occur inside hollow elements of the building construction, such as stud-formed voids, cavities in masonry, or the insides of hollow mullions in curtain wall systems. In these situations it is sometimes possible to simply let the

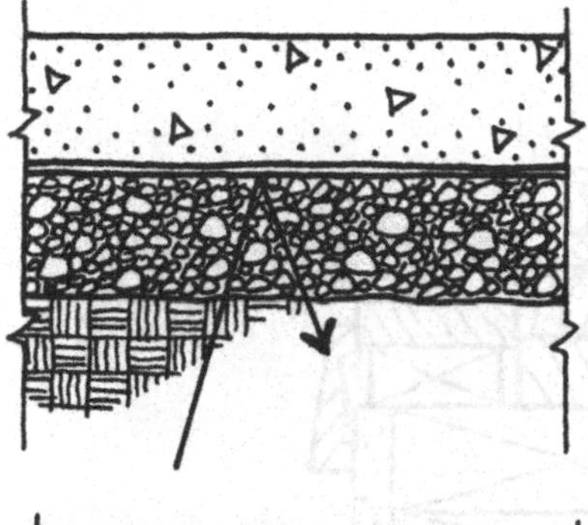

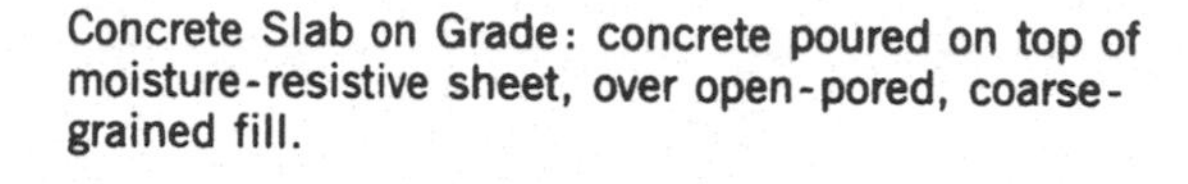
Concrete Slab on Grade: concrete poured on top of moisture-resistive sheet, over open-pored, coarse-grained fill.

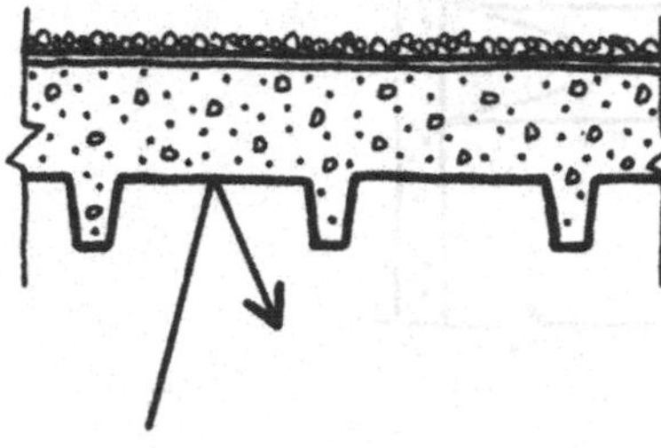

Metal Deck Roof: joints in deck units sealed to provide moisture barrier.

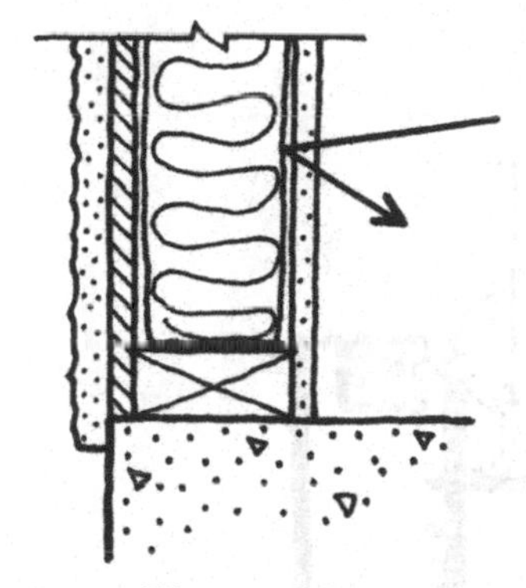

Surfaced Insulation Units: glass fiber batts with interior facing of moisture-resistive material (aluminum foil, etc.).

Figure 8.3 Barriers to the passage of moisture through the construction; as moisture barrier in a floor slab and vapor barriers (to airborne moisture) in roofs and walls.

moisture condense and provide for its drainage with the necessary shaping and sloping for flow control and use of flashing and weep holes to control its passage out of the construction.

This is a major issue of concern in development of the construction for the building enclosure, and there is a great deal of information available for study and design development.

8.4 WATER PENETRATION

Water from rain or melting snow and ice can flow into the building at many locations. The general problems for roofs in this regard are discussed in Sections 5.3 and 7.5. However, there are also many concerns for exterior walls as well as for windows and doors.

As discussed for roofs, there are two issues here—preventing penetration into the construction and generally getting rid of the water. For exterior walls in a vertical position, developing the drainage flow is not a general concern—obviously water will move rapidly down the vertical face. However, various effects can make it linger or

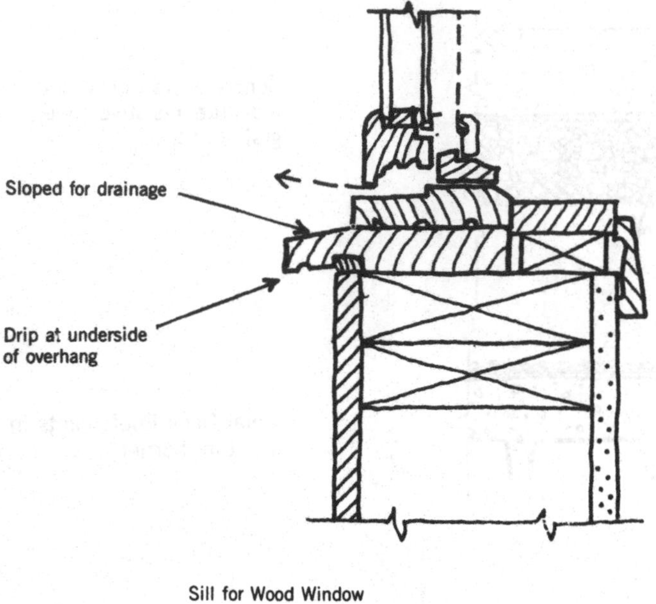

Sill for Wood Window

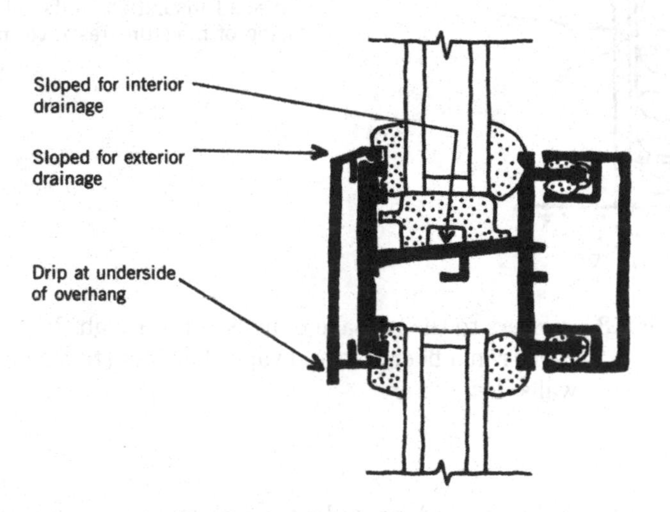

Horizontal Mullion in Curtain Wall

Figure 8.4 Development of horizontal joints for resistance to penetration of water on wall surfaces.

sneak into the building. Figure 8.4 shows some basic concerns for the detailing of horizontal elements and joints—a particularly vulnerable situation.

As with water on the roof, the water that runs down the face of a wall must also go somewhere. This may require some consideration for what happens at the bottom of a wall or at the heads of doors or windows that can be opened. At various places it may be necessary to use ledges, awnings, canopies, or other elements to divert flows.

A special problem is that of the sloped wall surface, which becomes transitional in many regards between a wall and a roof. It may need to be developed to serve many wall functions, but if it is large, it may also need to deal with roof problems, including those of water flow and disposal.

In some situations, particularly in tight urban spaces, the water from roofs and that running down walls may need to be collected and not dumped off of the building.

Roofs of tall buildings generally present this situation, whatever the roof form. The control of flow on surfaces, diversion to points of collection, means for collection and piping into the building, and ways to deal with it after it is in the building must all be considered in the development of a disposal system and detailing of all the construction elements involved. Relations to the exterior building form and spatial accommodation of interior drain elements must be studied as architectural design issues.

Water must usually be allowed to flow by gravity in drain systems. This is especially critical where there is need to provide for long horizontal movement—a general situation to be avoided whenever possible. Horizontal drains within the construction must have the ability to change their position vertically as they pass through the structure and other parts of the building system.

On the outside of the building, water may flow other than by simple downward paths due to gravity. Windblown rain and water flowing on surfaces may be moved along surfaces so that the flow is even upslope due to wind direction. This must be considered for various details; shingles, for example, must sustain some upward flow and wind uplift effects as well as the normal downslope fast-draining action.

8.5 SOUND CONTROL

One of the most elusive design factors is that of sound control. Goals for this differ for different occupant groups and for individual interior spaces. Typical major concerns are the following:

1. Privacy: This generally means the keeping of sound—notably conversations within a contained situation; frequently simply inside a room. Normal conversation has a specific range of sound level (decibels) and frequency (cycles per second), so that either transmission through boundaries (walls, floors) or masking with other sounds (music, etc.) can be reasonably managed in design.
2. Noise intervention: Noise is essentially unwanted sound—it could be beautiful music, but intrusive if it drowns out a lecture or conversation or keeps someone awake who is trying to sleep.
3. Room acoustics: This has to do with the management of sound within a bounded space, involving sustained, ambient sound levels (enduring sound), reverberation, echoes, and amplifications of specific frequencies.

Controlling any or all of these, or other special problems, may be reasonably established by careful design. Effecting it in the actual construction is another matter. Like air or water, sound can leak through tiny openings. It also travels by paths you don't anticipate; such as out one window and in the next, instead of through the strong barrier you have created between rooms.

Control of sound is less often critical for the building enclosure than for interior construction. However, incoming noise can be a consideration in busy urban areas, near freeways or railroad lines, near airports, and in other situations where there is some regular disturbing noise source outside the building. The usual concern is simply for use of the enclosure as a sound transmission barrier.

Various typical forms of construction are rated for their ability to modify sound levels between the spaces for which they serve as separators. In the usual situation barriers also vary in terms of their relative effectiveness with regard to sound pitch and total energy level. For a full design, the exact nature of the incoming sound and the form of activity inside the building must be given consideration. With all of that data and some investigation, a judgment can be made as to whether some modification or enhancement of the construction is indicated, or possibly a different basic form of construction is indicated. This issue is being given increasing attention in the design of curtain walls, as they occur quite frequently in urban areas with high volume of traffic of all kinds.

Effective sound management requires a lot of cooperation; between architect, acoustical engineer, builder, and even building occupants. That is quite a bit to ask for, but any one of these parties can subvert the effort.

Many simple things can be done in choosing surfacing materials, detailing of the construction, shaping of interior spaces in planning, isolating of noisy equipment, and other situations to generally improve sound conditions. Classically bad situations can be carefully avoided by minor attention in planning and in detailing of the construction.

It is possible to spend considerable money for sound control and miss the target. Major efforts must be made for serious situations, like those of auditoriums and recording studios. For most buildings, however, some simple attention to potential problems and the easy things to do to avoid them or compensate for them are the usual course.

One simple trip through a basic reference on good sound management in buildings will sensitize the average designer sufficiently to the simple issues. A lifetime of experience will then produce some judgment about what is really possible to achieve in ordinary circumstances.

REFERENCE

Mechanical and Electrical Equipment for Buildings, 8th edition, Stein and Reynolds, and McGuinness, Wiley, 1992. Chapters 26 and 27.

8.6 SECURITY

The exterior, enclosing construction of most buildings must offer some degree of security; referring mostly to the prevention of intrusion by unwanted persons. This varies considerably, of course, from the situation of a residence in a peaceful community to a bank in an urban area. Security involves the basic construction of exterior walls and roofs, but also the nature of doors, windows, skylights, utility tunnels, and all other potential points of entry.

Individual interior spaces or levels in buildings may also need to be made secure; the basic need and degree of security required, depending on the building usage. Determined burglars are hard to thwart. Most security achieved by the construction alone is in the order of preventing casual entry. Serious security—in construction terms—is usually reserved for jails, bank vaults, and other special situations.

As with design for other building responses, some simple considerations can improve security without major restriction on the general construction or general building planning. Good lighting at entries, secure locks for windows and doors, and other measures can improve the situation. In the end, however, alarms and surveillance are probably the real effective means to establish security, and effective, economically feasible measures within the scope of building design are of limited value.

REFERENCE

Mechanical and Electrical Equipment for Buildings, 7th edition, Stein, Reynolds, and McGuinness, Wiley, 1986. Chapter 22.

8.7 SIGNAGE AND TRAFFIC CONTROL

Buildings are essentially inanimate, although operation of various building services—elevators, HVAC systems, water heaters, etc.—involves some dynamic concerns. People, on the other hand, come and go, move about inside the building, and exit—usually leisurely, but occasionally in a rush, when there is a fire alarm or bomb scare.

Longtime users of a building know the premises and how to get around in the building. Infrequent users or first time visitors often need help, starting with finding the entrance.

Design of the building construction is not the way to deal with traffic management, but various considerations for the building usage may well affect the general planning of the building and the choice of some elements of the construction. Doors, for example, are major elements in traffic control, and many of their features derive from these concerns.

Some designers measure really effective planning by the lack of need for signs. This may be a good ideal, but some signage is actually required by codes, and good signage is never a negative thing for building users.

Visibility within the building interior may be a means of assisting users to find places. Walls of glass or doors with glazing may be effective for various purposes. More subtly, continuity of wall construction or simply wall or floor finishes may keep people on a traffic path by inference. A sudden change—in the size or shape of bound spaces, in lighting, in visibility of surrounding spaces, in floor finish, etc.—can grab peoples' attention and alert them to something.

Effective exiting of the building in emergencies shoud be carefully planned. This may begin with minimum code requirements, but should extend to deeper concerns for practical likelihoods of the responses of users. Fire exits may be legally marked by signs, but if nobody enters the building through them, they are not going to rush to them in an emergency. Safety and general security of interior construction should preferably be progressively more stringent as exiting occupants proceed along logical exit paths.

8.8 BARRIER-FREE ENVIRONMENTS

Increasing attention has been given in recent times to the development of buildings that are safe and are accessible by persons with limited capabilities. Design of public facilities has been most affected, but access to work places has also been promoted, so that many basic elements of construction are now affected by requirements in this category.

Early on in this effort a principal concern was for access for persons confined to wheel chairs. The basic concept and general interest is now of a broader scope and tends to spill over into general concern for a much wider range of building users; beginning with those with serious limitations, such as blindness, but extending to the elderly, the very young, illiterate persons, or others incapable of reading signs, and so on.

Specific details and features of buildings can be made less hostile or otherwise more accommodating to users—those with limited capabilities or not. Elimination of steps where possible, widening of doors, provision of grab bars in toilets and showers, and use of large, clear letters and identifying symbols on signs, represent efforts to create a generally better environment for everyone; but surely do indeed widen the range of potential building users.

▪8.9▪ GENERAL LIFE SAFETY

Attention to life safety has a simple goal: the reduction of the potential for loss of life or for serious injury to building users. This is a major factor in the general development of building codes, which are government ordinances when enforced—providing essentially for the public health and welfare.

Building codes are, however, basically meant to establish threshold, minimal levels of acceptability. *Optimizing* for life safety, without necessarily creating economically unfeasible or otherwise distorted designs, is a relatively new design emphasis that is gaining strength. Viewing all aspects of design decision making in the building design process from this point of view, and having data to input to a value analysis procedure, is a major research field in building design.

9

Construction Systems

System is a much overused word, but generally it describes any group of component elements that exists in some ordered relationship. Most often the group is composed to relate to some specific task. A building as a whole entity can be described as a system, although we more frequently use the term to describe some subunit, or subsystem, of a building. In this chapter the issue of building systems will be discussed with reference to the idea of a particular, distinctly definable means of construction. In that context there are a small number of very ordinary systems that are repetitively used to create buildings.

9.1 HIERARCHIES

Systems with many parts often have the parts arranged in some ranked order or hierarchy (see Figure 9.1). For building construction a hierarchy may be defined in terms of the value of parts to some task or issue. Structural systems may be reasonably treated in this manner, referring to the relative importance of individual parts to the integrity or the basic determination of the overall system. Parts may be viewed as major or minor, primary or secondary, and so on.

Considering construction from a design viewpoint, it is desirable to be able to identify the fundamental, primary elements of a system in order to understand their major roles in determining the general character of the system. Modifications of the system should be done in a manner that respects the roles of the primary elements of the system.

Modifications in the form of design variation may be more easily done when they deal only with secondary or minor elements of the basic system. Thus with the light wood frame, the studs are primary determiners of the system, while the surfacing is secondary to the basic system. Many types of materials may be used for surfacing—effecting major changes in the appearance of the construction—without essentially altering the basic nature of the system.

In considering the various major systems described here, the reader should attempt to visualize the hierarchies in them and the degree to which significant architectural design modification can be made without altering the basic character of the systems.

9.2 DESIGN PROCESS AND TASKS

The nature of the building design process and the various issues and tasks are described in Chapter 2. A major early design task is the selection of the basic systems

a

b

Figure 9.1 Articulated, hierarchical framing systems: (a) Steel system with rolled columns and beams, light prefabricated trusses, and formed sheet steel decking. (b) Strongly-expressed, custom-formed wood frame; Gamble House, Pasadena, CA., Greene and Greene, architects.

for the construction. Selections may well be limited by code requirements, building size and form, or other factors inherent in the design criteria or situation. However determined, the choice of basic systems at an early stage will greatly simplify progress of the design.

With systems defined, and their hierarchies understood, the forms of design modification that are most easily achieved can be explored. If the major design goals cannot be achieved within the practical limits of the basic systems, it may be necessary to make a major shift to other systems. This kind of design change has major effects on the progress of the work, and the earlier it happens, the better.

Although design work seldom proceeds in so orderly a fashion, the following is a logical process for determination of the construction during the general progress of a building design project.

1. Determine the basic systems. This means the selection of the general form and materials of the basic system. (Masonry bearing wall with CMU, for example.)
2. Determine specific features of the basic elements. With the general system defined, each part is now independently defined (mortar type and basic unit dimensions for the CMU wall).
3. Consider relations to other building elements. This is the general task of integration. (How do unit dimensions affect window sizes?)
4. Develop all necessary details for use of the system. Some details are inherent or basic to the system and many must be specifically developed for all situations that exist in the building. (Detail all the variations of wall intersections, wall openings, wall corners, and so on.)

As described earlier, the considerable work required in steps 2 through 4 will be wasted if a system change is made late in the process.

9.3 MIXTURES

Buildings consist of large collections of individual parts and subsystems. There are very few manufacturers of whole buildings; in the main each manufacturer produces only selected parts. The task of collecting parts to make a building is usually up to the building designer.

Economy generally dictates that the parts of buildings are mostly obtained as manufactured products, selected from the catalogs or samples of the producers and suppliers. The designer of a building in this situation thus has major tasks of catalog shopping and mixture making.

While the building represents a giant mixture, it is usually a mixture of subsystems and components, each of which may also be a mixture. For example, a designer may choose to use a light wood frame (2 × 4 wood studs, etc.) for a wall. This is a well-defined basic construction component and it can be "mixed" with various other components for an appropriate building assemblage. But the stud wall itself is a mixture, with the studs produced by one maker, interior paneling by another, exterior sheathing by a third, exterior finish by a fourth, connectors for the frame by a fifth, insulation by a sixth, and so on. Each of these items can represent a design choice not necessarily implied by the general choice to use a wood stud wall system.

With building materials and products largely produced by an industry that focuses on selected, individual items, the designer (shopper) faces the chore of comparative shopping, as was discussed at length in earlier chapters of this book. Selection between the available products for a single use (sheathing for the wood frame wall, for example) requires comparative value analyses that individual suppliers may not be fully relied on to provide assistance for an unbiased view. The larger view, as well as an unbiased one, is required. One type of sheathing may be the best in its class, with many superior qualities, but those qualities may be of minor concern for a particular application, and a lower quality, less expensive product may be perfectly adequate for the task at hand.

Designers (shoppers, buyers) must look diligently for the best choice for each individual part for the building. But the whole building is more important than any individual part. The best roofing material, placed on a poor roof structure, will not produce a good roof.

9.4 APPROPRIATENESS OF SYSTEMS

The achievement of good construction begins with the choice of the best materials, products, and systems for a particular use. Add in careful detailing and competent construction work, and a well-built building just might result. But a well-built building in the wrong place, for the wrong use, will still be inappropriate.

It is one thing to put plaster on a wall in the right way; it is another to select plaster as the proper surfacing for the wall for the situation at hand. Product information and design specifications can deal with the correctness of the plaster itself, but the larger question of whether or not to use plaster is in the hands of the designer.

Building codes provide some guidance for the selection of materials and systems for appropriate situations. This occurs mostly as a negative form of criteria—prohibiting certain uses, such as that of wood (or combustible construction in general) for hazardous or major life-threatening situations with regard to fire.

Manufacturers or industry associations sometimes provide guides to the appropriate use of their products. However, they have in general a major interest in promoting the use of their products, so they do not much like pointing out all the situations in which these products are inappropriate.

9.5 CHOICES: THE SELECTION PROCESS

Design of the building construction is generally simpler and faster to achieve when choices can be made in large bites. Choosing preestablished (or essentially predesigned) systems is one means for doing this. If the choice is good in all regards, the size of the bite is irrelevant.

Preestablished systems offer the potential advantage of some demonstrated success. If the demonstrated successful cases have a reasonable match with the situation for a proposed design, all the parties concerned—designer, builder, owner, investors, insurers, and code-enforcing agencies—may take some comfort in an assurance of success. This is, of course, the actual case, making any real innovative design difficult to sell in many quarters. Designers should not be totally discouraged, but must understand the realities in achieving any significant changes in the "good old ways" of doing things.

Exciting, uplifting architecture can be achieved with very ordinary means of construction, but can also—on occasion—come from real innovation or dramatic usage of the construction. In either case, the designer usually has a considerable understanding of, and an appreciation for, the construction details and processes. If ordinary means are used, the designer understands their limits and potential as well as what significant architectural design variations can be achieved within the basic system. If ordinary means are rejected, the designer usually does so in full knowledge of what exactly is being discarded and what the needs are for any replacements in terms of new means. Otherwise, innovation will inevitably be naive and unsatisfactory in some performance aspects.

9.6 SYSTEM DEVELOPMENT

Current construction of buildings in the United States occurs with a broad array of materials and systems. Some are modern versions of very old methods and uses; some are very recent additions to the inventory of technology.

The majority of buildings—modest in size, but extensive in total number—are built with a few common, basic systems. Many others are built with only minor variations of basic systems. The larger the building, the higher the total building budget, or the more prominent its place in the community, region, or country, the more its design budget may support some real innovative design work. Otherwise, innovation becomes mostly a determined personal effort of individual designers, probably mostly on their own time.

9.7 BASIC STRUCTURAL SYSTEMS

The following discussions present some of the most common methods for achieving the basic shell of the building. Specific examples of most of these are also shown in the building case studies in Chapter 10.

9.7.1 Light Wood Frame

The 2 × 4 stud wall and closely spaced joist floor and raftered roof system is firmly entrenched as the most common building system for modest-size buildings in the United States. Some typical details for the system and some of its characteristics are described in Figure 9.2.

Usage of this system for a single-family house is illustrated in Building 1 in Chapter 10. Another form of usage, for a one-story commercial building, is shown in Building 2 in Chapter 10. These are two very common uses, but the system can be used for just about any building with modest wall heights, spans, and general overall size. In addition to size limits, the other major source of limitation is that of code or zoning requirements for fire resistance. Total floor area, number of stories, type of or total number of occupants, or location in locally established fire zones may prohibit the use of this class of construction.

Variation in building appearance is mostly achieved by choice of finish materials for walls, floors, ceilings, and roofs. The basic frame system can also produce a great variety of building forms and accommodate many different windows and doors. This gives a great range for design manipulation within the basic system.

The frame itself can be modified by changes in the size or the spacing of frame members or by inclusion of heavier elements—beams or columns—within the ordinarily hollow voids of the structure. Provisions for exceptional loads or longer spans may thus be made with no apparent major modification of the basic system.

Light wood framing may also be used in combination with other systems. Buildings with structural masonry walls often have floor and roof structures of wood; if spans are modest, these will use typical light joists and rafters (see Figure 9.5). In low-rise

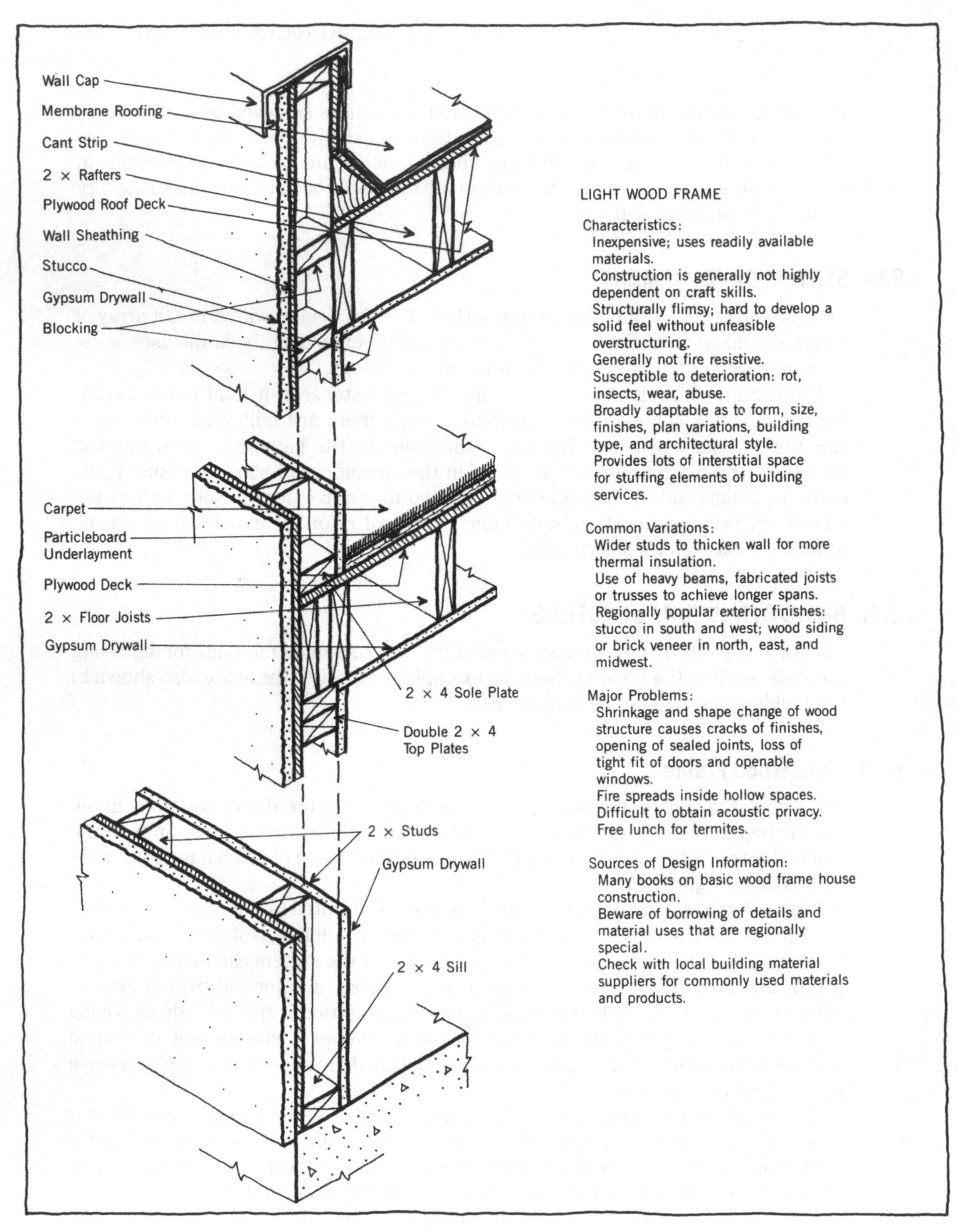

Wall Cap
Membrane Roofing
Cant Strip
2 × Rafters
Plywood Roof Deck
Wall Sheathing
Stucco
Gypsum Drywall
Blocking
Carpet
Particleboard Underlayment
Plywood Deck
2 × Floor Joists
Gypsum Drywall
2 × 4 Sole Plate
Double 2 × 4 Top Plates
2 × Studs
Gypsum Drywall
2 × 4 Sill
LIGHT WOOD FRAME
Characteristics:
Inexpensive; uses readily available materials.
Construction is generally not highly dependent on craft skills.
Structurally flimsy; hard to develop a solid feel without unfeasible overstructuring.
Generally not fire resistive.
Susceptible to deterioration: rot, insects, wear, abuse.
Broadly adaptable as to form, size, finishes, plan variations, building type, and architectural style.
Provides lots of interstitial space for stuffing elements of building services.
Common Variations:
Wider studs to thicken wall for more thermal insulation.
Use of heavy beams, fabricated joists or trusses to achieve longer spans.
Regionally popular exterior finishes: stucco in south and west; wood siding or brick veneer in north, east, and midwest.
Major Problems:
Shrinkage and shape change of wood structure causes cracks of finishes, opening of sealed joints, loss of tight fit of doors and openable windows.
Fire spreads inside hollow spaces.
Difficult to obtain acoustic privacy.
Free lunch for termites.
Sources of Design Information:
Many books on basic wood frame house construction.
Beware of borrowing of details and material uses that are regionally special.
Check with local building material suppliers for commonly used materials and products.

Figure 9.2 Light wood frame.

office buildings, where clear spans of 25 ft or more are desired, a popular solution consists of a steel column and column-line beam framework with light wood infill for the walls, roof, and floors. In fact, where fire requirements permit it, the light wood frame will typically be used for infill in just about any building, regardless of the principal structural system.

Information Sources

National Forest Products Association

American Institute for Timber Construction

9.7.2 Steel Frame

The most common steel frame building structure is one that uses I- or H-shaped rolled steel shapes (mostly W shapes) for beams and columns in a fully developed, three-dimensional arrangement (see Figure 9.3). Some details for such a system are shown in Building 6 in Chapter 10.

The range of sizes available in rolled steel shapes makes it possible to extend this basic system from the smallest buildings to high-rise structures. Most shapes are produced in a single, common grade of steel (currently ASTM A36), but higher grades are obtainable, making additional variation possible.

As with any framework assembled from individual pieces, the means used for achieving joints are a major concern. Steel frames are ordinarily assembled by direct connection of the members, using a few common types of connecting devices. Most widely used is the so-called framed connection, using a pair of angles as an intermediate device.

Connections of endless variety are possible, but are mostly used only when the ordinary, standard ones do not suffice for some reason. Reasons may include concerns for appearance when the frame is exposed to view or functional considerations, such as special loads, members not at right angles, and so on.

With or without intermediate devices, connecting is usually achieved with either welding or high-strength bolts. In general, welding is favored for connections made in the fabricating shop and bolting for those made at the building site. Welding can be done at the site, but is more reliably achieved in the controlled situation and environment in the shop. Equipment for bolting is also easier to use at the site, especially high up on the skeleton framework.

This system can achieve considerable variety in building form and size, either as the entire major structure or in conjunction with other systems. For special purposes, the ordinary straight, linear, rolled shapes can be bent or curved, or cut and welded into various shapes.

As with other frames, other surfacing elements must be used to produce walls, roofs, and floors. Depending on functional requirements, fire codes, and architectural design concerns, there is a wide range of possibilities for surfacing materials. In various situations, just about anything can be used—wood, masonry, concrete, plastics, steel, and other metals.

While the W shape is the most frequently used element, various other shapes can also be used. Other rolled shapes—channel, angle, tee, etc.—can be used, as well as pipe or tubular elements. And, if a special form is needed or a standard shape larger than those available is required, elements can be built up from steel plate or combinations of rolled shapes, using welding or bolting.

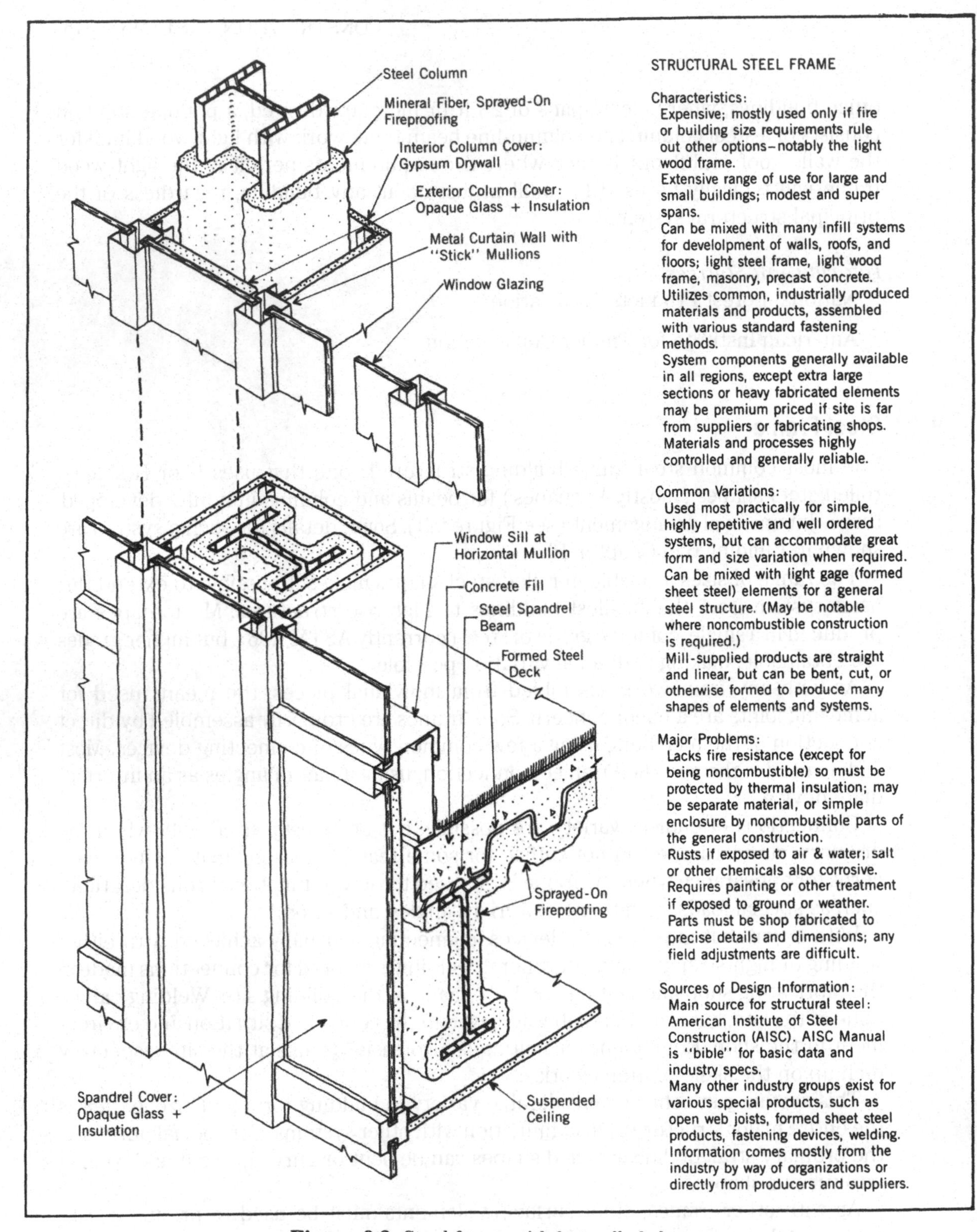

Steel Column
Mineral Fiber, Sprayed-On Fireproofing
Interior Column Cover: Gypsum Drywall
Exterior Column Cover: Opaque Glass + Insulation
Metal Curtain Wall with "Stick" Mullions
Window Glazing
Window Sill at Horizontal Mullion
Concrete Fill
Steel Spandrel Beam
Formed Steel Deck
Sprayed-On Fireproofing
Spandrel Cover: Opaque Glass + Insulation
Suspended Ceiling
STRUCTURAL STEEL FRAME
Characteristics:
Expensive; mostly used only if fire or building size requirements rule out other options–notably the light wood frame.
Extensive range of use for large and small buildings; modest and super spans.
Can be mixed with many infill systems for develolpment of walls, roofs, and floors; light steel frame, light wood frame, masonry, precast concrete.
Utilizes common, industrially produced materials and products, assembled with various standard fastening methods.
System components generally available in all regions, except extra large sections or heavy fabricated elements may be premium priced if site is far from suppliers or fabricating shops.
Materials and processes highly controlled and generally reliable.
Common Variations:
Used most practically for simple, highly repetitive and well ordered systems, but can accommodate great form and size variation when required.
Can be mixed with light gage (formed sheet steel) elements for a general steel structure. (May be notable where noncombustible construction is required.)
Mill-supplied products are straight and linear, but can be bent, cut, or otherwise formed to produce many shapes of elements and systems.
Major Problems:
Lacks fire resistance (except for being noncombustible) so must be protected by thermal insulation; may be separate material, or simple enclosure by noncombustible parts of the general construction.
Rusts if exposed to air & water; salt or other chemicals also corrosive.
Requires painting or other treatment if exposed to ground or weather.
Parts must be shop fabricated to precise details and dimensions; any field adjustments are difficult.
Sources of Design Information:
Main source for structural steel: American Institute of Steel Construction (AISC). AISC Manual is "bible" for basic data and industry specs.
Many other industry groups exist for various special products, such as open web joists, formed sheet steel products, fastening devices, welding.
Information comes mostly from the industry by way of organizations or directly from producers and suppliers.

Figure 9.3 Steel frame with hot-rolled shapes.

Major concerns include those for deterioration by rusting and vulnerability to fire. Elements must usually be protected for both these effects, either by other parts of the building construction or by specially applied protective coatings.

Information Source
American Institute of Steel Construction

9.7.3 Heavy Timber Frame

Before the light wood frame, there was the timber frame (see Figure 9.4). While the rest of the world still uses the word *timber* to describe structural wood in general, in the United States the term is reserved for lumber elements of a size beyond that normally used for the light wood frame. In that context, a timber frame begins with columns 6 × 6 or larger and beams mostly 6 in. nominal width and larger.

The larger size of the members implies a wider spacing than that in the light frame. Thus, surfacing or infill must be achieved with either a light wood frame system (a possibility) or with heavier elements. Plank decks (1.5 in. or thicker) can be used and walls are sometimes created with prefabricated panels.

Although still popular in some regions (mostly highly forested ones) or for some buildings (mostly seafood restaurants or any business whose name includes the word *warehouse*), this system has been considerably displaced by others. Good timbers in other than green lumber condition are hard to get, and the craft for handling them is vanishing.

Still, the architectural styles that use the forms of the timber structure endure in popularity, so the timber structure continues in some use, as real timber structural systems or as faked ones for appearance only.

One advantage of the timber frame is its fire resistance; it takes a bit of time to get a thick piece of wood to burn, as campers can testify. Building codes recognize this by giving some fire rating to this form of construction, which can heighten its feasibility for some situations. Used with masonry walls, for example, a reasonably fire-resistive construction can result.

The heavy timber truss is also a possibility for roof construction, usually in a gabled form. Although a real truss is possible, this may be more feasible if developed essentially as a tied gable beam system, using only the sloping roof frame members and the horizontal member. Interior members would mostly function only to support the horizontal member to keep it from sagging. There is nothing really wrong with such a structure, and many old timber "trusses" are actually of this nature.

Glued laminated members may be used for some parts of the heavy timber system, especially for very large sections and very long span beams. In addition to an unlimited size, advantages for the use of glued lamination include higher structural strength of the wood members and great dimensional stability. The latter refers to a major problem with wood in general, but with large sections especially. Warping, curling, splitting, and other dimensional changes are normal as the wood dries out from its green condition. This drying out takes a long time for thick pieces, and some of it is likely to occur after the members are installed in a building. Glued laminated members, although large as assembled, are mostly made up from nominal 2 in. elements, usually cured considerably from a green condition before being glued together.

One problem with using glued laminated members is that their dimensions do not coordinate with those of standard lumber. Multiples of the nominal 2 in. laminations

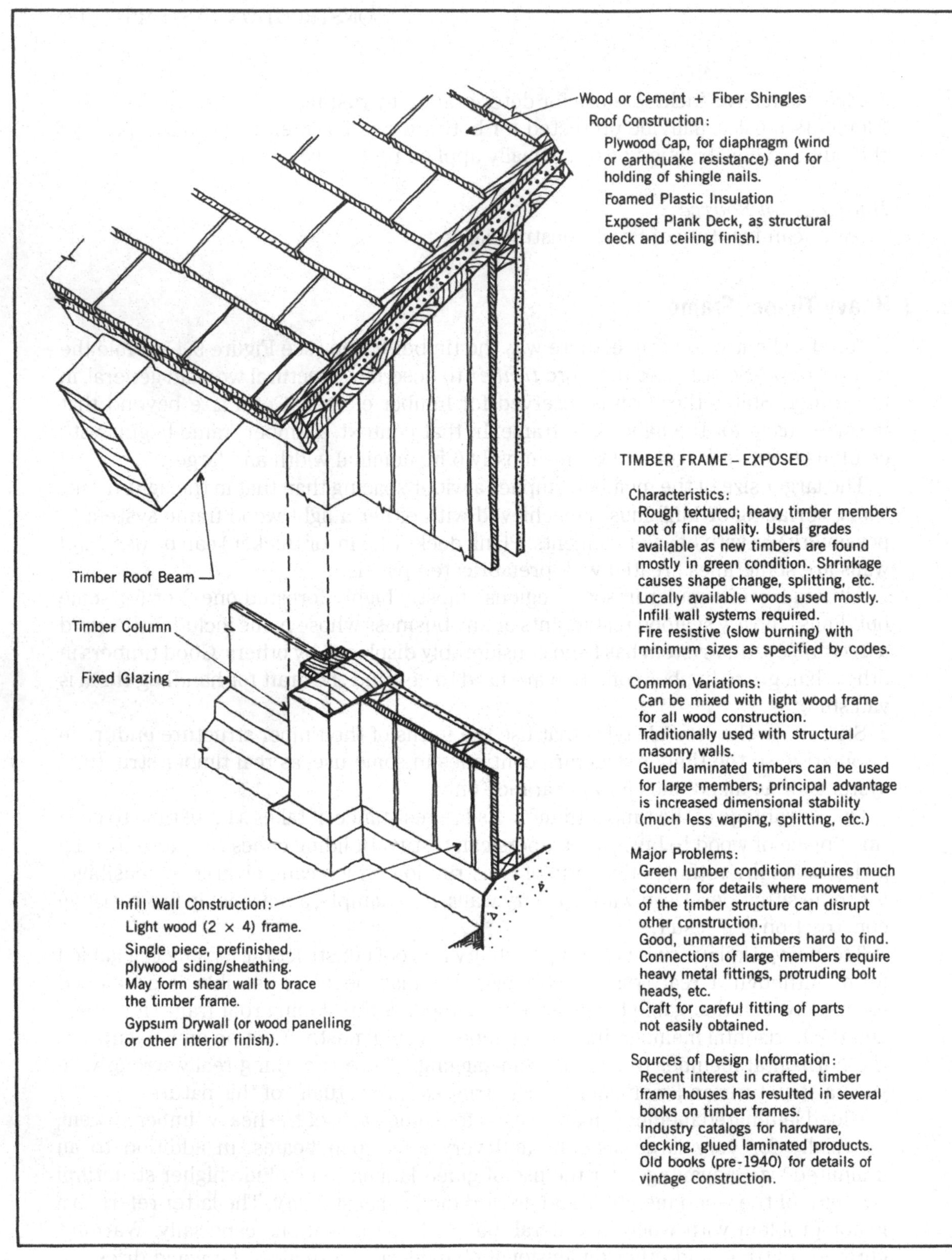

Wood or Cement + Fiber Shingles
Roof Construction:
Plywood Cap, for diaphragm (wind or earthquake resistance) and for holding of shingle nails.
Foamed Plastic Insulation
Exposed Plank Deck, as structural deck and ceiling finish.
TIMBER FRAME – EXPOSED
Characteristics:
Rough textured; heavy timber members not of fine quality. Usual grades available as new timbers are found mostly in green condition. Shrinkage causes shape change, splitting, etc.
Locally available woods used mostly.
Infill wall systems required.
Fire resistive (slow burning) with minimum sizes as specified by codes.
Common Variations:
Can be mixed with light wood frame for all wood construction.
Traditionally used with structural masonry walls.
Glued laminated timbers can be used for large members; principal advantage is increased dimensional stability (much less warping, splitting, etc.)
Major Problems:
Green lumber condition requires much concern for details where movement of the timber structure can disrupt other construction.
Good, unmarred timbers hard to find.
Connections of large members require heavy metal fittings, protruding bolt heads, etc.
Craft for careful fitting of parts not easily obtained.
Sources of Design Information:
Recent interest in crafted, timber frame houses has resulted in several books on timber frames.
Industry catalogs for hardware, decking, glued laminated products.
Old books (pre-1940) for details of vintage construction.
Timber Roof Beam
Timber Column
Fixed Glazing
Infill Wall Construction:
Light wood (2 × 4) frame.
Single piece, prefinished, plywood siding/sheathing.
May form shear wall to brace the timber frame.
Gypsum Drywall (or wood panelling or other interior finish).

Figure 9.4 Timber frame: wood structure with thick wood pieces.

produce one dimension that is a multiple of 1.5 in. The other dimension is slightly less than standard, since members are shaved to clean up the faces and produce a smoother surface. When mixed with solid sawn timbers, the connections and general detailing must be developed with this in mind.

Information Source

American Institute for Timber Construction

Many regional timber products organizations

9.7.4 Masonry Walls

As discussed in Chapter 6, masonry covers a range of materials and uses. As a structural material, masonry is used mostly for walls, with roofs, floors, and foundations achieved with other materials. The example shown in Figure 9.5 uses walls of precast concrete blocks (CMUs) resting on a concrete foundation and supporting a light wood frame roof and floor construction.

The illustration also shows the form of a type of masonry called *reinforced masonry*. This construction utilizes steel rods placed both vertically and horizontally at intervals in the wall and encased in concrete poured into the voids in the blocks. A special form of block is used for this with flat ends and only two large voids. In staggered courses (horizontal rows) this produces a vertical alignment of voids of significant size for casting what is a small reinforced concrete column inside the masonry. Notched cross walls are used in blocks to permit the horizontal rods to be placed and the horizontal reinforced concrete strips to be created.

For the reinforced masonry as shown, the reinforced concrete frame inside the wall constitutes a major contribution to the wall as a structure, making for a somewhat lower concern for the structural integrity of the concrete masonry units and mortar. Thus the two-celled block with relatively thin cell walls is used, allowing for the greatest size of the encased concrete elements. This form of construction is used most extensively in regions with high risk of windstorms or earthquakes, where the use of unreinforced structural masonry is not permitted.

Where it is permitted by codes, unreinforced masonry is still used extensively. Without the internal reinforced concrete frame, the integrity of the masonry itself becomes more critical, and a block form used is usually one with more cells of smaller size, resulting in more cross walls which are also thicker.

Applications of structural masonry walls are shown in Buildings 3 and 4 in Chapter 10.

Structural masonry—reinforced or not—can also be achieved with brick construction or walls made with combinations of brick and CMUs. Favored types of masonry construction are quite regionally determined, based on various traditions as well as primary concerns for structural properties (wind resistance versus earthquake resistance) and climate conditions (thermal expansion and freeze-thaw cycles.) Thus brick masonry is favored in the colder climates (mostly less concern for seismic response; more for rusting out of the reinforcing), while reinforced CMU construction is favored in the west and south (earthquakes and/or hurricanes; not much thermal range or freezing).

Masonry structures (mostly concrete block) can also be covered with various materials, including other masonry as veneer. One means for installing insulation is to place it on the inside surface of the wall, with rigid foam plastic units attached by

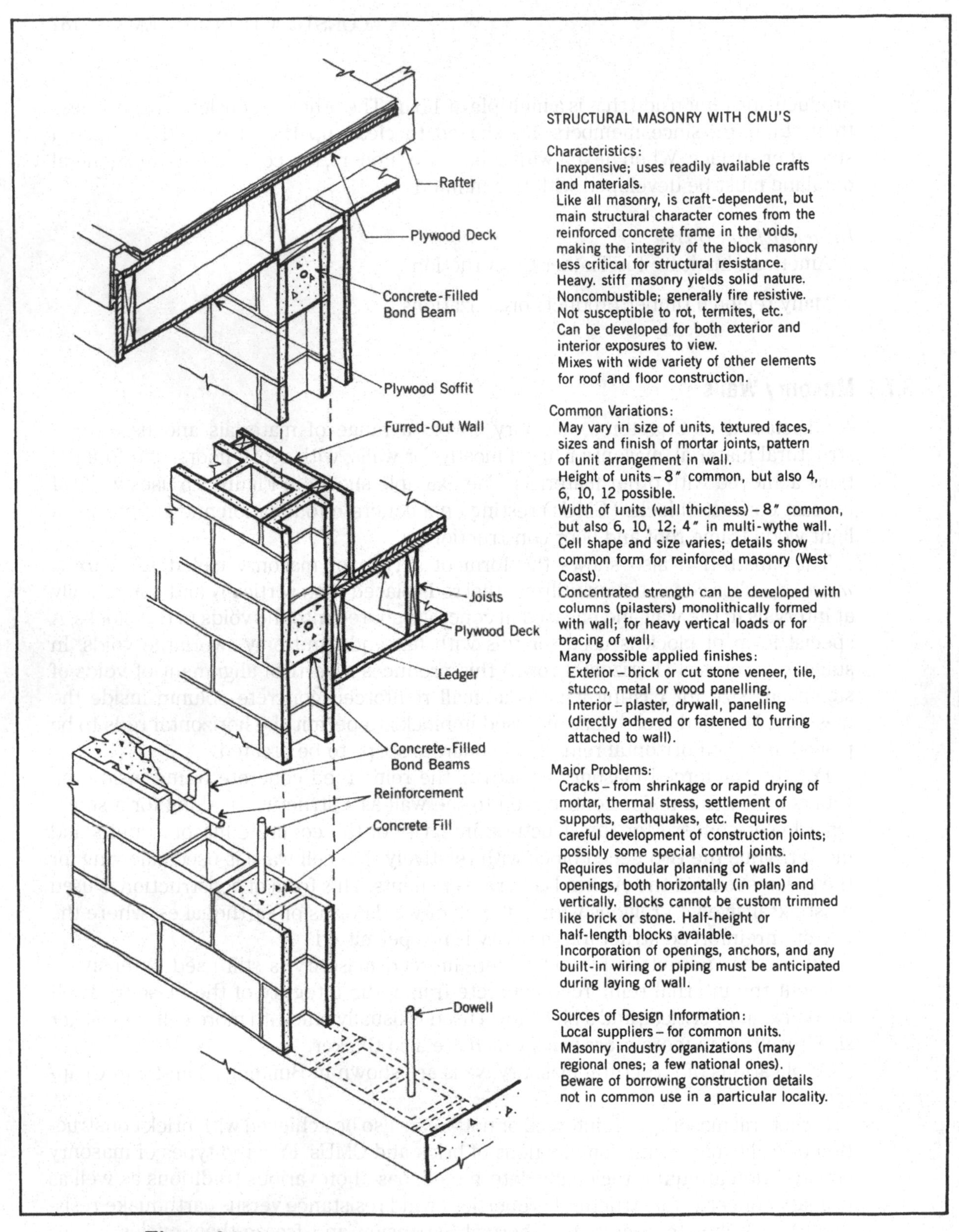

STRUCTURAL MASONRY WITH CMU'S
Characteristics:
Inexpensive; uses readily available crafts and materials.
Like all masonry, is craft-dependent, but main structural character comes from the reinforced concrete frame in the voids, making the integrity of the block masonry less critical for structural resistance.
Heavy. stiff masonry yields solid nature.
Noncombustible; generally fire resistive.
Not susceptible to rot, termites, etc.
Can be developed for both exterior and interior exposures to view.
Mixes with wide variety of other elements for roof and floor construction.
Common Variations:
May vary in size of units, textured faces, sizes and finish of mortar joints, pattern of unit arrangement in wall.
Height of units - 8" common, but also 4, 6, 10, 12 possible.
Width of units (wall thickness) - 8" common, but also 6, 10, 12; 4" in multi-wythe wall.
Cell shape and size varies; details show common form for reinforced masonry (West Coast).
Concentrated strength can be developed with columns (pilasters) monolithically formed with wall; for heavy vertical loads or for bracing of wall.
Many possible applied finishes:
Exterior - brick or cut stone veneer, tile, stucco, metal or wood panelling.
Interior - plaster, drywall, panelling (directly adhered or fastened to furring attached to wall).
Major Problems:
Cracks - from shrinkage or rapid drying of mortar, thermal stress, settlement of supports, earthquakes, etc. Requires careful development of construction joints; possibly some special control joints.
Requires modular planning of walls and openings, both horizontally (in plan) and vertically. Blocks cannot be custom trimmed like brick or stone. Half-height or half-length blocks available.
Incorporation of openings, anchors, and any built-in wiring or piping must be anticipated during laying of wall.
Sources of Design Information:
Local suppliers - for common units.
Masonry industry organizations (many regional ones; a few national ones).
Beware of borrowing construction details not in common use in a particular locality.
Rafter
Plywood Deck
Concrete-Filled Bond Beam
Plywood Soffit
Furred-Out Wall
Joists
Plywood Deck
Ledger
Concrete-Filled Bond Beams
Reinforcement
Concrete Fill
Dowell

Figure 9.5 Structural masonry with hollow concrete masonry units (CMUs, or good old concrete blocks).

adhesive, or fiber batts inside a furred-out stud space. Any form of interior paneling can then be used.

Information Sources

National Concrete Masonry Association

Brick Institute of America

American Concrete Institute

9.7.5 Sitecast Concrete

Concrete is used most extensively for foundations and pavements. However, there are many forms of concrete building systems, some of which are unique to the material. Domes, shells, folded plates and waffle slabs can be built in forms that only concrete can accomplish. The structure shown in Figure 9.6 is a basic frame of columns and beams with roof and deck systems that can be generally developed in wood or steel. What makes it unique as a concrete structure is the ability for the members of the system to be cast in continuous form rather than all produced as separate elements that need to be joined together.

In addition to eliminating the need for the extensive connections required for most frames, there are some other attributes of the concrete frame, as follows:

Natural lateral resistance: Jointed frames need special help for the resistance of lateral forces, in the form of shear walls trussing, or special connections with moment resistance. The sitecast concrete structure is naturally constituted as a rigid frame, although reinforcement must be provided in the joints for proper development of this action.

Natural fire resistance: The exposed concrete structure does not need protection for fire, although adequate concrete cover on the reinforcement is necessary for enhanced resistance.

Resistance to deterioration: Concrete is generally resistant to the effects of water, insects, fungus, and so on. Exposure to the weather and contact with soil are possible, although some extra provisions may be made, such as additional cover for reinforcement, admixtures in the concrete for enhanced resistance to freezing, or use of special coatings for water resistance.

Sitecast concrete structures are usually quite slow to erect and typically quite expensive. Their cost needs to be justified by some of the advantages to be gained from their use. Lack of need for fire or weather protection can sometimes be cost-effective by eliminating the need for additional materials to achieve these attributes.

Major economic factors for the sitecast concrete structure are typically the cost of reinforcement and the various costs involved in the casting operation. The latter include the costs for forming materials, structural supports for the forms, and the trucking, handling, depositing, finishing, curing and testing of the concrete. The total of these costs is usually greater than the cost of the concrete itself; thus design efforts to conserve the amount of concrete required or to use lower quality concrete are usually less effective than those aimed at reducing reinforcement, forming, or finishing costs. Reuse of forms for members of similar size is one common savings, making design of the fewest different member sizes or shapes a practical idea.

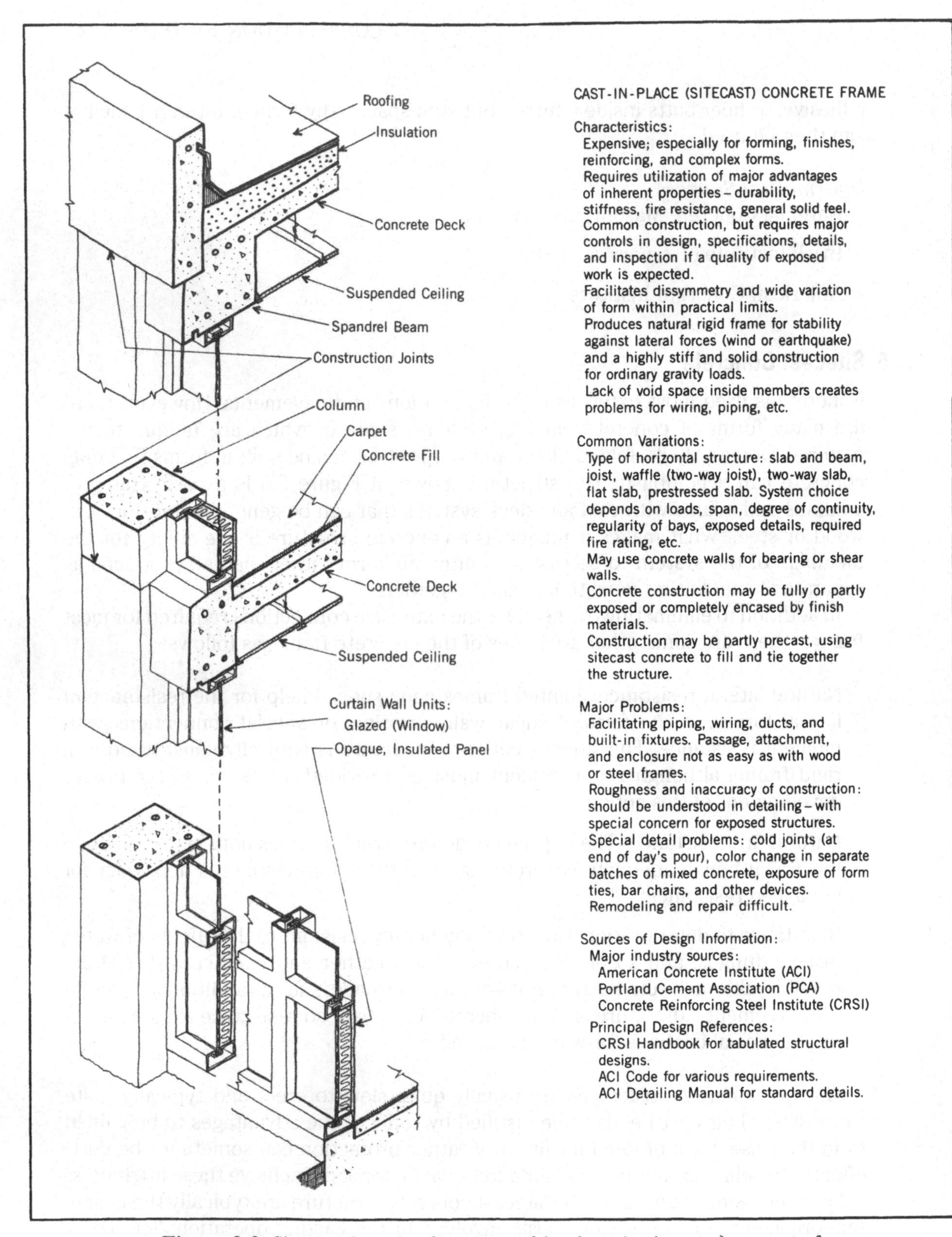

Roofing
Insulation
Concrete Deck
Suspended Ceiling
Spandrel Beam
Construction Joints
Column
Carpet
Concrete Fill
Concrete Deck
Suspended Ceiling
Curtain Wall Units:
Glazed (Window)
Opaque, Insulated Panel
CAST-IN-PLACE (SITECAST) CONCRETE FRAME
Characteristics:
Expensive; especially for forming, finishes, reinforcing, and complex forms.
Requires utilization of major advantages of inherent properties – durability, stiffness, fire resistance, general solid feel.
Common construction, but requires major controls in design, specifications, details, and inspection if a quality of exposed work is expected.
Facilitates dissymmetry and wide variation of form within practical limits.
Produces natural rigid frame for stability against lateral forces (wind or earthquake) and a highly stiff and solid construction for ordinary gravity loads.
Lack of void space inside members creates problems for wiring, piping, etc.
Common Variations:
Type of horizontal structure: slab and beam, joist, waffle (two-way joist), two-way slab, flat slab, prestressed slab. System choice depends on loads, span, degree of continuity, regularity of bays, exposed details, required fire rating, etc.
May use concrete walls for bearing or shear walls.
Concrete construction may be fully or partly exposed or completely encased by finish materials.
Construction may be partly precast, using sitecast concrete to fill and tie together the structure.
Major Problems:
Facilitating piping, wiring, ducts, and built-in fixtures. Passage, attachment, and enclosure not as easy as with wood or steel frames.
Roughness and inaccuracy of construction: should be understood in detailing – with special concern for exposed structures.
Special detail problems: cold joints (at end of day's pour), color change in separate batches of mixed concrete, exposure of form ties, bar chairs, and other devices.
Remodeling and repair difficult.
Sources of Design Information:
Major industry sources:
American Concrete Institute (ACI)
Portland Cement Association (PCA)
Concrete Reinforcing Steel Institute (CRSI)
Principal Design References:
CRSI Handbook for tabulated structural designs.
ACI Code for various requirements.
ACI Detailing Manual for standard details.

Figure 9.6 Sitecast (cast-in-place, poured-in-place, in situ, etc.) concrete frame structure.

Concrete can also be used for structural elements in a mixed system. Concrete walls, piers, or columns can be used to support other structures of different materials. Concrete slabs or framing systems can be used with supports of masonry or steel. A common situation is one in which a lower structure of concrete provides support for an upper structure of some other material—wood, steel, or masonry. The lower structure may be a basement or simply one at ground level for some use that favors the use of concrete. Low-rise apartments frequently have ground-level concrete structures for parking, where the exposed conditions in a partly open structure and the fire resistance for separation of the occupancies make the concrete a natural choice.

Information Sources

American Concrete Institute

Portland Cement Association

Concrete Reinforcing Steel Institute

9.7.6 Precast Concrete

Precasting refers to the process in which concrete elements are cast somewhere other than where they are needed, and then must be moved into place. This may occur on site, as in the base of tilt-up construction, in which large exterior wall panels are cast flat on the ground at the site and then moved into position by lifting or by simply tilting them up (see Figure 9.7).

Many precast elements, however, are produced in a factory or an outdoor casting yard at a plant that also produces the concrete. These may be custom-designed elements, but they also include standard shapes for deck units or beams. Factory casting can usually result in a higher quality product than sitecasting. Forms can be more carefully and accurately made, with steel or fiberglass for a smoother formed surface and cleaner, sharper edges and corners. Higher quality concrete can usually be produced in the more tightly supervised and better-equipped situation of the plant. Concrete design strengths may be as much as twice those for sitecast concrete, producing concrete that is stronger, denser, harder, and generally more resistant to wear or moisture penetration.

Factory production of multiple castings may also simply be a means for achieving economy by reducing the general casting cost and speeding up the work at the site. However, transportation and erection costs must be added, making the distance from the plant to the site and any problems in using heavy erection equipment at the site possible additional considerations.

Figure 9.8. shows a structure that uses a mixture of sitecast and precast elements —not an unusual situation. In this case it is sometimes possible to use the sitecast material to help effect the connection of some precast parts, reducing the cost and extra effort required to assemble the separate precast units. Joining with cast concrete may also help to restore some of the natural lateral resistance character that is essentially lost by precasting.

However, entire structures can also be precast, and members can be connected as needed for both gravity and lateral loads. The precast elements may produce only a skeleton framework or may also develop floor and roof decks and walls in panel forms.

It is also possible to produce so-called architectural concrete elements that are not major components of the building's structural system. Walls may be produced and

Figure 9.7 Tilt-up construction; walls sitecast in horizontal position and lifted into place.

used with frames of steel or sitecast concrete. At a smaller scale, individual units may be used as spandrel covers and column jackets in a curtain wall system.

Of course, the concrete block (CMU) is a precast concrete unit. Using it in the masonry structure shown in Figure 9.5, with internally developed cast concrete and reinforcement, is basically quite similar to the usage shown here with a blend of precast and sitecast elements.

Information Source
Prestressed Concrete Institute

9.7.7 Hard Surface Structures

There are generally three basic forms of structure: solid, framed, and surface. Solid structures use mass, as with large gravity dams. They are used in buildings mostly only for foundations or large piers and abutments. Framed structures are the most common form for buildings; made from framing members of wood, steel, or concrete, as described previously. Surface structures are those that use a relatively thin surface as the basic structural device (see Figure 9.9).

A general distinction is made between surfaces that are hard and can resist significant shear and compressive buckling of the surface, and those that are soft and resist basically only tension in the plane of the surface. Some examples of hard surface structures are the following:

Figure 9.8 Concrete structure that is partly precast and partly sitecast. Precast units form the exterior, exposed columns and spandrels and are filled with sitecast concrete as the sitecast horizontal structure is created.

Figure 9.9 Surface-type structure with hard surface material (concrete).

Slabs; The planar slab (plywood panel, concrete deck) is a simple surface structure that is commonly used. Slabs are usually parts of a framed structure, but are constituted as surface elements, even within the framed system. Large concrete slabs may form single structures by themselves, capable of unlimited thickness and a surface size proportionate to their thickness.

Structural walls: A planar wall is simply a vertical slab, but the vertical orientation allows for masonry as well as other materials. Walls can be framed, or—if thick enough—may approach the solid-structure category. But relatively thin walls are generally in the surface category for many considerations, structural and otherwise.

Concrete shells: Concrete can be used to form eggshell-type surfaces of various three-dimensional geometries. Shell surfaces are curved, either in single curvature—as in a cylinder—or in double curvature—as in a sphere. Double-curved surfaces are further classified as synclastic—with curvatures all in the same direction, as in a spherical dome—or anticlastic—with opposed curvatures in opposite directions, as in saddle shapes. The shell is a true surface structure only if it is relatively thin with respect to its overall size (as for an eggshell); if it is very thick, it becomes more like a solid structure (as for an igloo). Shells can be made of sheet metal, plastic, or other materials, but for building structures they are mostly concrete.

Folded plates: This is the world of origami with an endless variety of forms developed with multiple folds of flat surfaces that produce collectively some general form. A simple form is that produced by multiple parallel folds of a flat sheet into a set of parallel pleats. Some shell-like surfaces may be produced, with the folds of the surface adding some stiffness to the thin surface elements.

For construction considerations, a principal aspect of surface structures is that they are *surfaces*, thus producing enclosures as well as structures. In some situations surface structures may constitute the entire construction, exposed on the inside and outside. Such is the case for an exposed masonry wall or a plastic domed skylight. In such a case, the characteristics of the surface must be considered for all the enclosure functions as well as the structural ones. Interior appearance, exterior appearance, weather resistance, fire resistance, thermal transmission, solar effects, acoustic response—whatever is critical must be considered for a full design of the unenhanced surface.

But, of course, the surface (inside or outside) may be enhanced. Roofing, insulation, fire protection, moisture barriers, solar reflectors, and decorative or protective finishes of various kinds can be added.

Except for flat slabs, such as decks or walls, surface structures are usually of forms that create some interest but a number of potential problems. Some of these are:

Forming: Nonflat surfaces present special forming problems with any material. Doubly curved surfaces of concrete are especially challenging. Use of repetitive shapes with multiple surfaces is highly advisable.

Functional application: Floors need to be horizontally flat in most cases; nonflat geometries are usually only usable for roofs.

Roof drainage: Water flows by gravity; the nonflat surface for a roof defines a drainage pattern that must be considered—for flow on the surface as well as discharge off the surface.

Architectural effect: The structural implications of a surface must be coordinated with some concerns of the architectural effects, as the surface may dominate the building exterior—especially roofs of highl, complex, three dimensional form.

Information Source
Portland Cement Association

9.7.8 Soft Surface Structures

Soft surfaces are produced with flexible materials, such as membranes of woven fabric or sheet film (see Figure 9.10). A membrane surface has significant structural resistance only to tension in the plane of the surface. To maintain stability in a given position, the membrane surface must be made taut; that is prestretched. If adequately stretched and supported, it may resist additional forces, but only ones that can be resisted by development of additional tension in the membrane surface, in most cases. An exception is that where some compression in the surface may be developed, if it does not exceed the prestretching tension.

For buildings, soft surfaces are usually in the form of tents or pneumatically sustained surfaces. Many geometries are possible, but the forms must be honest to the equilibrium of the internal tension forces, the loading conditions, and the nature of the supports and restraints that sustain the surface. Design of the supports and restraining elements is a critical aspect of design of soft surface structures.

Figure 9.10 Surface-type structure with soft surface material (plastic film).

Soft surfaces constitute some of the lightest weight and thinnest of structures. Aspect ratios (span to thickness) for soft surfaces may easily be in the four-figure range, while those for solid slabs are typically much lower. (Plywood roof deck, 3/8 in. thick on a 24-in. span, for example: 24/0.375 = 64.) Even eggshell-like hard surfaces must have sufficient thickness to resist some degree of compressive buckling.

Natural characteristics of the soft surface establish some specific limitations for utilization of this form of construction for buildings. Concentrated (pinpoint) loads must generally be avoided, although some techniques may be used to develop them, such as local reinforcement of the surface (stitched-in patches, grommets, etc.) or spreaders that dissipate the loading to a larger surface. These are methods used to accommodate tent poles, for example.

Softness of the surface usually means it is relatively vulnerable to tearing, cutting, penetration, and other forms of destruction. Some tough fabrics or laminates can be used, but this remains a problem to some degree for most such surfaces.

The thickness of soft surfaces precludes the ability to offer much resistance to thermal flow. Heat loss in cold weather and solar heat gain in warm weather create major problems to be dealt with for building enclosures. For temporary structures—especially ones that do not effect total enclosure (awnings, canopies)—this may not be a critical concern, but for permanent construction it must be considered. Double-thick elements may be used, with open voids (as for double glazing) or some enclosed insulation; or separate surfaces, such as suspended ceilings, may be supported from the structural surface.

Pneumatically sustained surfaces require the added consideration of how the pneumatic pressure is developed and sustained to keep the surface in its pre-stretched condition. This is no small matter, and must deal with the situation of the nature of the occupancy and the problems of entering and exiting the building.

As with other structures that have a highly developed geometry, tension-sustained structures are mostly limited to roofs or total, single-unit enclosures (roof and exteriors walls in one). Issues of roof drainage and architectural applications, as discussed for the hard surface, are equally of concern here.

The tension-sustained surface element, in combination with some system for support and stretching, can be used to develop some special structures for other than permanent installation. Portable, rapidly deployable and demountable, temporary structures can be created in a range of sizes and forms for various purposes. Such structures have wide application for recreation, special events, or emergency shelter purposes.

9.7.9 Curtain Walls

The term *curtain wall* may be extended to include all exterior wall construction that does not involve structural walls. That is, the wall exists only as a wall providing enclosure for the building. Thus a light wood frame wall, or even a masonry wall, if used only as an infill with a frame structure, is technically a curtain wall.

However, the term is most often reserved to describe the form of construction developed for multistory buildings with frame structures. (See Figure 9.11.) As such, the curtain wall takes various forms, a common one being a system that has the following components:

A support system, consisting of the attachments of the wall to the building frame—most often to spandrel beams or the edge of roof and floor decks.

A major wall structure system, which may take various forms relating to the form of the construction of the wall. Typical examples are systems that use continuous vertical mullions (called a "stick" system) and ones that are developed with prefabricated panel units.

Secondary and accessory elements to develop the general system. For the "stick" system, there are many additional items required to complete the wall; for a panel system, the only additional items may be those required to join the panels and seal the wall surface.

Glazing, since most walls have windows.

Opaque infill or covering of the portions of the wall not occupied by windows.

A curtain wall for an individual building may be completely custom designed; and indeed such walls mostly were for many of the early buildings to use this system. Now, however, many of the actual components are likely to come directly from the catalogs of industry suppliers, many of whom have complete "systems" with the necessary elements to achieve all the necessary construction: structural mullions; heads, jambs, and sills; finish trim; joint sealing and glazing support materials; and so on.

In fact, whole walls may be essentially developed from catalog elements, with design work limited to choices of colors, mullion spacing dimensions, and attachments for the frame; and choices for glazing and opaque infill in the frame. However, for large building projects, suppliers can usually accommodate some amount of "customizing" of their products in the form of minor alterations.

Many manufacturers of curtain wall materials have product lines that include a range of different systems and components, responding to various situations of usage (low-rise, high-rise, storefront, etc.) and some range of architectural styles. Details for Buildings 6, 7 and 8 in Chapter 10 illustrate three different systems, all from a single manufacturer and supplier.

Because of their extensive use in large, tall buildings, curtain walls get a lot of attention, and their many problems get a lot of play in professional and technical journals. When the system for some large, prominent building develops a major flaw, necessary corrections may involve considerable wall surface and possibly a major liability case stretching over many years.

Some problems are enduring—having been encountered in early buildings—and still begging for good solutions. Others have been developed more recently, having to do with new concerns (energy use and pedestrian safety, for example). A few of the critical concerns are the following:

Thermal expansion: With walls of considerable extent—both vertically and horizontally—movements due to thermal change are critical: obviously they are more

a

Figure 9.11 Development of metal-framed curtain walls: (a) Metal stud system encasing the whole exterior. (b) Glass and metal window wall formed in one-story segments between the levels of the structure.

b

Figure 9.11 Continued.

serious when the outdoor temperature range is great. Assemblage of the construction typically involves a mixture of materials, each with its own coefficient of thermal expansion. How to hold the wall together, but still let it move as it wants to without damage is no small problem.

Water: This includes leaks of precipitation into the construction or through it into the building and condensation of moisture within the construction.

Structural movements: The curtain wall assemblage must tolerate the movements of the structure that supports it, including vertical deflection of spandrels and lateral movements from wind or seismic effects.

Glass breakage: This is first of all a safety problem if glass falls out of the frame—onto occupants if it falls inward, or onto pedestrians if it falls outward. Also, when glazing needs replacing, the frame must be partly disassembled, new glazing units installed, and the construction fully restored.

Thermal flow and sunlight control: The nature of these problems depends on climate, building orientation, and occupant needs. At issue for one is energy consumption, for which major concerns may be thermal flow, solar heat gain, or use of daylight for interior illumination. Sun conditions are different on north-facing walls, south-facing walls, and so on. It is nice to have sunlight at times, but seldom in a full-glare situation. It is hard to get the sunlight for illumination without the accompanying undesirable heat gain during warm weather. Insulation may be used for opaque wall surfaces, but large window areas are tough to make really thermally resistive.

Some of these concerns are discussed for the example buildings (2, 4, 6, and 7) in Chapter 10.

When columns are placed at the building perimeter (a common occurrence), their relationship to the building skin wall has a great bearing on the exterior appearance as well as on interior planning. Figure 9.12 shows four possible positions for the columns and spandrel beams relative to the exterior wall plane.

Although freestanding columns (Figure 9.12(a)) are the least desirable for interior planning, they might be tolerated if they are small (as in a low rise building) and are of an unobtrusive shape and size. In some cases they can be treated as features of the design. In framed structures of wood or steel the cantilevered edge of the horizontal framing usually presents a clumsy problem. In sitecast concrete structures, however, the cantilever is simply achieved and may even be an advantage in that it reduces span effects for interior beams and aids the transfer of forces to the columns.

Placing the columns totally outside the wall (Figure 9.12(d)) eliminates both the interior planning lump and the need for the cantilevered edge. A continuous exterior ledge is produced that can be used as a sun shield, for window washing, for balconies, or for an exterior circulation space. However, unless some such use justifies it, the ledge may be a nuisance in creating water runoff and dirt accumulation problems. The totally exterior column and spandrel also creates a potential problem with thermal expansion differences.

A common situation is that shown in Figure 9.12(b), where the skin wall is placed just at the face of the columns or incorporates them. With large columns this results in a lump at the inside wall surface, but permits a smooth building exterior surface. If the wall plane is moved in from the column's exterior face, the column lump moves steadily outside and may eventually produce a smooth interior wall surface.

The photographs in Figure 9.13 show three buildings with varying relationships for the positions of the skin wall and exterior columns. In Figure 9.13(a) the columns are just inside the wall, permitting a smooth exterior face with no real acknowledgment of the positions of the columns or the spandrels. The building in Figure 9.11(a) presents a variation of this, with a metal stud wall built essentially outside the building's frame structure.

In Figure 9.13(b) the wall plane is just at the interior face of the columns, with the

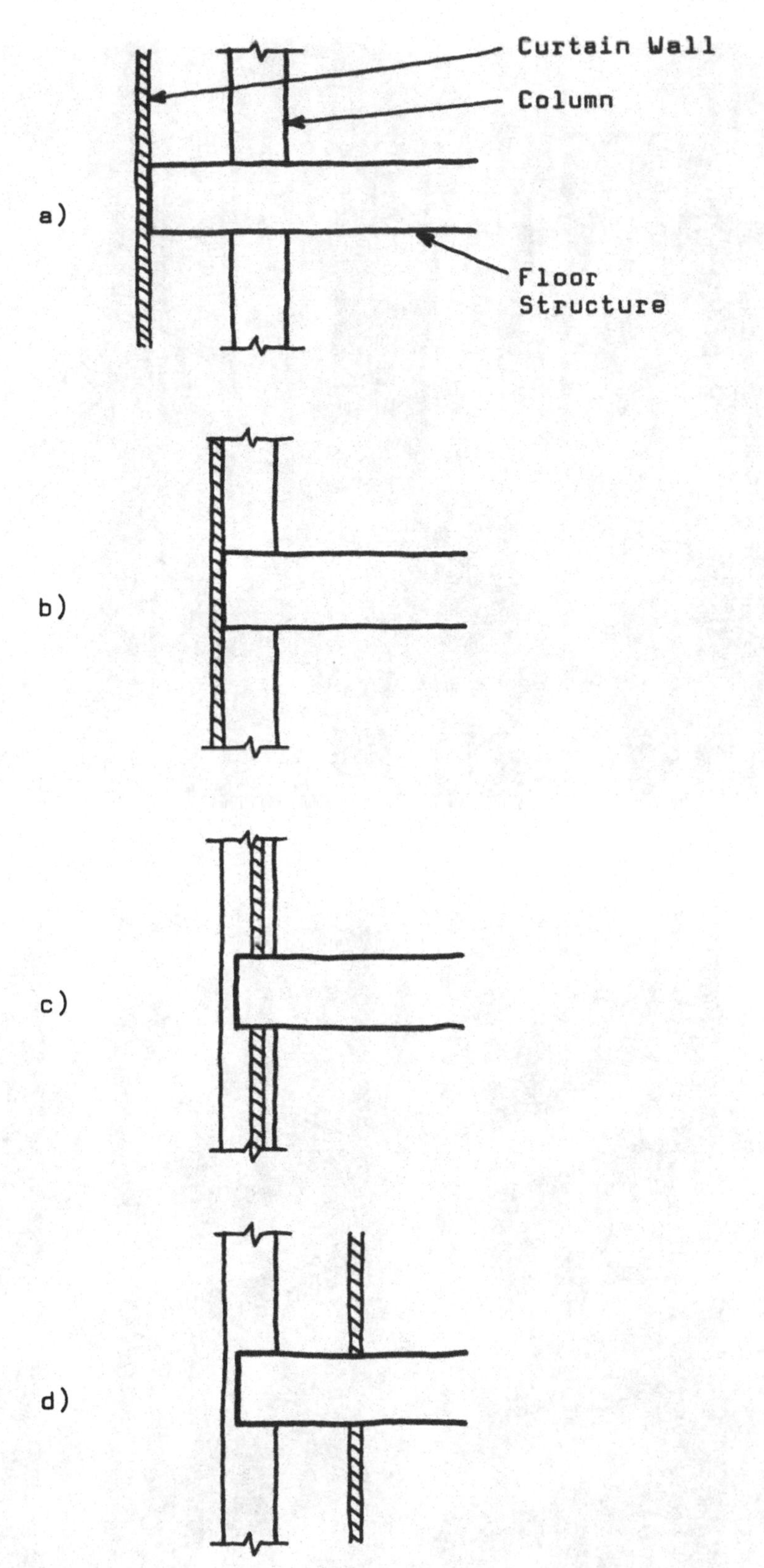

Figure 9.12 Relation of location of the curtain wall with respect to the building structure.

a

b

Figure 9.13 Positions of curtain walls: (a) Outside the structure. (b) In the same plane as the structure. (c) Inside the plane of the structure.

c

Figure 9.13 Continued.

columns and spandrels revealed on the exterior. The curtain wall in this case is built in the form of units that fit into the four-edged plane defined by the columns and the tops and undersides of the spandrels.

In the building shown in Figure 9.13(c) the building's structural edge and the columns are fully freestanding on the exterior. In this case the curtain wall is essentially constructed as a one-story wall in each story level—a much simpler system in general from that shown in Figure 9.13(a) and (b). Figure 9.11(b) shows another example of the freestanding exterior column system.

Information Sources

Wall Systems: Analysis by Detail, by Herman Sand, McGraw-Hill, 1986

The Building Envelope: Applications of New Technology Cladding, by Alan J. Brookes and Chris Grech, Butterworth, 1990.

9.7.10 Package Systems

Buildings consist mostly of individual collections of many separate parts, the parts coming from many sources. There are, however, some companies in the business of producing larger units, such as whole wall panels—or even whole buildings, possibly including the HVAC, lighting, and other service systems.

Buying packages, where they are appropriate, can reduce design time, simplify construction, and make some costs more easy to predict. Performance may be more predictable and reliable for an industrially produced package, subject to product testing and the quality control in a factory.

Packages can exist in a variety of forms. The following are some examples:

Windows: Most individual windows are bought as package units, with the window frame, operating sash, glazing, hardware, and possibly screens, storm sash, and blinds assembled for single placement in the wall structure of the building.

Roofing: Various priority roofing exists as a system, with all components provided as a package, possibly including some flashing and expansion joint materials. Single-ply membranes are often marketed as a whole system in this manner. So are some forms of metal roofing.

Curtain walls: Storefront and curtain wall systems are usually complete packages with framing, trim, seals, connectors, and other needed parts supplied in a kit by the manufacturer. (See examples in Buildings 2, 4, 6, and 7 in Chapter 10.) The package frame may accommodate various glazing and spandrel cover materials, or it may also use ones that the manufacturer provides with the frame.

Exterior wall finishes: Many kinds of exterior wall treatments are available in packages, such as preformed metal panel systems, exterior stone veneer systems, and multilayered insulated systems with acrylic finishes.

Metal building systems: Many prefabricated metal structural systems or whole building systems with metal skin walls are manufactured and sold as package systems. These are mostly used for industrial or agricultural buildings, or other utilitarian functions (see Figure 9.14). These range from small, backyard, home owner-assembled sheds to large buildings with longspan structures. Portions of packages can also be used to develop major parts of otherwise custom-designed buildings, reducing design costs and tightening construction cost control.

Whole buildings can also be prefabricated or simply factory assembled. The mobile home industry produces a significant amount of low-cost housing in the usual actual "mobile" form, or simply as large component units for site assembly.

Packaging in general is a growing concept in building, and more and bigger parts of buildings are likely to become mostly used as packages in the future.

9.8 SYSTEM DESIGN CONCERNS

Design of any system with many interacting parts requires a number of basic considerations. For building systems in particular, the following are some major concerns.

9.8.1 Integration of Construction

A great deal of the work done in design and research tends to focus on selected elements or issues, rather than on buildings as whole systems. The building industry

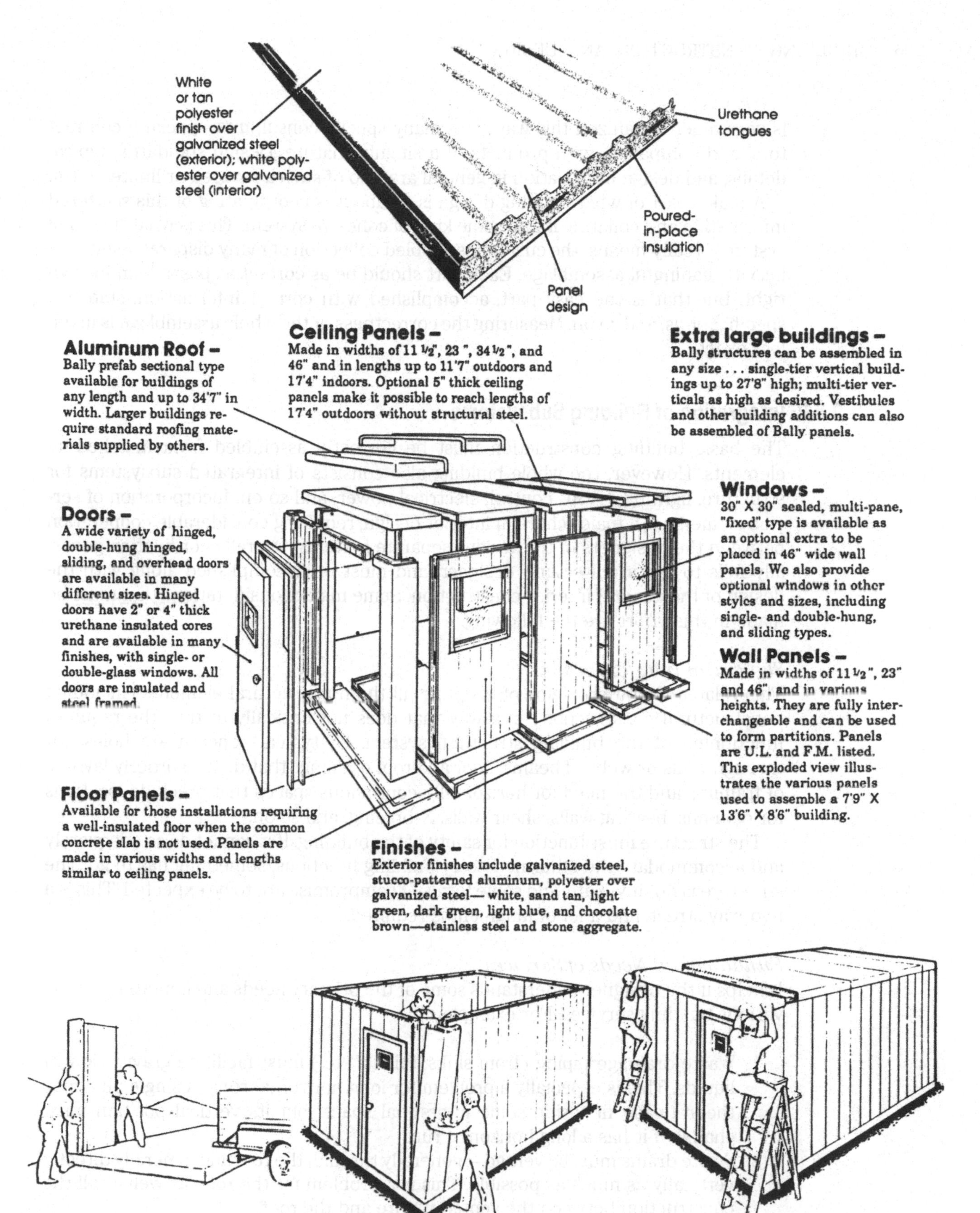

White or tan polyester finish over galvanized steel (exterior); white polyester over galvanized steel (interior)
Urethane tongues
Poured-in-place insulation
Panel design
Ceiling Panels – Made in widths of 11 1/2", 23", 34 1/2", and 46" and in lengths up to 11'7" outdoors and 17'4" indoors. Optional 5" thick ceiling panels make it possible to reach lengths of 17'4" outdoors without structural steel.
Aluminum Roof – Bally prefab sectional type available for buildings of any length and up to 34'7" in width. Larger buildings require standard roofing materials supplied by others.
Extra large buildings – Bally structures can be assembled in any size . . . single-tier vertical buildings up to 27'8" high; multi-tier verticals as high as desired. Vestibules and other building additions can also be assembled of Bally panels.
Doors – A wide variety of hinged, double-hung hinged, sliding, and overhead doors are available in many different sizes. Hinged doors have 2" or 4" thick urethane insulated cores and are available in many finishes, with single- or double-glass windows. All doors are insulated and steel framed.
Windows – 30" X 30" sealed, multi-pane, "fixed" type is available as an optional extra to be placed in 46" wide wall panels. We also provide optional windows in other styles and sizes, including single- and double-hung, and sliding types.
Wall Panels – Made in widths of 11 1/2", 23" and 46", and in various heights. They are fully interchangeable and can be used to form partitions. Panels are U.L. and F.M. listed. This exploded view illustrates the various panels used to assemble a 7'9" X 13'6" X 8'6" building.
Floor Panels – Available for those installations requiring a well-insulated floor when the common concrete slab is not used. Panels are made in various widths and lengths similar to ceiling panels.
Finishes – Exterior finishes include galvanized steel, stucco-patterned aluminum, polyester over galvanized steel— white, sand tan, light green, dark green, light blue, and chocolate brown—stainless steel and stone aggregate.

Figure 9.14 Priority prefabricated metal building system utilizing light-gauge formed sheet steel elements and custom-formed steel frame units. Basically produced for industrial, agricultural, and general utilitarian use as an economical, quickly erected, bare-bones shelter. (Courtesy of Bally Engineered Structures, Bally, PA.)

is very much organized this way, with many special consultants, specialty contractors, and companies with products of a singular nature. Design standards, typical details, and design information in general are also of such a focused or limited form.

A major part of what building design is all about is coordinating of this scattered information and collating it into some kind of cohesive system. This is what "building systems" really means: the custom-assembled collection of many disparate elements into a meaningful assemblage. Each part should be as correct as possible in its own right; but that is the easy part, accomplished with correct information, standard specifications, and so on. Measuring the correctness of the whole assemblage is much less exact.

9.8.2 Integration of Building Subsystems

The basic building construction must be correctly assembled from matched-up elements. However, the whole building also consists of integrated subsystems for structure, lighting, HVAC control, electrical power, and so on. Incorporation of service elements is a major chore in interior design, requiring considerable cooperation between the designers of the various separate systems. Overall coordination generally falls to the architectural designer, and must be accomplished along with the design of the general interior construction. Some major considerations that must be made in this effort are the following.

Structural Interference

This relates to the problems of inserting all the nonstructural elements (including nonstructural construction) in a way that does not critically disturb the required functioning of the building structural system. Of typical concern are holes cut through walls or webs of beams, floor and roof openings that disturb orderly layouts of framing, and the need for horizontally-continuous spaces that preclude locations for columns, bearing walls, shear walls, X-bracing, and so on.

The structure must function for safety of the building. If it cannot do so adequately and accommodate the other necessary building functions, something has to go. The structure is not inviolate, and some forms of compromise are to be expected. This is a two-way street, and a lot of judgment is required.

Fundamental Needs of Services

It helps if the designer understands some of the primary needs and limitations of the various building services. For example:

1. Waste drainage piping (from sinks, toilets, etc.) must facilitate gravity flow of liquids. This is especially a problem for long horizontal runs. A single pipe may theoretically fit inside some interstitial space, but its vertical position must change if it has a long horizontal run.
2. Waste drains must be vented—generally through the roof and generally directly vertically as much as possible. This is a problem for the roof, as well as all the construction between the vented fixture and the roof.
3. Hot water pipes and heating ducts get hot; chilled water pipes and air conditioning (cooled air) ducts get cold. Surrounding construction can be heated or cooled. Thermal expansion and contractions can create problems. Chilled objects can sweat and create moisture problems.
4. Electrical wiring must frequently be altered and various other elements should

be accessible for adjustments, repairs, alterations, and so on. Embedding elements in the permanent construction can create many problems.

5. Equipment with moving parts—notably motors, fans, pumps, and chillers—create both noise and direct, physical vibrations that can be annoying. They should be isolated; possibly by good planning of their locations, but also by use of vibration isolators, sound insulated partitions, or other special construction.

As noted previously, compromise is the name of the game. Ideally, all building systems should operate or exist with optimal conditions, but economic feasibility, technical limitations, and some priority of design values and goals must be recognized.

9.8.3 Facilitation of Modification

A peculiar requirement of modern buildings is their general need to facilitate modification. The dynamic nature of our society makes this imperative. People relocate regularly; businesses reorganize; the urban fabric continuously changes—growing, decaying, regrouping; and rapid changes in technology bring many sudden demands for facilitation of new opportunities.

Playing against this rapidly shifting scene is the great expense of building costs, which makes the writing off of initial construction over a long period mandatory for most investors. The net effect is that buildings must be built for permanence but somehow bc easily, feasibly, and preferably rapidly modifiable.

Ease of modification is a design goal and involves both basic planning and the final decisions for choices of materials and details of the construction. It also needs to permeate the whole design effort, including that of the structure and all the building services where possible. This is relatively easy for some things, such as wiring, ceilings, and plumbing fixtures; not so easy for sewers, foundations, and shear walls.

Designing for special requirements—such as ease of modification—may require extra effort and extra cost of the construction. In some ways, however, it merely means picking between competing alternatives with a view to those that facilitate ease of change. Bolted joints can be undone, for example, while glued or welded ones cannot.

10

Case Studies

The purpose of this chapter is to present a number of examples of building types and a range of construction usages and to illustrate the development of the general construction for each example. The major intent is to present construction in terms of whole systems for specific building cases. Design of the whole system thus becomes the dominant concern, and the variability of individual elements can be viewed in the context of their role in the general construction.

The presentations here are meant to compliment the materials in preceding chapters. Critical isolated issues are dealt with more extensively in those chapters, and that material should be used as a resource for the more detailed information on individual items shown in these examples.

However, building designers must ordinarily face the problem of the whole building, even though the actual design may be worked out by concentrating on one part at a time. Here, the presentations begin with consideration of the whole building, followed by presentations of some of the primary increments of the construction with all the significant parts displayed.

The building designs presented here are not meant as illustrations of superior architecture. The interest here is directed mostly to setting up reasonably realistic situations that cover a range of circumstances in regard to design of construction. It is not the intention to present boring, ugly buildings, but the use of a limited number of examples to display a broad range of construction use is the dominant concern.

An attempt has been made to show construction usage and the form of details that are reasonably correct. This is, however, a matter for much individual judgment and is often tempered by time and location. There are not many situations in which only one way is correct and all others are wrong.

Reasonably correct alternatives are often possible, and the particular circumstances for an individual building can make a specific choice better on different occasions. Differences in climate, building codes, local markets, building experience, or community values can produce a way of doing things that is preferred. Over time, changes in any of those areas of influence will produce variety and hopefully some evolution toward a better way of doing things.

▪10.1▪ BUILDING 1

Single-Family Residence
Light Wood Frame
Sloping Site

Building 1 is a split-level, single family residence; an all-American suburban house (see Figure 10.1). The light wood frame (Sec. 9.7.1) is extensively used for this type of building in all parts of the United States. The long endurance of the timber industry, the continued availability of stands of timber in North America, and the highly developed means for product distribution keep this method of construction in a position of dominance, as discussed in Chapter 6.

Cape Cod cottages used this basic system hundreds of years ago and are still imitated as a style in housing developments. The style may stick with us, but the products and details of the construction are considerably changed. What does not change much over time or region is the basic structure and general form of the light wood frame system—using studs, joists, rafters, sheathing, and decking.

Other than exterior and interior cosmetics, what varies mostly are items that relate to local and regional concerns. Roofing and exterior finishes vary in this regard, as do the types of window construction, need for insulation, and general level of concern for the weather tightness of the enclosure systems.

The construction shown here is mostly typical and with minor modifications could be built just about anywhere in the United States.

Notes on Figures 10.2 through 10.5 identify basic elements of the construction and, in some cases, also mention basic issues and possible variations. The discussion that follows relates to the illustrations, but generally considers the basic building type.

Basic Structure

The light wood frame is still predominant in use for this type. The 2 × 4 stud, 2 × 6, 8, 10, or 12 joists and rafters, and the 4-ft by 8-ft panel size for sheathing and decking set a modular system of dimensioning in place. Frame spacings of 12, 16, 24, 32, 48, and 96 in. are thus obvious and the 3.5-in.-wide void space in walls is typical. Some dimensioning off of the basic module is to be expected, but the pressure to reduce waste makes use of the popular, readily available units of materials important. A major cost factor—as for all building construction—is the need to reduce to a minimum the need for on-site labor, particularly of the skilled crafts.

Despite the modular dimensions of standard products, this structural system has immense potential for the accommodating of variations of building form, size, and detail. The materials are relatively easy to work with—on or off the site—and custom forming can be done virtually without limit. This adds considerably to the enduring popularity of this building type with designers and builders.

There are, of course, other possible structures for this type of building. The light wood frame can be emulated in steel, using light-gage elements. Use of steel offers the principal advantage of reducing the mass of combustible material in the building. This may be a desirable feature in some situations, or an actual building or zoning code requirement in others. It is hardly ever, however, an economic advantage; nor does it make the design or construction easier or faster to achieve.

In times past, the wood structure was more frequently developed as a heavy timber frame, often with infilling utilizing elements of light wood framing. While the elegantly crafted, exposed wood frame can be very attractive and dramatic in appearance, it is not economically competitive for most ordinary buildings. The lack of available craft

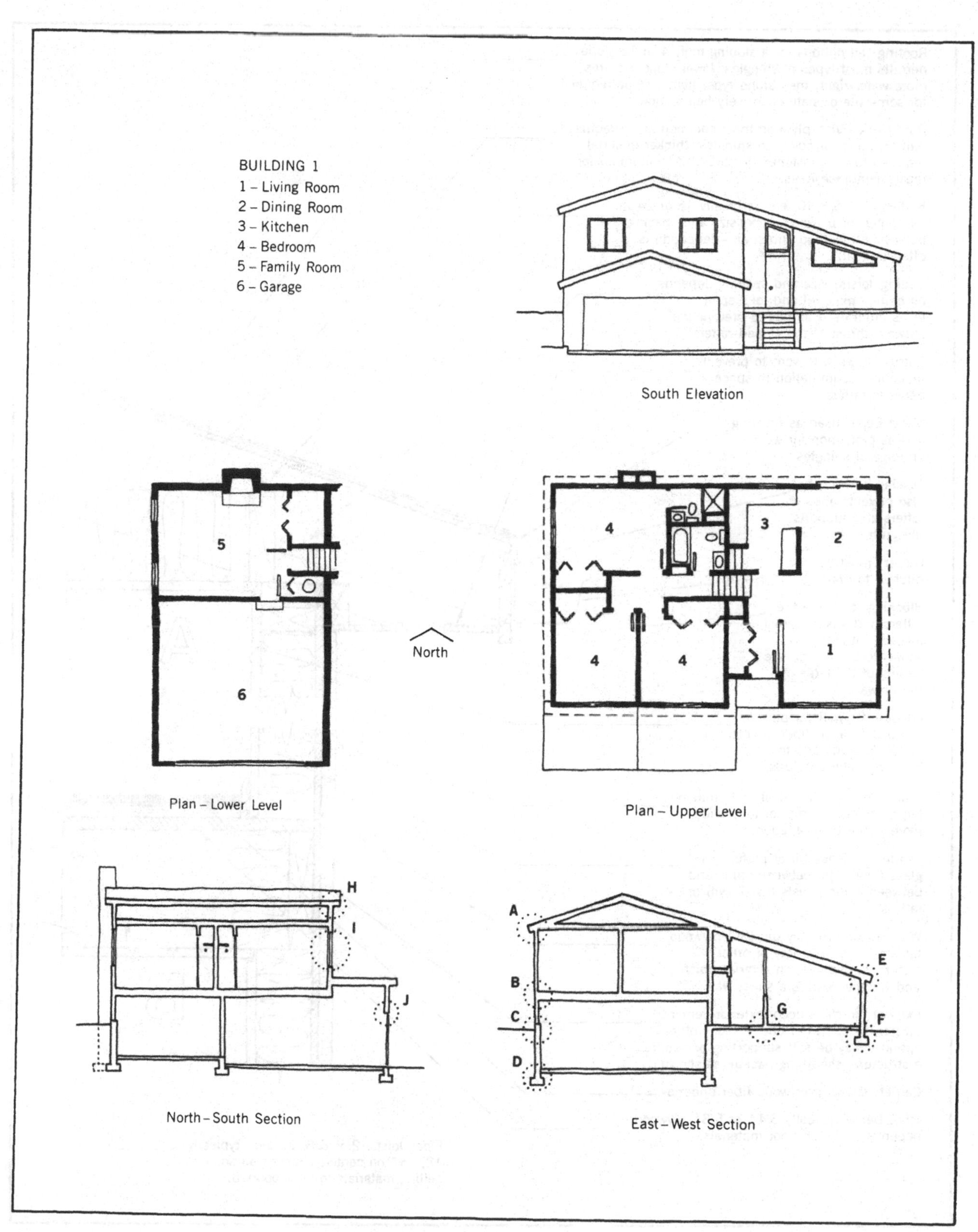

Figure 10.1 Building 1.

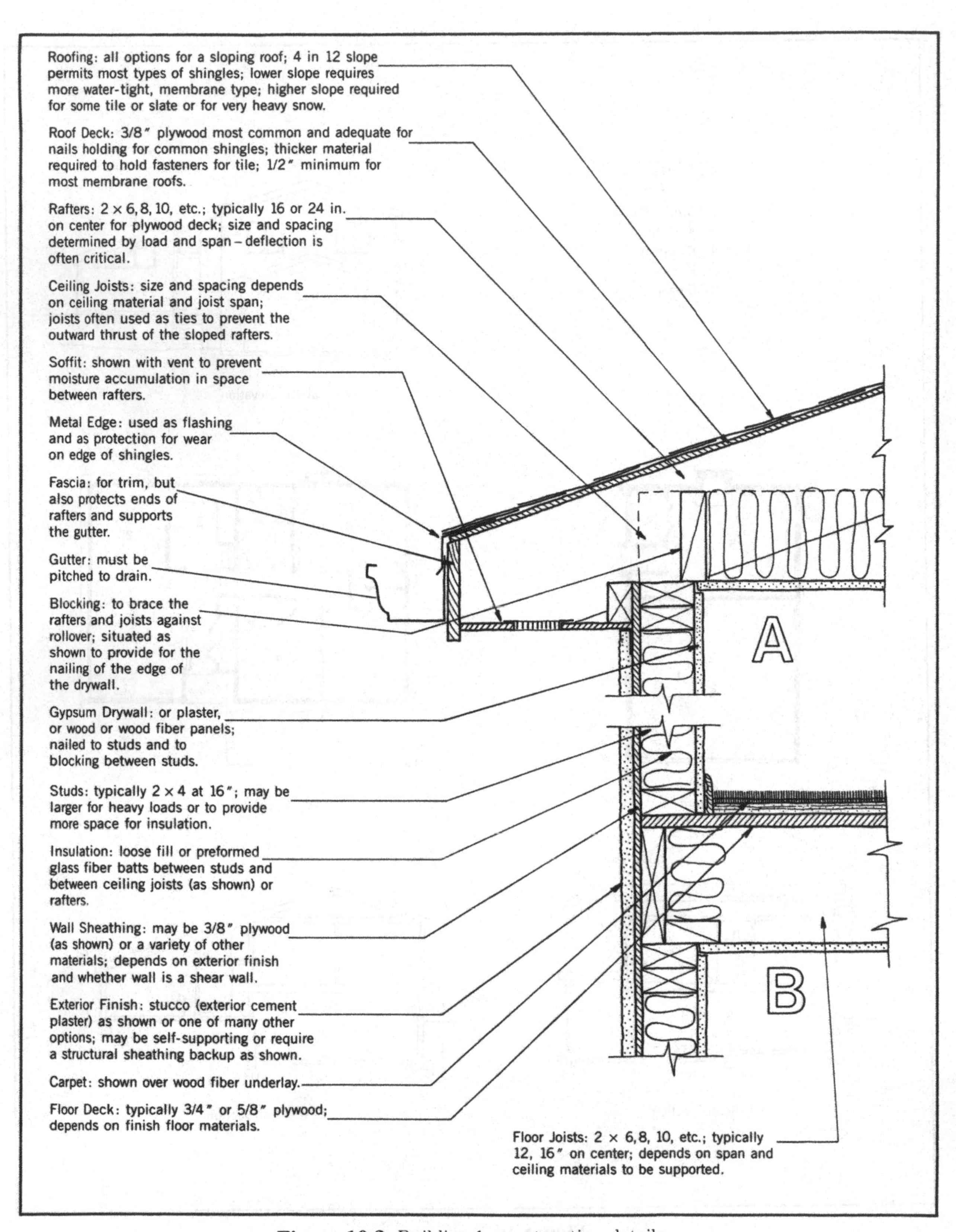

Roofing: all options for a sloping roof; 4 in 12 slope permits most types of shingles; lower slope requires more water-tight, membrane type; higher slope required for some tile or slate or for very heavy snow.
Roof Deck: 3/8" plywood most common and adequate for nails holding for common shingles; thicker material required to hold fasteners for tile; 1/2" minimum for most membrane roofs.
Rafters: 2 × 6, 8, 10, etc.; typically 16 or 24 in. on center for plywood deck; size and spacing determined by load and span – deflection is often critical.
Ceiling Joists: size and spacing depends on ceiling material and joist span; joists often used as ties to prevent the outward thrust of the sloped rafters.
Soffit: shown with vent to prevent moisture accumulation in space between rafters.
Metal Edge: used as flashing and as protection for wear on edge of shingles.
Fascia: for trim, but also protects ends of rafters and supports the gutter.
Gutter: must be pitched to drain.
Blocking: to brace the rafters and joists against rollover; situated as shown to provide for the nailing of the edge of the drywall.
Gypsum Drywall: or plaster, or wood or wood fiber panels; nailed to studs and to blocking between studs.
Studs: typically 2 × 4 at 16"; may be larger for heavy loads or to provide more space for insulation.
Insulation: loose fill or preformed glass fiber batts between studs and between ceiling joists (as shown) or rafters.
Wall Sheathing: may be 3/8" plywood (as shown) or a variety of other materials; depends on exterior finish and whether wall is a shear wall.
Exterior Finish: stucco (exterior cement plaster) as shown or one of many other options; may be self-supporting or require a structural sheathing backup as shown.
Carpet: shown over wood fiber underlay.
Floor Deck: typically 3/4" or 5/8" plywood; depends on finish floor materials.
A
B
Floor Joists: 2 × 6, 8, 10, etc.; typically 12, 16" on center; depends on span and ceiling materials to be supported.

Figure 10.2 Building 1: construction details.

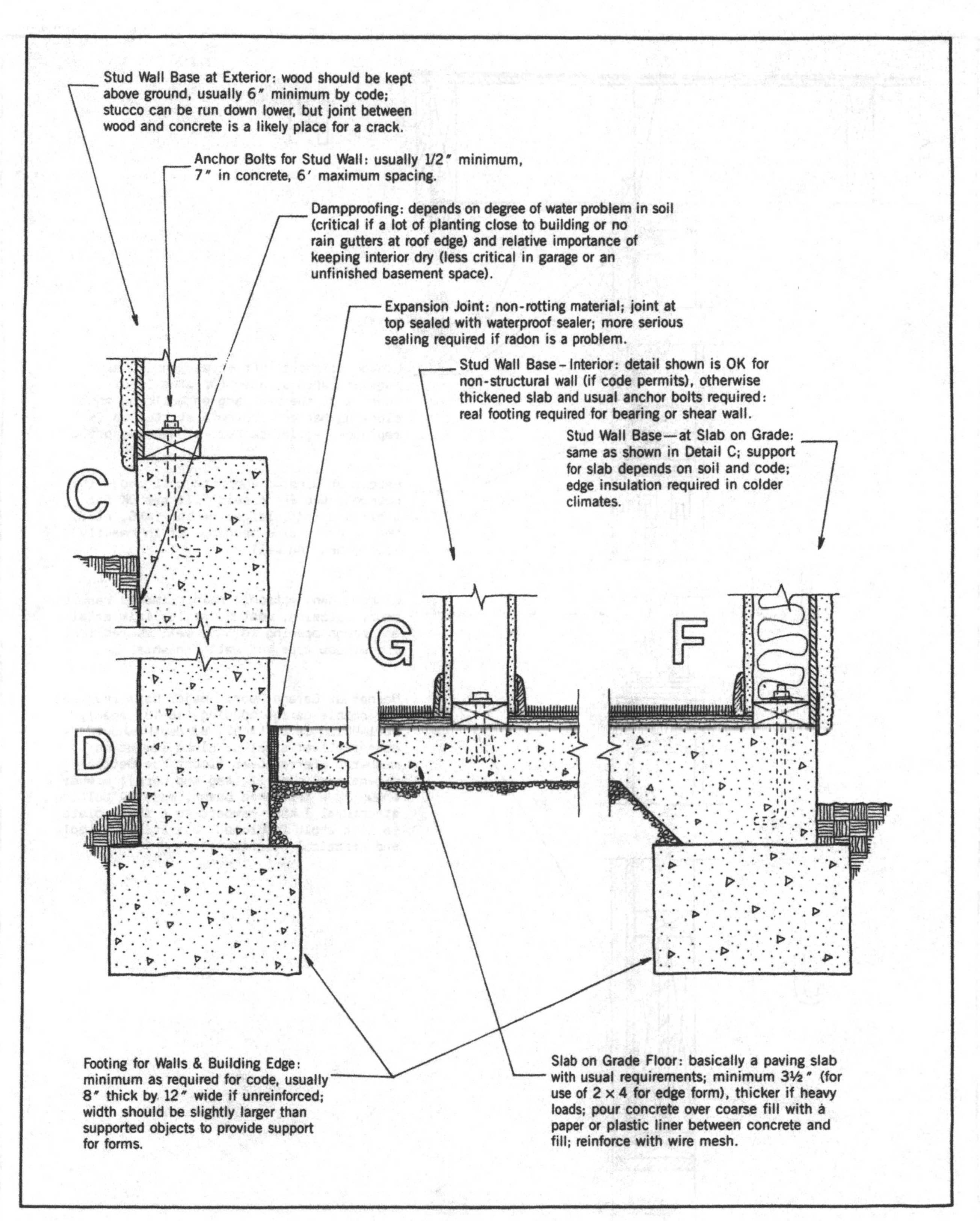

Stud Wall Base at Exterior: wood should be kept above ground, usually 6" minimum by code; stucco can be run down lower, but joint between wood and concrete is a likely place for a crack.
Anchor Bolts for Stud Wall: usually 1/2" minimum, 7" in concrete, 6' maximum spacing.
Dampproofing: depends on degree of water problem in soil (critical if a lot of planting close to building or no rain gutters at roof edge) and relative importance of keeping interior dry (less critical in garage or an unfinished basement space).
Expansion Joint: non-rotting material; joint at top sealed with waterproof sealer; more serious sealing required if radon is a problem.
Stud Wall Base - Interior: detail shown is OK for non-structural wall (if code permits), otherwise thickened slab and usual anchor bolts required: real footing required for bearing or shear wall.
Stud Wall Base—at Slab on Grade: same as shown in Detail C; support for slab depends on soil and code; edge insulation required in colder climates.
C
D
G
F
Footing for Walls & Building Edge: minimum as required for code, usually 8" thick by 12" wide if unreinforced; width should be slightly larger than supported objects to provide support for forms.
Slab on Grade Floor: basically a paving slab with usual requirements; minimum 3½" (for use of 2 × 4 for edge form), thicker if heavy loads; pour concrete over coarse fill with a paper or plastic liner between concrete and fill; reinforce with wire mesh.

Figure 10.3 Building 1: construction details.

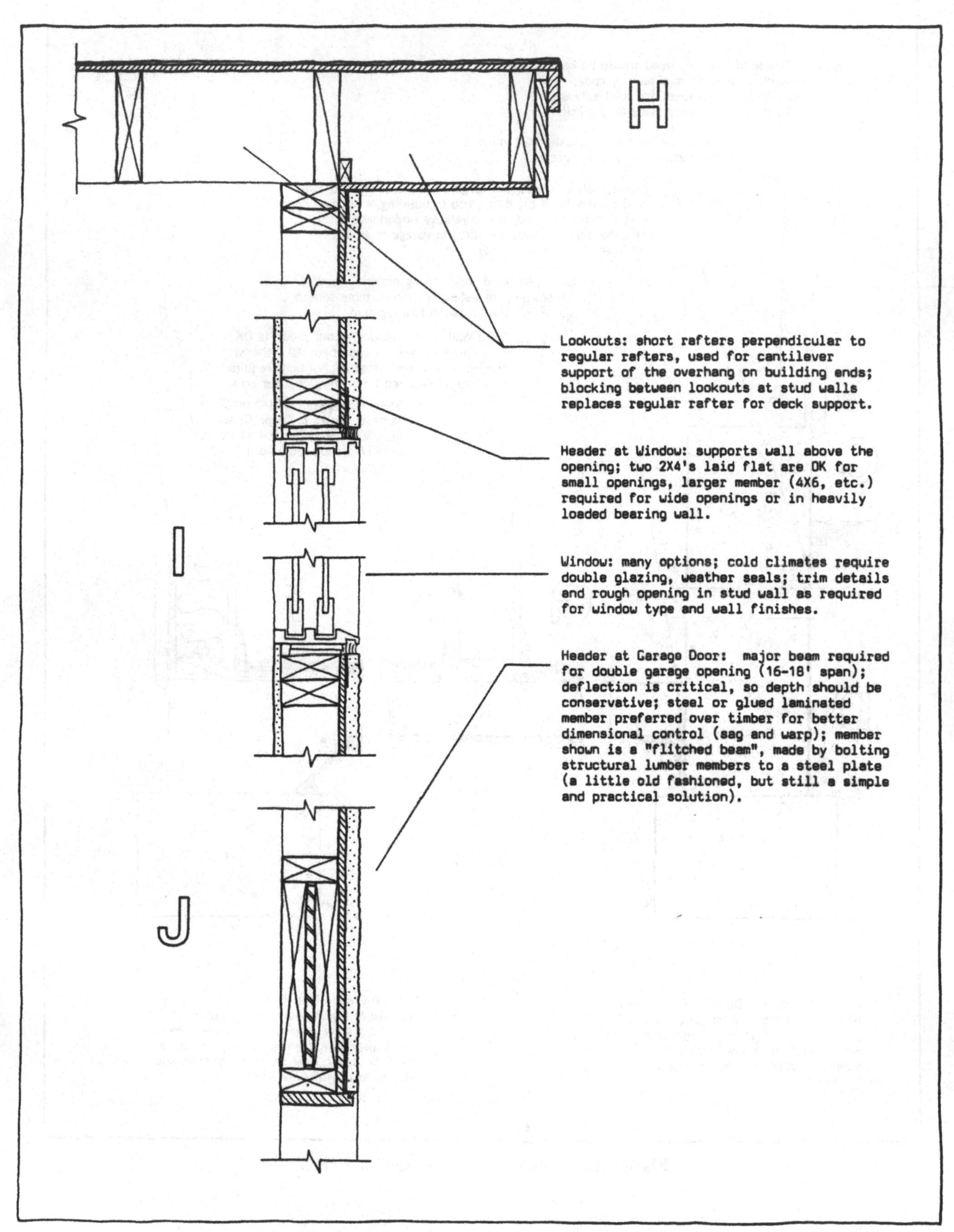

H
I
J
Lookouts: short rafters perpendicular to regular rafters, used for cantilever support of the overhang on building ends; blocking between lookouts at stud walls replaces regular rafter for deck support.
Header at Window: supports wall above the opening; two 2X4's laid flat are OK for small openings, larger member (4X6, etc.) required for wide openings or in heavily loaded bearing wall.
Window: many options; cold climates require double glazing, weather seals; trim details and rough opening in stud wall as required for window type and wall finishes.
Header at Garage Door: major beam required for double garage opening (16-18' span); deflection is critical, so depth should be conservative; steel or glued laminated member preferred over timber for better dimensional control (sag and warp); member shown is a "flitched beam", made by bolting structural lumber members to a steel plate (a little old fashioned, but still a simple and practical solution).

Figure 10.4 Building 1: construction details.

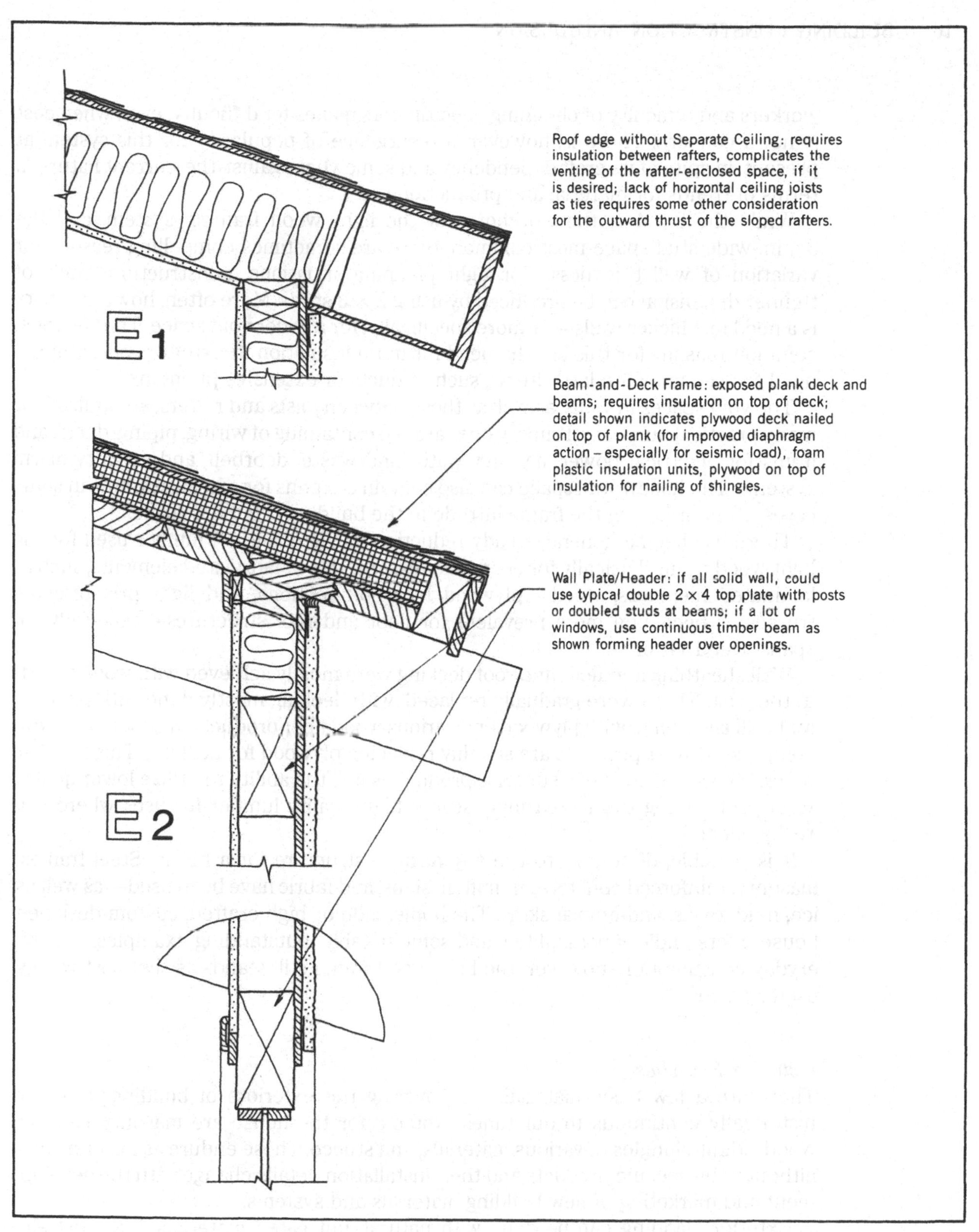

Roof edge without Separate Ceiling: requires insulation between rafters, complicates the venting of the rafter-enclosed space, if it is desired; lack of horizontal ceiling joists as ties requires some other consideration for the outward thrust of the sloped rafters.
E1
Beam-and-Deck Frame: exposed plank deck and beams; requires insulation on top of deck; detail shown indicates plywood deck nailed on top of plank (for improved diaphragm action – especially for seismic load), foam plastic insulation units, plywood on top of insulation for nailing of shingles.
Wall Plate/Header: if all solid wall, could use typical double 2 × 4 top plate with posts or doubled studs at beams; if a lot of windows, use continuous timber beam as shown forming header over openings.
E2

Figure 10.5 Building 1: construction details.

workers and difficulty of obtaining good timbers makes for difficulty, even when cost is not a concern. There is, however, a resurgence of popularity for this system at present, in spite of its craft dependency and somewhat against-the-current nature in terms of trends for material and production use.

While the 2 × 4 is the workhorse of the light wood framing system, and the 3.5-in.-wide stud space most common, there are sometimes compelling reasons for variation of wall thickness. For tight planning situations, nonstructural walls of thinner dimension can be produced by using 2 × 3 studs. More often, however, there is a need for thicker walls—or more specifically, for a wider void space. The two most common reasons for this are the need for more insulation in exterior walls and the need to accommodate large items, such as ducts or extensive plumbing.

The void spaces in walls, as well as those between joists and rafters, are utilized for many practical purposes. Primary ones are the containing of wiring, piping, ducts, and items for electrical power, lighting, water and waste, doorbell, and security alarm systems. The hollow wall space can also contain columns for a heavier frame in some cases, without having the frame intrude in the building occupied space.

There has been a general, steady reduction in the quality of lumber used for the light wood frame, basically for economic reasons. Use of fabricated elements, such as laminated members, wood + plywood I or box sections, and light prefabricated trusses, is becoming more prevalent for floor and roof structures—especially for spans over 15 ft or so.

Wall sheathing and floor and roof decking were mostly achieved with wood boards in the past. These were gradually replaced, with decking mostly done with plywood and wall sheathing, with plywood and various wood fiber products in panel form. Now compressed-fiber products are steadily replacing plywood for decking. This is not so much the ascendancy of a superior product as it is the ability to utilize lower quality wood, conserving the increasingly scarce high-quality lumber for uses where it is really needed.

It is possible, of course, to use any form of structure for a house. Steel frames, masonry, reinforced concrete, aluminum skins, and fabric have been used—as well as ice, mud, twigs, and animal skins. The homemade or high crafted, custom-designed house offers endless possibilities and some notable, outstanding examples. For everyday consumption, however, the light wood frame still stands as the most widely used system.

Exterior Finishes

There are a few basic materials for covering the exteriors of buildings that are historically continuous to our times. Notable for the house are masonry veneers, wood siding, shingles of various materials, and stucco. These endure as major means, although the specific products and their installation details change with the development and marketing of new building materials and systems.

Exterior cladding can be done with natural, untreated materials, but most elements used to surface the exteriors of buildings now get some form of coating. Finish treatments may be factory applied or site applied; may be single coat or multiple coat; and may be long enduring or need frequent renewal or recoating.

Good old wood siding can be imitated with other materials: vinyl, aluminum, fiberglass, or coated steel. Brick veneers can be produced with real bricks and mortar, or with thin tiles adhered with mastic and joints caulked. Spanish tile roofing can be done with real clay tiles, but is more often done with "tiles" of fiber concrete or various compounds that have less weight and more structural resistance to cracking than the clay units. Name just about any "real" product or material and there is an imitator that most likely outperforms the original.

Stucco—as shown in the details for Building 1—is still a popular material, mostly in the south and on the west coast in the United States. Extreme cold weather and freeze-thaw cycles do not treat it well. However, much that now passes visually for stucco is really a plastic-coated, fiber-reinforced, sandwich system with foamed plastic insulation and a backup of some sheathing element or structural masonry. In extreme climates, this is usually very superior to real stucco.

Appearance and practical performance over time are obvious major concerns for exterior finishes. However, choices are also strongly affected by local availability, cost (including installation), product and manufacturer reliability, and practical concerns for installation (available crafts, specialty contractors, etc). As the potential owners of new homes move steadily up the national economic status level, the idea of owner-maintained buildings becomes less viable. Doctors, lawyers, dentists, and yuppies in general are not likely to repaint the wood trim and siding every few years.

Roofing

As discussed in Sec. 5.6, there are two very distinct problems in roofing—the fast-draining surface and the slow-draining surface, with an intermediate condition that has features of both. The roof shown for Building 1 is definitely fast draining, so this discussion is limited to that problem.

Typically, the simplest, cheapest solution to the fast-draining surface is the shingle. The cheapest shingles are usually the bituminous, felt, fiberglass, or sheet form type, available in a wide variety of form, texture, and finish appearance in general. Wood shingles were quite popular until the popular conception of their vulnerability to fire—justified or not—brought other choices forward.

There are minimum angles of slope required for shingles, tiles, and various treatments for fast-draining surfaces. Some of these are established by building codes or recommended by manufacturers, while others are simply long-standing "good practice." All shingles require a significant slope, a bare minimum usually being 2-1/2 in 12, or approximately 12 degrees. That low a slope, however, is usually considered more in the intermediate class, and 4 in 12 or so is a more comfortable slope for most shingles.

Tile, of real clay or otherwise, usually requires at least a 6 in 12 slope. Real clay tile is one of the heaviest roofing systems, approaching 15 psf in deadweight in most installations. It is somewhat ironic that this is such a popular roofing system in California, where placing great weight on the roof presents a major seismic concern.

For the building designer, a fast-draining roof typically presents a situation in which the roof surface is highly visible. Thus, while there are many serious concerns for the physical character of the roofing, appearance is a major issue. This applies not only to the general roof surface, but to all the edges, ridges, valleys, and the general overall roof system. Careful detailing of the various discontinuities of the roof is critical to both its good appearance and its good performance. Leaks occur most frequently at points of discontinuity—ridges, valleys, intersections, and so on.

Sheet metal roofing is an option for both the fast-draining and intermediate-slope roof, and offers a very attractive and secure roofing surface. Its cost is generally high, so it is less often used for houses, although it may be used for only a portion of a roof and achieve some major design emphasis as an architectural feature.

One of the most difficult items to handle in roof design is the control of location and form of the various roof penetrations that are required for plumbing vents, heater vents, chimneys, skylights, roof ventilators, and the general array of roof "junk." This is typically not handled well with regard to the architectural design and also presents a heightened concern for potential leaks. The necessity of these items should be carefully determined and the details for achieving them carefully controlled if a good roof is to be obtained—both visually and for good construction.

Interior Construction and Finishes

From the point of view of construction, the development of the interior of this building is considerably predictable. Interior form, room arrangements, and finish materials can vary endlessly, but the basic construction and the general pallet of materials is about as common as the 2 × 4 stud.

Development of the interior begins with the consideration of the overall construction as determined by the basic shell of the building enclosure. This presents the following:

Underside of the roof structure with sloping rafters and the plywood deck.

Inside of the exterior walls with the studs and the exterior wall sheathing.

Insides of below-grade walls; most likely, of sitecast concrete or CMU construction (concrete blocks).

Top of the concrete paving slabs.

None of this construction is likely to be very desirable in an exposed condition, with the following possible exceptions. Concrete pavements may be acceptable in the basement and garage spaces, if the concrete surface is finished to a very smooth, hard quality. Insides of below-grade walls may be acceptable in the garage; less likely in the basement when the space is really occupied, not just for storage and equipment. The exposed stud construction may be acceptable in the garage.

Mostly, however, the basic structure needs to be covered for development of interior spaces. For the roof, a first question is whether the sloping form is acceptable for a ceiling. If so, the ceiling may be developed as shown for the living room in the section in Figure 10.1 and the details in Figure 10.5. The minimum construction in this case would be the application of drywall (as shown) or some other surfacing, such as wood panelling or lath and plaster. An alternative here would be to use the timber frame and plank deck as an exposed structure, as shown in Detail E2 in Figure 10.5.

With the ceiling surfacing applied directly to the bottom of the rafters, the enclosed void space exists only between rafters. If ceiling-mounted lighting is used in this area, it may be necessary to run wiring perpendicular to the rafters; a common situation, but one that does require drilling holes through a number of rafters.

If an all-air, central HVAC system is used, it would be difficult to deliver air to the far side (east side) of the living room and dining room with the ceiling as shown. Ducts might be placed beneath the concrete pavement, but this is expensive construction. There are many options in this situation, but it is one consideration to be made in developing the interior design.

The high, sloping ceiling may be desirable in the living, entry, and dining areas, but questionable in the small kitchen—a trap for stale air, for one. In this case, a lower ceiling could be framed separately to the kitchen walls.

In the bedrooms in the upper level, it would also be possible to have the sloping ceiling, as in the living room. However, the sections in Figure 10.1 and the details show the use of a flat ceiling with an attic space above it. The roof and ceiling here could be framed with prefabricated trusses, with the top chords forming the roof and the bottom chords the ceiling. The same trusses could also be used in the front portion of the garage.

A major structural problem here is the achieving of a clear span across the garage. This is easy for the roofed-over portion in the front part of the garage, but the rear part is beneath the bedrooms, as shown in Figure 10.1. A possible solution would be to use a large beam beneath the wall at the front of the bedrooms, spanning across the garage to posts in the garage walls. Then relatively short floor joists could be used between this beam and the back wall of the garage.

The garage is shown with its floor at some distance below the room behind it. This may be developed in the site design, but also relates to a code requirement to have the floor lower so that gas fumes that lay at floor level do not creep into the occupied space through the door. Code requirements must also be satisfied for fire separation between the garage and the rest of the house, affecting the details of wall and floor construction and choice of the door materials.

Except for the baths, kitchen, and front entry area, the most commonly used floor finish these days is carpet, although many other options are possible. For the construction, the only concern is the anticipation of required thickness allowances. If thick tiles or solid wood flooring is used, the structure must be sufficiently depressed in these areas.

A problem that sometimes occurs is where flooring materials change in the same level. For example, suppose that carpet is used in the living and dining rooms, vinyl tile or linoleum in the kitchen, and thick, ceramic tile in the entry way. Keeping the wearing surfaces of these materials at the same level means the accommodating of three different thicknesses of flooring materials. Not to complicate the installation of the concrete pavement—nor preclude possible future changes of flooring—the top of the concrete should be set at the lowest level to accommodate the thickest anticipated materials. Then fillers can be used to build up to the necessary level for thinner materials. Details for this should follow recommendations of the suppliers of the flooring materials.

Special Interior Construction

The details indicate very ordinary solutions for the construction of the major portions of the house. There are also a number of special areas that must be considered, and various options for their construction.

Stairs. These may be built with ordinary wood framing and wood steps, with the necessary provisions of openings in the floor framing and through the walls as required. The steps to the lower level and to the garage, however, also sit on the concrete structure, and some options are possible for their construction of concrete, in conjunction with the general construction of the foundations, basement walls, and floor pavements. Possible details of this construction, showing some options, are given in Figure 10.6.

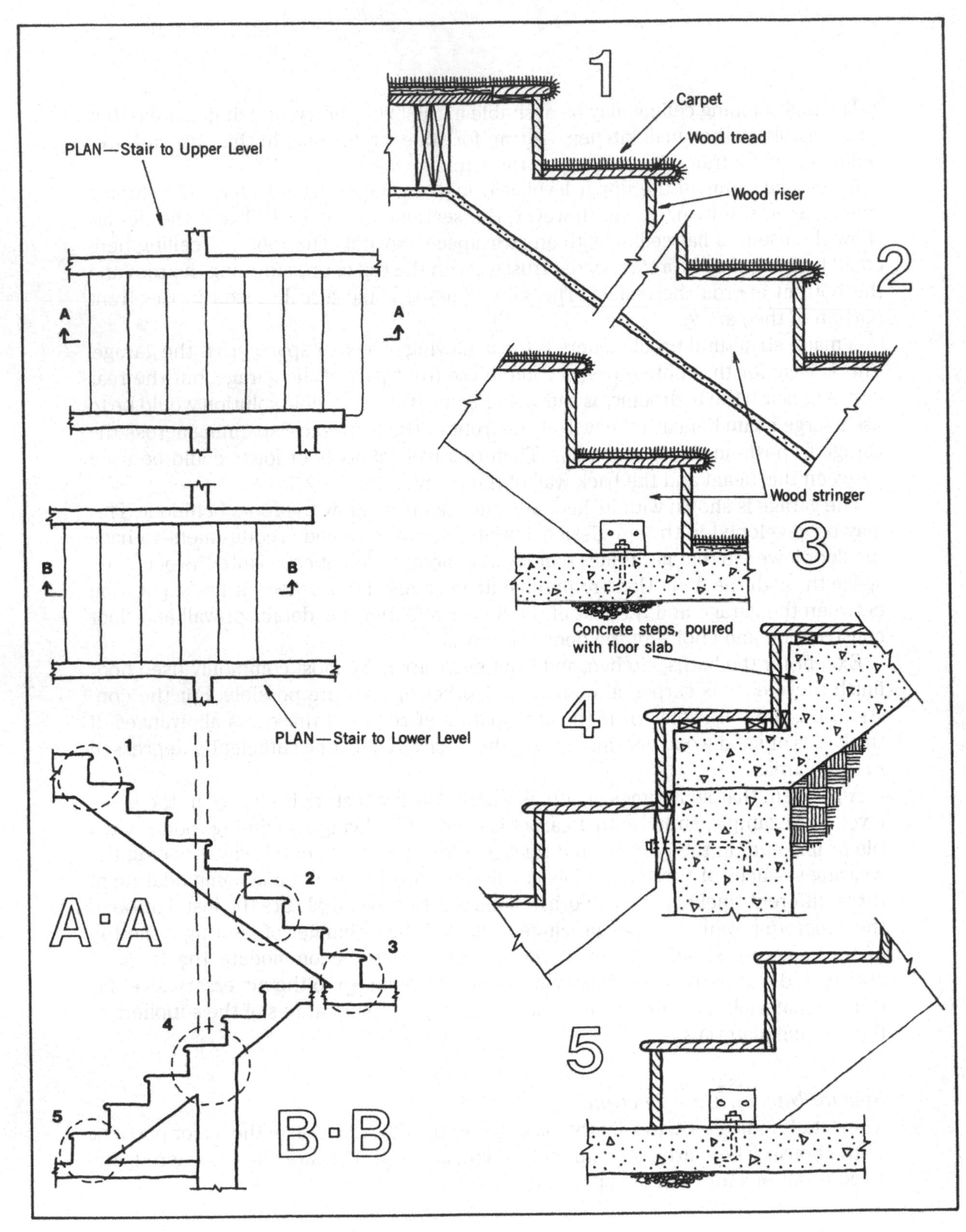

Figure 10.6 Building 1: stairs.

Fireplace. The fireplace, as shown in Figure 10.7, is developed as an all masonry structure in a very traditional manner. The chimney accommodates two flues—one for the fireplace inside the house, and one for the barbecue on the rear terrace. While this construction is still possible, it is very expensive, and prefabricated units are now available to achieve the fireplace itself and the chimney. If the prefabricated elements are used, a masonry shell could still be built around them, but would be much simpler in construction.

Kitchen. Although the image of the kitchen as a major activity center is fading (what with eating out, microwave ovens, two-career families, etc.), it is still likely that some basic elements would be provided here. These may include provisions for built-in cabinets and countertop units, built-in ovens and ranges, sinks, and an exhaust fan. The construction may accommodate these in various ways, but probably with little basic alteration of the system. In this building, the locations of the kitchen, baths, and lower storage room offer some possibilities which are developed in the details in Figure 10.8. This is all very tight planning and requires careful study for simplification of the construction and economizing of space.

Bathrooms. Figure 10.9 shows some details for the construction at the bathrooms in the upper level. This is a location of considerable plumbing, for three different services: cold water supply, hot water supply, and waste water (sewer) piping. Installation of the piping and the fixtures requires various considerations for the plumbing, including the following:

Space within the voids of walls and floors for the pipes and some portions of recessed fixtures.

Allowance for drainage of horizontal piping runs, especially for the sewer lines.

Provision for extension of vent piping through the roof, with vents for each drain for sinks, shower, bathtub, and toilets.

Cutouts as required in the floor deck for the toilets and the drains for the shower and bathtub.

In order to help accommodate the concentrations of vertical pipes, the walls between the two bathrooms and between the bathrooms and the kitchen are thickened slightly. This can be done by using larger studs (2 × 8?) or by using two rows of studs. The wall between the lower and upper portions at this location is quite chopped up with the plumbing, the stairs, and some provisions for the HVAC and hot water services. This would require some careful study, once the exact form of the equipment is determined.

Divider at Entry. This is an element that achieves some separation between these two areas. Due to the rather long span of the roof from the outside wall of the living room to the wall between the levels, this divider may also be used to support the rafters. This structural function would call for a somewhat sturdier form of construction; however, the divider is quite tall, and would need some substantial stiffness in any event. Some possible details for the divider are shown in Figure 10.10. These assume that the divider supports the roof (thus the heavier foundation detail, compared to that shown for Detail G on Figure 10.2). The low, solid portion of the divider wall allows some furniture in the living room to be placed along the wall. Many other details are possible for this semistructural, semi-decorative element.

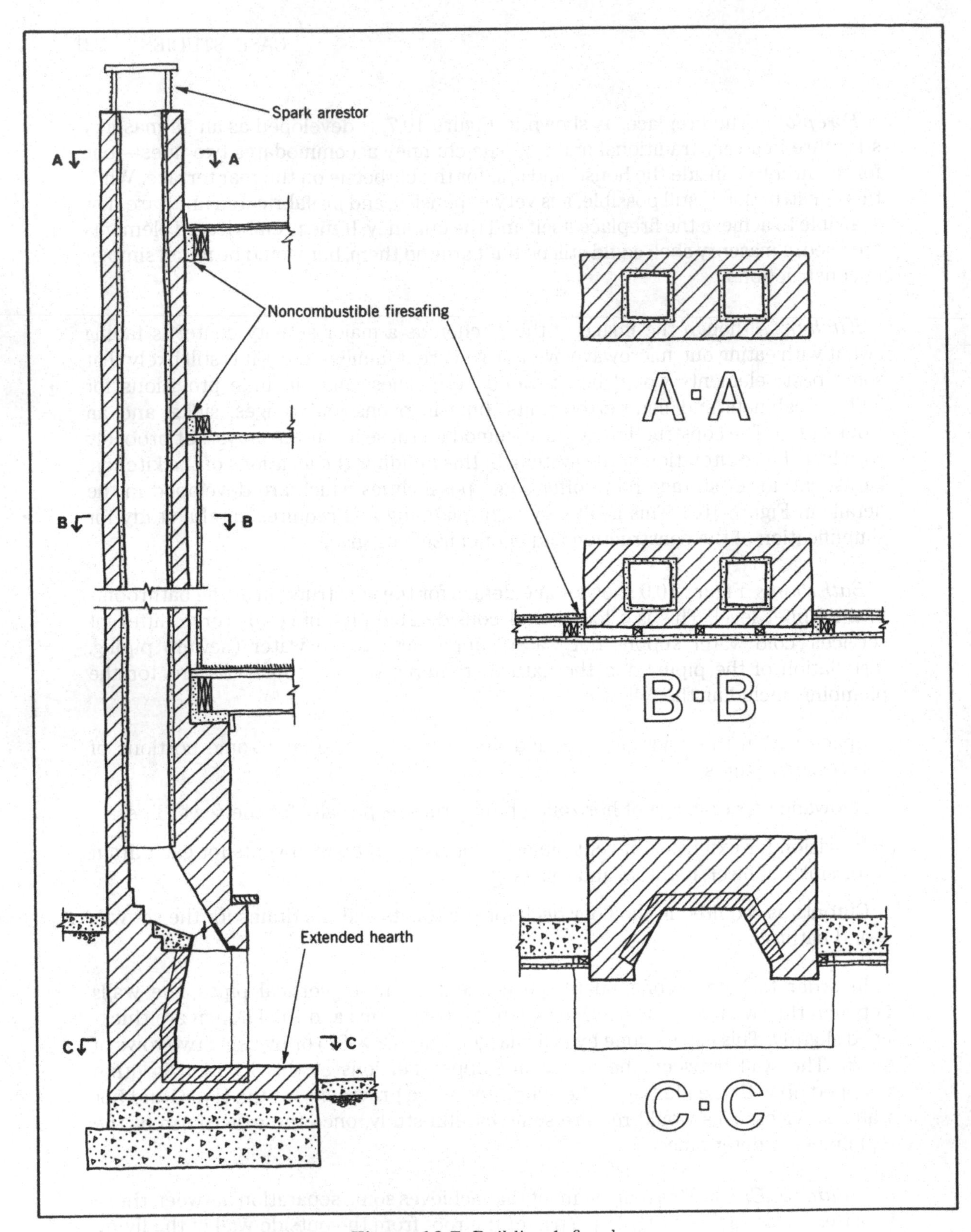

Figure 10.7 Building 1: fireplace.

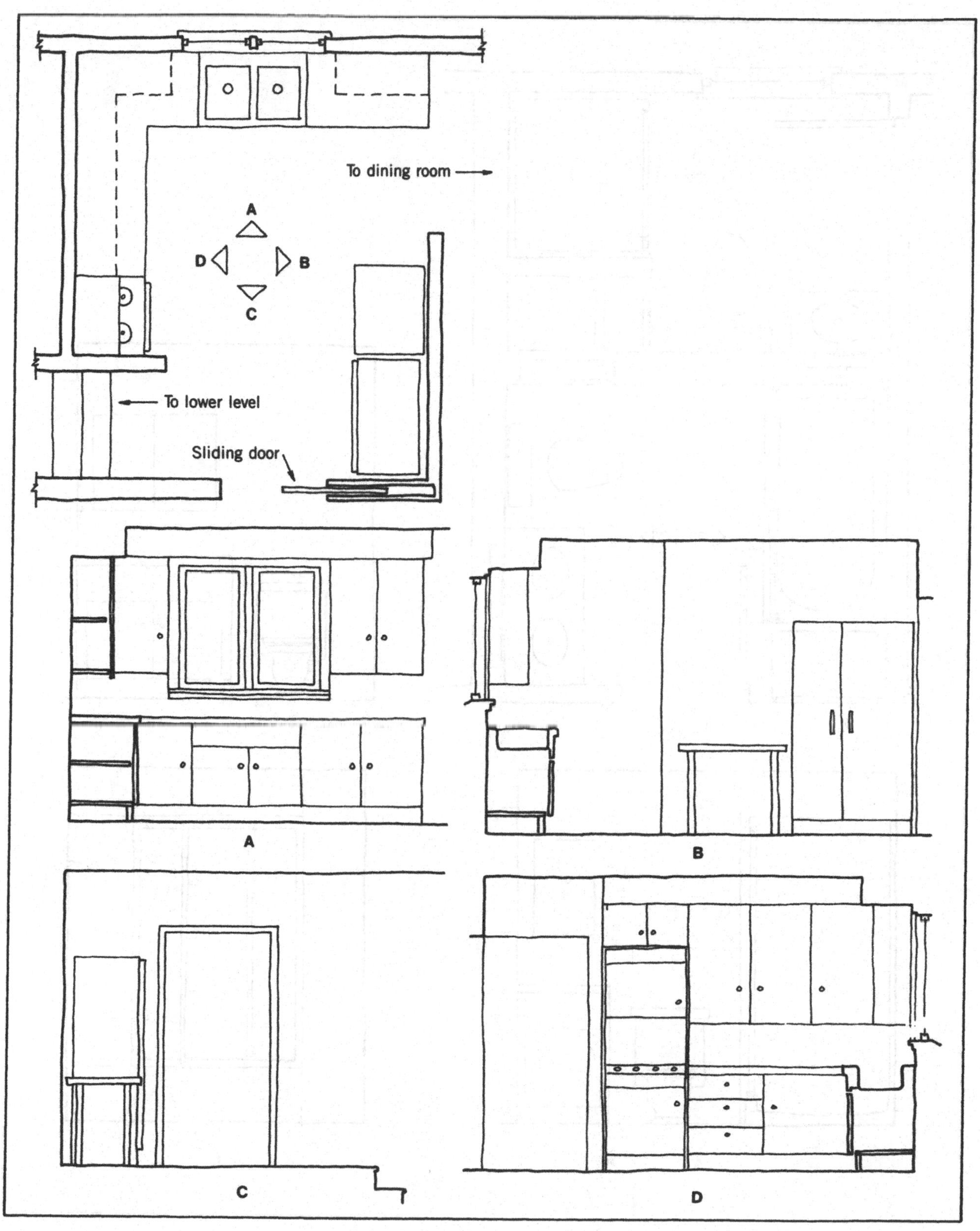

Figure 10.8 Building 1: kitchen.

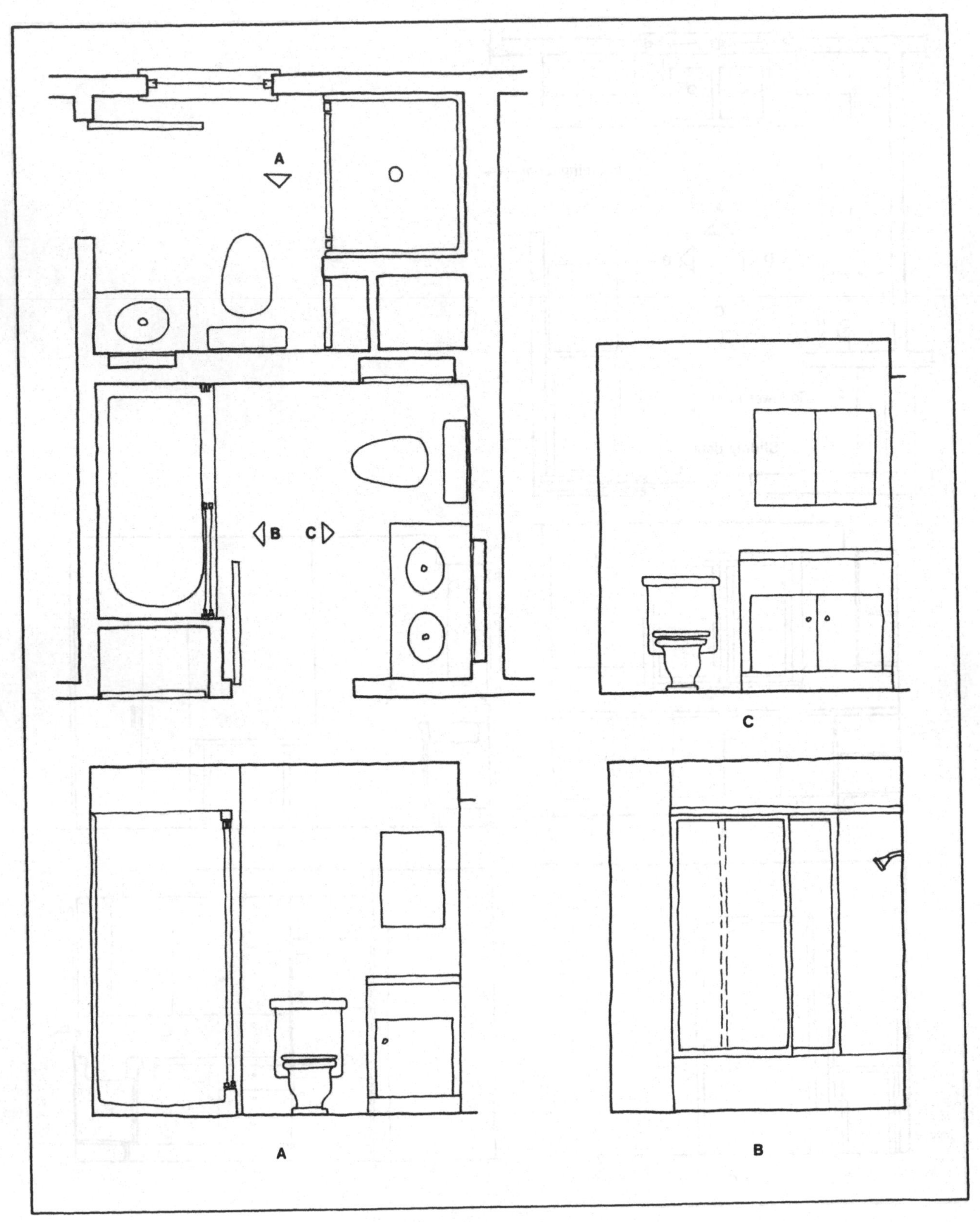

Figure 10.9 Building 1: bathrooms.

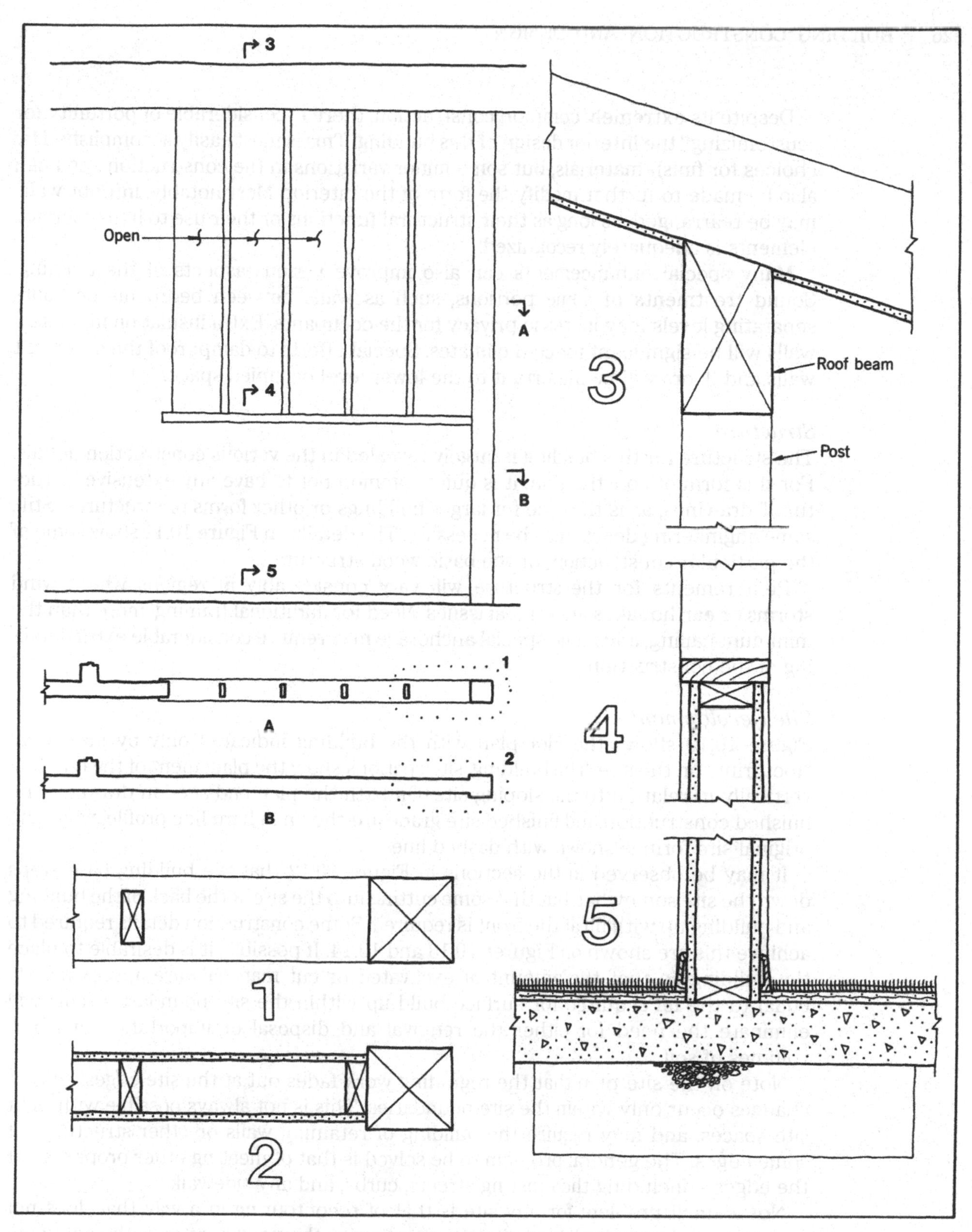

Figure 10.10 Building 1: divider at entry.

Despite its extremely common construction, there is considerable opportunity for "customizing" the interior design of this building. This is most easily accomplished by choices for finish materials, but some minor variations in the construction itself can also be made to further modify the form of the interior. Most notably, interior walls may be rearranged, as long as their structural functions or their use to house service elements is adequately recognized.

Many special enhancements can also improve various aspects of the building. Sound treatments of some portions, such as walls between bedrooms or floors separating levels may increase privacy for the occupants. Extra insulation in outside walls will be significant for cold climates. Special efforts to dampproof the basement walls and floors will be important to the lower level occupied space.

Structure

The structure for this building is largely revealed in the various construction details. For this form of construction, it is quite common not to have any extensive "structural" drawings, as is the case for larger buildings or other forms of structures. Still, some engineering design may be necessary. The details on Figure 10.11 show some of the particular construction for the basic wood structure.

Requirements for the structure will vary considerably in regions where wind storms or earthquakes are critical issues. Need for additional framing, more than the minimum nailing, and some special anchorage may require considerable extra detailing of the construction.

Site Development

Figure 10.12 shows the plot plan with the building indicated only by its overall "footprint" on the site. The building/site sections show the placement of the building vertically in relation to the sloping site. On both the plan and section drawings the finished construction and finished site grade are shown in hard line profile, while the original site form is shown with dashed lines.

It may be observed in the sections in Figure 10.12 that the building base steps down the site somewhat, but that some cutting into the site at the back of the building and building up with fill at the front is required. Some construction details required to achieve this are shown on Figures 10.13 and 10.14. If possible, it is desirable to place the building so that the amount of excavated or cut materials are approximately equal to the fill required for surface build-up within the site boundaries. This will eliminate the need for either the removal and disposal or importation of major volumes of soil.

Note on the site plan that the regrading work fades out at the site edges, so that changes occur only within the site boundaries. This is not always possible with tight site spaces, and may require the building of retaining walls or other structures at some edges. The general problem to be solved is that of meeting other properties at the edges—including the existing streets, curbs, and any sidewalks.

Not a small problem for any site is that of recontouring in a way that does not unintentionally channel precipitation runoff onto the neighbors or with some concentrated velocity onto the street. Covering the site with a large building and a lot of paving may well result in more total runoff than occurred with the barren site; further aggravating the problem. Once a preliminary site plan is developed, a careful study of the site drainage should be done.

The site drainage problem can also be aggravated by the installation of an extensive irrigation system, producing in some cases more total runoff than that occurring

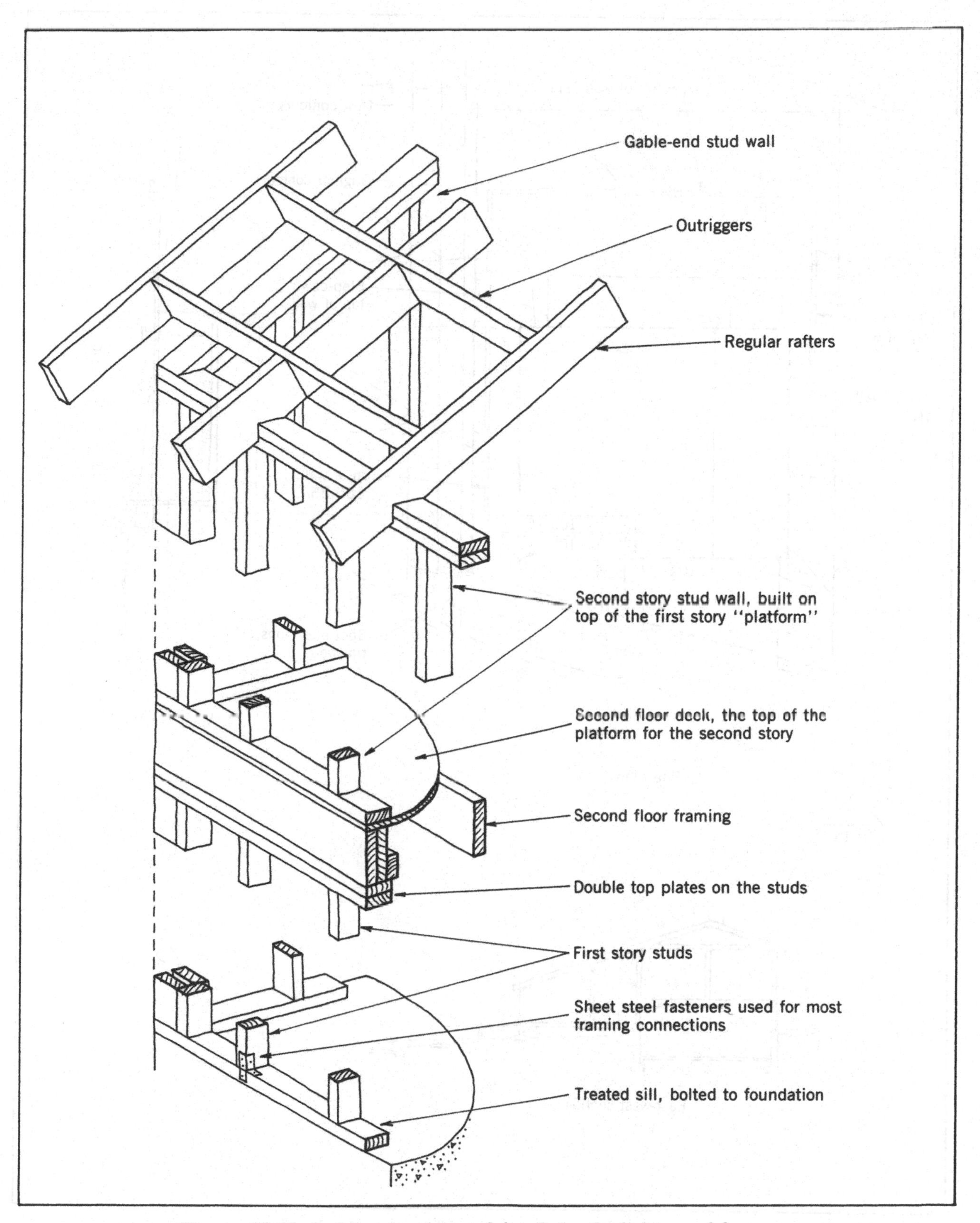

Figure 10.11 Building 1: structural details for the light wood frame.

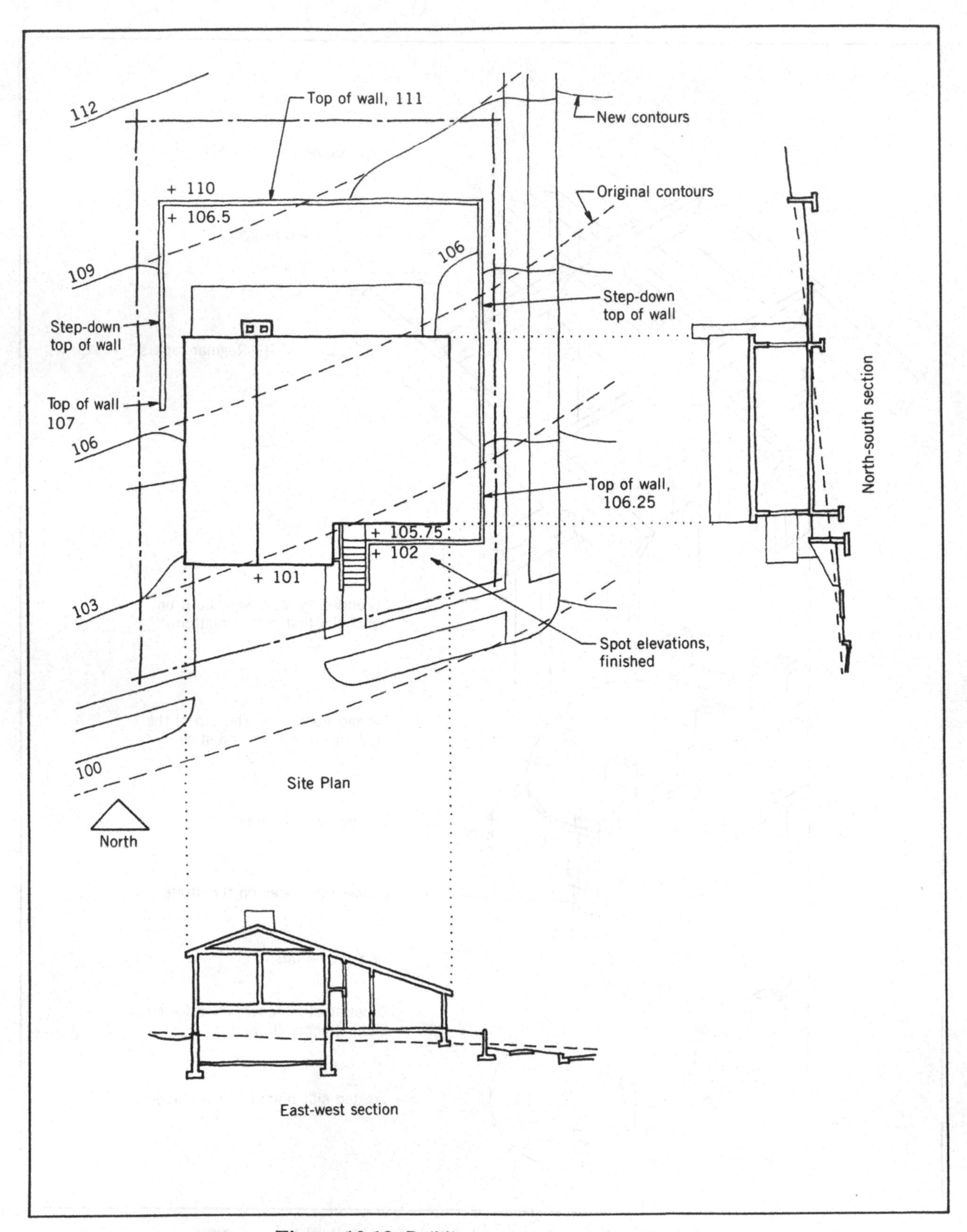

Figure 10.12 Building 1: site plan and sections.

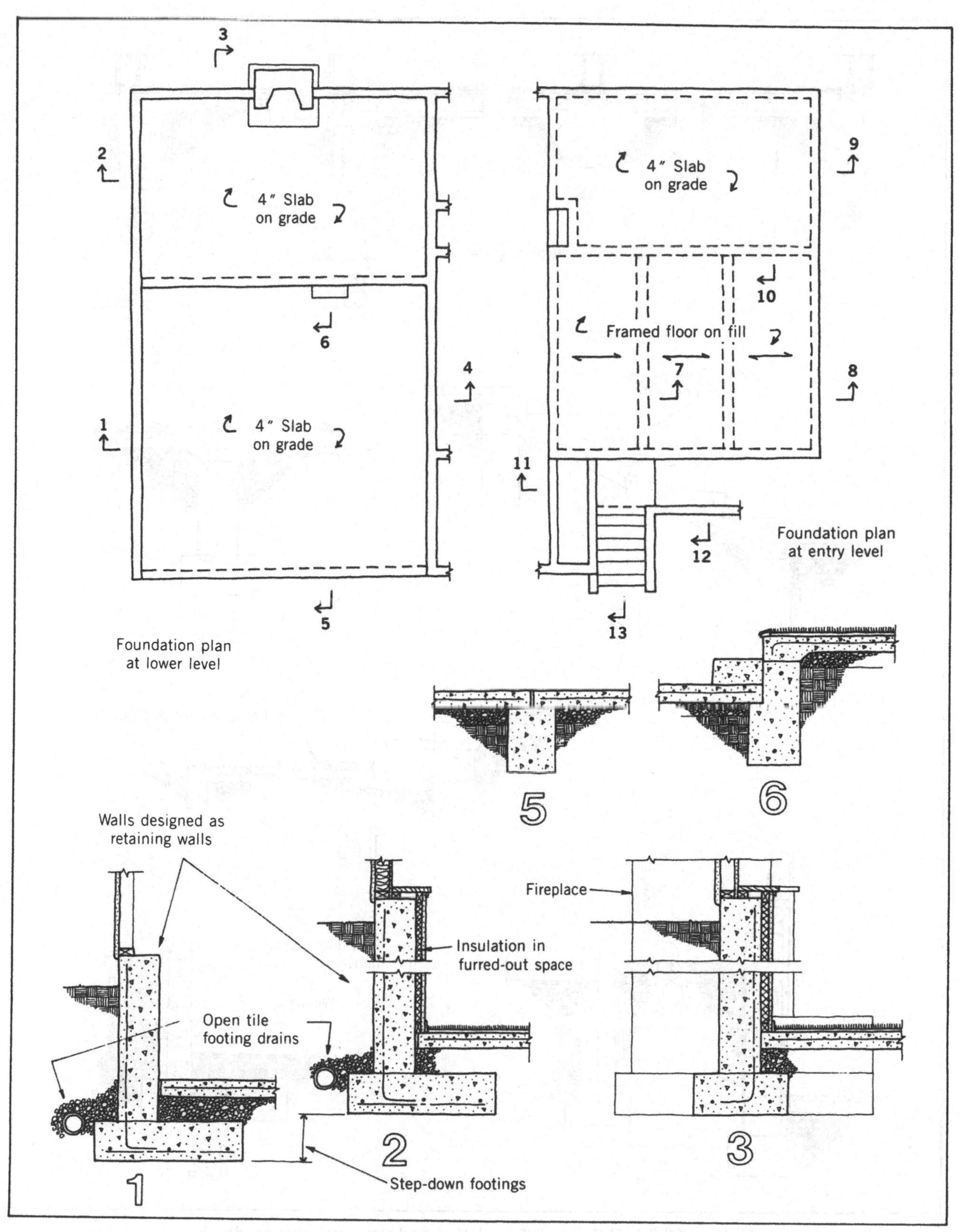

Figure 10.13 Building 1: basement and foundation plan and details.

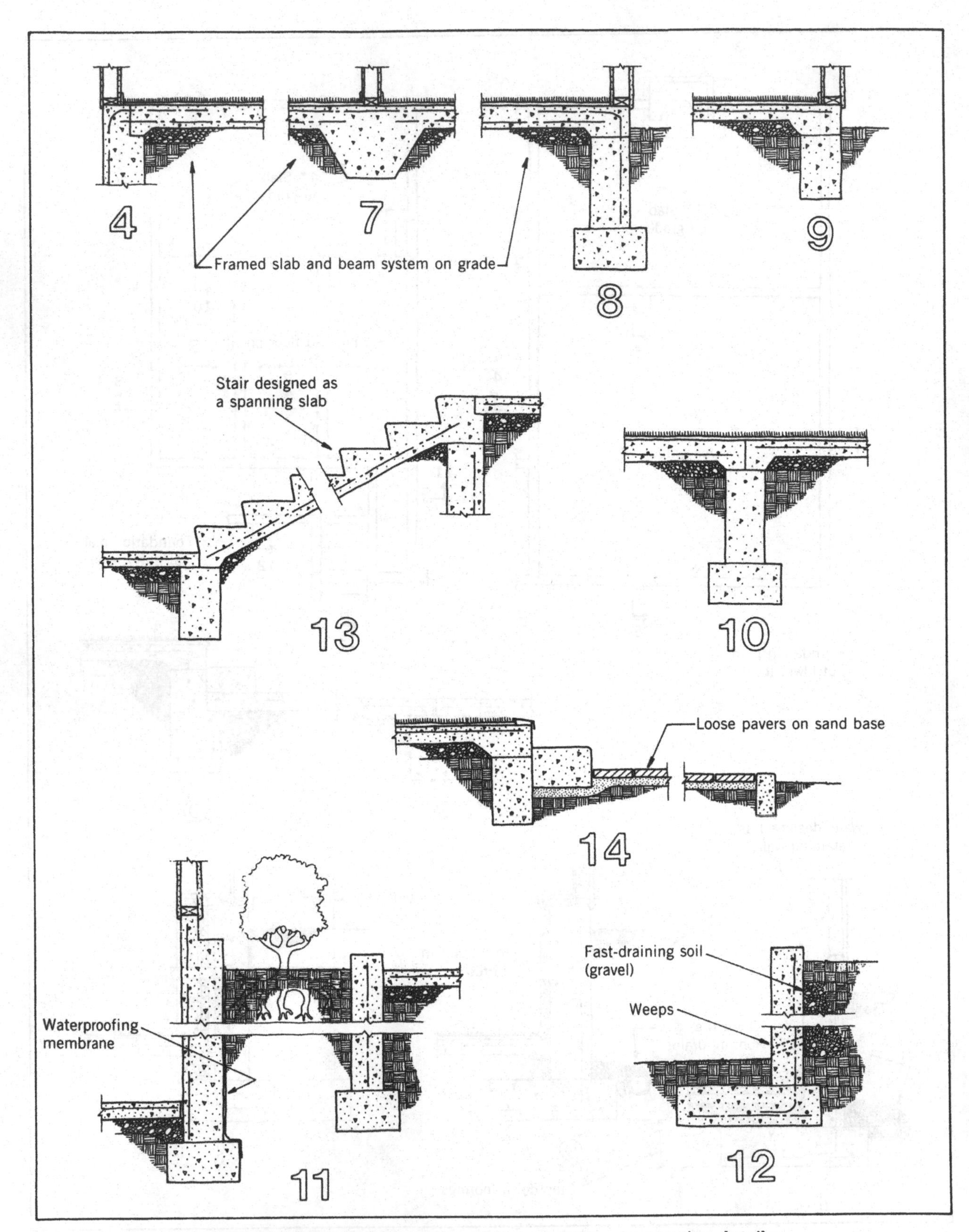

Figure 10.14 Building 1: foundation and site construction details.

with precipitation. Avoiding this requires careful design of the irrigation system, a study of site drainage from this source, and a lot of thought about the landscape design. If this is a major potential problem (which hillside houses typically present), the selection of water-conserving plantings, use of trickle systems for watering, and other methods should be used.

The plan of the building base is shown in Figure 10.14. This indicates the bounding foundation walls in the lower part of the house, the concrete floor paving slab in the entry level part, and the plans (in dashed line) of the foundations. Inspection of this drawing together with the site plan and sections will reveal the various conditions with regard to the positions of construction elements relative to both the original site surface and the finished grades.

The walls at the sides of the garage and on the north and west sides of the lower level must retain some soil pressures against their outer sides. This presents a special problem because the concrete portions of these walls are not supported laterally at their tops in most cases. This makes it necessary to design these walls as cantilever retaining walls or as horizontal-spanning ones in a lateral direction. The details in Figure 10.14 show the use of a form of retaining wall construction, although the bracing by the floor slabs is also critical to the wall stability.

At the front end of the house, the floor in the living room and entry areas, the front walk, and the exterior stair are placed on considerable fill. The plans and details show the use of spanning concrete construction at this location. Note that all foundations are taken down to a level that is slightly below the original site surface.

At the rear of the house the opposite problem occurs, and it is necessary to cut into the original grade to provide the flat patio area. This also requires the construction of some form of retaining structure on three sides of the patio. As the details in Figure 10.14 indicate, a short concrete cantilever wall is placed across the site at the point of the deepest cut. This wall is returned a short distance at its ends, but the remainder of the retaining is done with a loose-laid stone wall, sloped into the cut. Various other options for this type of structure are discussed in Sec. 7.15.

Building Service Systems

As shown here, the light wood frame construction system provides considerable hollow, interstitial space which can be extensively used for installations of out-of-view piping and wiring and some built-in elements of the various service systems. Some elements of the latter type that are routinely accommodated in this fashion are power outlets, light switches, wall and ceiling HVAC registers, and some recessed ceiling light fixtures.

There are two situations shown in the construction details, however, where burying of service elements is somewhat of a problem. Details C, D, and F (Figure 10.3) show the concrete slab floor and foundations. These solid concrete elements offer less accommodation to burying of elements, although some wiring is occasionally so treated, as well as small items such as power outlets.

Another situation of this type is shown in Detail E2 in Figure 10.5. With the plank roof deck, any elements for services at the ceiling must likely be surface-mounted or suspended; whereas the use of the light wood construction shown in Detail E1 would permit relatively easy concealing of any wiring, recessed lights, or even small ducts.

Figure 10.15 shows the general scheme for development of an all-air HVAC system for Building 1. There are several possible variations on this system, with certain forms more appropriate for particular climatic conditions. Shown here is a sys-

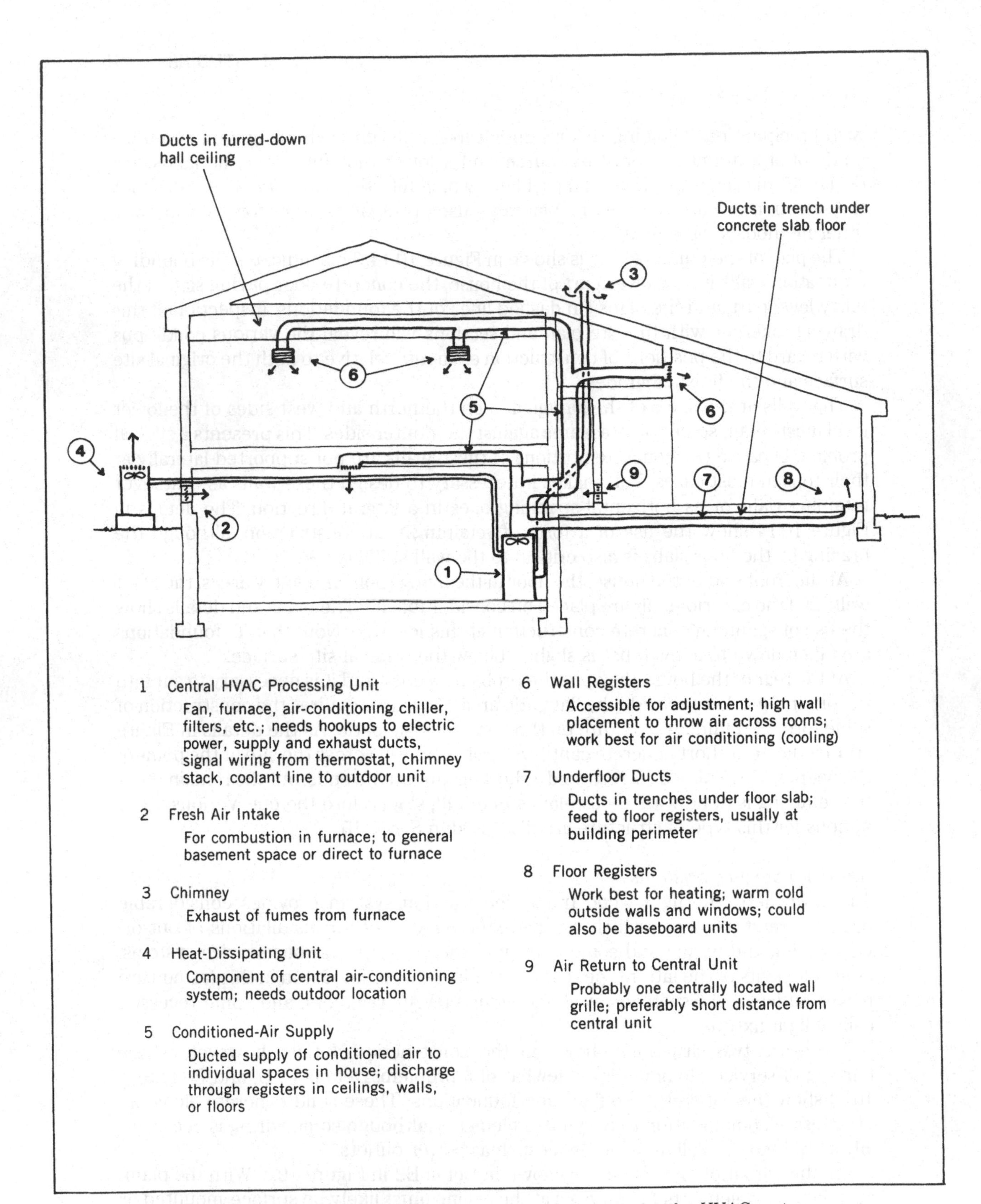

Ducts in furred-down hall ceiling
Ducts in trench under concrete slab floor
1
2
3
4
5
6
7
8
9
1 Central HVAC Processing Unit
Fan, furnace, air-conditioning chiller, filters, etc.; needs hookups to electric power, supply and exhaust ducts, signal wiring from thermostat, chimney stack, coolant line to outdoor unit
2 Fresh Air Intake
For combustion in furnace; to general basement space or direct to furnace
3 Chimney
Exhaust of fumes from furnace
4 Heat-Dissipating Unit
Component of central air-conditioning system; needs outdoor location
5 Conditioned-Air Supply
Ducted supply of conditioned air to individual spaces in house; discharge through registers in ceilings, walls, or floors
6 Wall Registers
Accessible for adjustment; high wall placement to throw air across rooms; work best for air conditioning (cooling)
7 Underfloor Ducts
Ducts in trenches under floor slab; feed to floor registers, usually at building perimeter
8 Floor Registers
Work best for heating; warm cold outside walls and windows; could also be baseboard units
9 Air Return to Central Unit
Probably one centrally located wall grille; preferably short distance from central unit

Figure 10.15 Building 1: general scheme and details for the HVAC system.

tem that favors air conditioning (cooling) through use of high-wall registers; a system more appropriate for warm climates where heating is not a major concern.

In cooler climates, especially if air conditioning is not developed by the central air handling system, it would probably be better to use a system with perimeter registers in the floor. For the construction, this would be easily achieved within the light wood frame for the upper level bedroom portion of the house. However, the floor registers and feeder ducts for the portions with slab floors would be somewhat more difficult. This might favor the use of crawl space construction for the living room portion of the house. Use of the crawl space would also make it somewhat easier to assure warm floors and a better response to the problems of edge heat loss at the ground level.

Separately developed—although with some similar construction details—are the ventilation exhaust systems for the bathrooms and kitchen. Use of the dropped ceilings in these areas greatly simplifies these installations.

Figure 10.16 shows the general nature of the plumbing systems for water supply and waste. Again, the light wood frame is reasonably accommodating for these installations. In fact, it is almost too easy to hack up the light wood frame, resulting in some serious losses of the structural frame in some cases. Generally, the structure has considerable redundance and load sharing by the closely-spaced members, but some careful inspection should be made to assure the structural integrity of the frame after the plumbers and electricians get through with it.

Careful coordination of the construction work must be made where the plumbing systems pass through or under the concrete slabs, walls, and foundations. If other services are also delivered underground to the site, these must also be developed in cooperation with the scheduling of other site and foundation construction work. Tearing up cast slabs, cutting holes through concrete walls, or decimating the finished landscape work should be avoided. The construction work of several different trades and subcontractors and the design work of several different professionals may be involved. This is where the architects, construction managers, or general contractors earn their money, if serious conflicts are avoided.

Also shown in Figure 10.13 is the use of a perimeter footing drain for the lower portion of the house where the floor is below grade. Roof drainage downspouts (vertical leaders) may also be fed into this system, if good site drainage away from the below-grade walls cannot be assured.

Some details for the basic electrical power system are shown in Figure 10.17. The typical system for a residence of this size would provide for both 110 and 220 volt service. The 110 volt system is used for lighting and the general outlets providing for random hookup of moveable items, such as TVs and floor lamps. The higher voltage system would be used for the HVAC system, clothes dryers, electric range/ovens, and possibly for attachment of heavy shop equipment if a workshop is provided.

Small transformers are usually used to step voltage down for built-in systems for doorbells, smoke alarms, thermostats (for the HVAC system), intercoms, and possibly for some outdoor lighting. Figure 10.18 shows some details for lighting elements that are built into the construction. Wiring for lighting is integrated with that for the general power distribution, while the wiring for low voltage systems (including phones and cable TV) is separately, and somewhat more easily, developed in the construction.

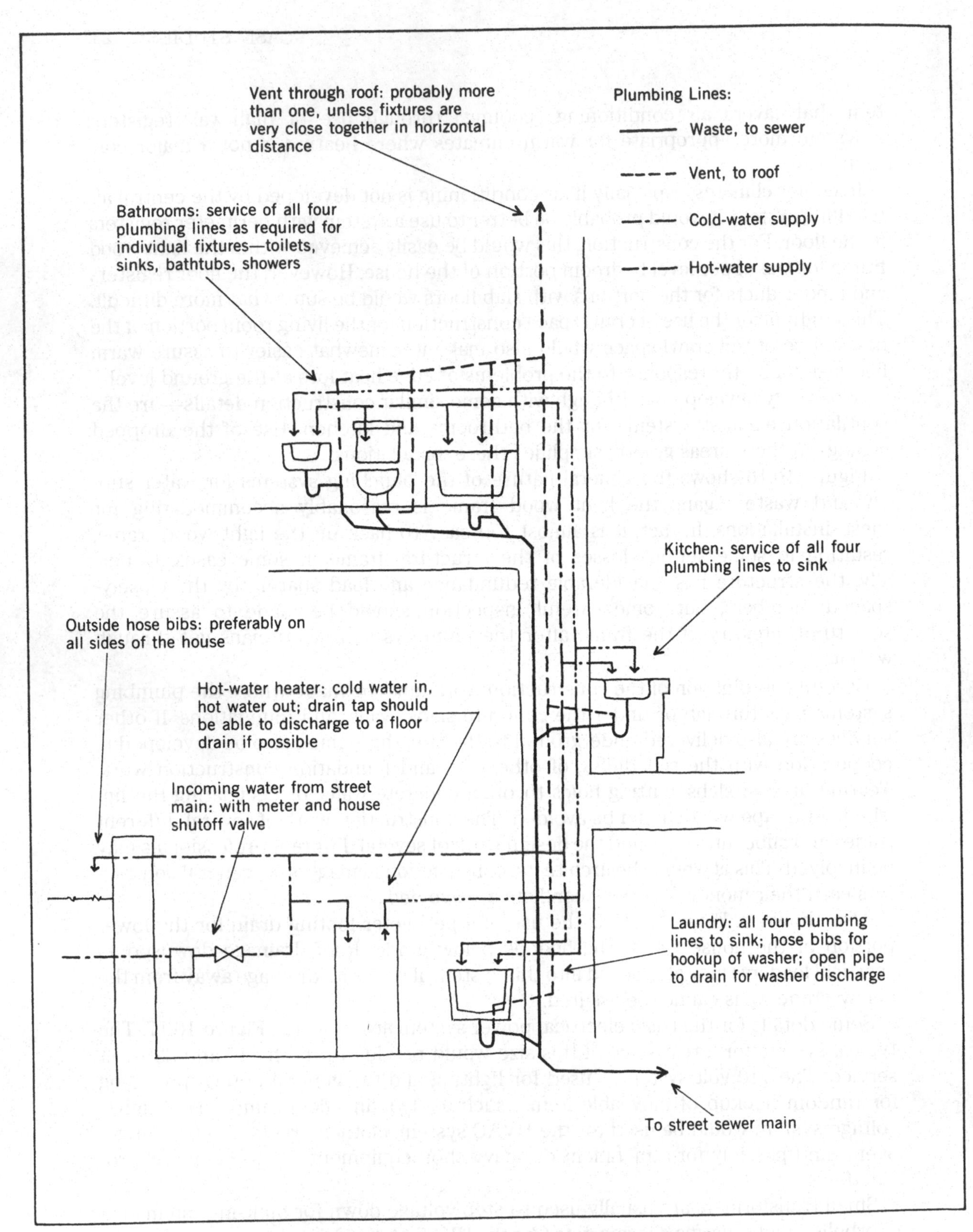
Vent through roof: probably more than one, unless fixtures are very close together in horizontal distance
Plumbing Lines:
Waste, to sewer
Vent, to roof
Cold-water supply
Hot-water supply
Bathrooms: service of all four plumbing lines as required for individual fixtures—toilets, sinks, bathtubs, showers
Kitchen: service of all four plumbing lines to sink
Outside hose bibs: preferably on all sides of the house
Hot-water heater: cold water in, hot water out; drain tap should be able to discharge to a floor drain if possible
Incoming water from street main: with meter and house shutoff valve
Laundry: all four plumbing lines to sink; hose bibs for hookup of washer; open pipe to drain for washer discharge
To street sewer main

Figure 10.16 Building 1: plumbing system details.

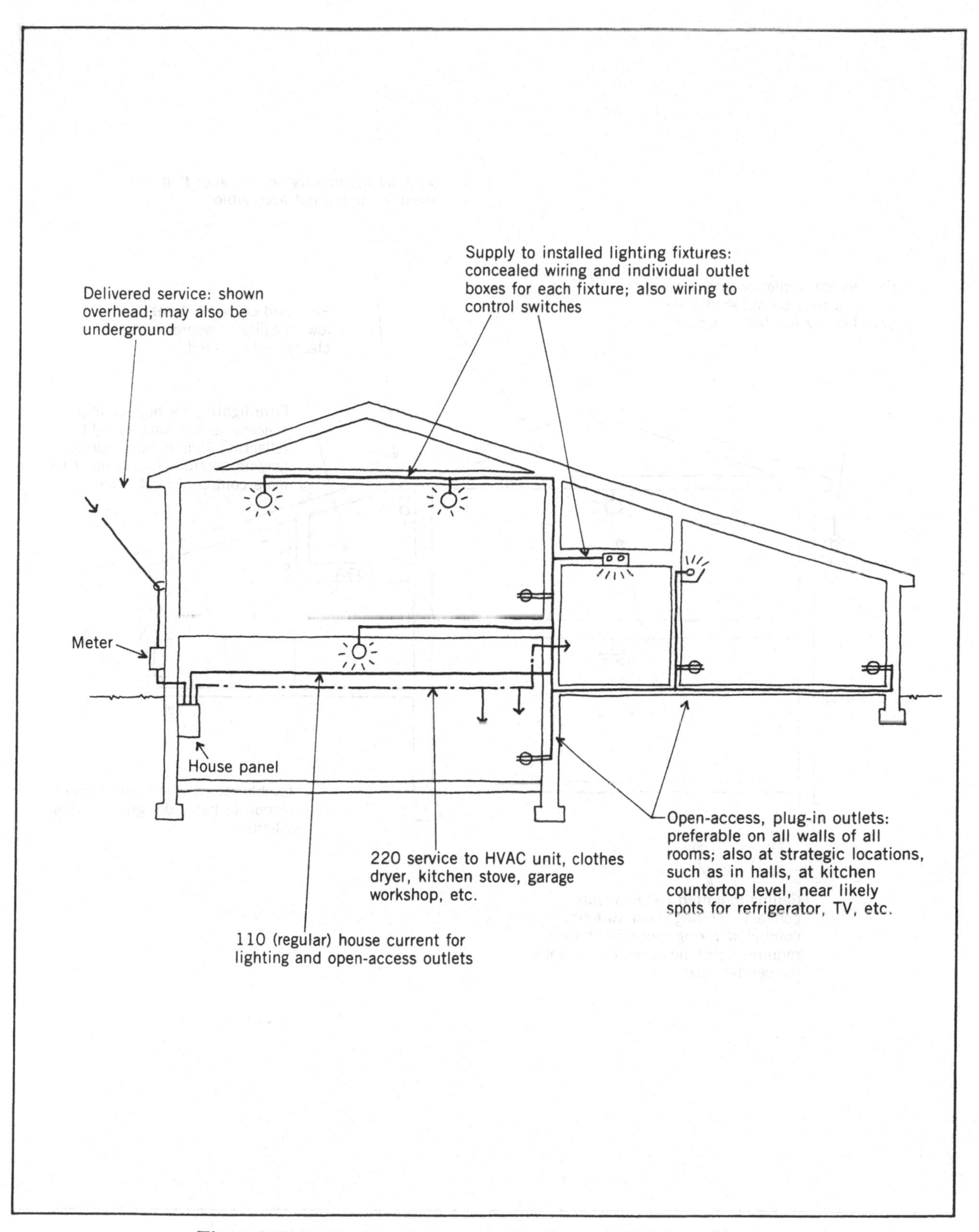

Figure 10.17 Building 1: details of the electrical power system.

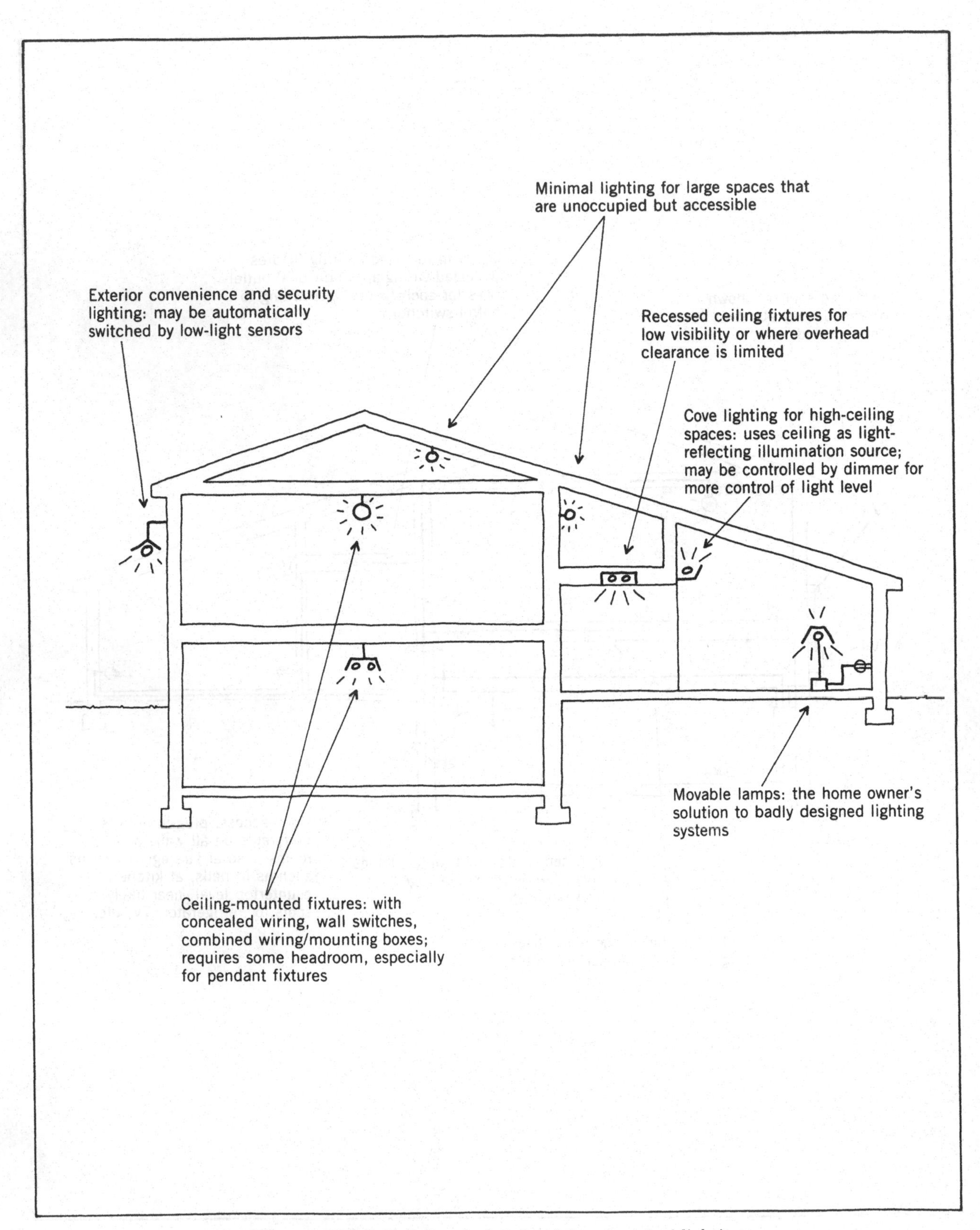
Minimal lighting for large spaces that
are unoccupied but accessible
Exterior convenience and security
lighting: may be automatically
switched by low-light sensors
Recessed ceiling fixtures for
low visibility or where overhead
clearance is limited
Cove lighting for high-ceiling
spaces: uses ceiling as light-
reflecting illumination source;
may be controlled by dimmer for
more control of light level
Movable lamps: the home owner's
solution to badly designed lighting
systems
Ceiling-mounted fixtures: with
concealed wiring, wall switches,
combined wiring/mounting boxes;
requires some headroom, especially
for pendant fixtures

Figure 10.18 Building 1: details of the electrical lighting.

10.2 BUILDING 2

One-Story Commercial
Light Wood Frame
Flat Site

This is a small, one-story building with a flat roof and a parapet wall of uniform height on all sides (see Figure 10.19). Although the building plan is generally symmetrical about a central corridor, the building exterior is developed to relate to a site orientation that results in one long side being developed as the front (street-facing) and the opposite long side as the rear. A short cantilevered canopy is provided on the front side and the roof surface is sloped to the rear. The most likely method for draining the roof surface in this situation would be by scruppers through the rear parapet and vertical leaders on the rear wall.

The details, as shown on Figures 10.20 through 10.23, reveal the structure to be light wood frame (Sec. 9.7.1) with structural plywood siding on the rear and ends and brick veneer on the front. The floor is a concrete slab on grade and the roof structure consists of clear-span trusses.

Use of the clear-span roof structure permits interior walls to be placed at will, with remodeling for other desired arrangements in the future easily achieved. This form of structure is common for commercial buildings built as speculative rental properties.

The prefabricated trusses can be formed to have a flat bottom chord but a sloping top one to accommodate the roof drainage. For the simplest ceiling construction, framing may be applied directly to the bottom chords of the trusses. In the void space between the roof deck and the ceiling, ducts, piping, and wiring may be placed, running through the spaces between the truss diagonal web members.

Lateral bracing for wind or seismic effects will be achieved by using the horizontal plane of the roof deck as a diaphragm, with vertical bracing from the plywood siding acting to form shear walls. Some detailing of the roof-to-wall joints, the framing of the roof deck and walls, and the foundations for the walls must respond to these lateral bracing functions.

It is assumed that heating, cooling, and air conditioning is to be achieved with an all-air-system, with ducts and ceiling registers in the void space between the roof and ceiling. Equipment for the HVAC system may be in the equipment room shown on the plan or may be placed on the rooftop, as is common in mild climates.

Plumbing is of minor concern in this building, consisting mostly of service for the rest rooms at one corner of the plan. However, there may also be a piped, underground drain system for the roof drains and a fire sprinkler system in the ceiling. In colder climates it may also be desirable to have a perimeter baseboard heating system to augment the air system in the ceiling.

This is a small building in total floor area, permitting the use of the light wood frame by most building codes. However, the construction may have to develop a one-hour rating, depending on the occupancy or on zoning requirements. If the plan were larger, the use of the sprinkler system or placing of a separating fire wall may still allow the light wood frame to be used.

The basic structure for this building is quite typical, with only minor variations to be expected for regional concerns. The exterior appearance can be endlessly varied by using different finish materials, windows, roof edge treatments, and detail features, such as the canopy shown here. A canopy on all four sides, for example, can create the appearance from ground level of a building with a shingled or tiled, fast-draining roof form.

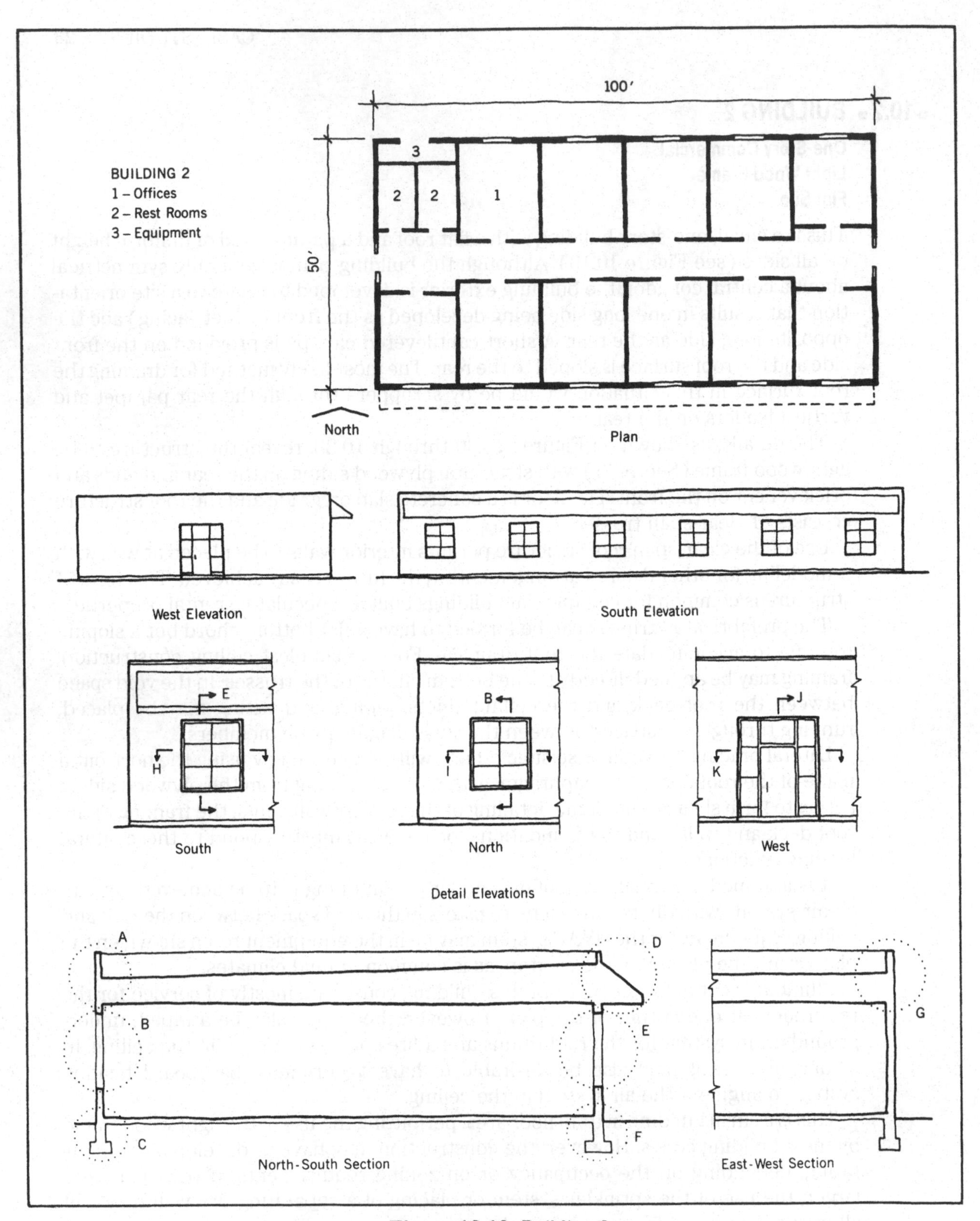

Figure 10.19 Building 2.

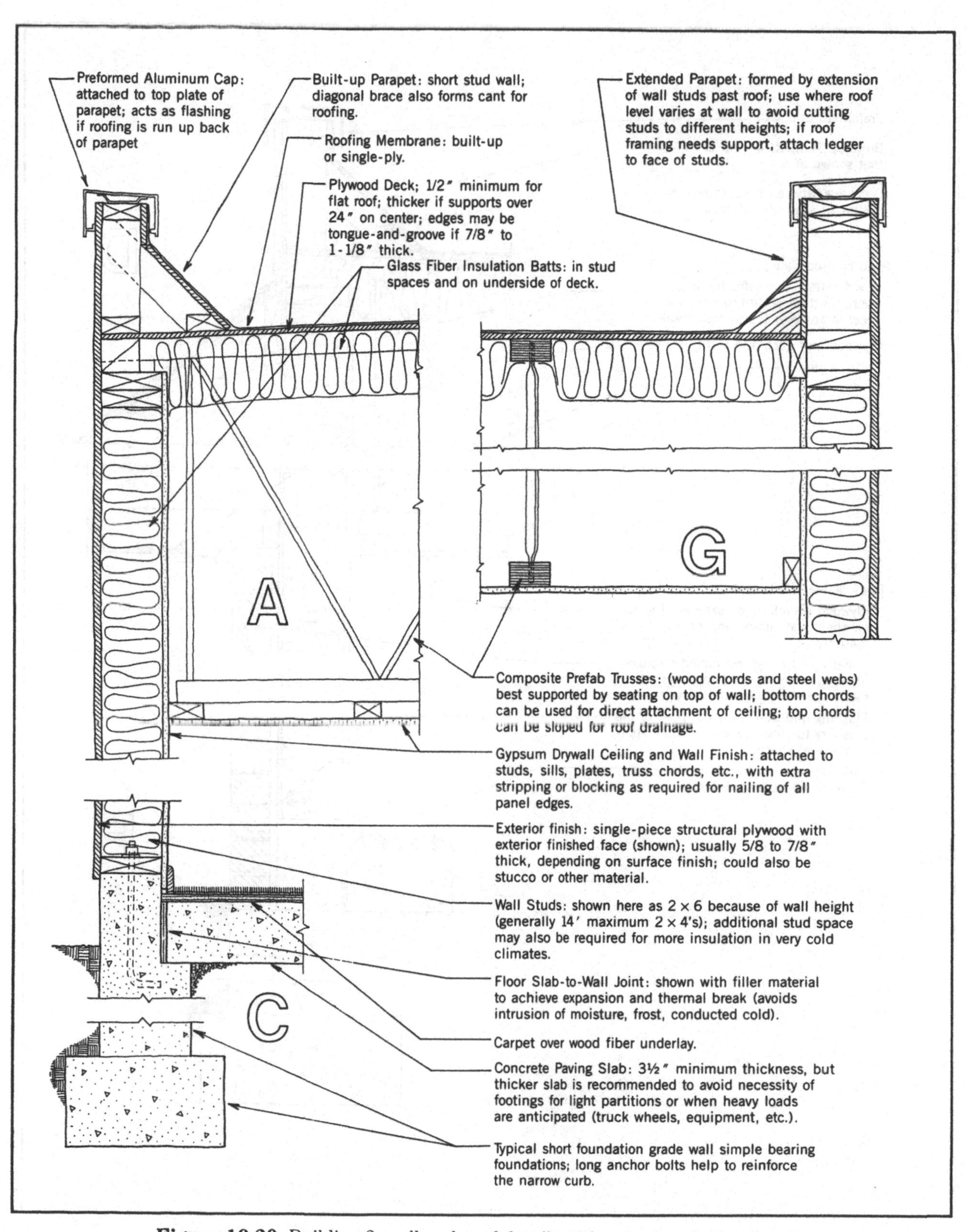

Preformed Aluminum Cap: attached to top plate of parapet; acts as flashing if roofing is run up back of parapet
Built-up Parapet: short stud wall; diagonal brace also forms cant for roofing.
Roofing Membrane: built-up or single-ply.
Plywood Deck; 1/2" minimum for flat roof; thicker if supports over 24" on center; edges may be tongue-and-groove if 7/8" to 1-1/8" thick.
Glass Fiber Insulation Batts: in stud spaces and on underside of deck.
Extended Parapet: formed by extension of wall studs past roof; use where roof level varies at wall to avoid cutting studs to different heights; if roof framing needs support, attach ledger to face of studs.
A
G
C
Composite Prefab Trusses: (wood chords and steel webs) best supported by seating on top of wall; bottom chords can be used for direct attachment of ceiling; top chords can be sloped for roof drainage.
Gypsum Drywall Ceiling and Wall Finish: attached to studs, sills, plates, truss chords, etc., with extra stripping or blocking as required for nailing of all panel edges.
Exterior finish: single-piece structural plywood with exterior finished face (shown); usually 5/8 to 7/8" thick, depending on surface finish; could also be stucco or other material.
Wall Studs: shown here as 2 x 6 because of wall height (generally 14' maximum 2 x 4's); additional stud space may also be required for more insulation in very cold climates.
Floor Slab-to-Wall Joint: shown with filler material to achieve expansion and thermal break (avoids intrusion of moisture, frost, conducted cold).
Carpet over wood fiber underlay.
Concrete Paving Slab: 3½" minimum thickness, but thicker slab is recommended to avoid necessity of footings for light partitions or when heavy loads are anticipated (truck wheels, equipment, etc.).
Typical short foundation grade wall simple bearing foundations; long anchor bolts help to reinforce the narrow curb.

Figure 10.20 Building 2: wall and roof details with exterior plywood walls.

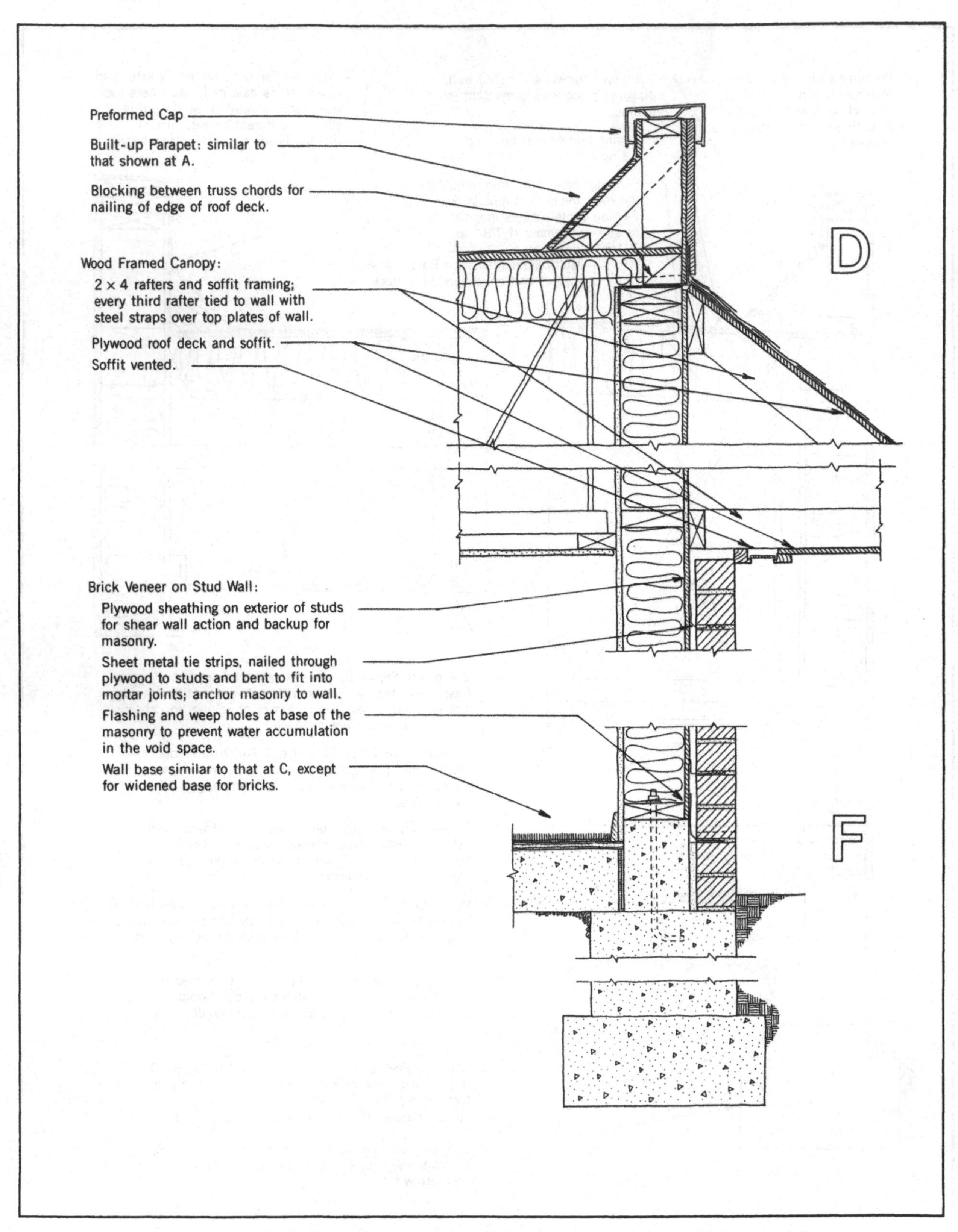

Preformed Cap
Built-up Parapet: similar to that shown at A.
Blocking between truss chords for nailing of edge of roof deck.
Wood Framed Canopy:
2 × 4 rafters and soffit framing; every third rafter tied to wall with steel straps over top plates of wall.
Plywood roof deck and soffit.
Soffit vented.
D
Brick Veneer on Stud Wall:
Plywood sheathing on exterior of studs for shear wall action and backup for masonry.
Sheet metal tie strips, nailed through plywood to studs and bent to fit into mortar joints; anchor masonry to wall.
Flashing and weep holes at base of the masonry to prevent water accumulation in the void space.
Wall base similar to that at C, except for widened base for bricks.
F

Figure 10.21 Building 2: wall and roof details with canopy and brick veneer.

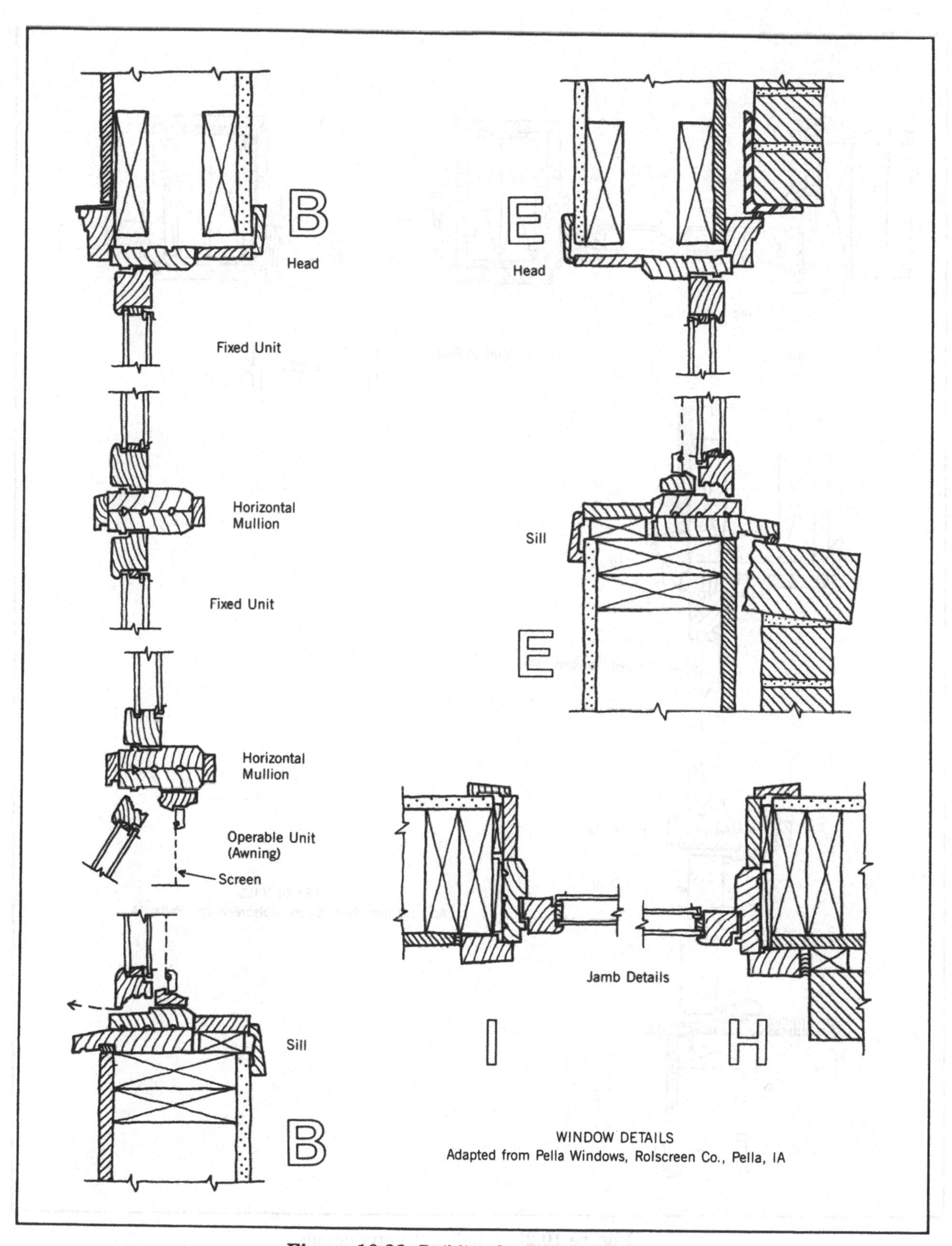

Figure 10.22 Building 2: window details.

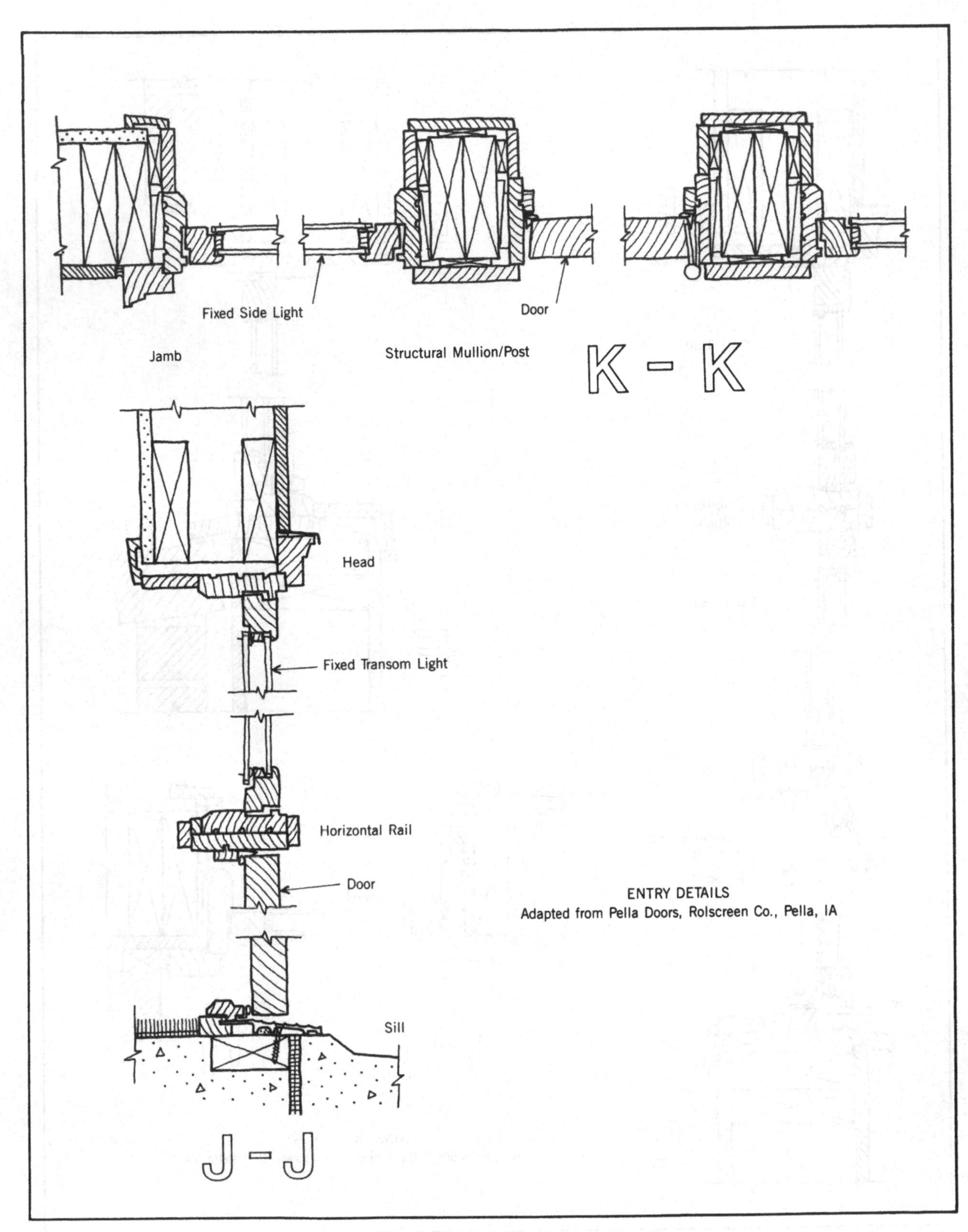

Figure 10.23 Building 2: entry details.

This is probably the most architecturally manipulatable form of construction with regard to exterior appearance. Other forms of construction can be used, of course, such as tilt-up concrete walls (see Building 3), exposed steel frame, structural masonry walls, or heavy timber frame. Used most directly, all of those systems will produce some specific forms and details that will affect the building appearance.

The light wood frame is generally quite anonymous with regard to regional characteristics. Some modifications are required for structural responses to severe wind conditions or high seismic risk, but will often not have much effect on the building appearance. Increased insulation in exterior walls may require a wider void in the stud space, which can be achieved with wider studs (2×6 or 2×8) or by simply furring out on the face of ordinary 2×4 studs. Wider studs may also be required because of the wall height, as shown here with 2×6 studs.

As with any building, the choice of window materials and details, depth of foundations below grade, need for insulation of the floor slab at the building edges, and other items may require special consideration for extreme climate conditions. Choice of exterior finish materials may also be a regional concern, although just about any type of appearance can be achieved anywhere by one means or another—not always with the same materials, but still with a similar appearance.

Despite the accommodation of interior rearrangements, certain features become relatively fixed and not subject to easy change. Primary elements in this regard are the building foundations and major structure. Also difficult to change are windows and doors, making some reasonably accommodating locations the most desirable. Plumbing—particularly sewer lines built into the pavement—are also hard to change.

Changes will be made easier if most services are installed in walls or in the space above the ceiling. This will allow for major reworking of the distribution of wiring, ducts, and various items for the building power, lighting, communication, and HVAC systems. The development of details for the construction of both permanent elements and speculatively demountable elements should be made with these provisions in mind.

As previously mentioned, the prefabricated trusses can be fabricated to have a flat bottom chord but a sloping top one to accommodate the roof drainage and a ceiling applied directly to the bottom chords of the trusses. With the open webs of the trusses, this still allows for major elements, such as ducts, to run perpendicular to the trusses. This is a case where a deeper than average truss might be used to place the ceiling at a desired level or to provide a maximum of space between truss web diagonals.

An option for the windows, using simple wood-framed elements is shown in Figure 10.22. These windows are generally not intended to accommodate the intersection of partitions, so the arrangements of any interior partitioning should avoid having the partitions meet the exterior walls except between windows.

Some options for ceiling construction are shown on Figure 10.24. Details are shown for a drywall ceiling at two positions: directly attached to the trusses inside the rooms, and suspended with a separate framing at the hallway. The hall ceiling could also be framed by joists supported by the hallway walls, as the span is quite short. Other forms of ceilings could also be used; supported in any of the three basic ways.

To assure some greater sound privacy for the tenants, the wall between the rest rooms and the adjacent office might be built with some acoustic enhancement. Some possibilities for this are shown on Figure 10.25. Real sound privacy requires that the total construction be carefully studied during design and inspected during construc-

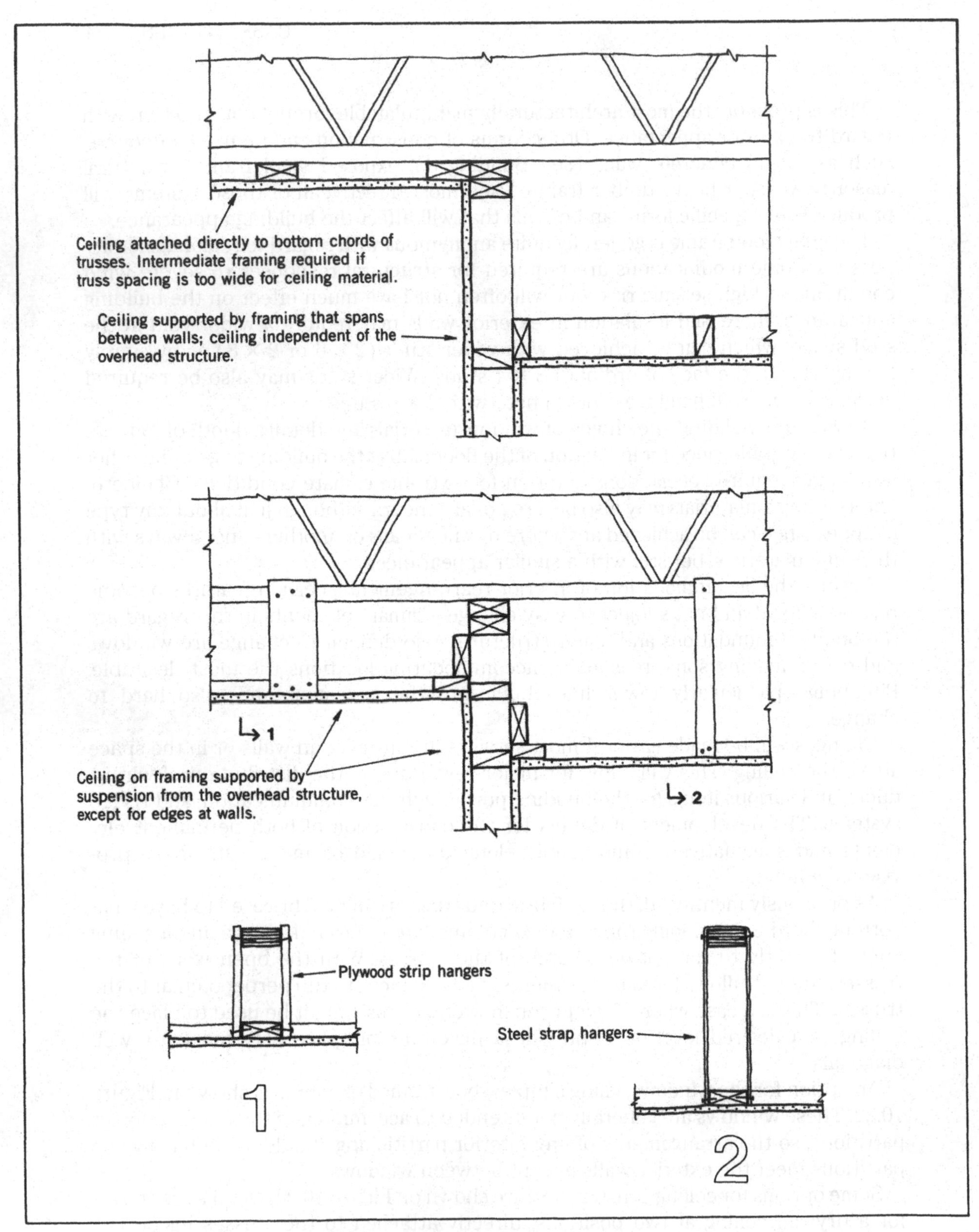

Ceiling attached directly to bottom chords of trusses. Intermediate framing required if truss spacing is too wide for ceiling material.
Ceiling supported by framing that spans between walls; ceiling independent of the overhead structure.
1
2
Ceiling on framing supported by suspension from the overhead structure, except for edges at walls.
Plywood strip hangers
Steel strap hangers
1
2

Figure 10.24 Building 2: ceiling details.

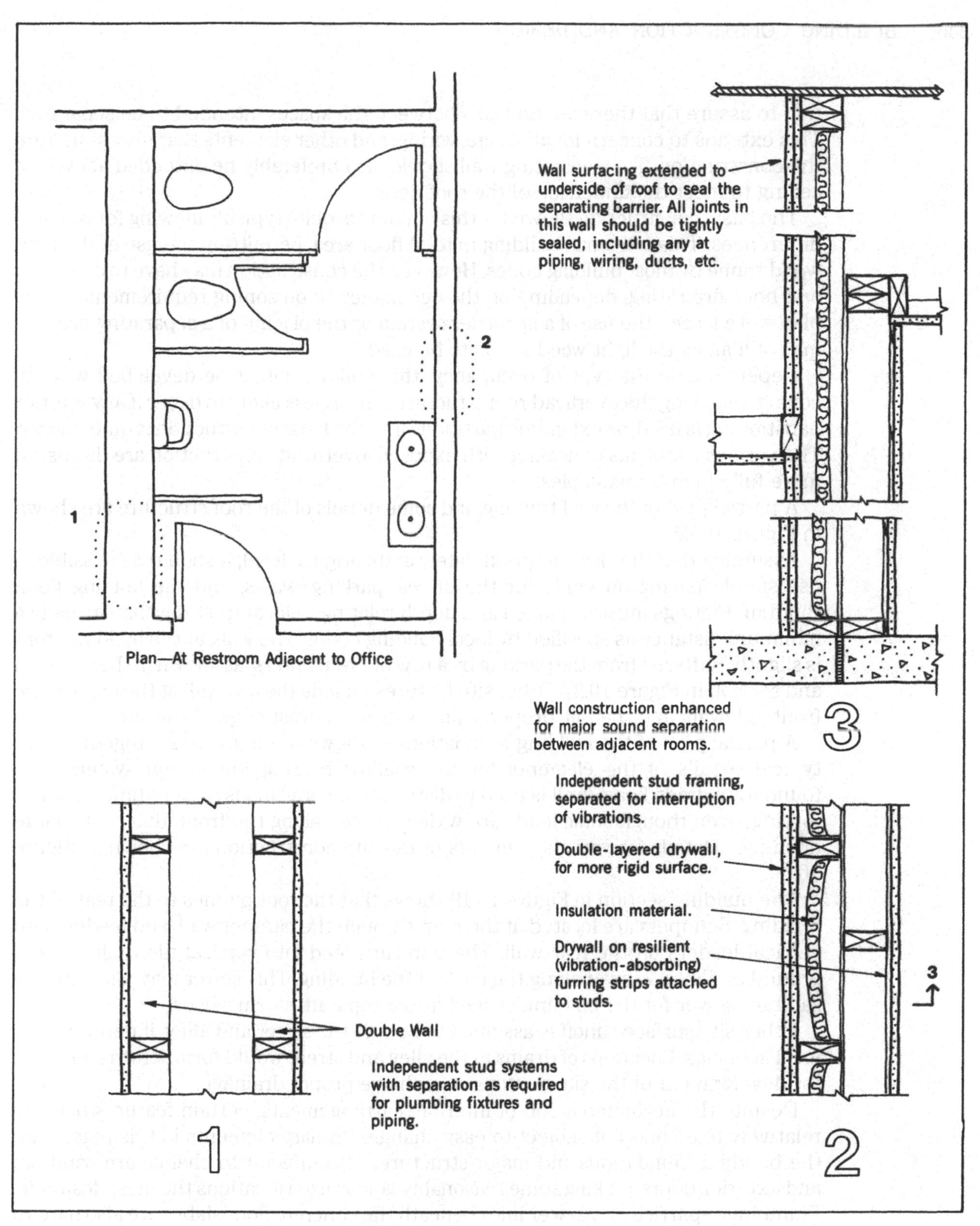
Wall surfacing extended to underside of roof to seal the separating barrier. All joints in this wall should be tightly sealed, including any at piping, wiring, ducts, etc.
2
1
Plan—At Restroom Adjacent to Office
Wall construction enhanced for major sound separation between adjacent rooms.
3
Independent stud framing, separated for interruption of vibrations.
Double-layered drywall, for more rigid surface.
Insulation material.
Drywall on resilient (vibration-absorbing) furring strips attached to studs.
3
Double Wall
Independent stud systems with separation as required for plumbing fixtures and piping.
1
2

Figure 10.25 Building 2: wall details at restrooms.

tion to assure that there are no leaks between the spaces intended to be separated. This extends to concern for air ducts, wiring, and other elements that may penetrate the construction. The separating wall should also preferably be extended above the ceiling to meet the underside of the roof deck.

The basic construction shown for this building is quite typical, allowing for regional differences. This is a small building in total floor area, permitting the use of the light wood frame by most building codes. However, the construction may have to develop a one-hour fire rating, depending on the occupancy or on zoning requirements. If the plan were larger, the use of a sprinkler system or the placing of a separating fire wall may still allow the light wood frame to be used.

Depending on the type of occupancy, this building might be developed with no ceiling, exposing the overhead roof structure. This is less likely to occur if any interior partitioning is used, as extending partitions into the truss construction is quite messy. The various problems of dealing with exposed, overhead construction are discussed more fully in other examples.

A partial plan of the roof framing and some details of the roof structure are shown in Figure 10.26.

Assuming that the site is approximately at its original level, it should be possible to use simple paving on grade for the drives, parking, walks, and the buiding floor. Building footings must be placed at a depth relating to local frost depths—usually a minimum distance as specified by local building codes. The walk at the building front is slightly buffered from the parking by a row of low planters, as shown in the site plan and section in Figure 10.27. Other site features include the low wall at the edge of the front sidewalk and the tall, property-line wall at the west edge of the site.

A partial plan of the building foundations is shown in Figure 10.28, together with typical details of the elements for the shallow bearing foundation system. The foundation grade beam/wall is used to distribute the wall loads to a continuous width footing, even though some loads are widely spaced along the front, due to the large openings. Details for various elements of the site construction are shown in Figure 10.29.

The building section in Figure 10.19 shows that the roof pitches to the rear of the building. Scuppers are located at the rear, through the parapet wall, and feeding into vertical leaders on the rear wall. These in turn feed into vertical tile drains at the ground and into a sewer along the back of the building. This sewer may join with the sanitary sewer for the building or feed into a separate storm sewer.

Other site surface runoff is assumed to be into the street and alley, if permitted by local agencies. Locations of drains in the alley and street would form some references for development of the site contours to achieve proper drainage.

Despite the accommodation of interior rearrangements, certain features become relatively fixed and not subject to easy change. Primary elements in this regard are the building foundations and major structure. Also difficult to change are windows and exterior doors, making some reasonably generalized locations the most desirable. Plumbing—particularly sewer lines beneath the concrete floor slab—are also hard to change.

Changes will be made easier if most services are installed in walls or in the space above the ceiling. This will allow for major reworking of the distribution of wiring, ducts, and various items for the building power, lighting, communication, and HVAC systems. The development of details for the construction of both permanent elements and speculatively demountable elements should be made with these provisions in mind.

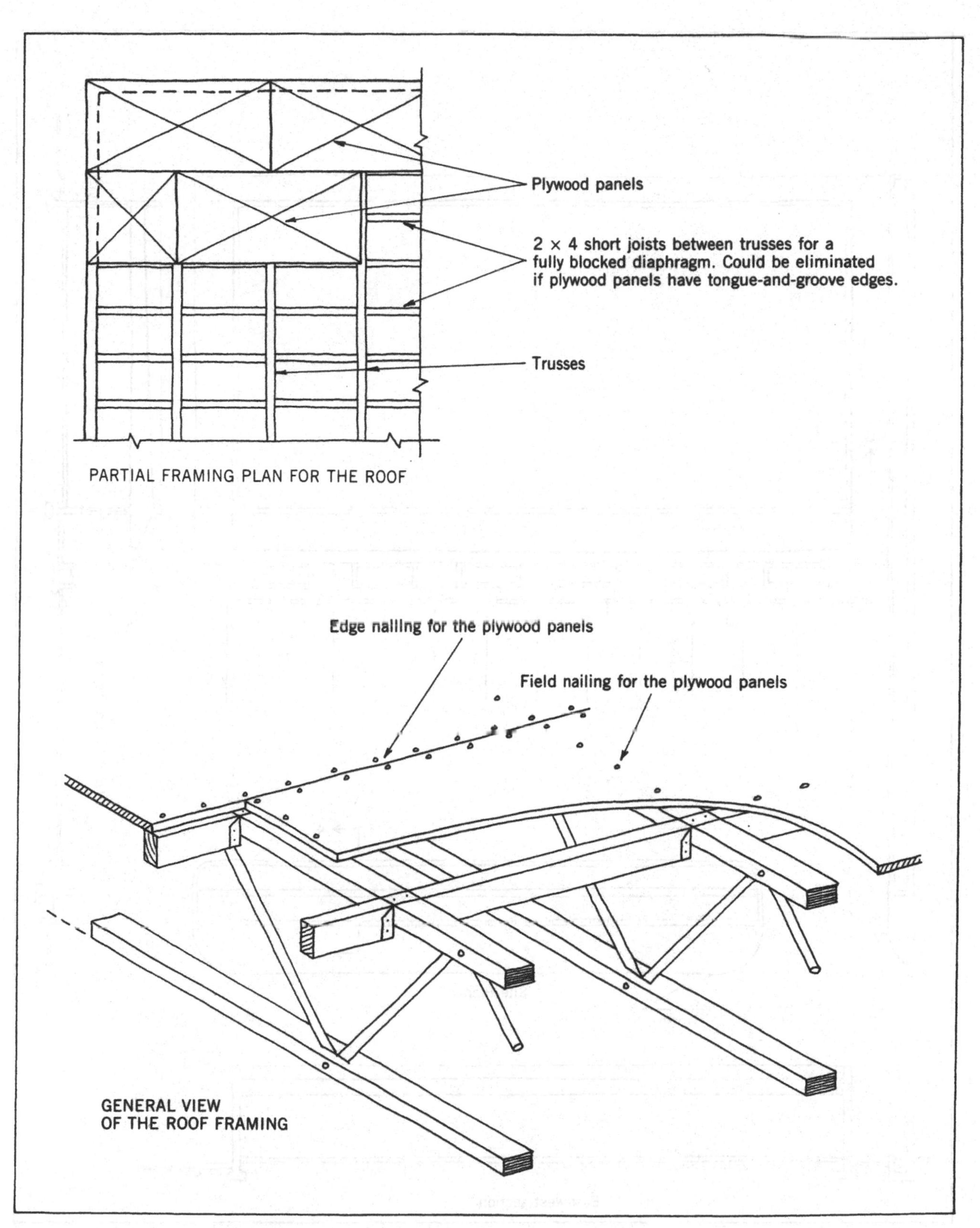

Figure 10.26 Building 2: roof framing plan and structural details.

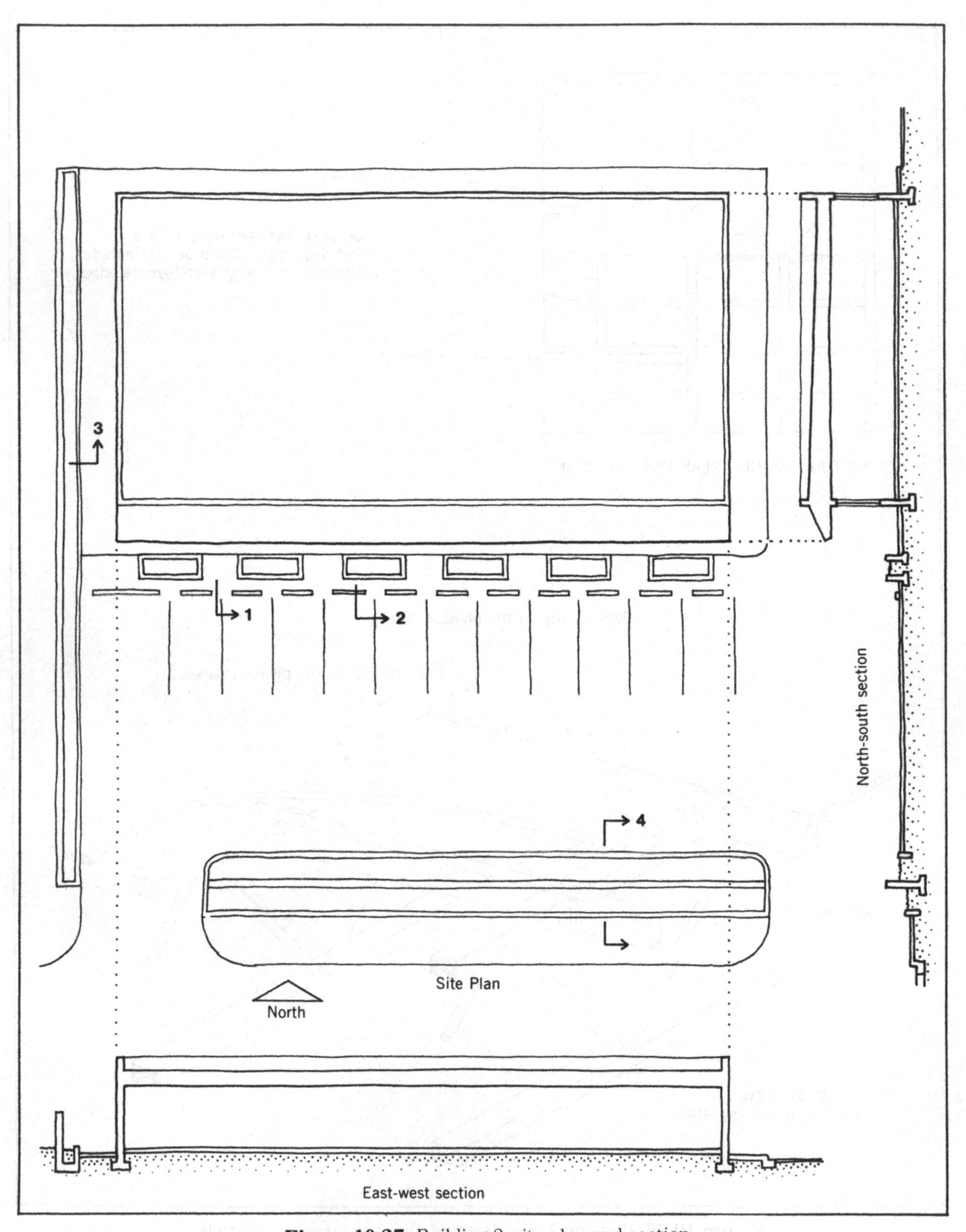

Figure 10.27 Building 2: site plan and section.

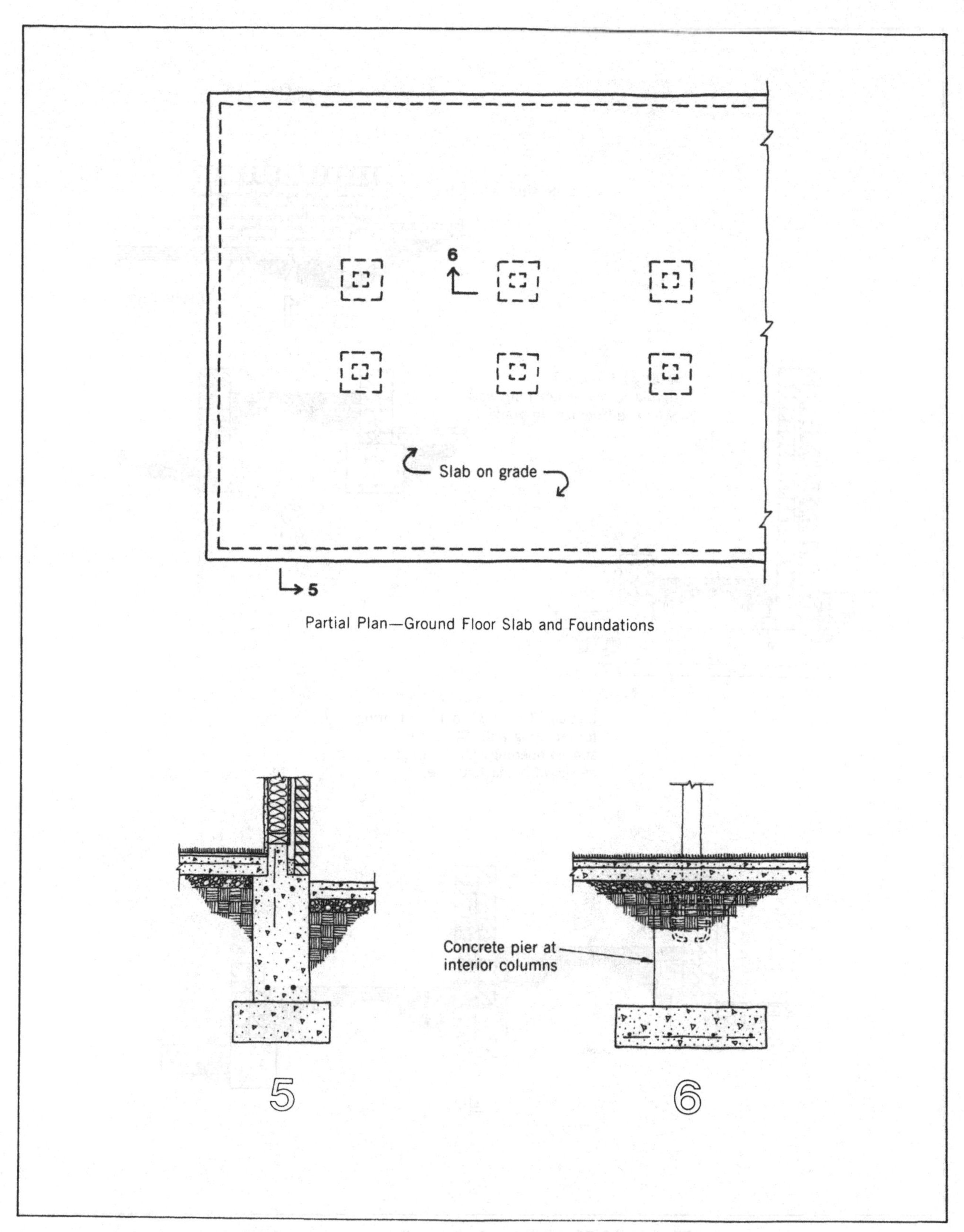

Figure 10.28 Building 2: foundation plan and details.

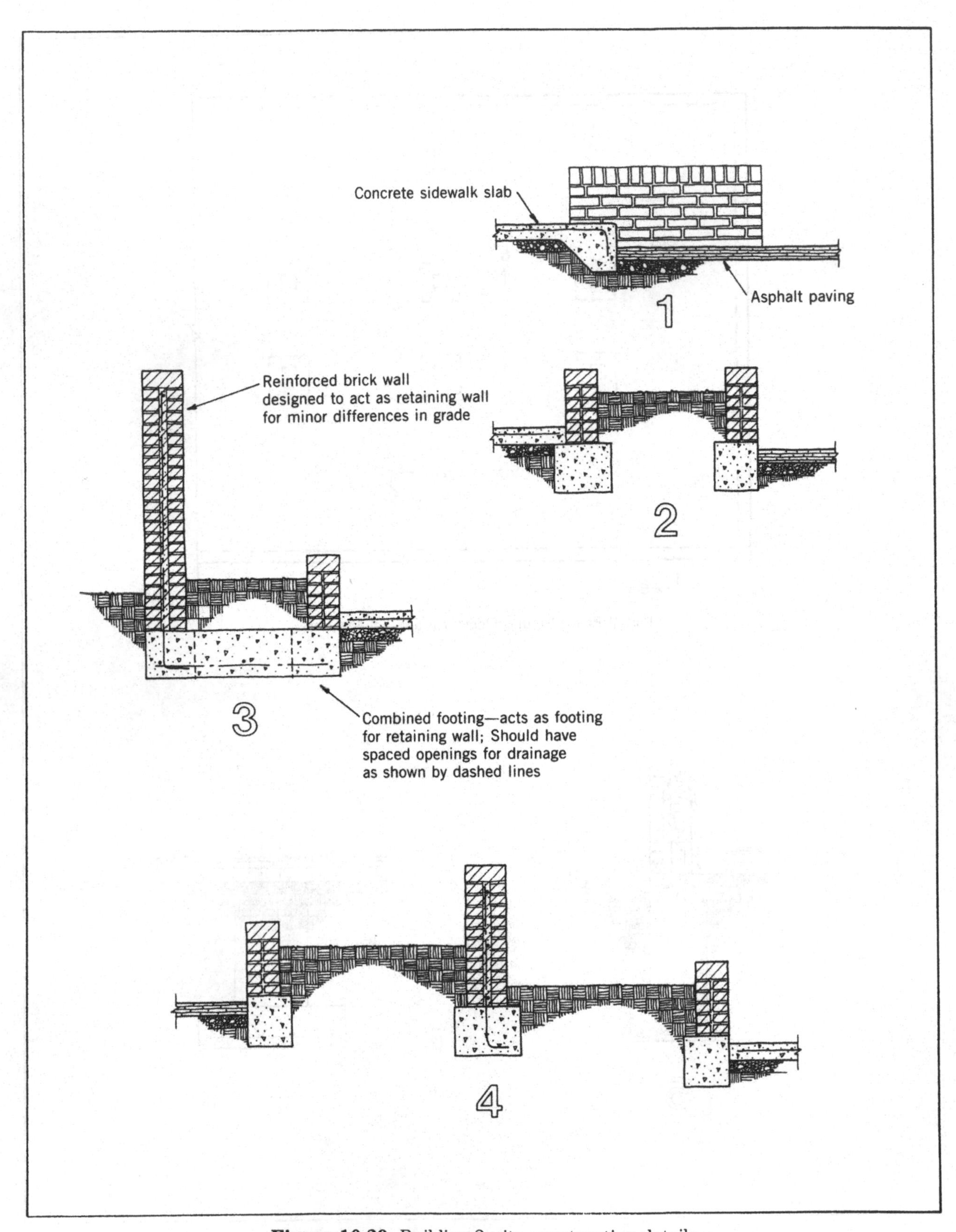

Figure 10.29 Building 2: site construction details.

The general form and some details for the HVAC system are shown in Figure 10.30. The system shown is an all air, single duct system, using the ceiling interstitial space as a plenum return. The air-handling equipment is installed on the roof. Heating is generated by a hot water boiler that supplies piped hot water to a fan coil system. Chilled water is similarly supplied by a refrigeration system.

In this building, the water heater and chilled water systems are housed in a room at floor level, which is accessed from the exterior. (See Figure 10.19.) The chilled water system normally uses a heat-dissipating unit that must be exposed to outdoor air; requiring it to be placed outside the building—on the roof or at ground level.

The selection of the basic HVAC system, the individual elements of the system, and the general operation of the system must respond to many concerns, principal of which are the following:

The local climate, concerning the range of temperatures, predominant concerns for heating or cooling, prevailing humidity, and wind conditions.

Planning concerns for the building, regarding the type of occupants, need for many individual controls (single interior spaces), concerns for noise and vibration of equipment, and the need for concealing of equipment—both inside and outside.

Specific form of construction, concerning particularly provision of wall and/or ceiling interstitial space, ease of access of hidden elements for maintenance or alteration, and natural thermal responses of the construction.

Desired locations for equipment and for ducts, registers, controls, and so on.

For the system as shown here, some specific concerns that relate to the construction are the following:

Support of the rooftop equipment and separation or isolation for control of vibration and noise. This becomes more critical when the roof structure is longspan and when the equipment is directly above noise-sensitive interior spaces.

Accommodation of the ducts, registers, and return grilles in the ceiling construction and the roof/ceiling interstitial space. This may critically affect the choice of the whole roof and ceiling construction systems.

Accommodation of the heating and chilling equipment in the building interior; as a space assignment concern and a noise separation issue.

Installation details for the piping of the hot and chilled water and for wiring for electrical power and electrical signal control systems.

Use of the light wood frame with its hollow stud walls and the trussed roof with suspended ceiling results in considerable potential for accommodation of hidden elements of the HVAC system. However, this construction does not provide much help for passive thermal response or noise and vibration separation; requiring some considerable enhancement where these are major concerns.

Some details for the installation of the plumbing system are shown in Figure 10.31. Reference should also be made to the plans and details for the construction at the rest room areas shown in Figure 10.25. Plumbing in this building is quite minimal in nature, serving only the rest rooms, the drinking fountain, and the hose bibs for the

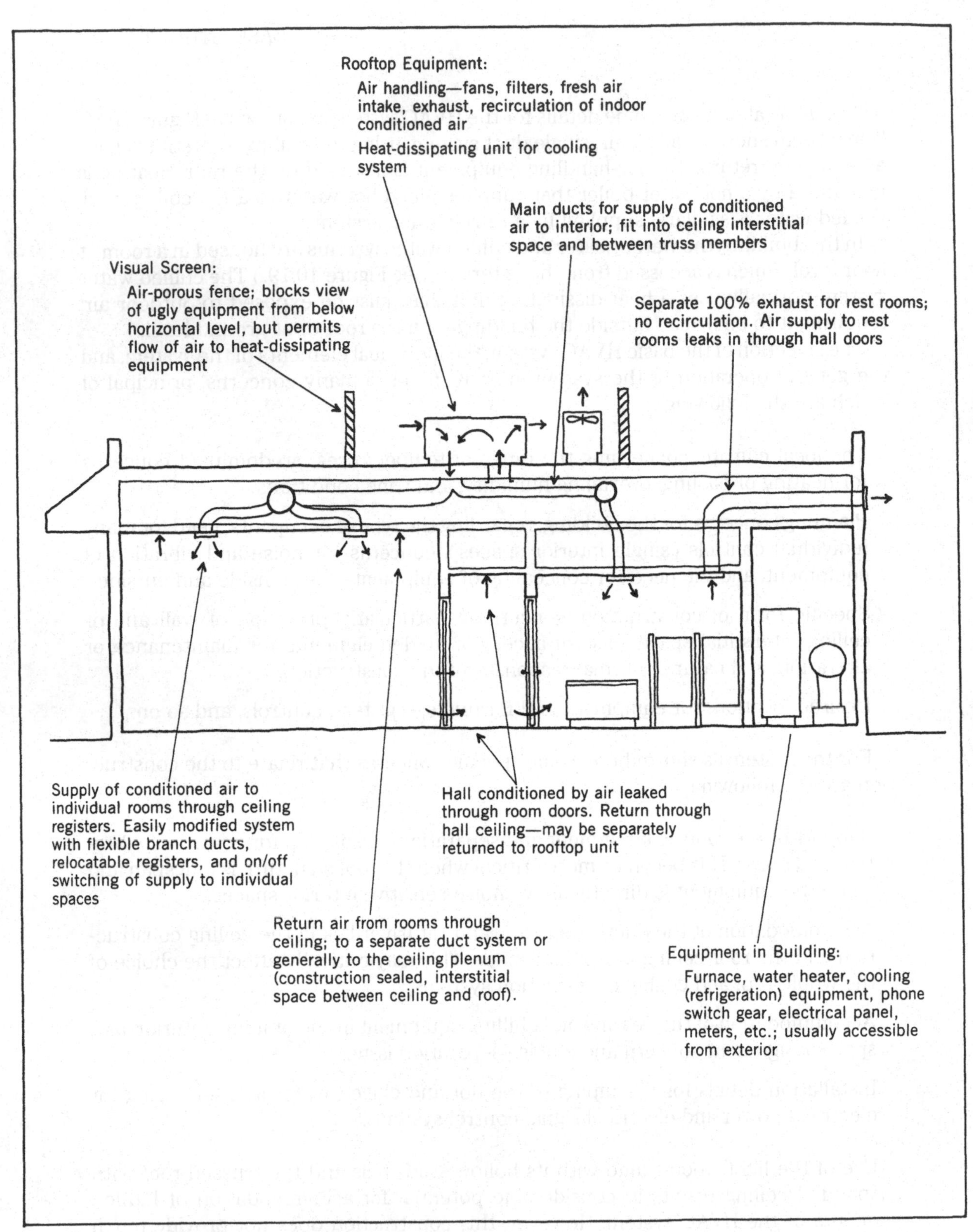
Rooftop Equipment:
Air handling—fans, filters, fresh air intake, exhaust, recirculation of indoor conditioned air
Heat-dissipating unit for cooling system
Main ducts for supply of conditioned air to interior; fit into ceiling interstitial space and between truss members
Visual Screen:
Air-porous fence; blocks view of ugly equipment from below horizontal level, but permits flow of air to heat-dissipating equipment
Separate 100% exhaust for rest rooms; no recirculation. Air supply to rest rooms leaks in through hall doors
Supply of conditioned air to individual rooms through ceiling registers. Easily modified system with flexible branch ducts, relocatable registers, and on/off switching of supply to individual spaces
Hall conditioned by air leaked through room doors. Return through hall ceiling—may be separately returned to rooftop unit
Return air from rooms through ceiling; to a separate duct system or generally to the ceiling plenum (construction sealed, interstitial space between ceiling and roof).
Equipment in Building:
Furnace, water heater, cooling (refrigeration) equipment, phone switch gear, electrical panel, meters, etc.; usually accessible from exterior

Figure 10.30 Building 2: general scheme and details for the HVAC system.

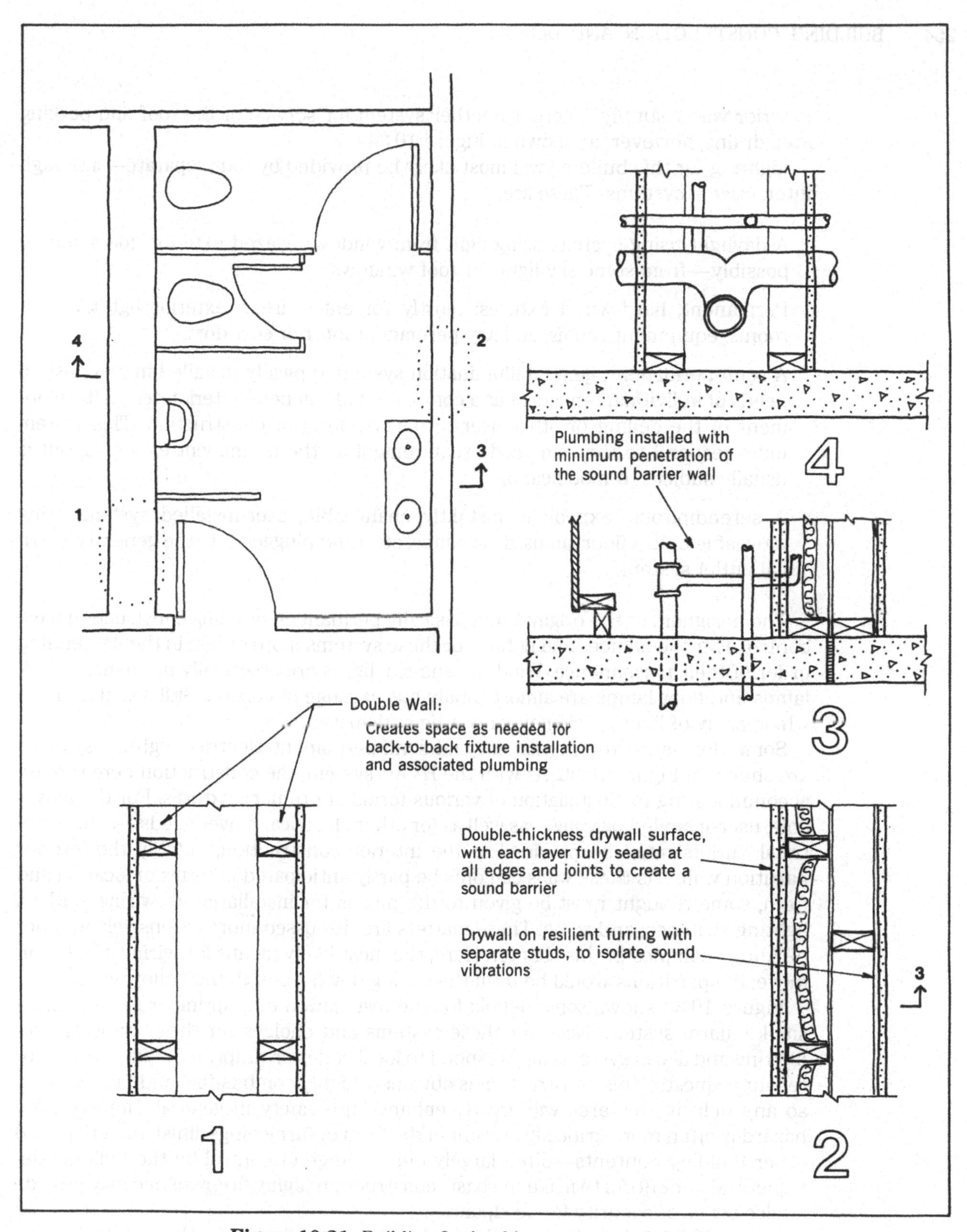

Figure 10.31 Building 2: plumbing systems and details.

exterior water supply. There is another system for service of the roof and parking area drains, however, as shown in Figure 10.34.

Lighting for this building will most likely be provided by four separate—although interactive—systems. These are:

A daylight-using system, using light from windows, glazed exterior doors, and—possibly—from some sky lights or roof windows.

Permanent, hard-wired fixtures; mostly for entry areas, exterior lighting, rest rooms, equipment rooms, and any permanent interior corridors.

A semi-permanent, general illumination system; typically installed in response to needs of individual tenants or as a component of a general interior design development of the ceiling or other user-responsive interior construction. This system must respond to the same code requirement as the permanent elements, but is usually subject to modification.

A serendipitous, expedient, instantly changeable, user-installed system, using moveable units (floor lamps, desk lamps, etc.) and plugged into the general electrical outlet system.

The designers of the original, and basically permanent, building construction have control over the choices and details for these systems more or less in the descending order of their permanence. Windows and sky lights are essentially permanent; desk lamps and floor lamps are almost totally out of range of control. Still, the use of the whole array of lighting elements must be anticipated.

Some details for the permanent and semi-permanent electrical lighting systems are shown in Figure 10.32. As with the HVAC system, the construction here is quite accommodating for installation of various forms of ordinary fixtures. For the moveable, user-installed lighting—as well as for other electrical power needs—some electrical outlets must be provided in the interior construction; notably the interior partition walls. As these walls can only be partly anticipated in terms of location and form, some thought must be given to the means for installation of wiring, outlets, lighting switches, and so on. These matters are discussed more extensively in other building examples in this chapter. Here, the most likely means for wiring of interior, moveable partitions would be from the ceiling down through the hollow walls.

Figure 10.33 shows some details for the installation of a sprinkler system and a smoke alarm system. Need for these systems and choices for these or other fire-fighting and alarm systems may respond to local codes or simply to building owner or tenant requests. This construction is obviously highly combustible and fast-burning, so any help in this area will greatly enhance fire safety in general. However, fire hazard is often more critically present in the form of furnishings, finish materials, and other building contents—often largely out of range of control by the building designers. Major efforts to make the basic construction highly fire-resistive may provide a false sense of security for occupants.

Figure 10.34 shows the general site plan and some details for the installation of a sewer system for storm water (rain) drainage from the roof and general lot spaces. Site surface drainage must relate to general conditions regarding neighboring streets and properties. Much of the runoff from paved areas near the site edges will likely feed into curb gutters and sewers on public land, so this is a highly constrained factor for site development.

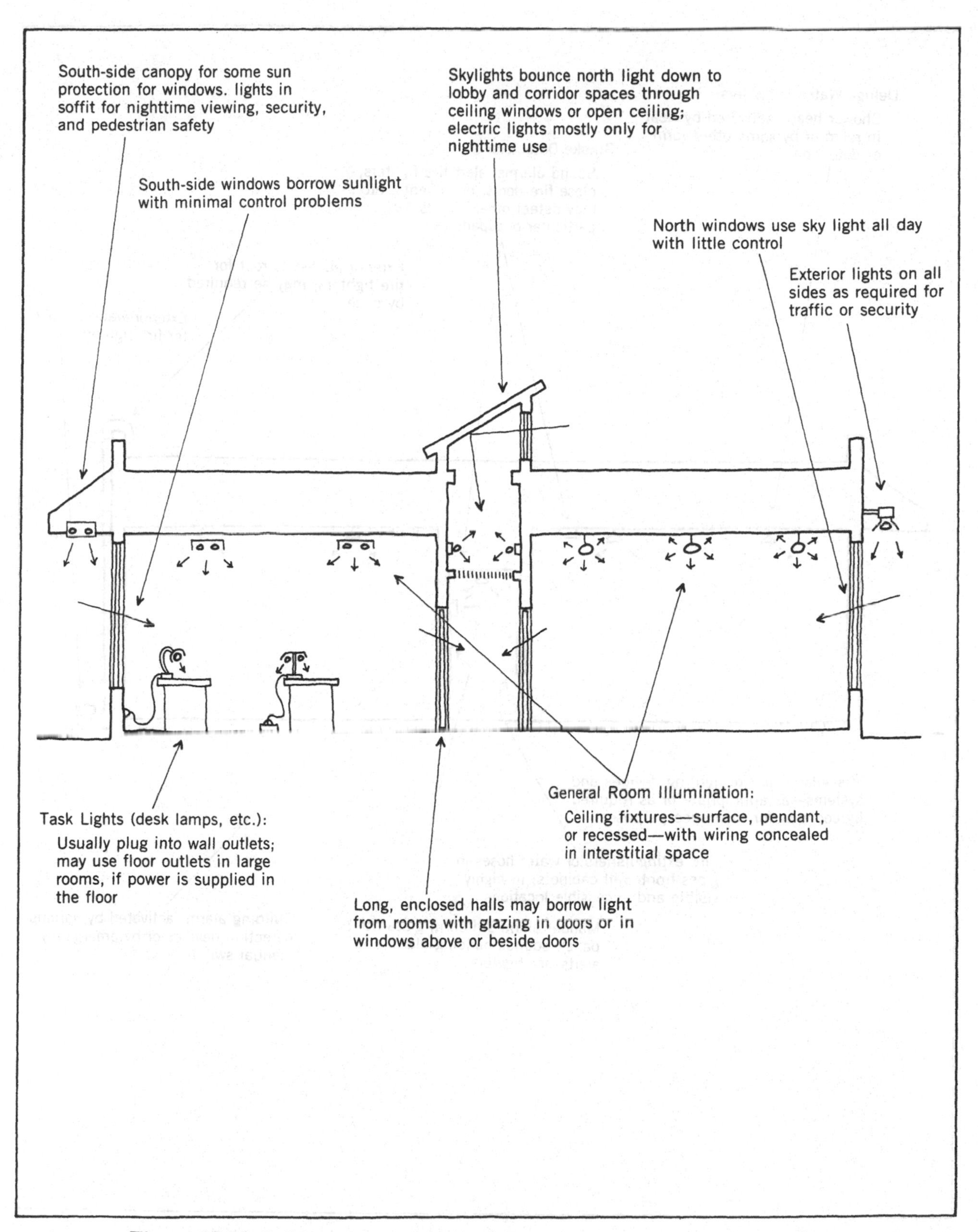

Figure 10.32 Building 2: details of the electric lighting for general illumination.

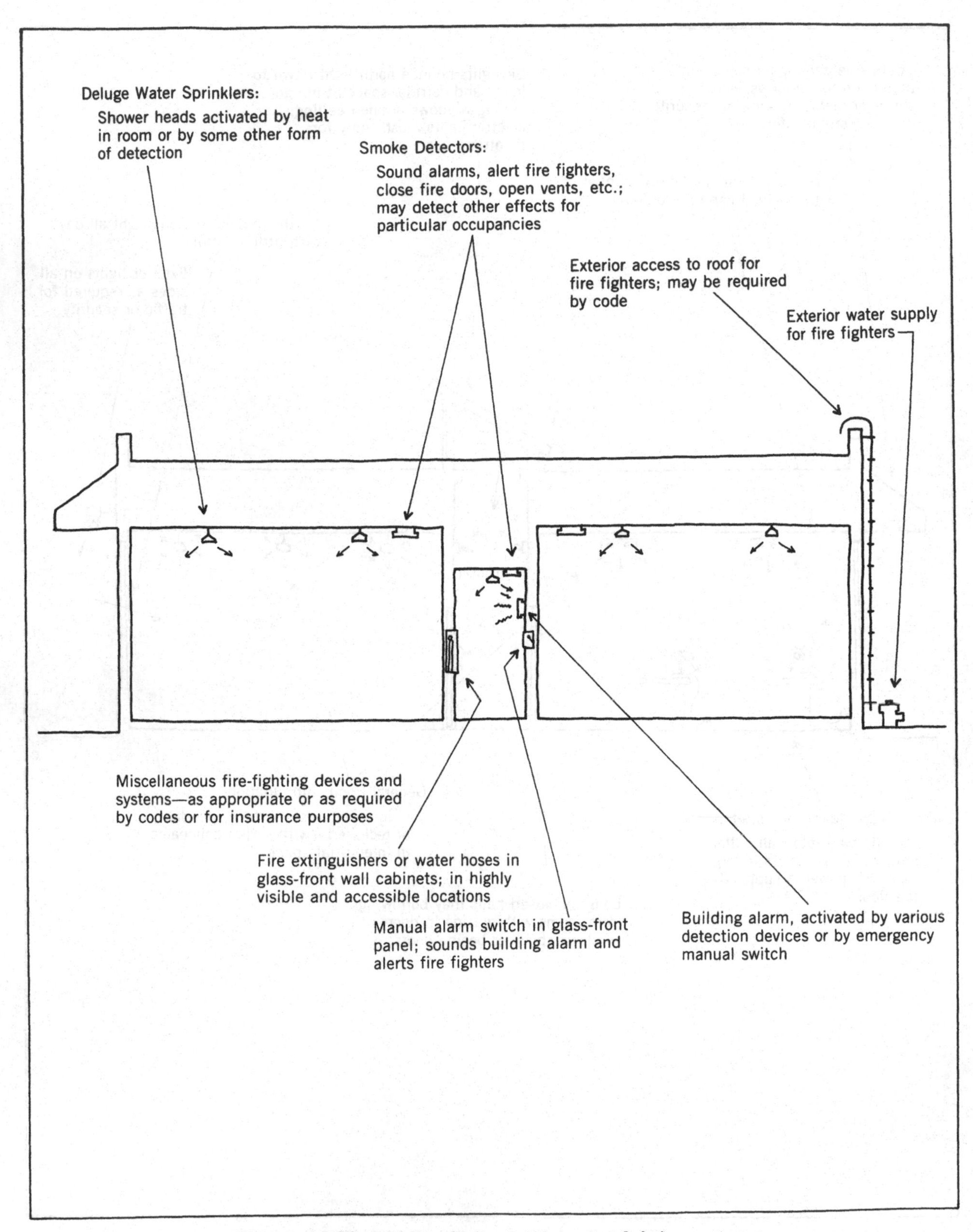
Deluge Water Sprinklers:
Shower heads activated by heat in room or by some other form of detection
Smoke Detectors:
Sound alarms, alert fire fighters, close fire doors, open vents, etc.; may detect other effects for particular occupancies
Exterior access to roof for fire fighters; may be required by code
Exterior water supply for fire fighters
Miscellaneous fire-fighting devices and systems—as appropriate or as required by codes or for insurance purposes
Fire extinguishers or water hoses in glass-front wall cabinets; in highly visible and accessible locations
Manual alarm switch in glass-front panel; sounds building alarm and alerts fire fighters
Building alarm, activated by various detection devices or by emergency manual switch

Figure 10.33 Building 2: details of the fire fighting systems.

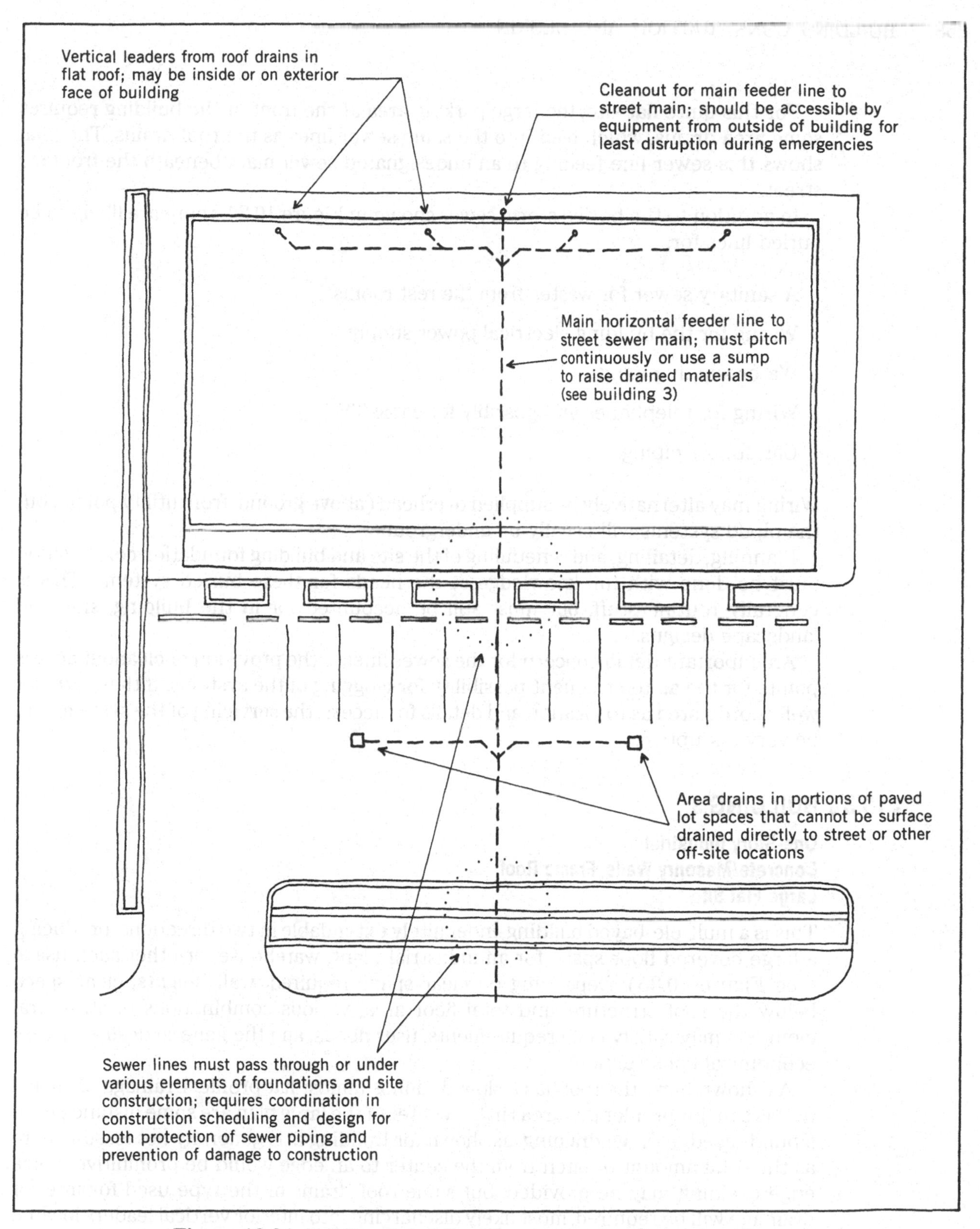
Vertical leaders from roof drains in flat roof; may be inside or on exterior face of building
Cleanout for main feeder line to street main; should be accessible by equipment from outside of building for least disruption during emergencies
Main horizontal feeder line to street sewer main; must pitch continuously or use a sump to raise drained materials (see building 3)
Area drains in portions of paved lot spaces that cannot be surface drained directly to street or other off-site locations
Sewer lines must pass through or under various elements of foundations and site construction; requires coordination in construction scheduling and design for both protection of sewer piping and prevention of damage to construction

Figure 10.34 Building 2: details of the storm water drainage system.

For this quite flat site, the large parking area at the front of the building requires some area drains, which feed into the same sewer lines as the roof drains. The plan shows this sewer line feeding to an undesignated sewer main beneath the frontage street.

In addition to the buried storm sewer shown in Figure 10.34, there are likely to be buried lines for:

A sanitary sewer for wastes from the rest rooms.

Wiring for the building electrical power supply.

Water supply piping.

Wiring for telephones and possibly for cable TV.

Gas supply piping.

Wiring may alternatively be supplied overhead (above ground, from utility poles), but the piped systems will usually be underground.

Planning, detailing, and scheduling of the site and building foundation construction must be done with full knowledge of the needs for these buried systems. This is generally routine stuff, but must still be accounted for in the building, site, and landscape designs.

An important detail concern for the sewer lines is the provision of cleanout access points for the all too frequent possibility for clogging of the systems. If these are not well coordinated as to location and details for access, the servicing of the systems can be very disruptive.

10.3 BUILDING 3

One-Story Industrial
Concrete/Masonry Walls, Frame Roof
Large Flat Site

This is a multiple-bayed building, indefinitely extendable in two directions, producing a large covered floor space for an industrial plant, warehouse, or other such usage (see Figure 10.35). Depending on clear spans required, wall heights, clear space below the roof structure, and total floor area, various combinations of structural elements may satisfy code requirements, user needs, and the general desire for very economical construction.

As shown here, the roof has a slow-draining, almost flat profile. Drainage of such a roof is a major problem, as areas in the center of the large plan are some distance from a building edge. Edge draining, as shown for Building 2, is generally not feasible here as the total amount of pitch from the center to an edge would be prohibitive. Some edge draining may be provided, but some roof drains of the type used for interior drainage will be required, most likely discharging into interior vertical leaders located at columns and in turn feeding into an underfloor piped drainage sewer system.

Roof slopes will most likely be achieved here by simply tilting elements of the roof framing. The nonflat bottom of such a frame is not of concern if no ceiling is provided. However, a suspended ceiling structure of the usual kind can be provided in some areas if one is desired.

Choice of wall materials will depend on fire requirements, local climate, interior

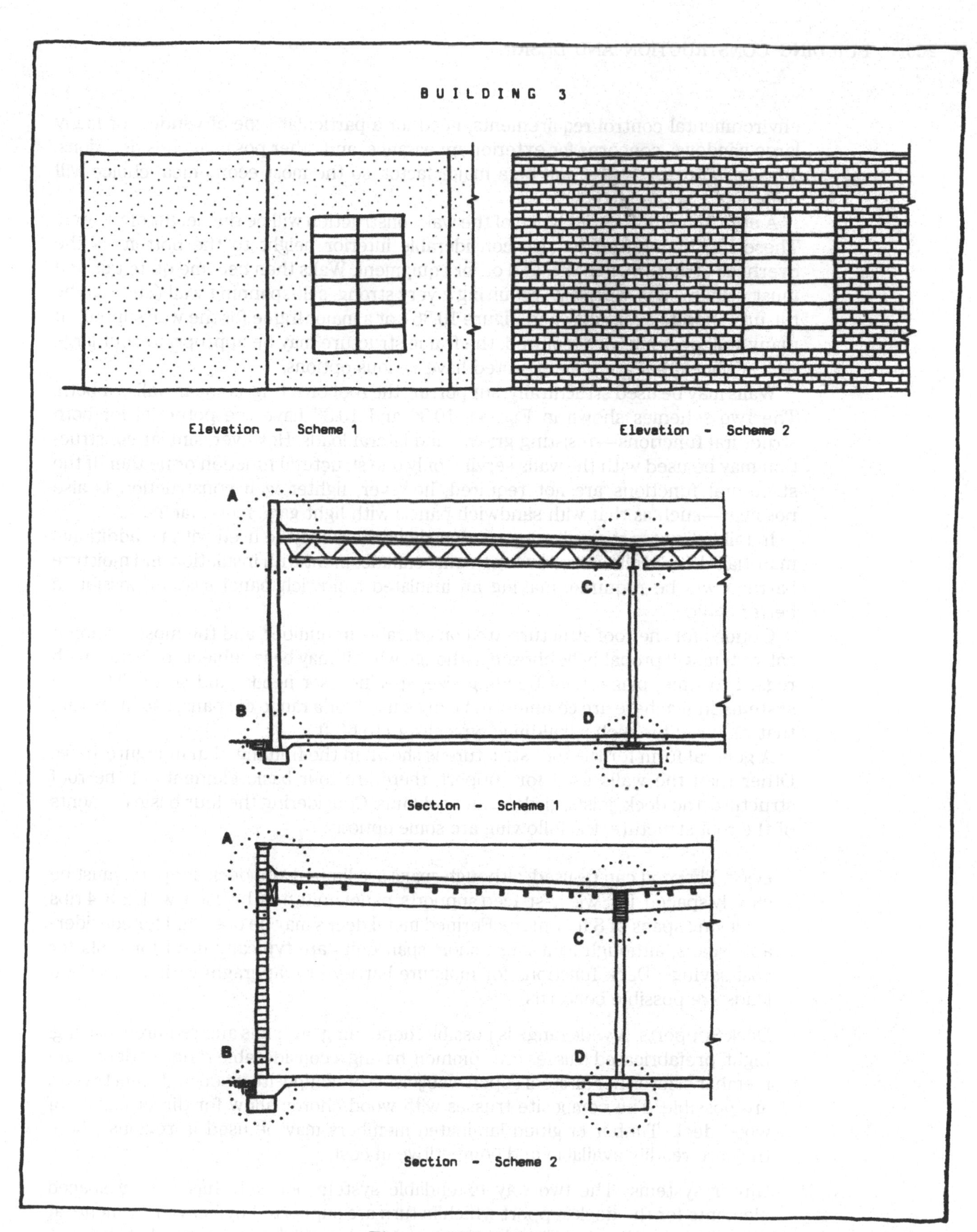

Figure 10.35 Building 3.

environmental control requirements, need for a particular type of window or many large windows, concerns for exterior appearance, and other possible considerations. Typically, however, cost will be a major factor, so the most economical choice will quite likely be preferred.

A critical factor for the choice of the wall construction will be the height of the wall. These buildings often require considerable interior height to the bottom of the overhead structure, lighting, and other equipment. Walls then become quite tall, and must usually span vertically, requiring a very strong, self-spanning wall (such as the tilt-up concrete slabs shown in Figure 10.36), or a major frame for the wall support. If many wall openings are required, the frame structure may be appropriate, but large wall surfaces that are uninterrupted have various options.

Walls may be used structurally, supporting the roof or acting as shear walls, or both. The two schemes shown in Figures 10.36 and 10.37 have the potential for both structural functions—resisting gravity and lateral loads. However, similar construction may be used with the walls serving only one structural function or neither. If the structural functions are not required, however, lighter wall construction is also possible—such as that with sandwich panels with light-gage metal facing.

In mild climates the wall construction shown here may be used with no additional materials. In cold climates, however, some enhancement with insulation and moisture barriers will be required, making an insulated sandwich panel system possibly a better choice.

Options for the roof structure are considerable in number, and the most economical system will probably be chosen, although what it may be is subject to change with regard to time, place, total building size, specific user needs, and so on. The two systems shown here are common and can be used for a range of spans, usually within that required for such a building (typically 40 to 60 ft).

A general form for the roof structure is shown in the framing plan in Figure 10.38. Other than the walls used for support, there are four basic elements of the roof structure: the deck, joists, girders, and columns. Considering the four basic elements of the roof structure, the following are some options.

Deck: Plywood can be used, although spans are limited and deck supports must be closely spaced. For wider-spaced supports, panel units may be used, with 2 × 4 ribs achieving spans of 8 ft or more. Formed metal decks may be obtained for considerable spans, although light-gage, short-span units are typically used for roofs for cost savings. Deck functions for moisture barrier or diaphragm action for lateral loads are possible concerns.

Deck Supports: A wide range is possible, depending on spans and required spacing. Light, prefabricated trusses are common, having a considerable range of depth and member size; either all steel (open web joists) or composite wood and steel trusses are possible. The composite trusses with wood chords allow for direct nailing of wood deck. Timber or glued laminated members may be used in regions where they are readily available and competitive in cost.

Girder systems: The two-way extendable system normally has closely spaced elements for the deck support which in turn are supported by heavier elements at right angles in the plan. The elements used for this will depend somewhat on spans required. The three types of members most often used are:

Glued laminated beams, available in almost unlimited size, although shipping, erection, and cost must be considered

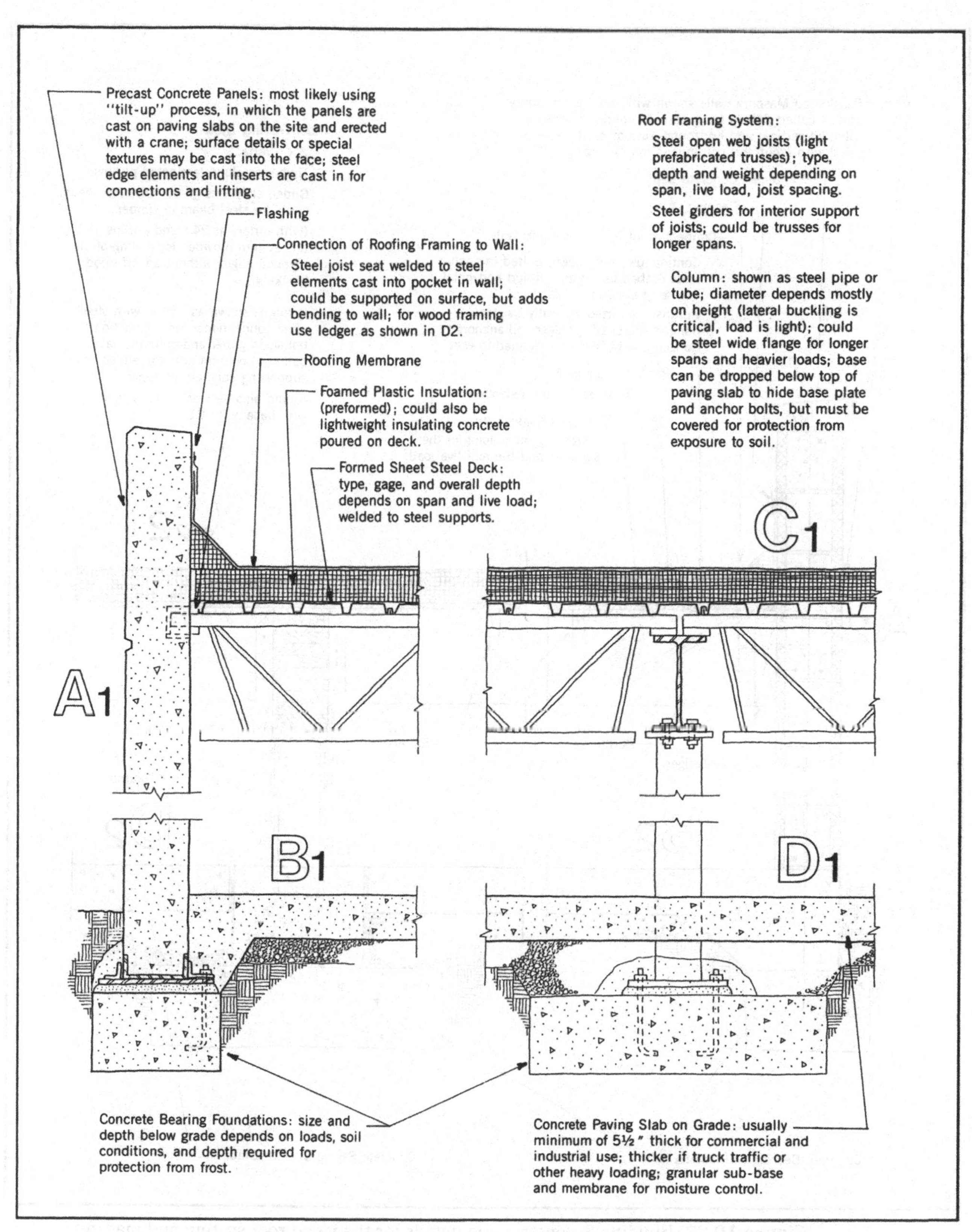

Precast Concrete Panels: most likely using "tilt-up" process, in which the panels are cast on paving slabs on the site and erected with a crane; surface details or special textures may be cast into the face; steel edge elements and inserts are cast in for connections and lifting.
Flashing
Connection of Roofing Framing to Wall: Steel joist seat welded to steel elements cast into pocket in wall; could be supported on surface, but adds bending to wall; for wood framing use ledger as shown in D2.
Roofing Membrane
Foamed Plastic Insulation: (preformed); could also be lightweight insulating concrete poured on deck.
Formed Sheet Steel Deck: type, gage, and overall depth depends on span and live load; welded to steel supports.
Roof Framing System:
Steel open web joists (light prefabricated trusses); type, depth and weight depending on span, live load, joist spacing.
Steel girders for interior support of joists; could be trusses for longer spans.
Column: shown as steel pipe or tube; diameter depends mostly on height (lateral buckling is critical, load is light); could be steel wide flange for longer spans and heavier loads; base can be dropped below top of paving slab to hide base plate and anchor bolts, but must be covered for protection from exposure to soil.
A1
B1
C1
D1
Concrete Bearing Foundations: size and depth below grade depends on loads, soil conditions, and depth required for protection from frost.
Concrete Paving Slab on Grade: usually minimum of 5½" thick for commercial and industrial use; thicker if truck traffic or other heavy loading; granular sub-base and membrane for moisture control.

Figure 10.36 Building 3: construction details for the steel roof system and tilt-up concrete walls.

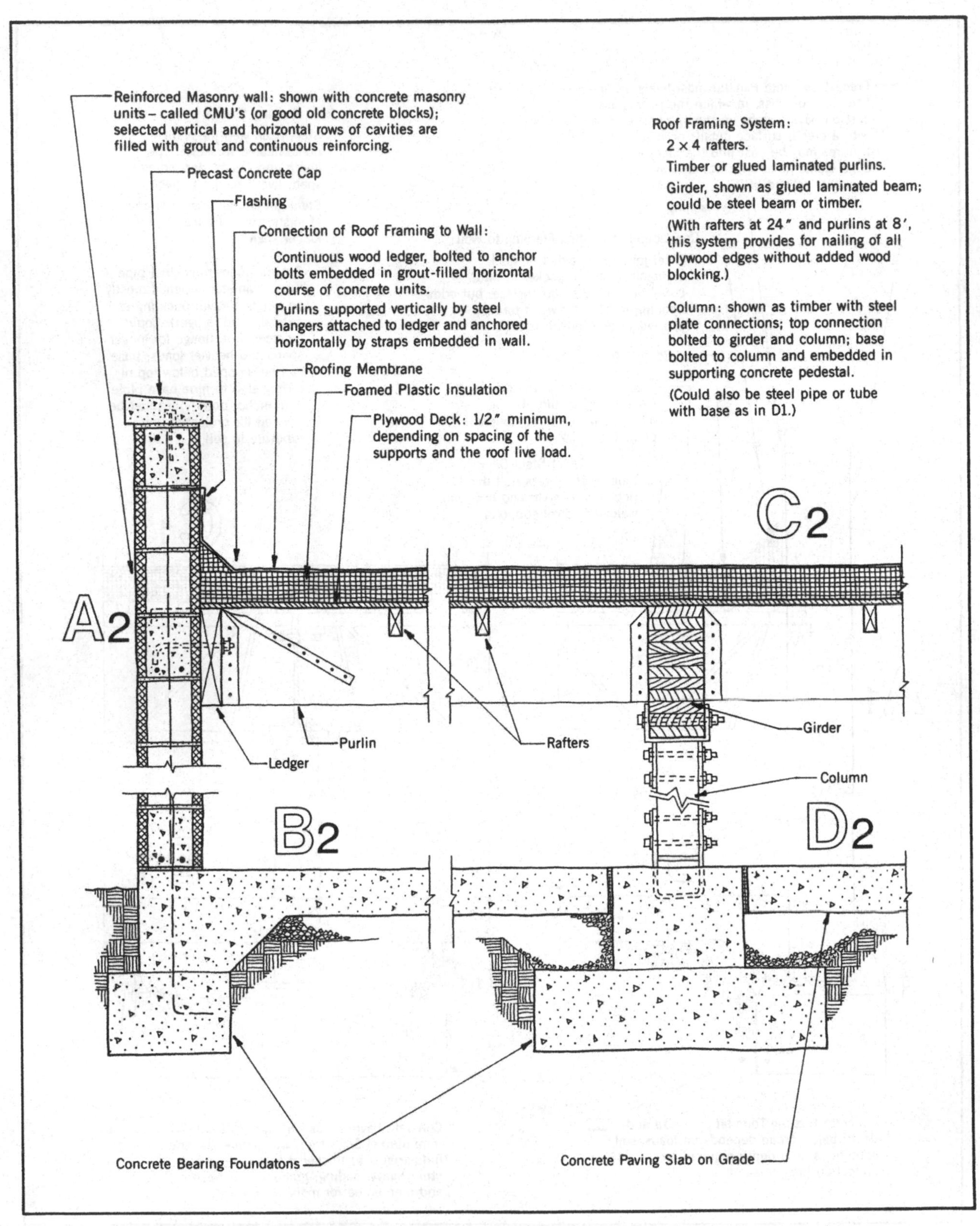
Reinforced Masonry wall: shown with concrete masonry units – called CMU's (or good old concrete blocks); selected vertical and horizontal rows of cavities are filled with grout and continuous reinforcing.
Precast Concrete Cap
Flashing
Connection of Roof Framing to Wall:
Continuous wood ledger, bolted to anchor bolts embedded in grout-filled horizontal course of concrete units.
Purlins supported vertically by steel hangers attached to ledger and anchored horizontally by straps embedded in wall.
Roofing Membrane
Foamed Plastic Insulation
Plywood Deck: 1/2″ minimum, depending on spacing of the supports and the roof live load.
Roof Framing System:
2 × 4 rafters.
Timber or glued laminated purlins.
Girder, shown as glued laminated beam; could be steel beam or timber.
(With rafters at 24″ and purlins at 8′, this system provides for nailing of all plywood edges without added wood blocking.)
Column: shown as timber with steel plate connections; top connection bolted to girder and column; base bolted to column and embedded in supporting concrete pedestal.
(Could also be steel pipe or tube with base as in D1.)
C2
A2
Ledger
Purlin
Rafters
Girder
Column
B2
D2
Concrete Bearing Foundatons
Concrete Paving Slab on Grade

Figure 10.37 Building 3: construction details for the wood roof system and masonry walls.

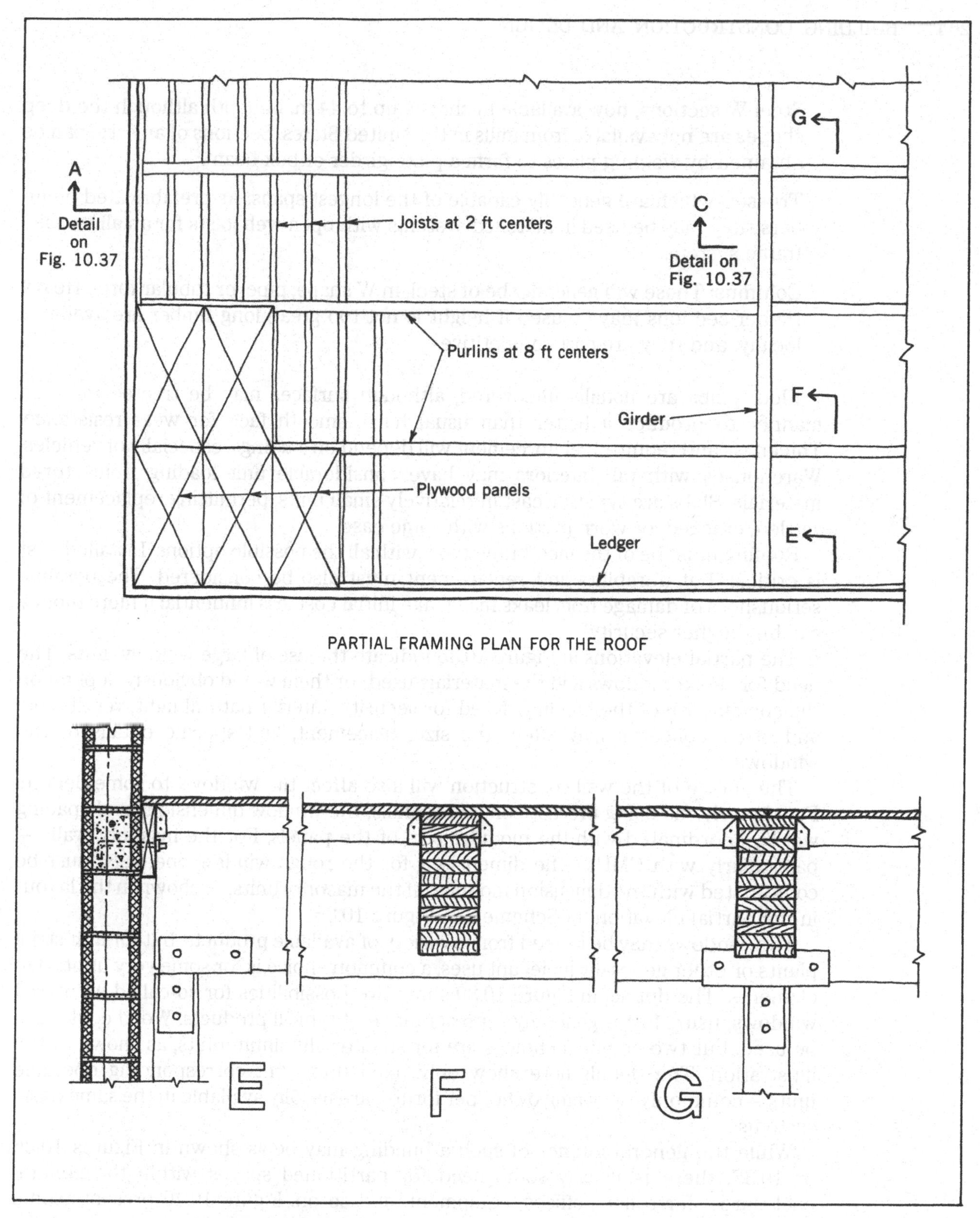

Figure 10.38 Building 3: roof framing plan and structural details.

Steel W sections, now available in shapes up to 44 in. in depth although the deep shapes are not available from mills in the United States. Sections of any size can be obtained by welding plates to form a plate girder or box beam.

Trusses, which are generally capable of the longest spans. As prefabricated members they may be used in matched systems with open web joists for an all trussed frame system.

Columns: These will generally be of steel, in W shape, pipe, or tubular form. Heavy timber sections may be used if height is not too great, long timber are available locally, and they are cost competitive.

Floor slabs are usually uncovered, although surfaces may be treated in some manner to produce a better than usual hard, smooth face for wear resistance. Thickness and required reinforcement will depend on loadings, especially of vehicles. Warehouses with tall interiors may have considerable unit loading from stored materials. Slabs are typically cast in relatively small units, permitting replacement of settled, cracked, or worn portions with some ease.

Roofing must be of the membrane type, with all the possible options. Installed cost is critical, but durability and replacement must also be considered. The possible seriousness of damage from leaks may make initial cost less influential if more money can buy higher security.

The partial elevations in Figure 10.35 indicate the use of large window units. The need for these windows and the materials used for them would obviously depend on the specific use of the building. Need for security, interior natural light, ventilation, and other concerns may affect the size, placement, and specific details of the windows.

The choice of the wall construction will also affect the windows to some degree. For the scheme using precast concrete panels, the window dimensions and spacing will be coordinated with the modular size of the panels. For the masonry walls—particularly with CMUs—the dimensions for the rough window openings must be coordinated with the dimension modules of the masonry units, as shown in the layout in the partial elevation for Scheme 2 in Figure 10.35.

The windows may be formed from a variety of available products, but for industrial plants or other generally inelegant uses, a common choice is for some very utilitarian elements. The details in Figure 10.39 show two possibilities for so-called industrial windows, using forms generally representative of typical products. Wood could also be used, but two common choices are for steel or aluminum units, as shown in the illustration. The details here show only fixed units, but corresponding operable units—commonly of awning or hopper form—are usually available in the same basic systems.

While the general interior of such a building may be as shown in Figures 10.36 or 10.37, there is usually some need for partitioned spaces within the general enclosure—for toilets, offices, equipment, and so on. Figure 10.40 presents some details for the creation of such spaces that includes the development of partitions, a suspended ceiling, and some enhancement of the inside surface of the exterior wall. Floor treatments may also be used in such a space, although the change in finished floor surface level will present a minor problem at doorways between the two types of spaces, which probably favors use of relatively thin flooring materials.

The site plan in Figure 10.41 indicates a flat site and shows some of the connections to services. Water, gas, and sewer lines must be underground, involving some careful

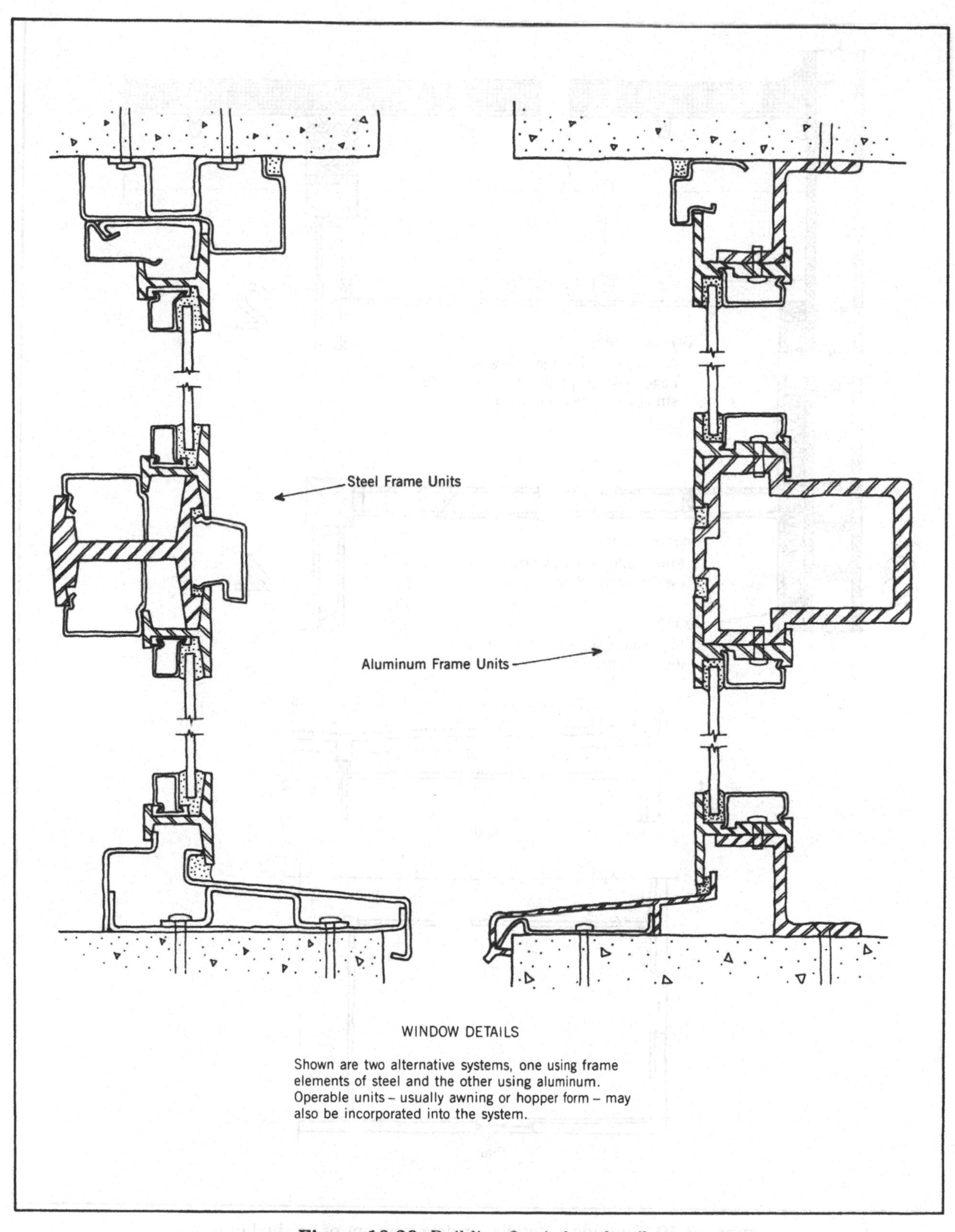

Steel Frame Units
Aluminum Frame Units
WINDOW DETAILS
Shown are two alternative systems, one using frame elements of steel and the other using aluminum. Operable units – usually awning or hopper form – may also be incorporated into the system.

Figure 10.39 Building 3: window details.

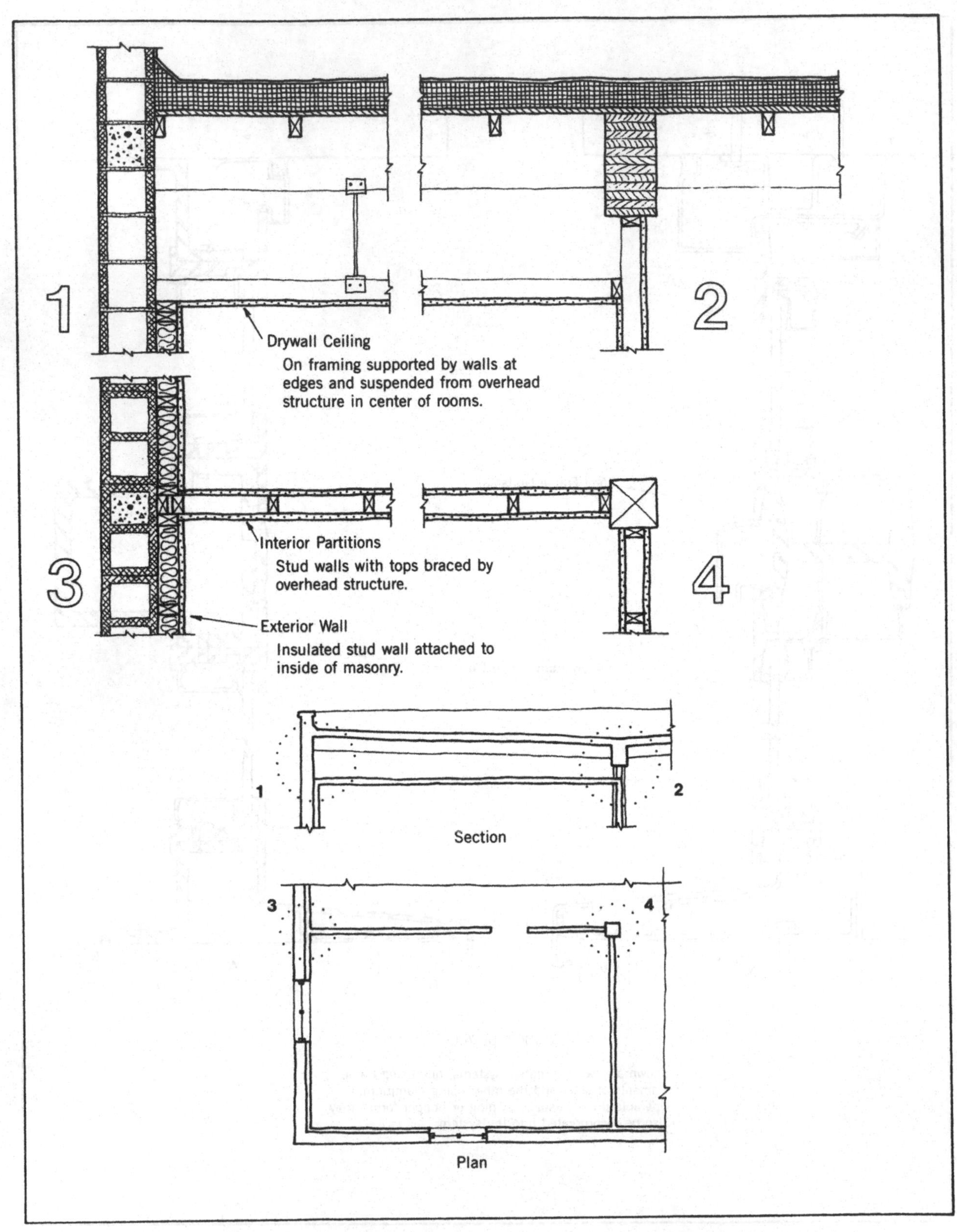

Figure 10.40 Building 3: details for development of finished interiors.

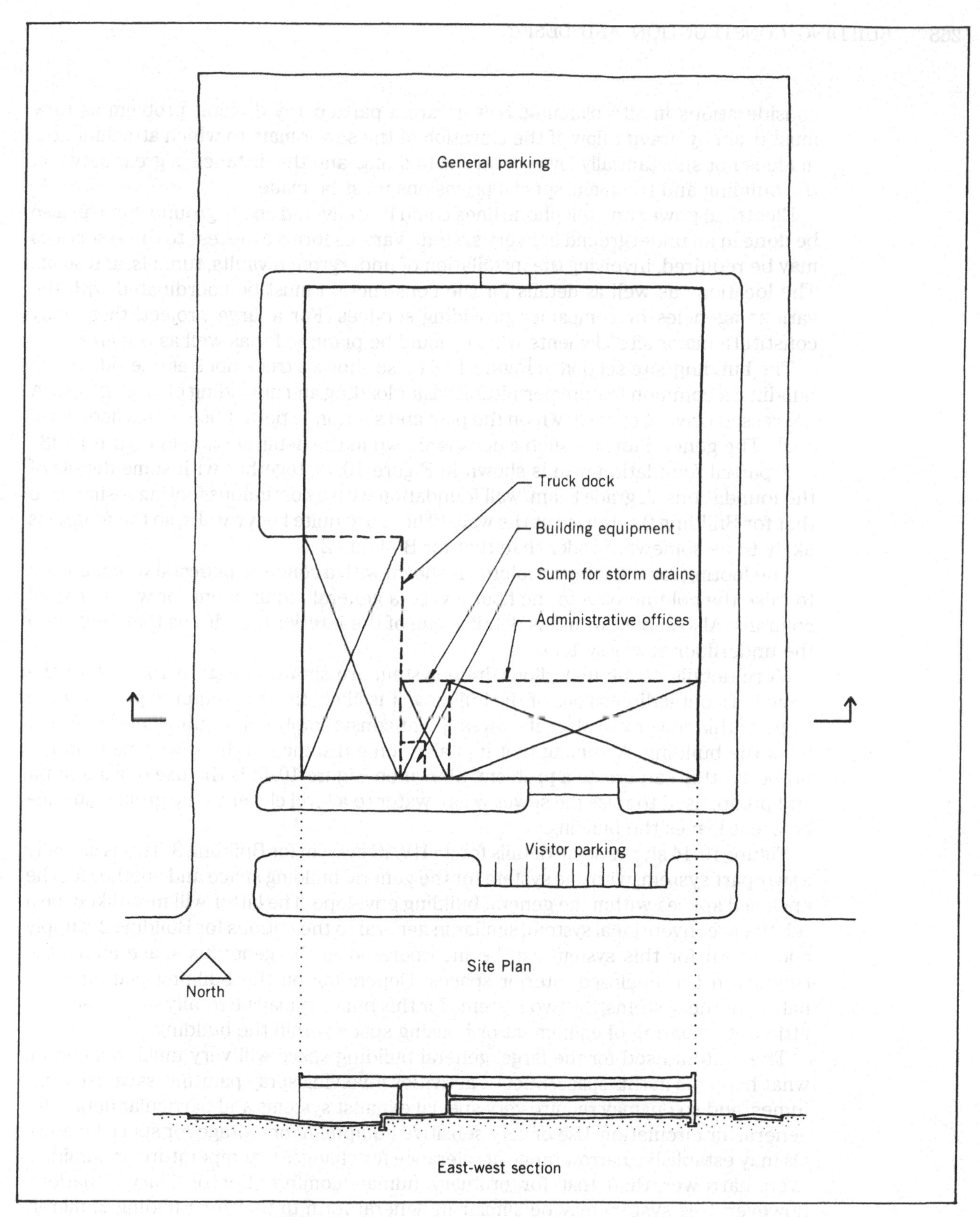

Figure 10.41 Building 3: site plan and section.

considerations in site planning. Sewers are a particularly difficult problem as they must drain by gravity flow. If the elevation of the sewer main to which attachment is made is not substantially lower than the building, and the distance is great between the building and the main, special provisions must be made.

Electrical power and telephone lines could be delivered above ground, but can also be done in an underground delivery system. Various forms of access to these services may be required, involving the installation of underground vaults, tunnels, and so on. The locations as well as details for the construction must be coordinated with the various agencies or companies providing services. For a large project, these may constitute major site elements which should be planned for as well as possible.

The building/site section in Figure 10.41 also shows a truck dock at one side of the building; a common feature permitting easier loading and unloading of large trucks. A depressed drive area is shown on the plan and section to permit use of this floor-level dock. The general form of such a dock is shown in the detail section in Figure 10.43.

A partial foundation plan is shown in Figure 10.42, together with some details of the foundations. A grade beam/wall foundation with a continuous footing—similar to that for Building 2—is used at the walls. These are quite heavy walls, so the footing is likely to be somewhat wider than that for Building 2.

The footing for the interior column is shown with a concrete pedestal which is used to raise the column base to the floor level—a general requirement for wood or steel columns. Also shown with this detail is one of the interior roof drains that feeds into the underfloor sewer system.

More details of the underfloor drain system are shown in Figure 10.43. With the large horizontal dimensions of the building, it is likely that the required gravity drain slope of this sewer will place its lowest point considerably below grade level before it exits the building. Assuming that it is also some distance to the sewer main off the property, this can create a problem. Shown on Figure 10.43 is the use of a sump pit and pump, used to raise the sewer waste water to a level closer to the ground surface before it leaves the building.

Figure 10.44 shows some details for an HVAC system for Building 3. This is actually a two-part system, with one system for the general building space and another for the enclosed spaces within the general building envelope. The latter will most likely be a relatively conventional system, similar in general to the options for Building 2. Supply and return for this system can be incorporated in the generous space above the ceilings in the enclosed interior spaces. Depending on the building plan and the nature of the systems, the two systems for this building may be totally separated with little or no sharing of equipment or housing space within the building.

The system used for the large, general building space will very much depend on what happens in this space. Special activities involving spray painting, sawdust, toxic fumes, and so on, may require very special exhaust systems and particular details for general air circulation. Use of very sensitive equipment or storage for special materials may establish a narrow range of tolerance for changes in temperature or humidity, even narrower than that for ordinary human comfort. For ordinary situations, however, this system may be similar in general form to that for Building 2, merely larger in all respects; with bigger ducts, fans, and other equipment elements, more registers and returns, and so on.

Instead of the single, centrally-operating and centrally-controlled system as shown in Figure 10.44, it may be possible to use a series of individual, rooftop systems, each serving a limited area of the building. This is more likely in some climates and when different needs occur at different locations in the building.

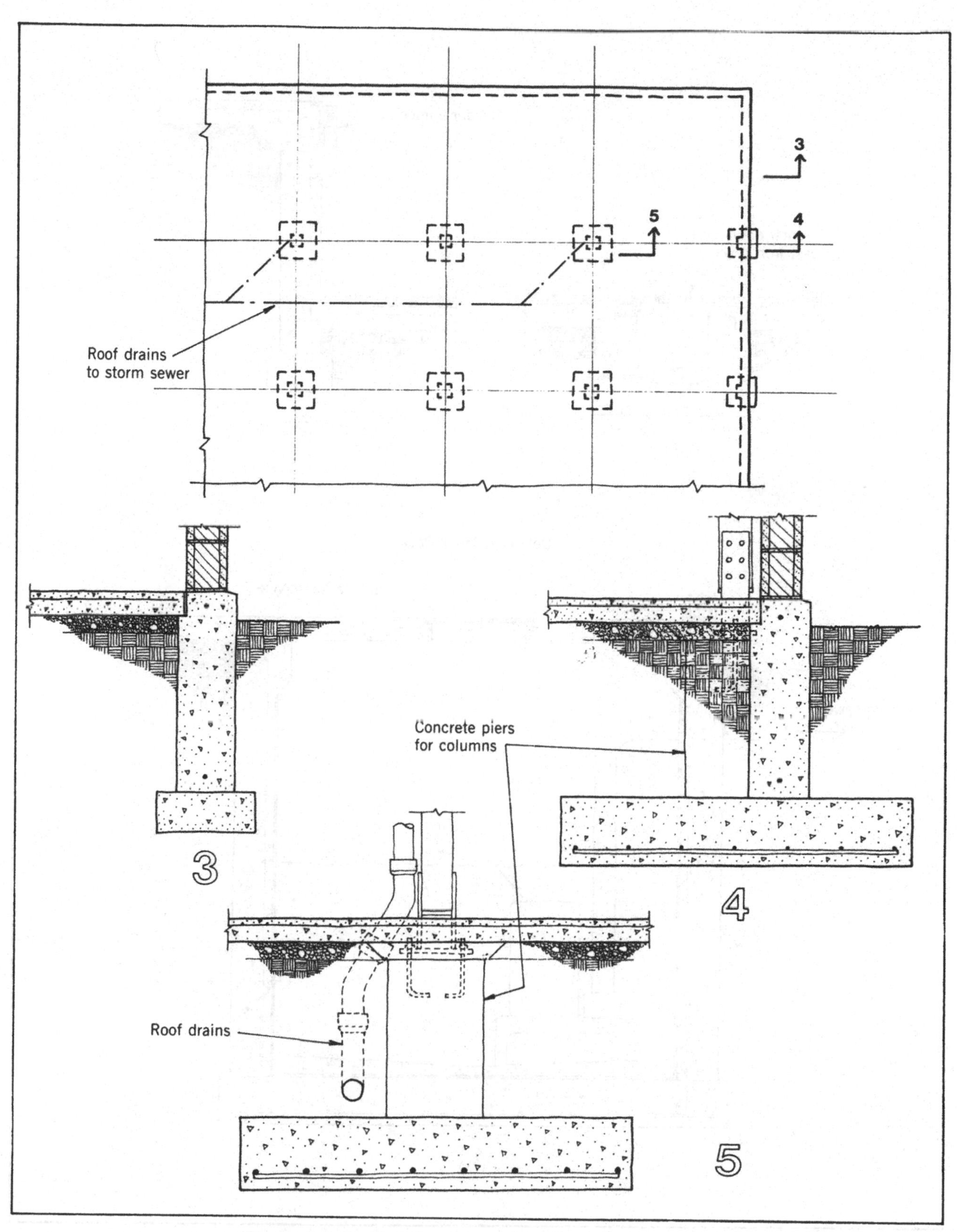

Figure 10.42 Building 3: foundation plan and details.

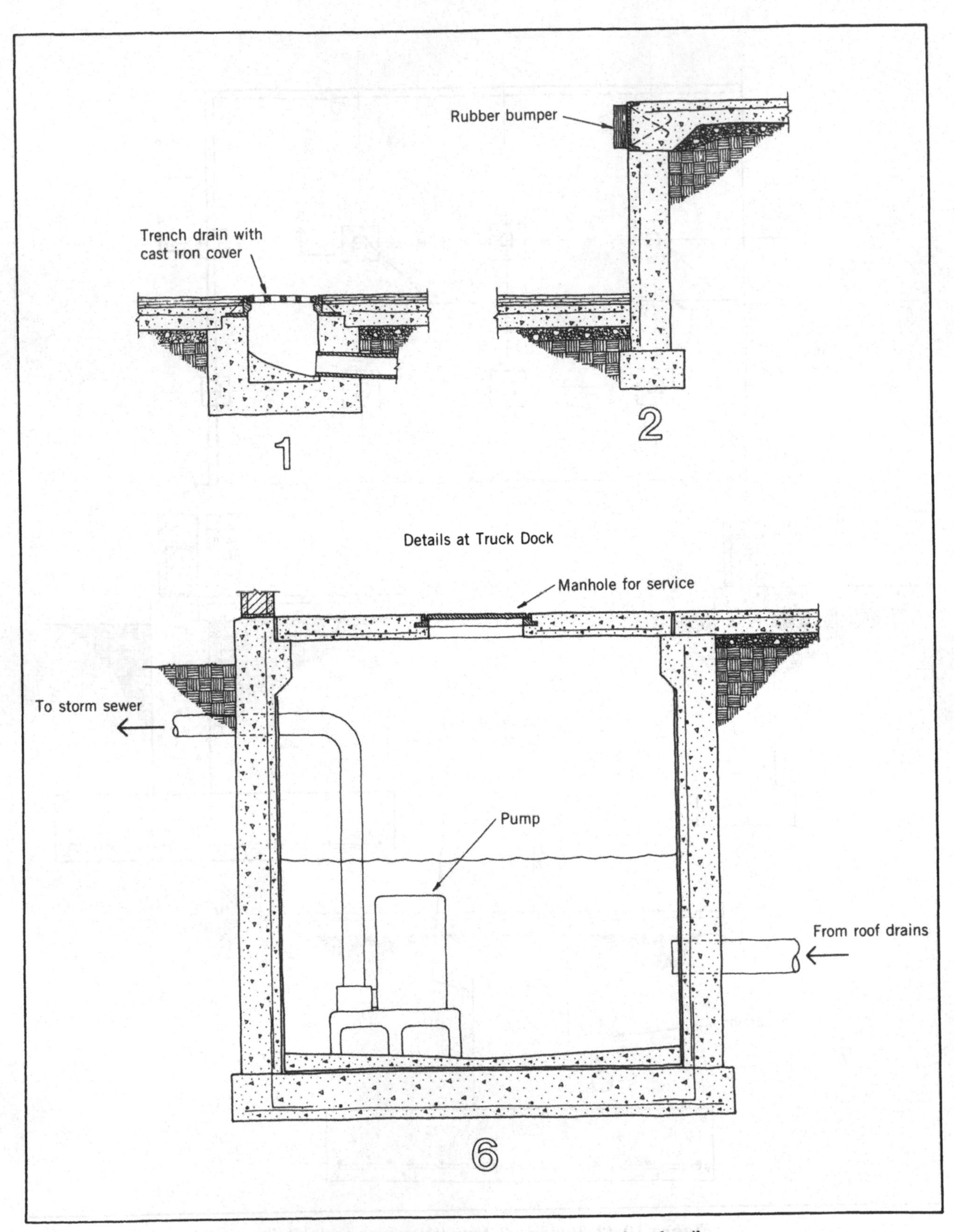

Figure 10.43 Building 3: special construction details.

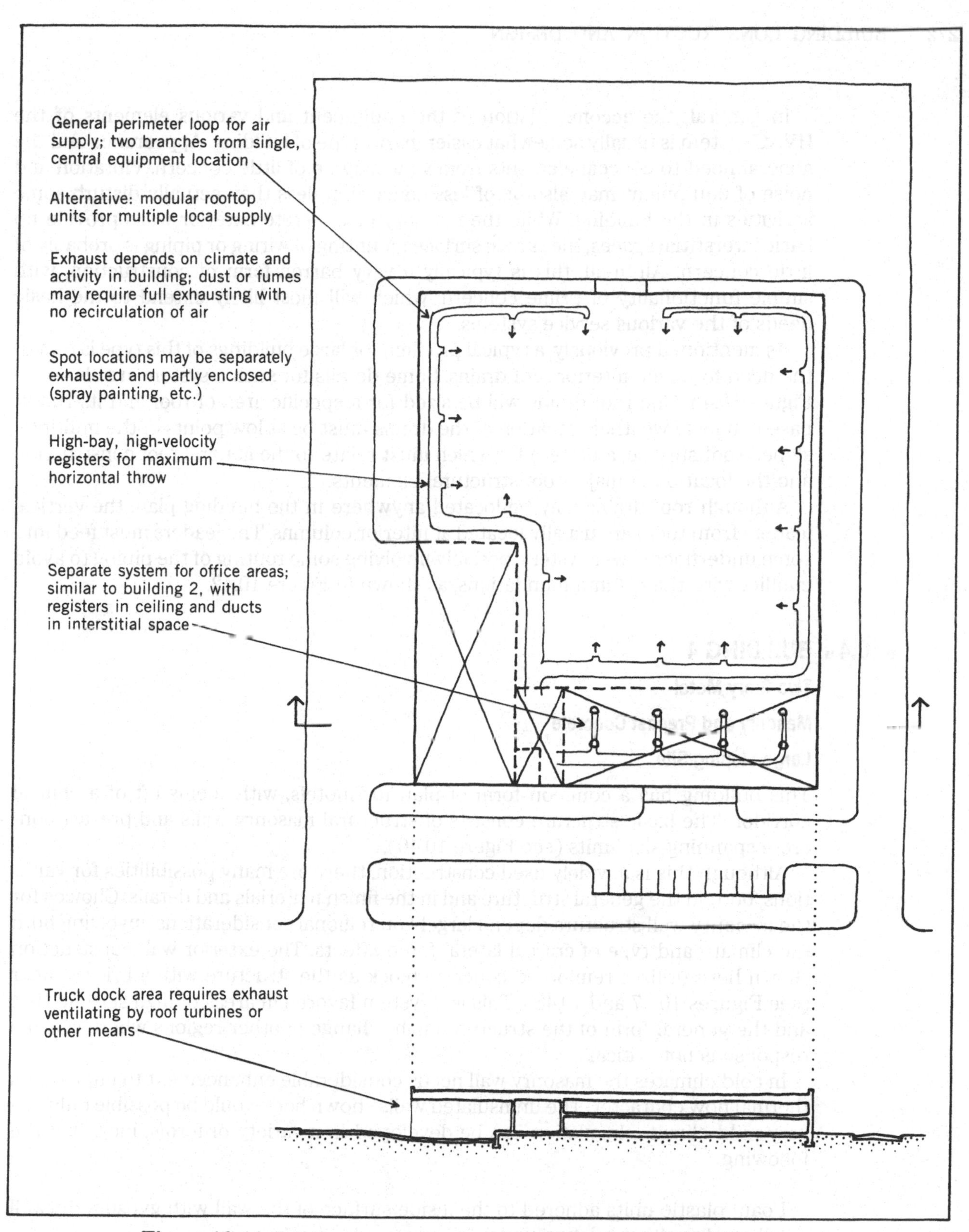
General perimeter loop for air supply; two branches from single, central equipment location
Alternative: modular rooftop units for multiple local supply
Exhaust depends on climate and activity in building; dust or fumes may require full exhausting with no recirculation of air
Spot locations may be separately exhausted and partly enclosed (spray painting, etc.)
High-bay, high-velocity registers for maximum horizontal throw
Separate system for office areas; similar to building 2, with registers in ceiling and ducts in interstitial space
Truck dock area requires exhaust ventilating by roof turbines or other means

Figure 10.44 Building 3: general scheme and details for the HVAC system.

In general, the accommodation of the equipment and various elements of the HVAC system is usually somewhat easier in this type of building. Appearance and the general need to conceal elements from view may be of little concern. Vibration and noise of equipment may also be of less concern, unless they actually disturb some activities in the building. While the masonry or concrete exterior walls provide no basic interstitial spaces, the simple surface mounting of wiring or piping is probably of little concern. All-in-all, this is typically a very barren form of construction, with simple functionality of prime concern; which will most likely extend to the basic needs of the various service systems.

As mentioned previously, a typical problem for large buildings of this type is that of the need for some interior roof drains. Some details for such a system are shown in Figure 10.45. The roof drains will be sized for a specific area of roof and its runoff based on local weather. Location of the drains must be at low points of the multiple-sloped roof surface; a pattern for which must relate to the general roof construction and the location of major roof structural elements.

Although roof drains may be located anywhere in the building plan, the vertical leaders from them are usually located at interior columns. The leaders must feed into some underfloor sewer system, possibly involving some routing of the piping to avoid conflict with the column foundations, as shown in Figure 10.42.

10.4 BUILDING 4

Two-Story Motel

Masonry and Precast Concrete

Large Sloping Site

This building has a common form of plan for motels, with rooms off of a central corridor. The basic structure consists of structural masonry walls and precast concrete spanning slab units (see Figure 10.46).

Although this is a widely used construction, there are many possibilities for variations, both in the general structure and in the finish materials and details. Choices for the masonry wall structure depend largely on regional considerations, involving both the climate and type of critical lateral force effects. The exterior wall construction shown here utilizes reinforced concrete block as the structure with a brick veneer (see Figures 10.47 and 10.48). This is a system favored in areas of high seismic risk, and the general form of the structure might change in other regions where seismic response is not critical.

In cold climates the masonry wall needs considerable enhancement to improve its thermal flow character. The uninsulated walls shown here would be possible only in a very mild climate. Insulation can be developed in a variety of forms, including the following:

Foam plastic units adhered to the inside surface of the wall with gypsum drywall adhered to the insulation units.

Batt insulation in a void space created by furring out on the inside surface of the masonry, with gypsum drywall attached to the furring strips.

Foam plastic insulation in the cavity space between the concrete block and the brick veneer.

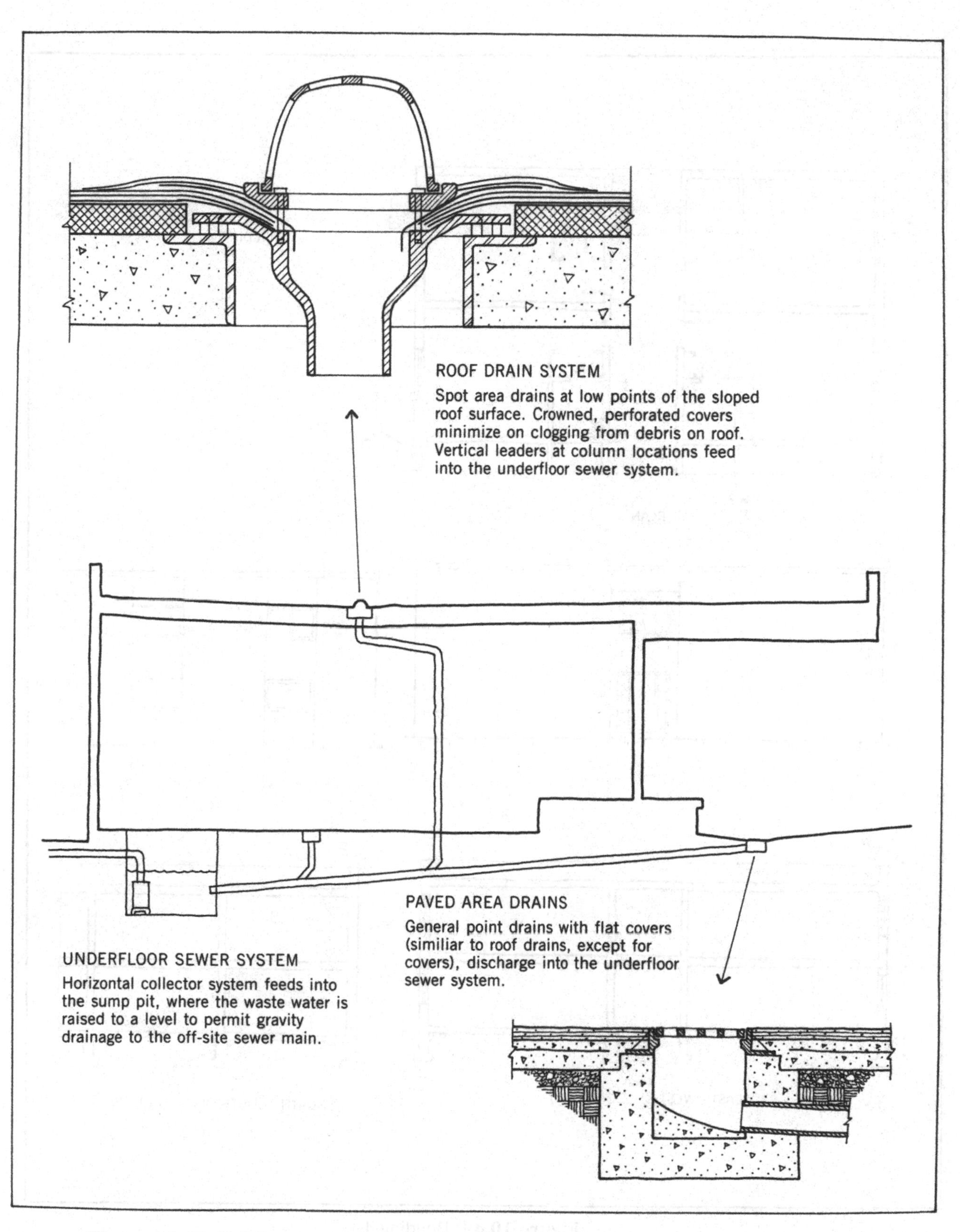

ROOF DRAIN SYSTEM
Spot area drains at low points of the sloped roof surface. Crowned, perforated covers minimize on clogging from debris on roof. Vertical leaders at column locations feed into the underfloor sewer system.
PAVED AREA DRAINS
General point drains with flat covers (similiar to roof drains, except for covers), discharge into the underfloor sewer system.
UNDERFLOOR SEWER SYSTEM
Horizontal collector system feeds into the sump pit, where the waste water is raised to a level to permit gravity drainage to the off-site sewer main.

Figure 10.45 Building 3: details of the interior roof drain and storm water sewer system.

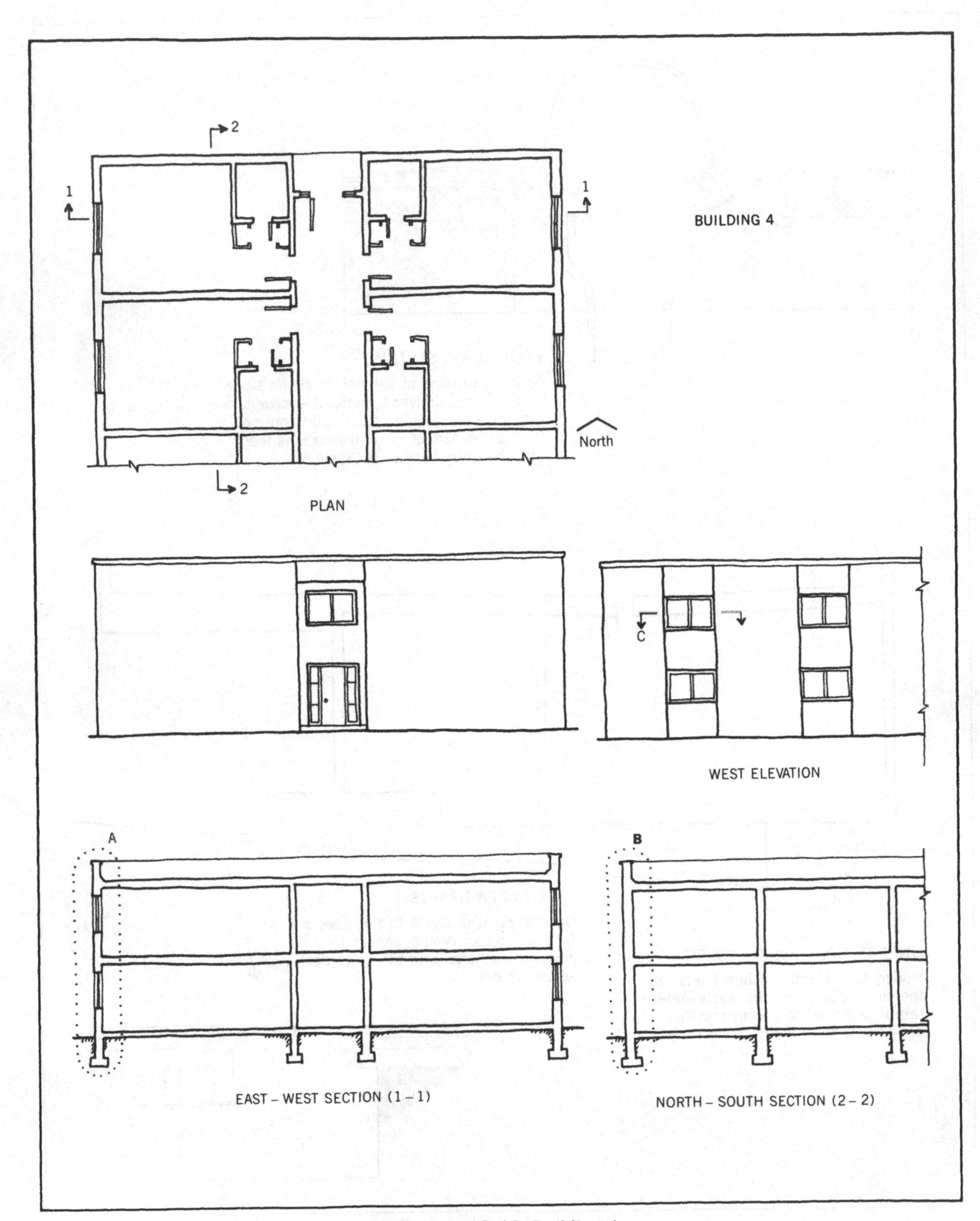

Figure 10.46 Building 4.

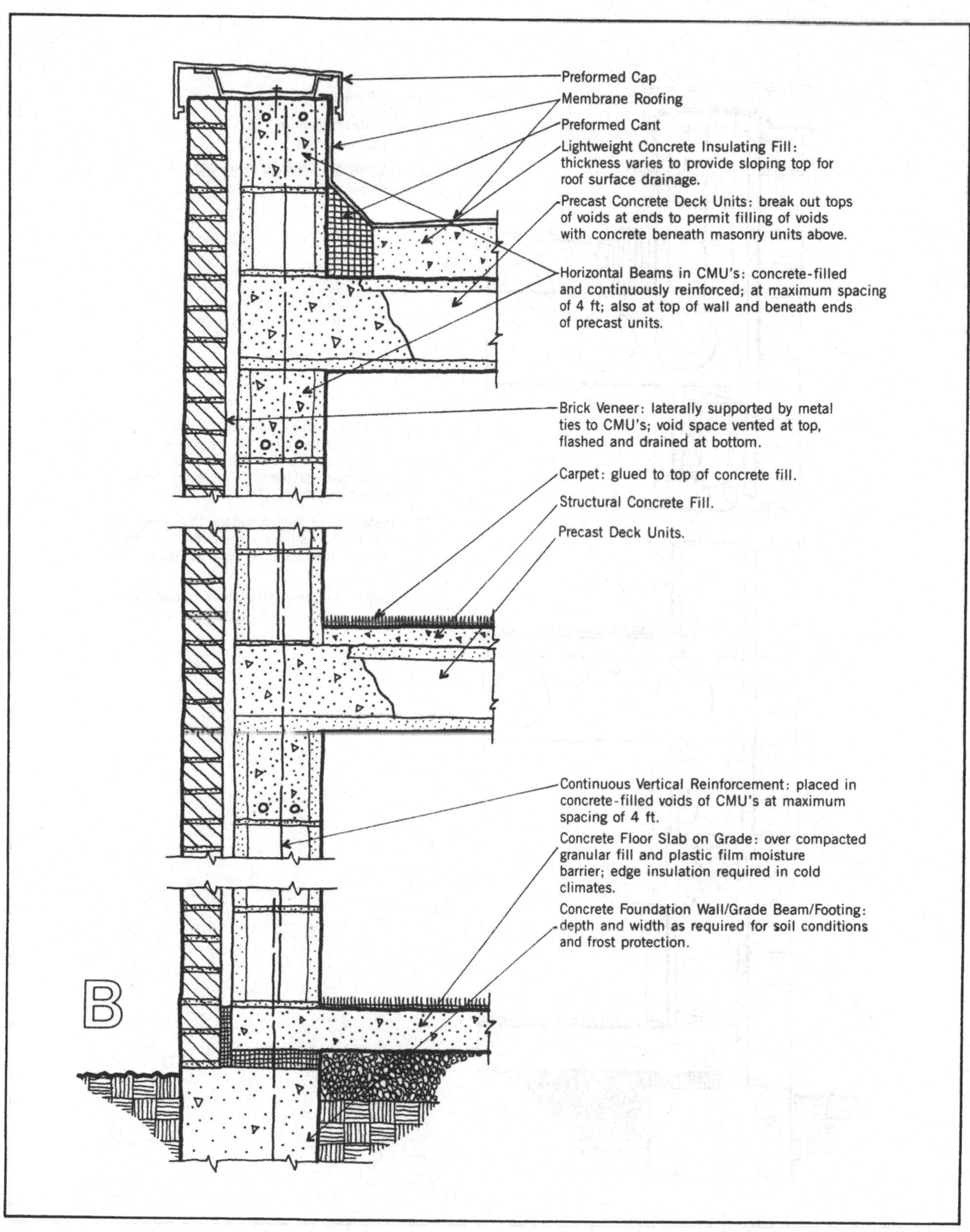

Preformed Cap
Membrane Roofing
Preformed Cant
Lightweight Concrete Insulating Fill: thickness varies to provide sloping top for roof surface drainage.
Precast Concrete Deck Units: break out tops of voids at ends to permit filling of voids with concrete beneath masonry units above.
Horizontal Beams in CMU's: concrete-filled and continuously reinforced; at maximum spacing of 4 ft; also at top of wall and beneath ends of precast units.
Brick Veneer: laterally supported by metal ties to CMU's; void space vented at top, flashed and drained at bottom.
Carpet: glued to top of concrete fill.
Structural Concrete Fill.
Precast Deck Units.
Continuous Vertical Reinforcement: placed in concrete-filled voids of CMU's at maximum spacing of 4 ft.
Concrete Floor Slab on Grade: over compacted granular fill and plastic film moisture barrier; edge insulation required in cold climates.
Concrete Foundation Wall/Grade Beam/Footing: depth and width as required for soil conditions and frost protection.
B

Figure 10.47 Building 4: construction details at brick veneer and structural masonry wall.

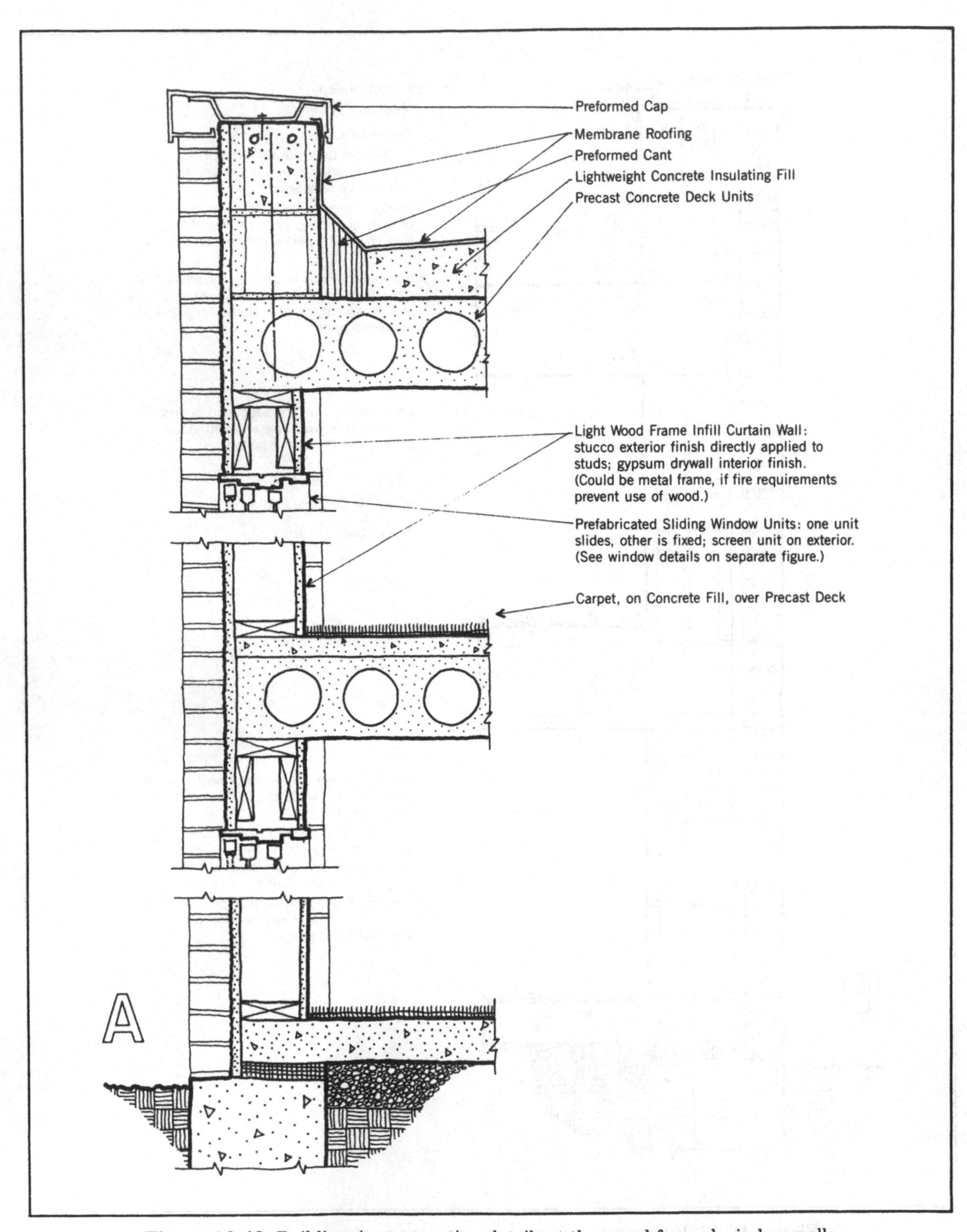

Figure 10.48 Building 4: construction details at the wood framed window walls.

Foam plastic units adhered to the outside of the masonry (without the brick veneer) with a vinyl stucco finish with fiberglass mesh reinforcement.

The roof may also be insulated in a variety of ways, depending on the type of roofing system selected. The details here show a conventional multilayered membrane on top of lightweight concrete insulating fill.

A cost-saving feature with this system is the use of the exposed surfaces of the concrete block walls and the underside of the precast floor and roof units. These may be sealed and painted, but are commonly not otherwise finished. Of course, finish paneling materials can be adhered to the walls or attached to furring strips, and ceilings can be finished in various ways—with sprayed-on coatings of thick plaster, or a conventional suspended ceiling. The latter is frequently used in halls, bathrooms, and room entry areas to accommodate ducting, piping, and wiring in horizontal distribution runs to the rooms.

The precast concrete deck units must be ones produced by some precasting facility within reasonable distance of the building site. Choice may thus be limited to a particular manufacturer's product in some regions. For the details shown here the Flexicore system has been used, as produced by the Flexicore Corporation.

The building end walls are shown here with no penetration, except for the corridor ends. There are thus two very long, uninterrupted masonry walls on the building ends. On the other sides, the walls must accommodate the room windows (and possibly ground-level entries and second-floor balconies).

For lateral bracing, the masonry walls must serve as shear walls. This is not a problem on the ends of the building, but must be studied with care on the room window sides. The drawings here show the use of wide strips of solid masonry between window units, resulting in an alternating series of masonry piers and window units on the elevation (see Figure 10.46). The masonry piers must be limited to a reasonable aspect ratio in terms of height-to-plan length in order to function as shear walls with reasonable overturn resistance. In this example the short end walls are much too slender to serve as shear walls, but the piers between rooms are of a reasonable form.

Details for the installation of horizontally sliding wood windows in the wood stud walls are shown in Figure 10.49. The windows shown are priority units that come as a complete package with glazing, frames, and sliding combination screen and storm windows. Just plop them into the rough opening; true them up in position; seal the outside edges to the plaster screed molding; and trim them up on the inside—not much demand for time and effort in the field.

Development of interior walls is shown in Figure 10.50. The walls between motel rooms are structural CMU construction, providing support for the roof and floor units. These are relatively heavy walls, and may well be left exposed, possibly receiving only some paint. However, for a higher quality finish, they may receive some surfacing with paneling or plaster. Their sound privacy quality may also be enhanced by using a furred out surface on one side; with an additional degree achieved by using resilient attachment of the surfacing to the furring.

The nonstructural partitions for the halls, bathrooms, and closets would most likely be achieved with simple light wood framing. A dropped ceiling is typically used in these areas to allow for encasing of air ducts, wiring, and piping. This form of construction is shown in the details in Figure 10.50.

The general form of the precast concrete structure for the floor and roof is revealed in the section details. (See Figures 10.47, 10.48, and 10.50.) This basic system of

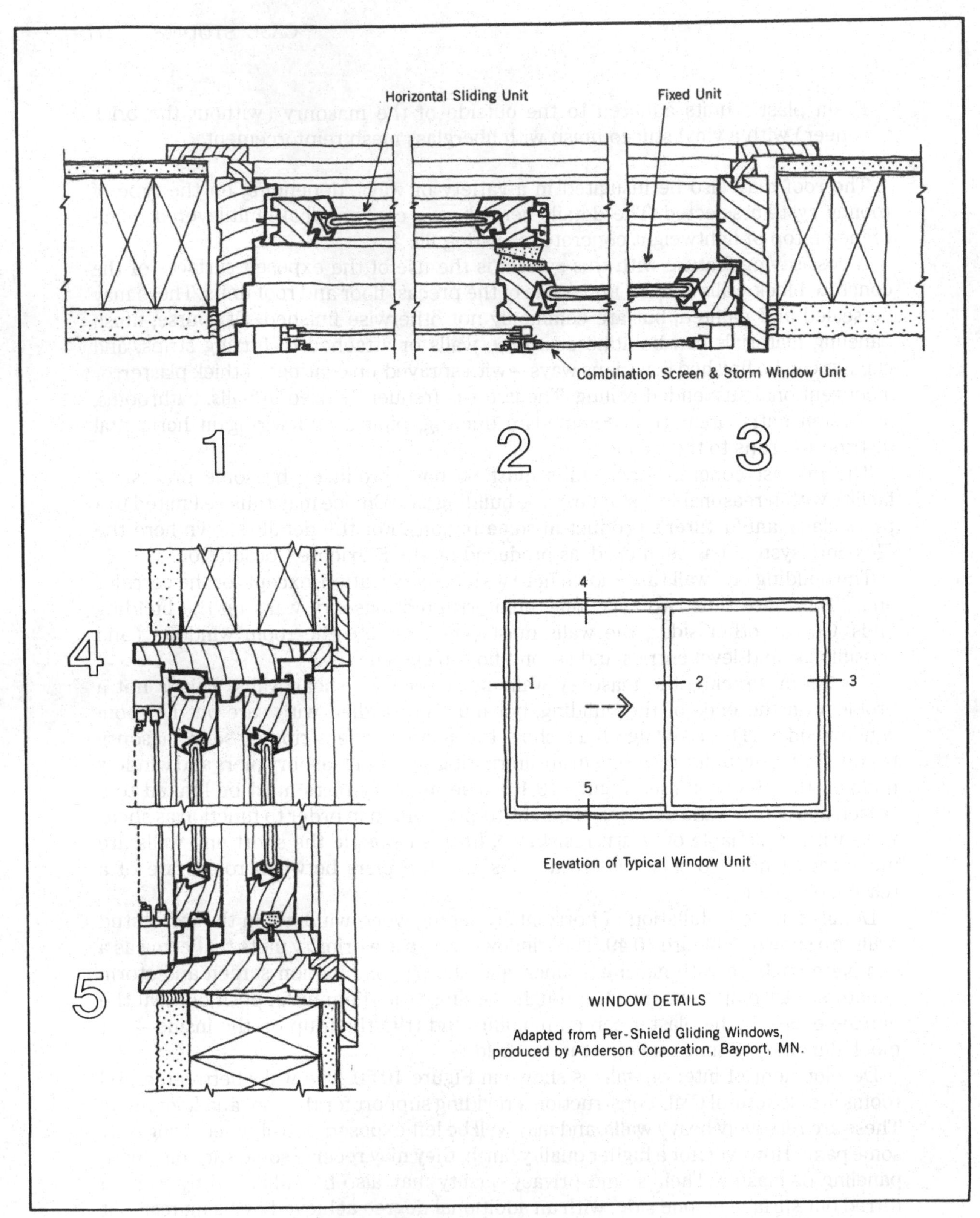

Figure 10.49 Building 4: window details.

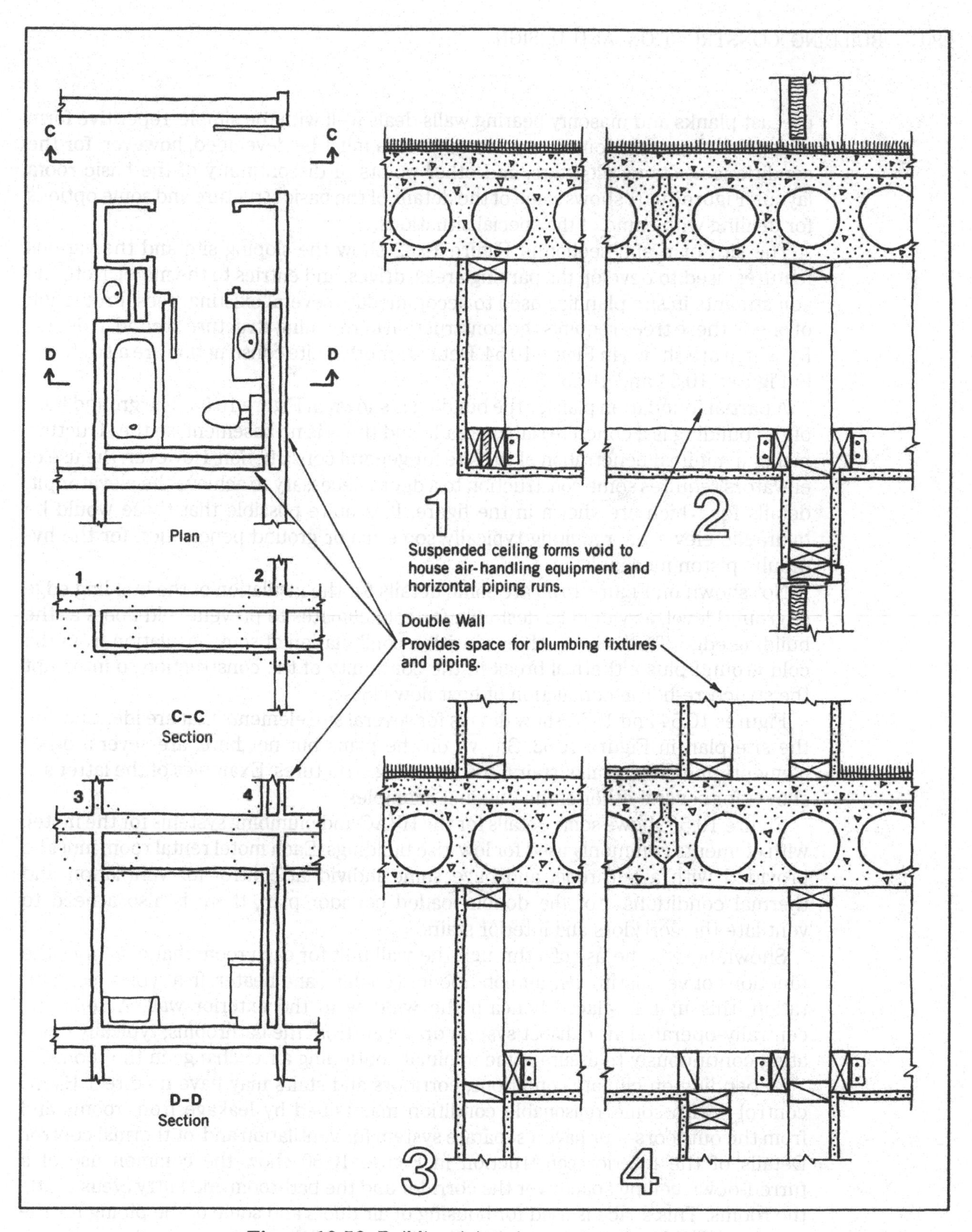

C
C
D
D
Plan
1
2
C–C
Section
3
4
D–D
Section
1
2
3
4
Suspended ceiling forms void to
house air-handling equipment and
horizontal piping runs.
Double Wall
Provides space for plumbing fixtures
and piping.

Figure 10.50 Building 4: interior construction details.

precast planks and masonry bearing walls deals well with the simple, repetitive form of the motel room layout. Some special details must be developed, however, for the conditions at stairs, elevators, and other points of discontinuity of the basic room layout. Figure 10.51 shows some of the details of the basic structure and some options for dealing with some of the special situations.

The site plan and section in Figure 10.52 show the sloping site and the various features used to develop the parking areas, drives, and entries to the motel. Note the adjustments in site planning used to accommodate several existing trees. Protection of one of these trees requires the construction of retaining structure around it; details for which are shown in Figure 10.54. Details for other site construction are also shown in Figures 10.54 and 10.55.

A partial foundation plan for the building is shown in Figure 10.53. The ground floor of the building is a concrete slab on grade and there is no basement, so the structure makes a minimal penetration of the site for general construction. However, the use of elevators requires some construction to a depth necessary to achieve the elevator pit; details for which are shown in the figure. It is quite possible that these would be hydraulic elevators, requiring typically some major ground penetration for the hydraulic piston mechanism.

Not shown on Figure 10.53 are some details for the insulation of the building edge at ground level, as would be desirable for cold climates to prevent cold floors at the building edge. This is best achieved with a combination of some insulation from the cold ground plus a thermal break in the continuity of the construction to interrupt the structure-borne conduction of heat flow (loss).

Figures 10.54 and 10.55 show details for several site elements that are identified on the site plan in Figure 10.52. Shown on the plans but not here, are several other elements, including walks, stairs, and retaining structures. Examples of the latter are shown for several other of the building examples.

Figure 10.56 shows some details for the HVAC and plumbing systems for the motel, with elements commonly used for low-rise buildings. Each motel rental room must be provided with a bathroom and have some individual control for ventilation and thermal conditions. For the double-loaded corridor plan, there is also a need to ventilate the corridors and interior stairs.

Shown here is the use of a through-the-wall unit for each room that combines the functions of ventilating fan, air conditioner (cooler), and heater. In a typical configuration, this unit is placed beneath the window in the exterior wall. A separate, centrally-operated air exhaust system draws air from the bathrooms; typically operating continuously to assure some minimal continuing air exchange in the rooms.

Depending on climate conditions, corridors and stairs may have no direct HVAC control—with some reasonable condition maintained by leakage from rooms and from the outdoors—or have a separate system for ventilation and/or thermal control. Details of the interior construction in Figure 10.50 show the common use of a furred-down ceiling space over the corridor and the bathroom and entry areas inside the rooms. This space is used for housing of air ducts and some of the piping for the bathrooms. Use of the dropped ceiling is essential here in conjunction with the precast concrete floor and roof systems, which provide for little useful interstitial space. If a non-exposed, wood or steel framed system were used, the dropped ceiling might not be necessary.

Figure 10.57 shows some details for the installation of the central elevator system, indicating the use of a hydraulic elevator—a common solution for the low-rise building.

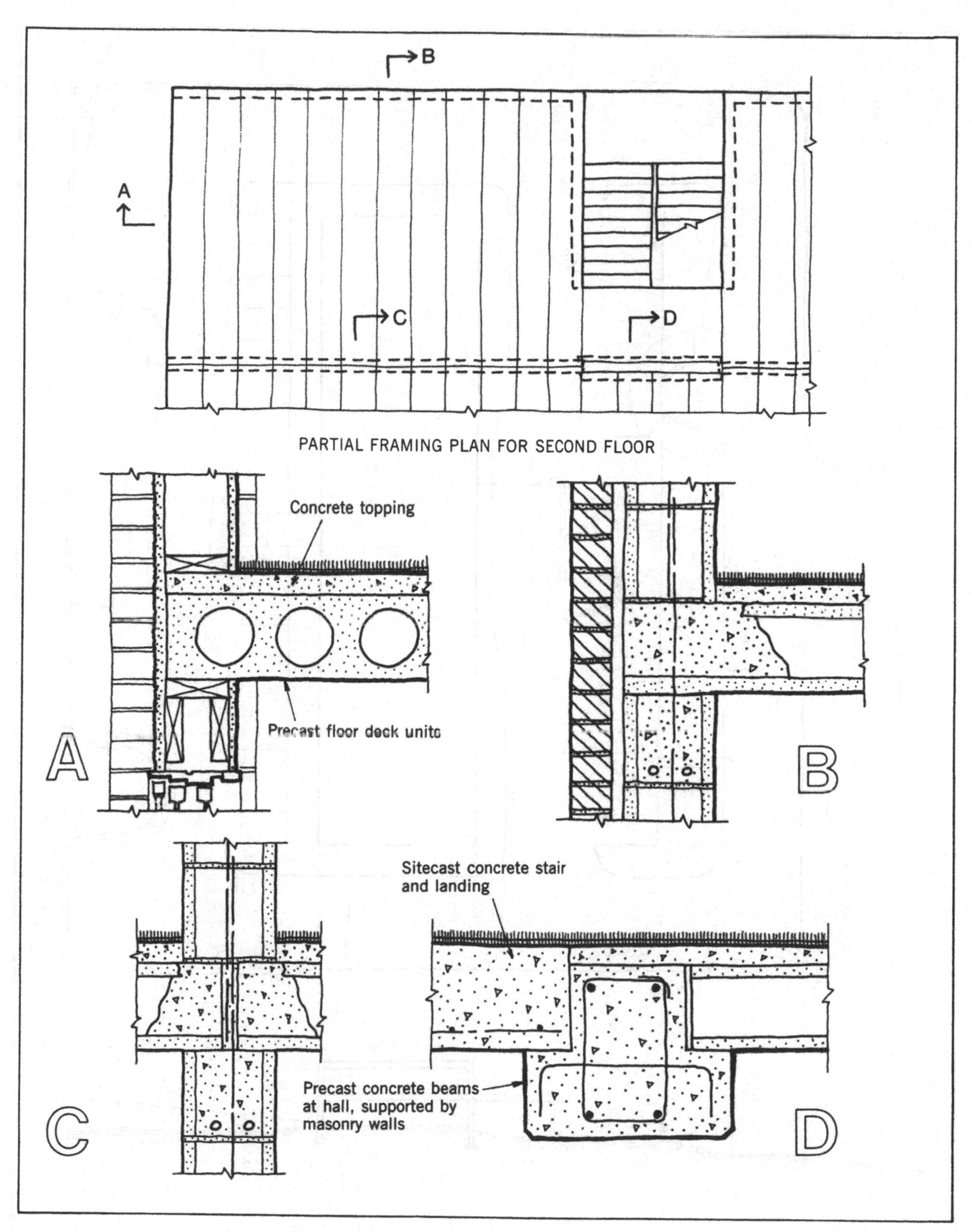

Figure 10.51 Building 4: structural details.

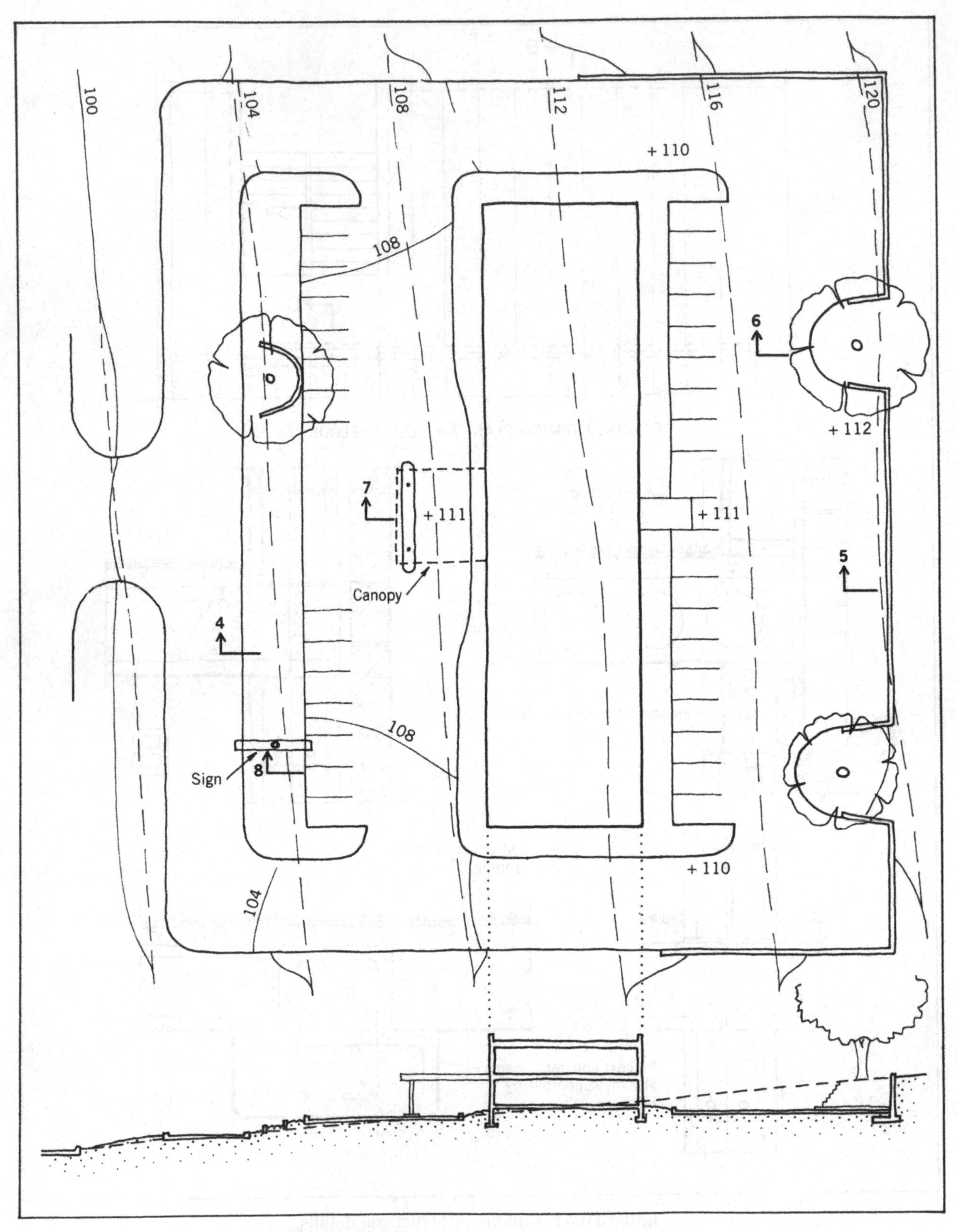

Figure 10.52 Building 4: site plan and section.

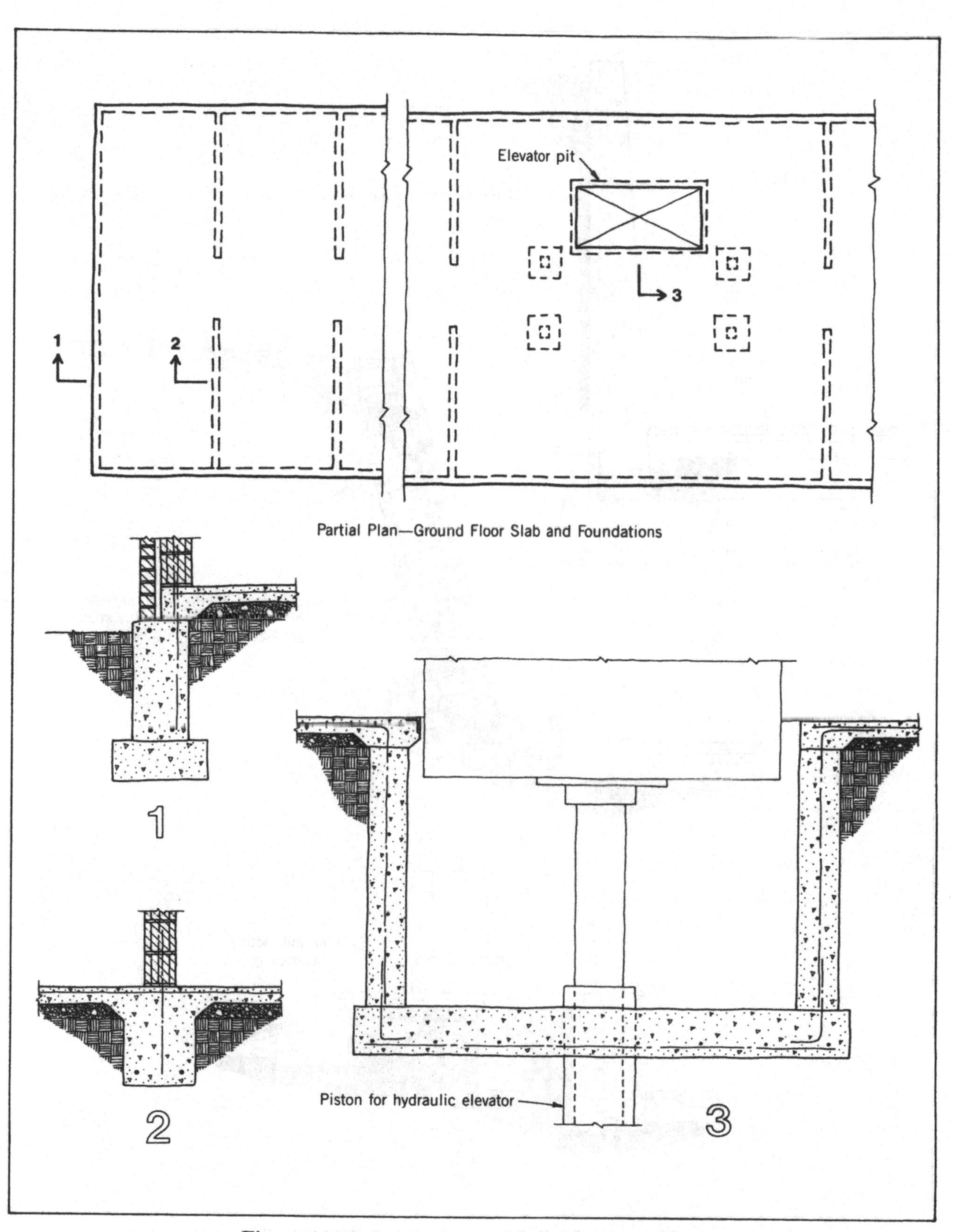

Figure 10.53 Building 4: foundation plan and details.

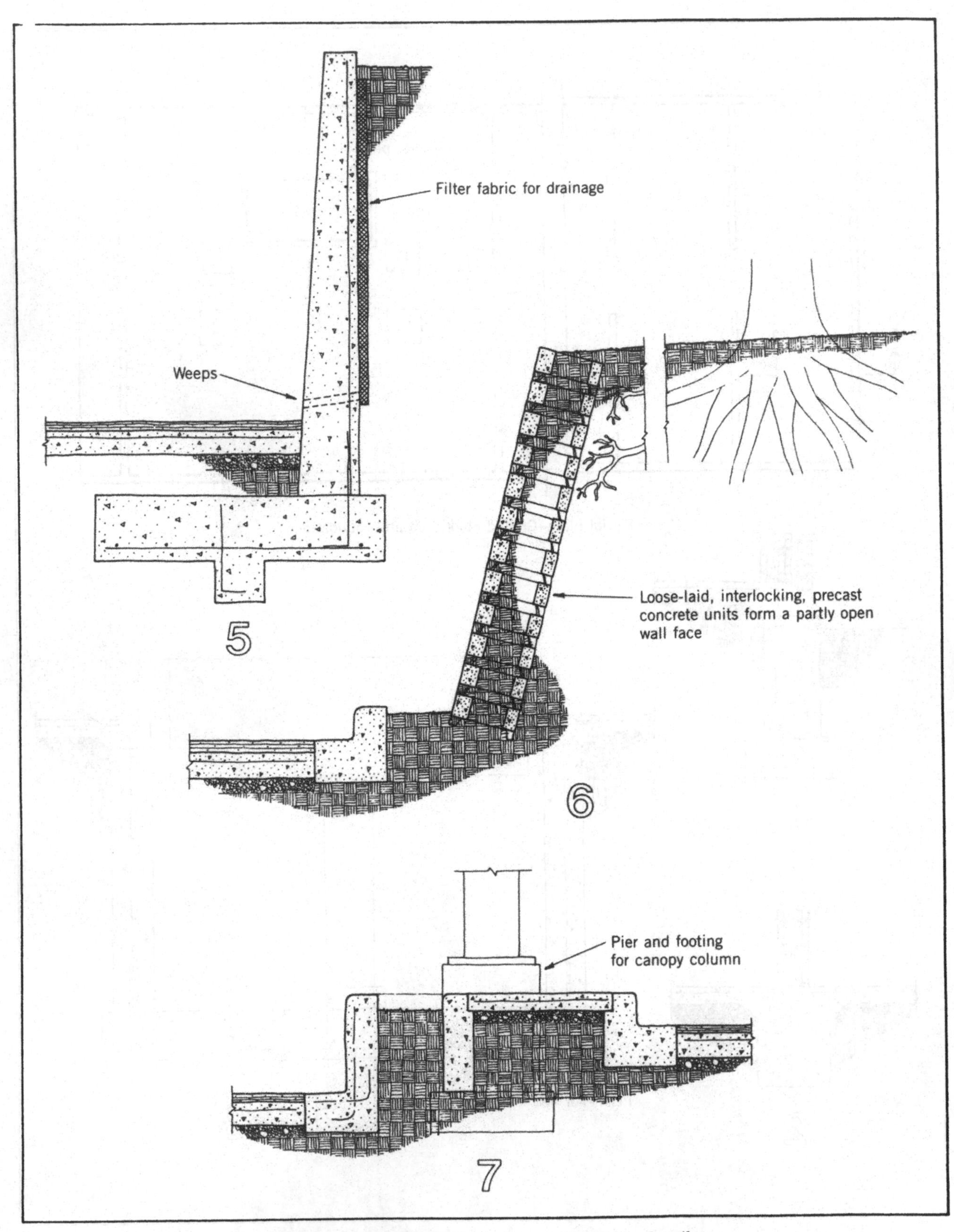

Figure 10.54 Building 4: site construction details.

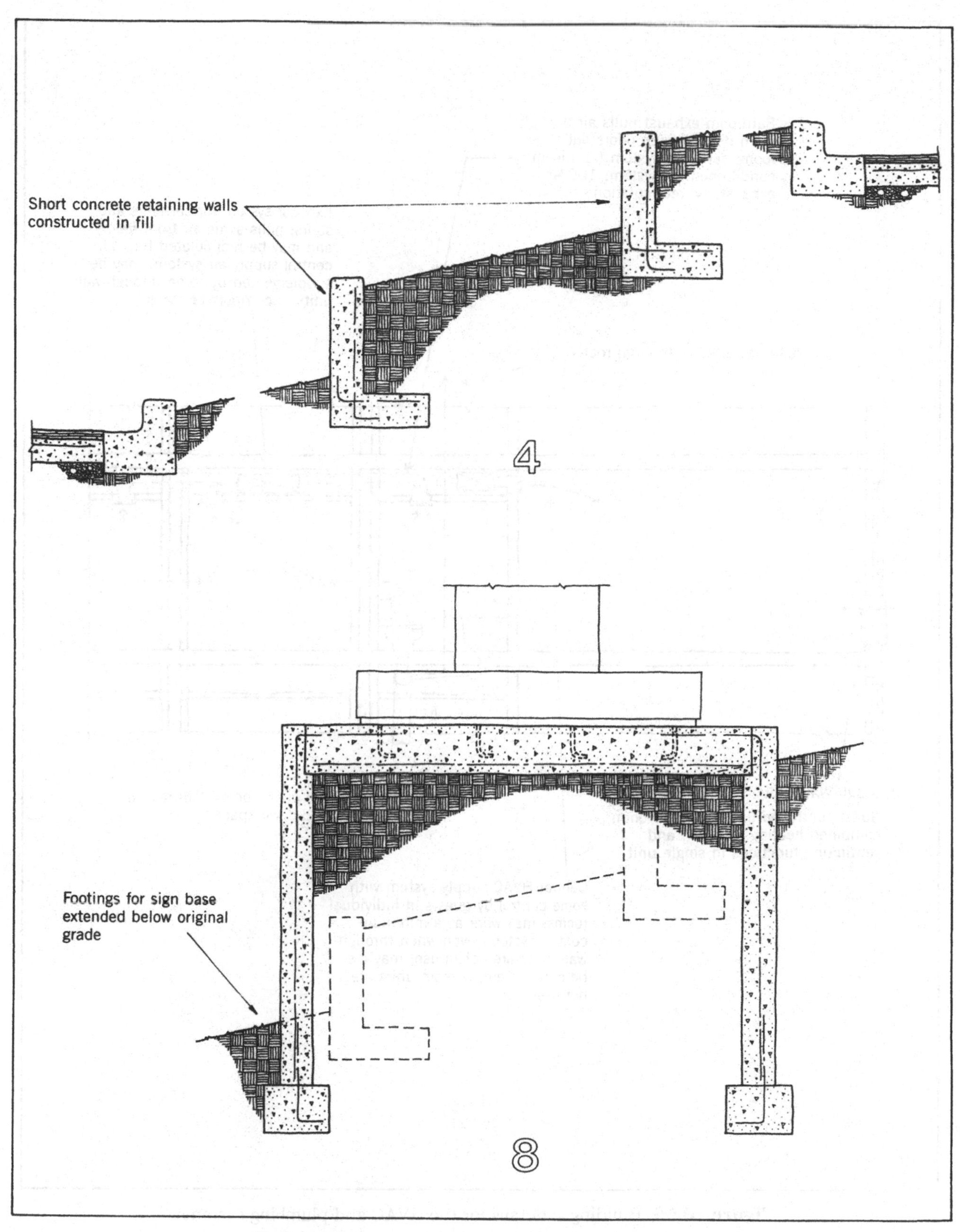

Figure 10.55 Building 4: site construction details.

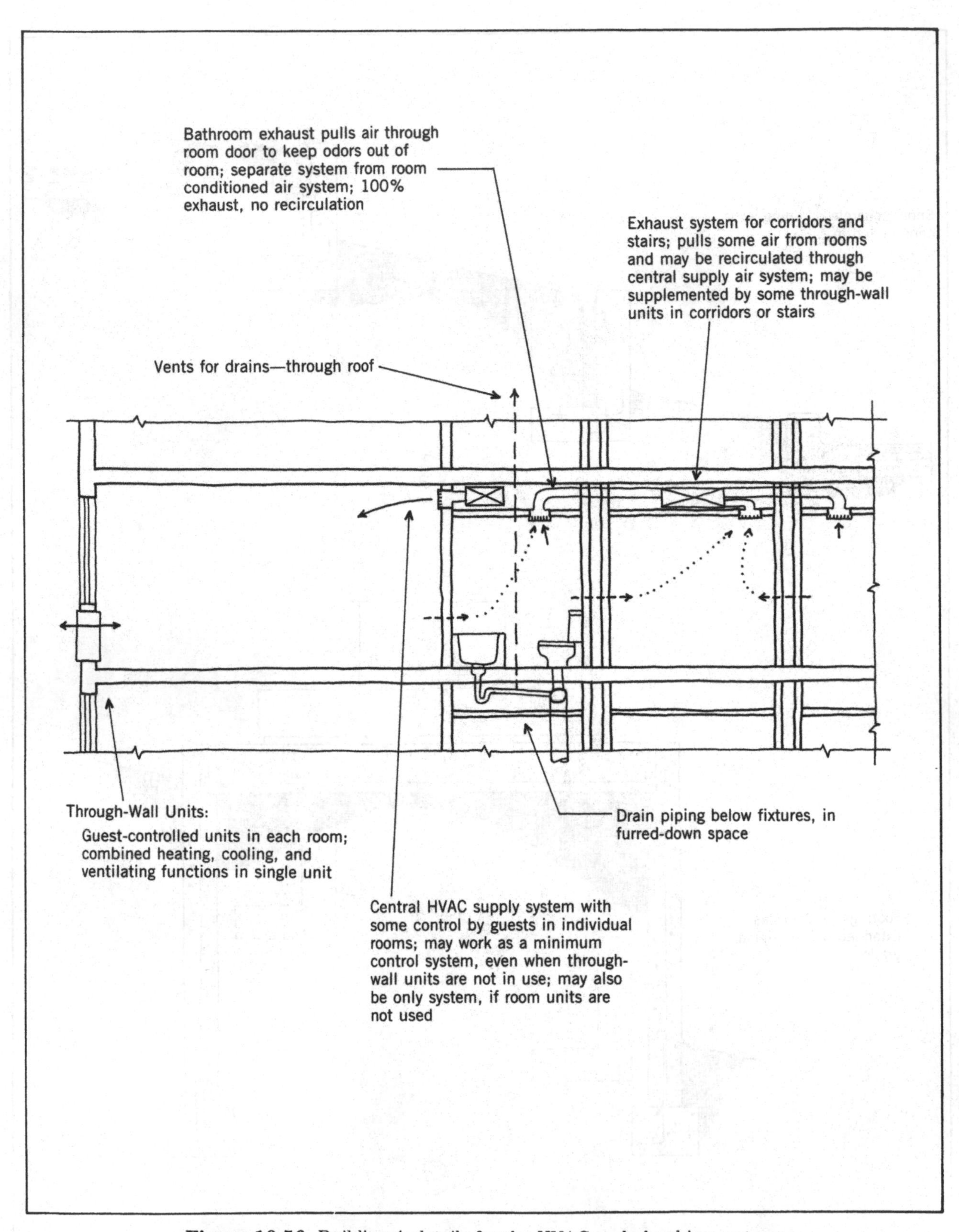

Bathroom exhaust pulls air through room door to keep odors out of room; separate system from room conditioned air system; 100% exhaust, no recirculation
Exhaust system for corridors and stairs; pulls some air from rooms and may be recirculated through central supply air system; may be supplemented by some through-wall units in corridors or stairs
Vents for drains—through roof
Through-Wall Units:
Guest-controlled units in each room; combined heating, cooling, and ventilating functions in single unit
Drain piping below fixtures, in furred-down space
Central HVAC supply system with some control by guests in individual rooms; may work as a minimum control system, even when through-wall units are not in use; may also be only system, if room units are not used

Figure 10.56 Building 4: details for the HVAC and plumbing systems.

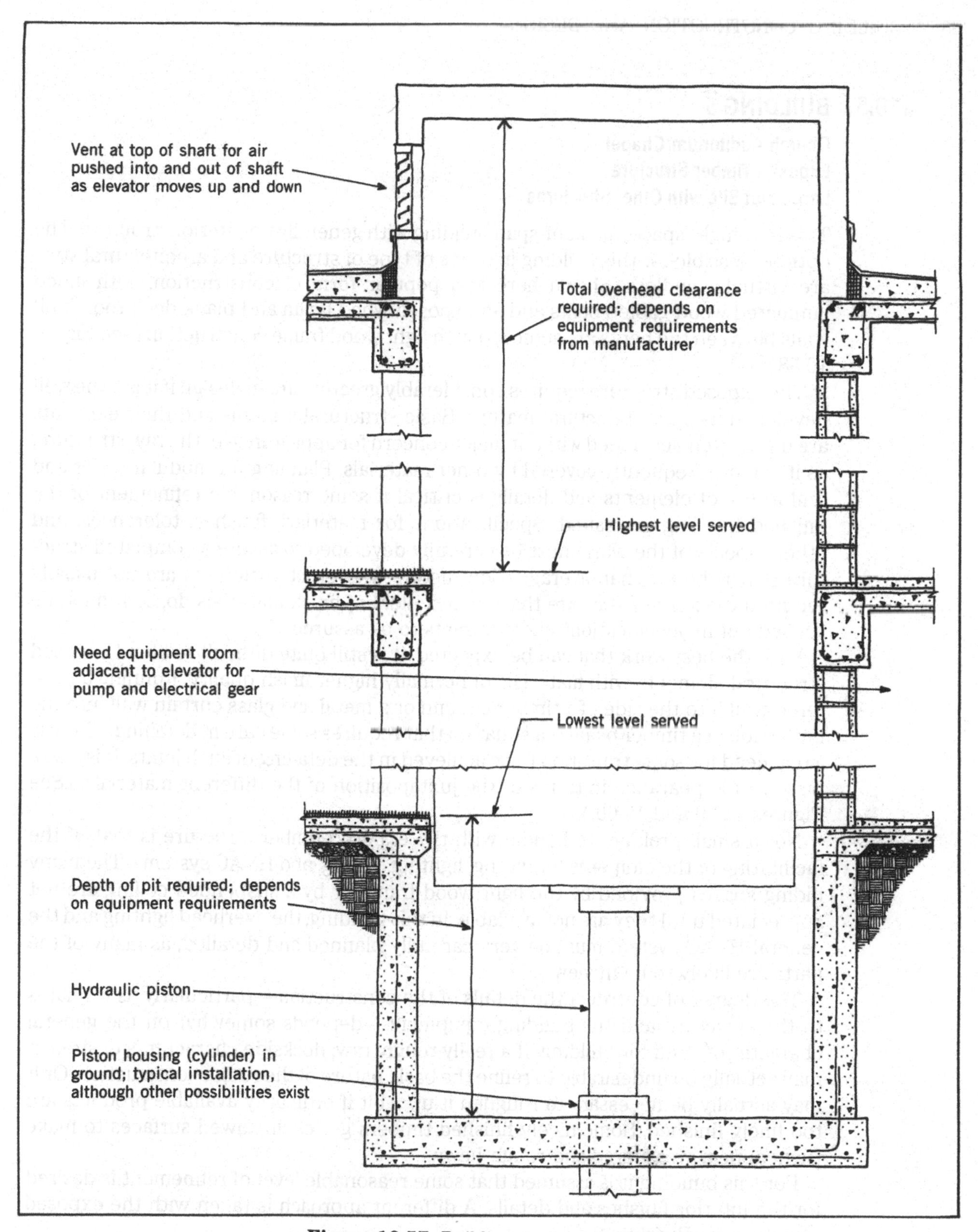

Vent at top of shaft for air pushed into and out of shaft as elevator moves up and down
Total overhead clearance required; depends on equipment requirements from manufacturer
Highest level served
Need equipment room adjacent to elevator for pump and electrical gear
Lowest level served
Depth of pit required; depends on equipment requirements
Hydraulic piston
Piston housing (cylinder) in ground; typical installation, although other possibilities exist

Figure 10.57 Building 4: elevator details.

10.5 BUILDING 5

Church Auditorium/Chapel
Exposed Timber Structure
Large Flat Site with Other Structures

This is a single-space, medium-span building with generally no interior structure. The options possible for this building in terms of type of structure and architectural style are virtually endless. Shown here is a popular form of construction, with glued laminated wood gabled bents and an exposed wood beam and plank deck roof. Infill walls between the bents are achieved with light wood frame construction (see Figure 10.58).

The exposed structure requires considerably greater care in design if it is to be well developed as an architectural feature. Basic structural systems and their elements are more often developed without major concern for appearance of the raw structure, as it is more frequently covered by other materials. Planning for modular order and uniformity of elements and details is critical if some reasonable refinement of the finished building is desired. Specifications for materials, finishes, tolerances, and other aspects of the work must be carefully developed to assure a completed structure that is better than average. Contractors who erect structures are not usually required to exercise the care that the installers of finish materials do, so it must be spelled out in specifications and contracts to be assured.

With the best work that can be expected, it is still quite difficult to blend exposed structural elements with materials of normally higher finish quality. Running a plastered wall into the side of a timber column or a metal and glass curtain wall into the underside of a timber beam is a situation that requires some care in detailing. There is a real need for some transition to be achieved in the detailing of such joints. It is also a matter of appearance in terms of the juxtaposition of the different materials. (See Figures 10.59 and 10.60.)

Not a small problem to handle with the exposed timber structure is that of the facilitating of the elements for wiring, lighting, piping, and HVAC systems. The many hiding spaces provided by the light wood frame or by a suspended ceiling are not appreciated until they are not available. In this building, the overhead lighting and the general HVAC system must be very carefully planned and detailed, as many of the parts are likely to be in view.

The degree of control of the details of the construction—particularly as it relates to the structure and the building equipment—depends somewhat on the general character desired for building. If a really rough, raw, dockside character is desired, it may actually be undesirable to refine the basic nature of the rough construction. Or it may actually be necessary to roughen it up a bit if ordinarily available products are too finely finished. Some nicely finished timbers get chain-sawed surfaces to make them appear weathered and rough-hewn.

For this building it is assumed that some reasonable level of refinement is desired for the interior finishes and details. A different approach is taken with the exposed structure for Building 9.

Development of the podium/altar area will depend very much on the form of services by the church group using the building. It may also need to respond to other uses for the building, if it doubles as a general purpose auditorium for meetings, concerts, plays, and so on. The steps and platform can be achieved in various ways; two common ones being those shown in Figure 10.61.

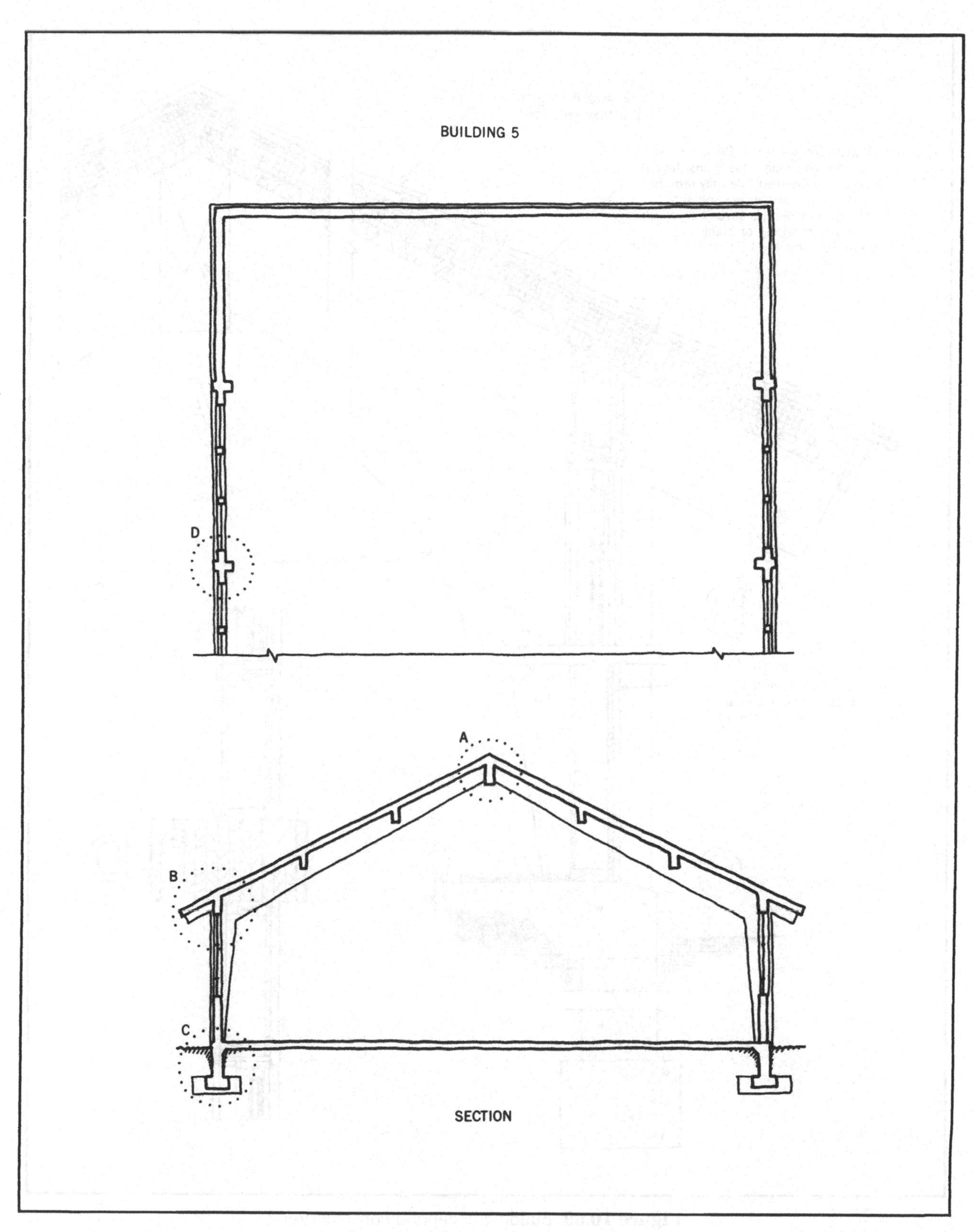

Figure 10.58 Building 5.

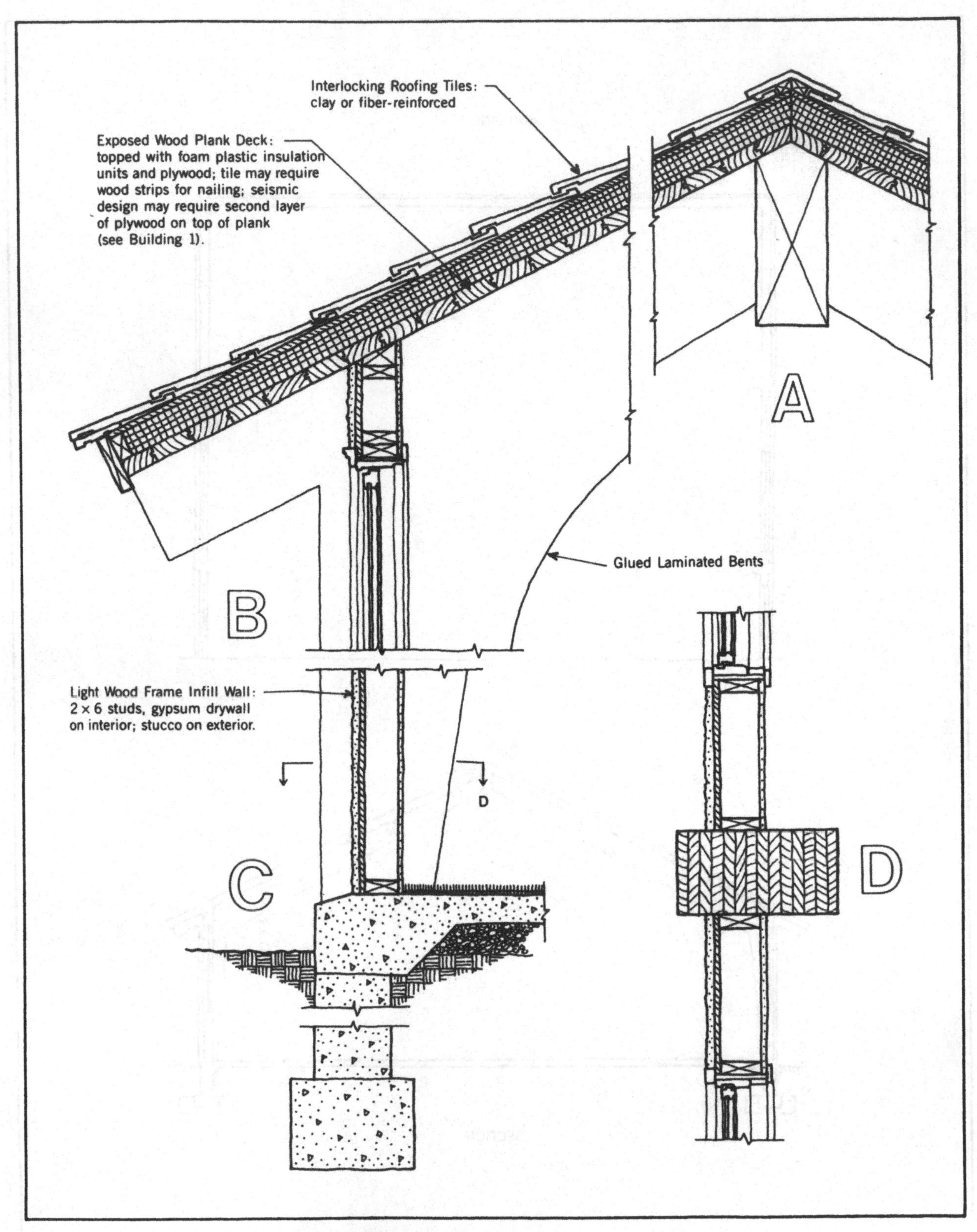

Figure 10.59 Building 5: general construction.

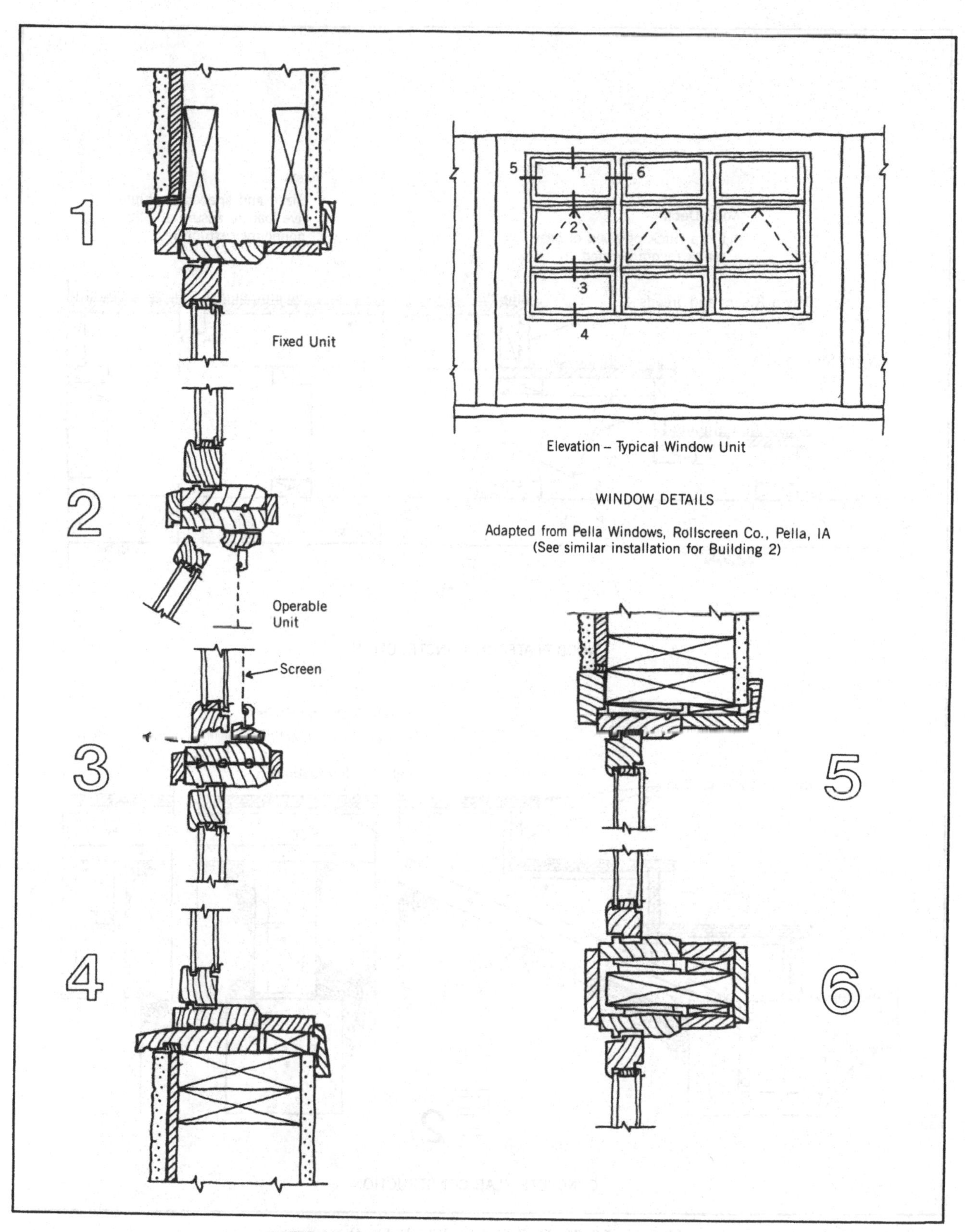

Figure 10.60 Building 5: window details.

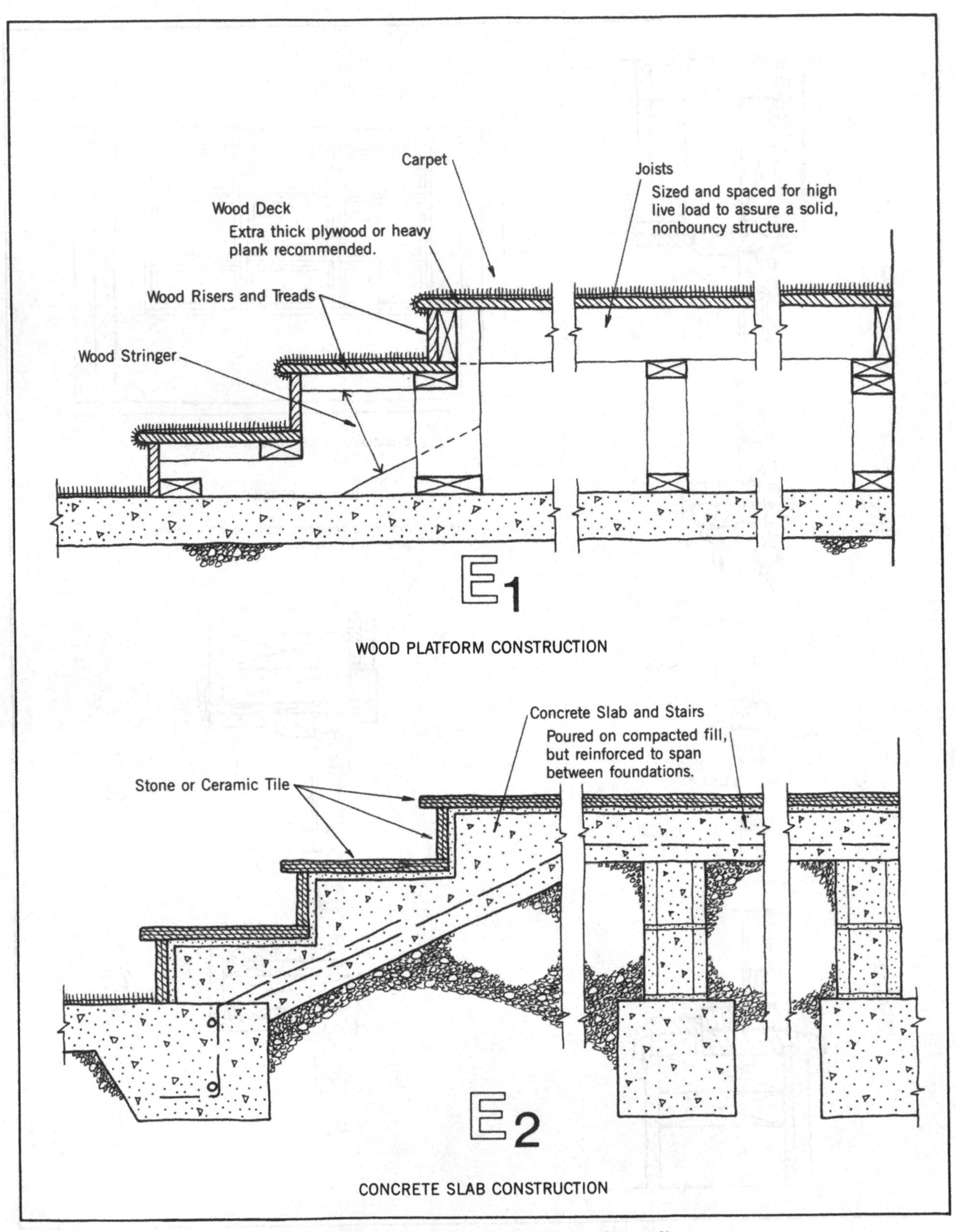

Figure 10.61 Building 5: details for the podium.

The wood platform is simply built up on top of the concrete paving slab. This is undoubtedly the cheap way to go, although construction can be made as sturdy as desired. An advantage for this construction is the possibility of future alteration with relative ease, should the users of the building develop different needs. The wood construction should be well anchored and joists should be extra stiff to eliminate bounciness. The deck should also be extra stiff (not pushing limits of recommended thicknesses) and should be held down with screws and glue to assure no squeaking.

The alternative construction shows a stepped concrete structure. This is a much sturdier, bounce-proof form of construction. One reason for choosing it might be the choice of tile floor surfacing, as shown in the details. This provides a much better support for the tiles, although either finish could actually be placed on either of the structures shown here.

This building presents an example of a situation where the basic construction and actual structural elements are hung out for display, and the designer of the interior must anticipate the various details and finishes of the materials which can normally be expected. Tolerance for the normal range of acceptable dimensional inaccuracies must be developed in the construction that abuts and infills the basic structure. For example, at the top of the wall in Detail B and the ends of the walls in Detail D (see Figure 10.59), some form of molding or reveal should be used to achieve the joint between the two planes of different materials; so as not to have just a blind collision.

Jointing of the timber frame, grade specifications for the timber elements, and side finish for the glued laminated bents should be done with good information on local products and fabricating practices. Joints may be boldly developed with exposed heavy steel elements, or more cleanly developed with some concealed fastening devices. The general impact of these details on the whole interior design should be carefully studied.

A general framing plan and some details for the roof structure are shown on Figure 10.62. The precise form of details of the framing of the glued laminated elements are quite likely to be developed by the fabricators of the timber structure, who will possibly also be involved in its erection.

This is the principal building of a small complex that includes a residence, a classroom/office building, and an extensive parking facility. The buildings are discrete structures, but are linked with outdoor spaces, structures, and landscaping.

The main church building itself is a single-space, medium-span building with essentially no interior spatial division—an enclosing shell for housing of singular events. These events may include a range of types—from very formal liturgical services, funerals, and weddings, to general meetings and social events for the church members.

To utilize the site space efficiently, there will be some pressure to economize on outdoor space not used directly for parking, buildings, and necessary walks and driveways. Still, some setting should be achieved for the church and some space available besides the parking lot for outdoor assembly and casual meeting before and after events. It may also be important to have the entry into the church be one that occurs as a procession through some transitory spaces. It is assumed here that the spaces between and around the buildings will be developed with these purposes in mind.

The character of exterior construction should relate to that of the church building. As shown in Figures 10.59 and 10.60, the church is constructed with an exposed timber frame, a tiled roof, and stucco walls. This rough, rich texture and the natural wood finish should be a reference for selection of exterior finishes.

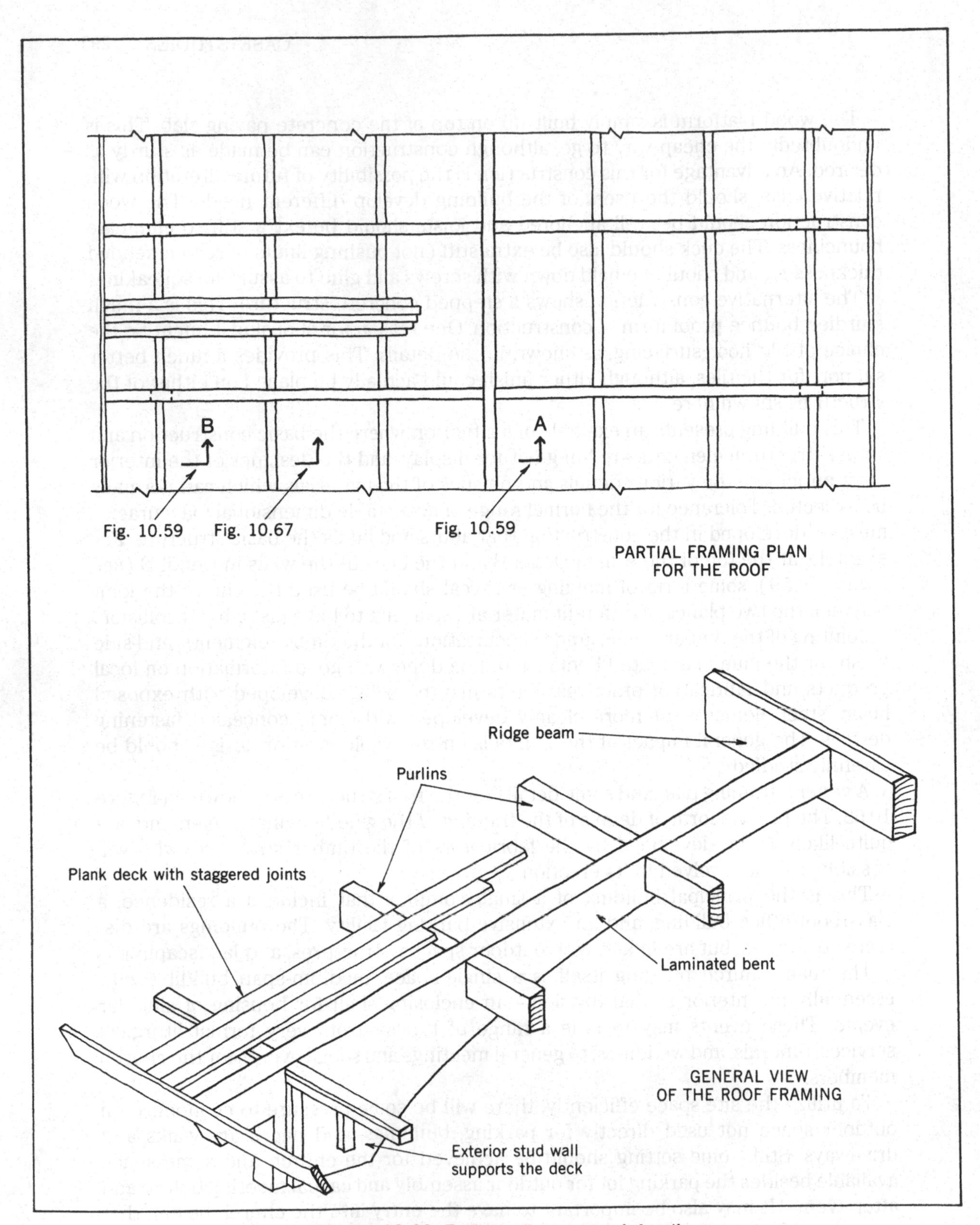

Figure 10.62 Building 5: structural details.

The site plan in Figure 10.63 shows the general arrangement of the complex. The church is featured centrally on the site and is further emphasized by raising it slightly on a built-up fill, a few feet above the rest of the flat site. Walks, landscaping, and an entry court also serve to create a transitional entry and visual spaciousness for the church building. The site section in Figure 10.63 shows both the horizontal and vertical form of this spatial arrangement.

Figure 10.64 shows a partial foundation plan for the church building. The concrete floor slab and exterior walls are supported by a grade beam and wall footing that fits between the piers for the frame bents. The details for the piers show the use of one method for developing the resistance to outward thrust at the base of the bents. In this case, the heavy floor slab is used to anchor reinforcing extended from the piers. Another solution is to provide a continuous tie across the structure, from pier to pier on opposite sides. The slab drag anchors shown are generally simpler and considered adequate for such structures when the span is modest, as it is here.

Since the floor of the church is considerably above the level of the original site, there is some judgment to be made about its structure. Use of a simple slab on grade would require very careful placing and compaction of the fill used to achieve the slab base, all the way up from the original cleared grade. If this is not considered to be feasible, a framed floor on grade—such as that used in the front area of Building 1—must be used to assure a stable floor structure.

Figure 10.65 shows the construction of some of the site elements outside the church. Carrying out the general nature of the rough, natural materials of the church, these are developed with stone, tile, and timber as much as possible. Low walls are developed with large stones in a strongly tapered, truncated pyramidal form; reminiscent of stone piles achieved without mortar or concrete by early builders. Even though modern concrete is used here to tie the construction together, the stone work should still be carefully developed so that the rock piles are as stable as possible without the concrete.

In keeping with other materials, the exposed concrete work here could be done with a concrete with a tan tinting admixture and possibly some surfaces brushed or air blasted to expose the aggregate.

Some walk and terrace surfaces are shown with paving of stone or tile. Paving in the open spaces around the church could be developed with loose-laid pavers over a compacted granular base. Since the base soil in this area is considerably built up, this form of paving can probably suffice without concern for surface drainage—relying on a very fast-draining sub-base for the pavers. If necessary, an open tile drain system could be placed beneath the paving, draining out to the lower parking areas.

A general concern for such a complex is that of continuous maintenance, which may be mostly by volunteers from the church membership. These workers may be motivated and industrious, but are likely not professional landscape maintenance people. Thus, the choice of plantings and general landscape elements should be made with simple maintenance as a major concern.

With the exposed form of construction for the roof, there is no suspended ceiling or other form of interstitial space in the upper portion of the auditorium. This poses some severe limitations on the systems for HVAC and lighting; especially if general concealment of the installation is desired. A general overhead ducted HVAC system is out of the question, unless the whole thing can be hung out for display. This is possible of course (see Figures 10.100 and 10.111), but does not fit with this interior and basic construction.

Other options for the HVAC system for the auditorium depend somewhat on local

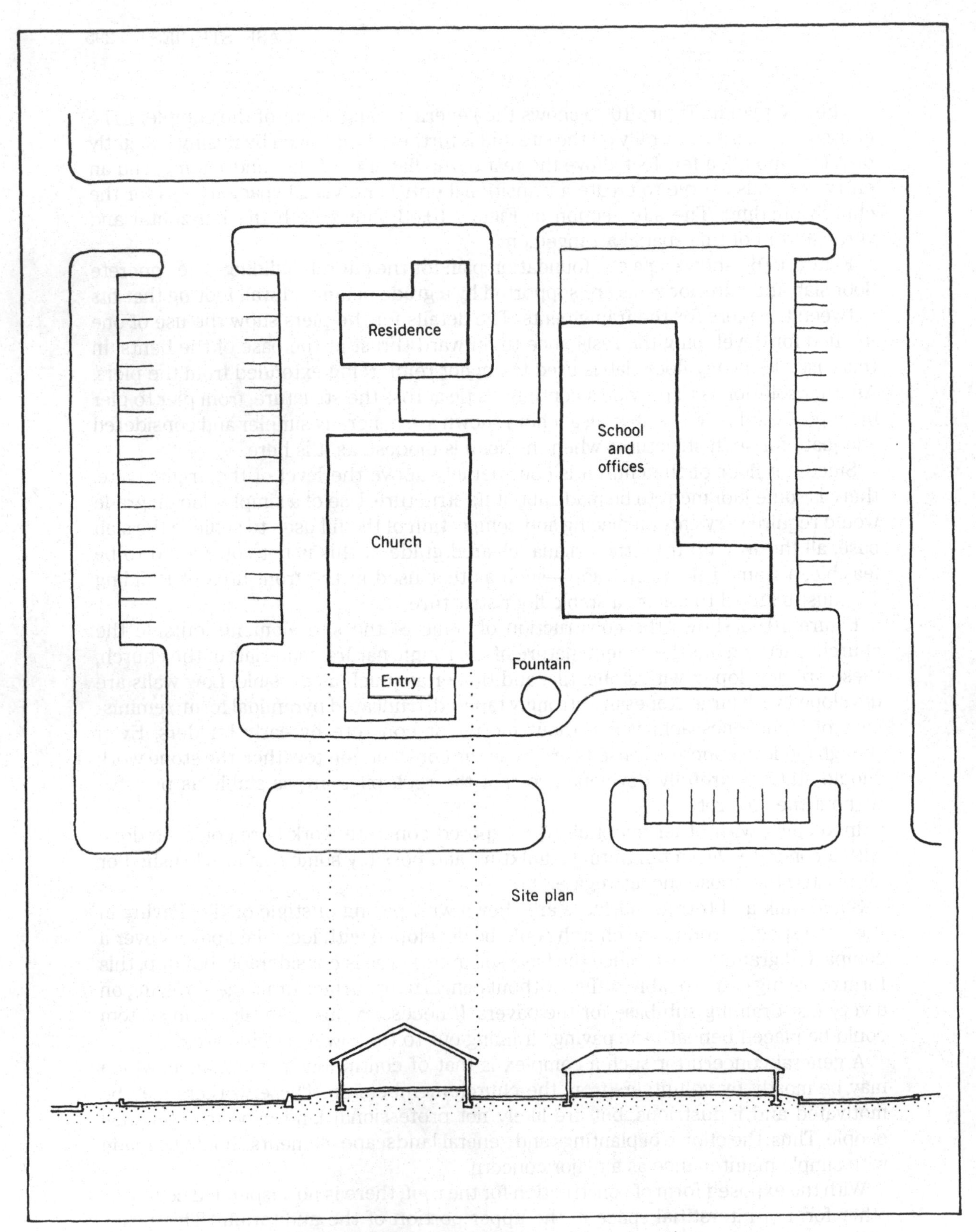

Figure 10.63 Building 5: site plan and section.

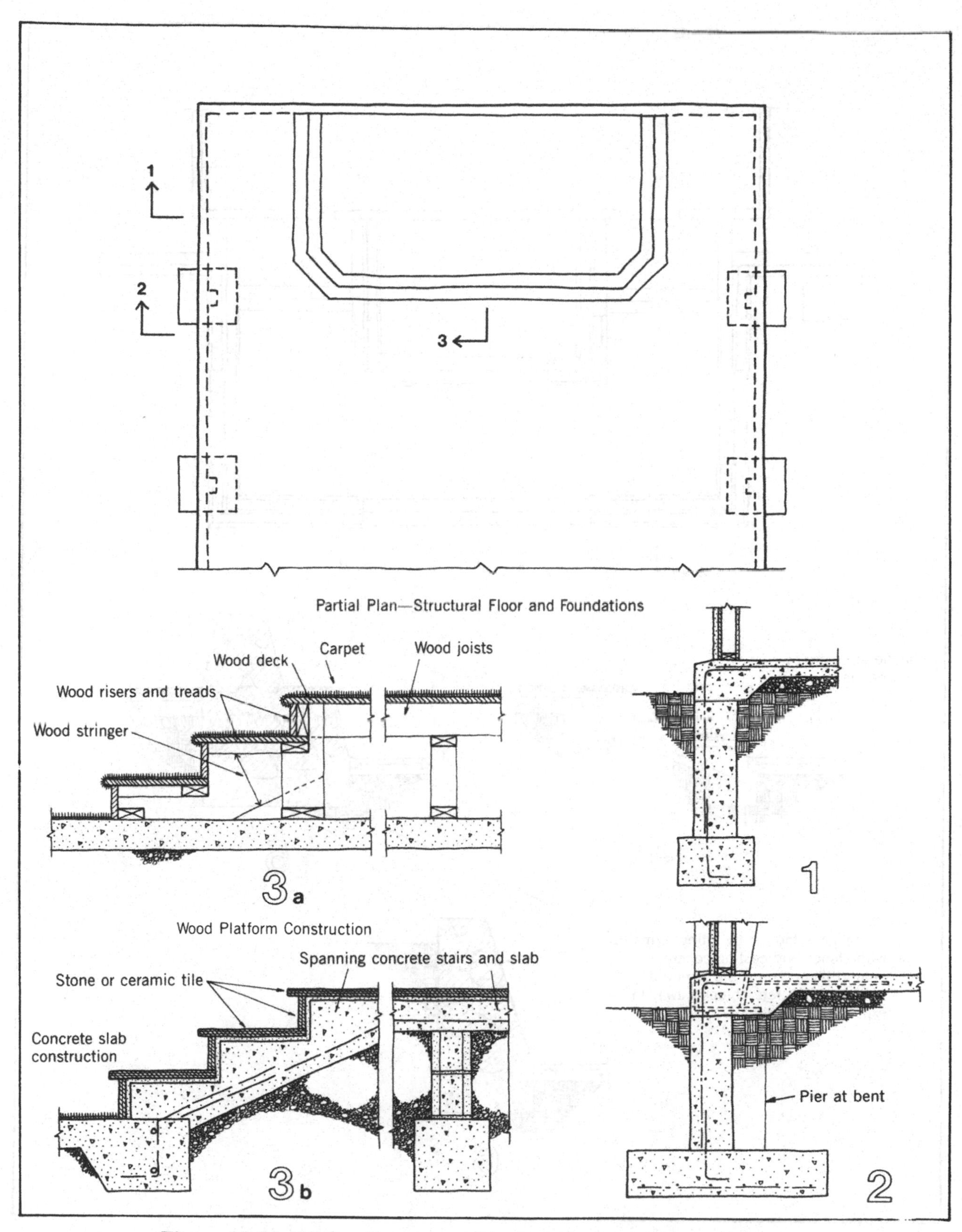

Figure 10.64 Building 5: structural floor and foundation plan and details.

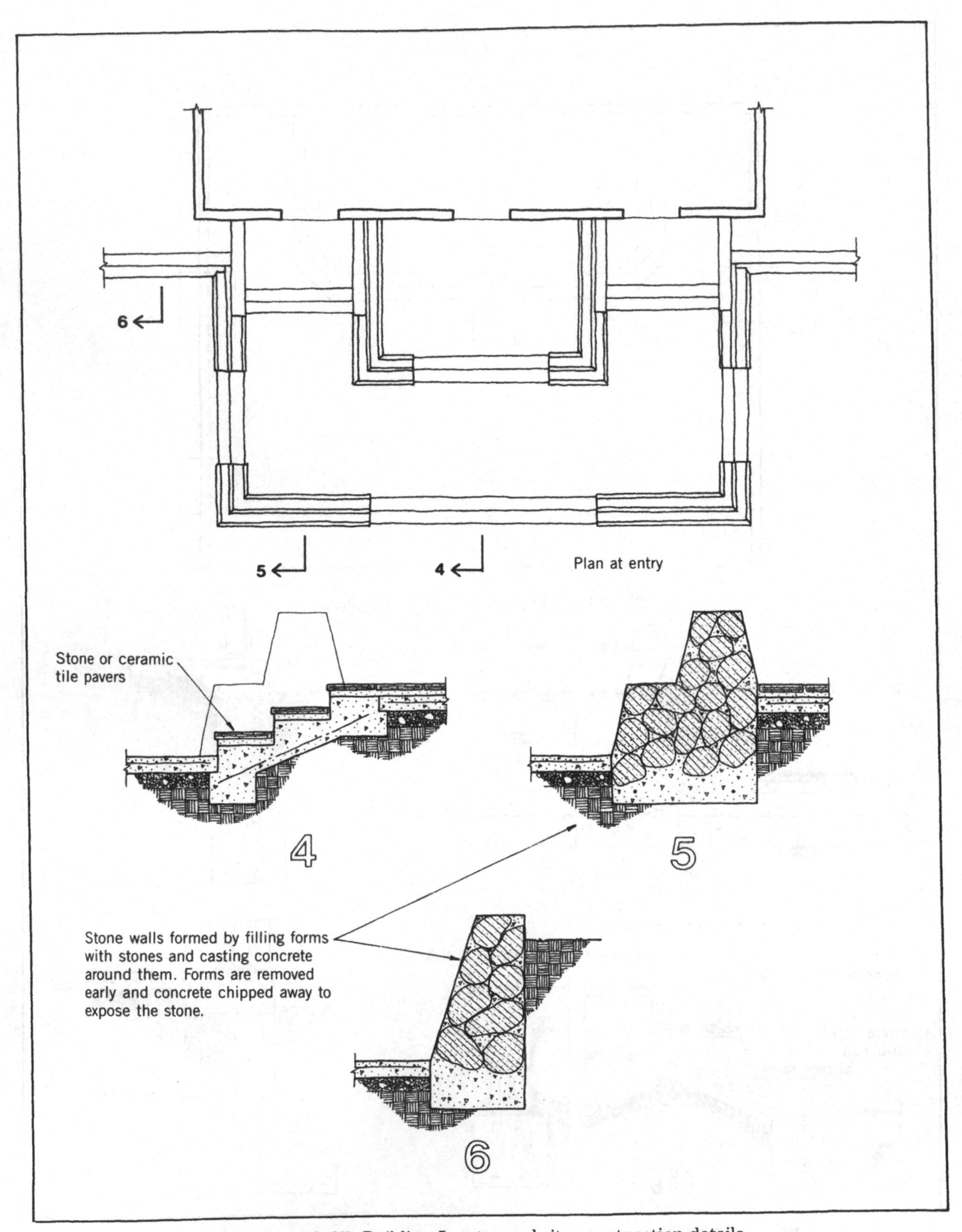
6
5
4
Plan at entry
Stone or ceramic
tile pavers
4
5
Stone walls formed by filling forms
with stones and casting concrete
around them. Forms are removed
early and concrete chipped away to
expose the stone.
6

Figure 10.65 Building 5: entry and site construction details.

climate and the overall use of the building. Churches are often used only at certain times, so a system may be able to be shut down—or at least operate at limited capacity—for extended periods. However, when the space is to be used, it is usually desired to bring it to a comfort level fairly quickly; a performance best had from an all air system. If the space sees a lot of ongoing use, however, (for church, school, and general social activities) it needs a system that is similar to that for other buildings with regular usage.

A possibility for an all air system is shown in Figure 10.66, employing underfloor ducts that feed perimeter floor-level registers. The registers may be in the floor, as shown, or possibly in the wall beneath the windows. A variation on this would be to have the registers high in the walls, possibly above the windows, which would work somewhat better for delivering air to the center of the space.

A consideration for very cold climates—here as in other buildings with concrete slab floors—is the possibility for heat loss and very cold floors at the building edges. Various forms of edge insulation may alleviate this, but the use of the perimeter heating system is one possible solution for warming up the floors and the bottoms of the exterior walls.

Another possibility for cold climates is the use of a radiant system in the floor; probably using piped hot water. This could be combined with an air system to provide warm floors and building edges, as well as a minimum general heat level between uses of the auditorium. The air system could then be turned on to quickly raise the temperature for use of the space.

Lighting of the auditorium would probably include a general illumination system and various special lighting for entries, for the podium area, and so on. The general lighting system may use wall-mounted fixtures with some bounce light from fixtures directed at the ceiling. However, if the ceiling is stained wood, this is less feasible. Figure 10.67 shows a more common solution using pendant fixtures suspended from the underside of the roof, with the wiring concealed in the roof construction.

If the suspended light fixtures are supported from the laminated bents, the wiring could be installed in channels cut into the top of the bents and drilled vertical holes to the individual fixtures. The bents would have to be oversized for this cutting and the cutting work carefully supervised to preserve the integrity of the exposed construction.

▪10.6▪ BUILDING 6

Three-Story Office Building
Steel Frame, Metal Curtain Wall
Large Sloping Site

This is a modest-size building, generally qualified as being low rise (see Figure 10.68). In this category there is a considerable range of choice for the construction, although in a particular place, at a particular time, a few popular forms of construction tend to dominate the field. Currently available products, market competition, current code requirements, popular architectural styles, and the general preferences of builders, craft people, developers, and designers usually combine to favor a few basic forms of general construction.

Times change, however, and the new fads, new building products, new construction methods, new paranoia for various issues (such as sun use, energy conservation, toxic substances, environmental pollution, etc.), and new community or national

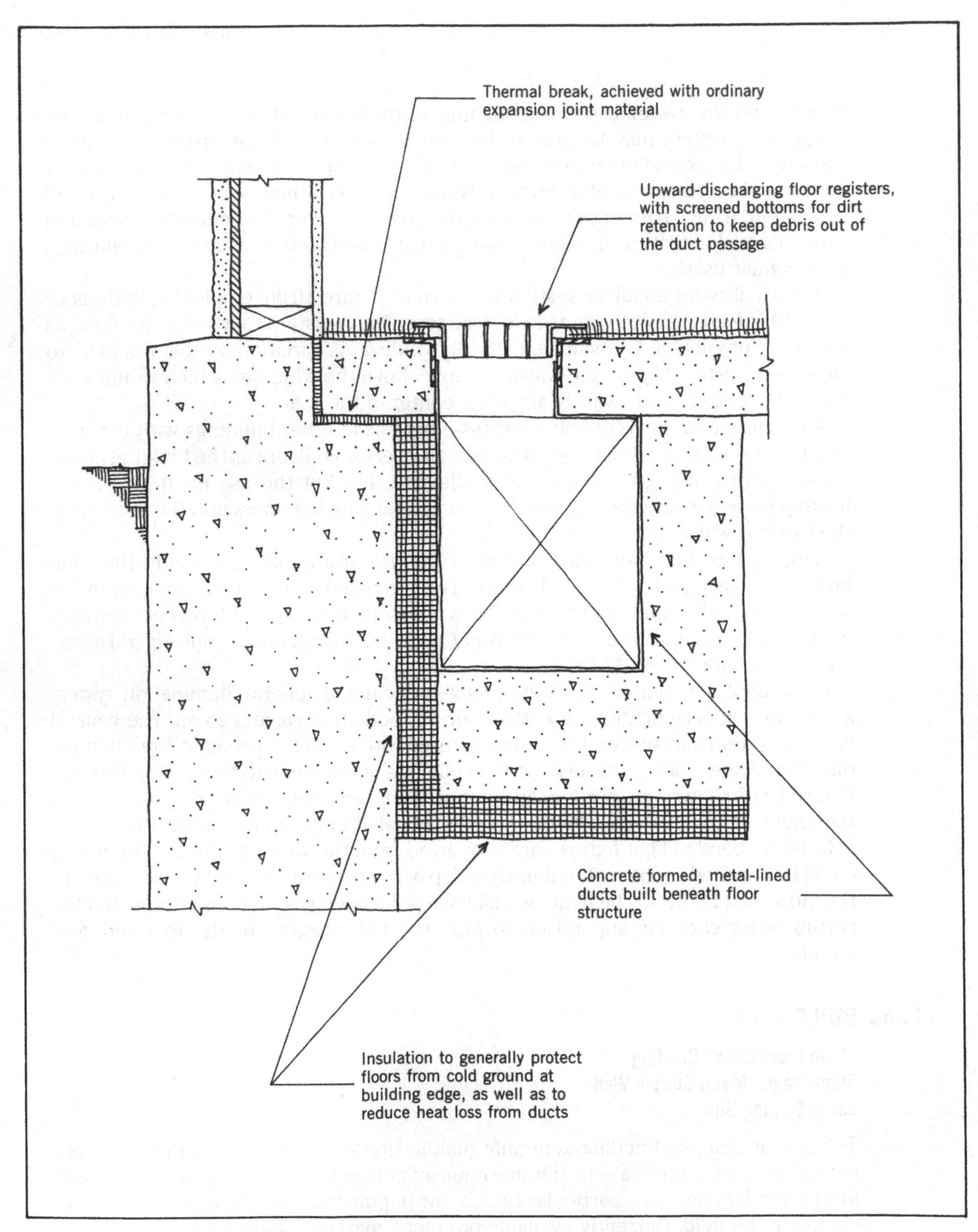

Figure 10.66 Building 5: details for the HVAC system.

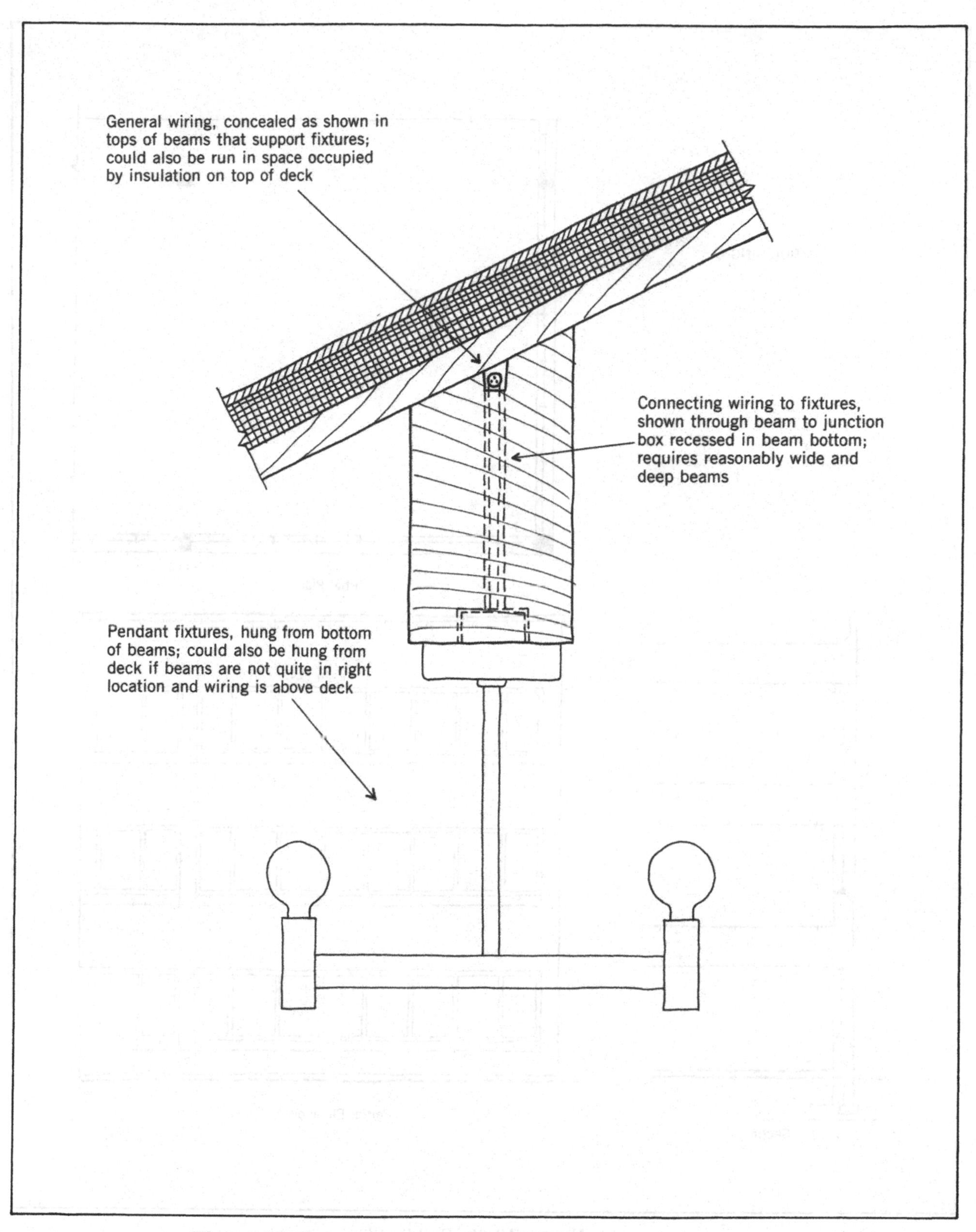
General wiring, concealed as shown in tops of beams that support fixtures; could also be run in space occupied by insulation on top of deck
Connecting wiring to fixtures, shown through beam to junction box recessed in beam bottom; requires reasonably wide and deep beams
Pendant fixtures, hung from bottom of beams; could also be hung from deck if beams are not quite in right location and wiring is above deck

Figure 10.67 Building 5: details for the general lighting system.

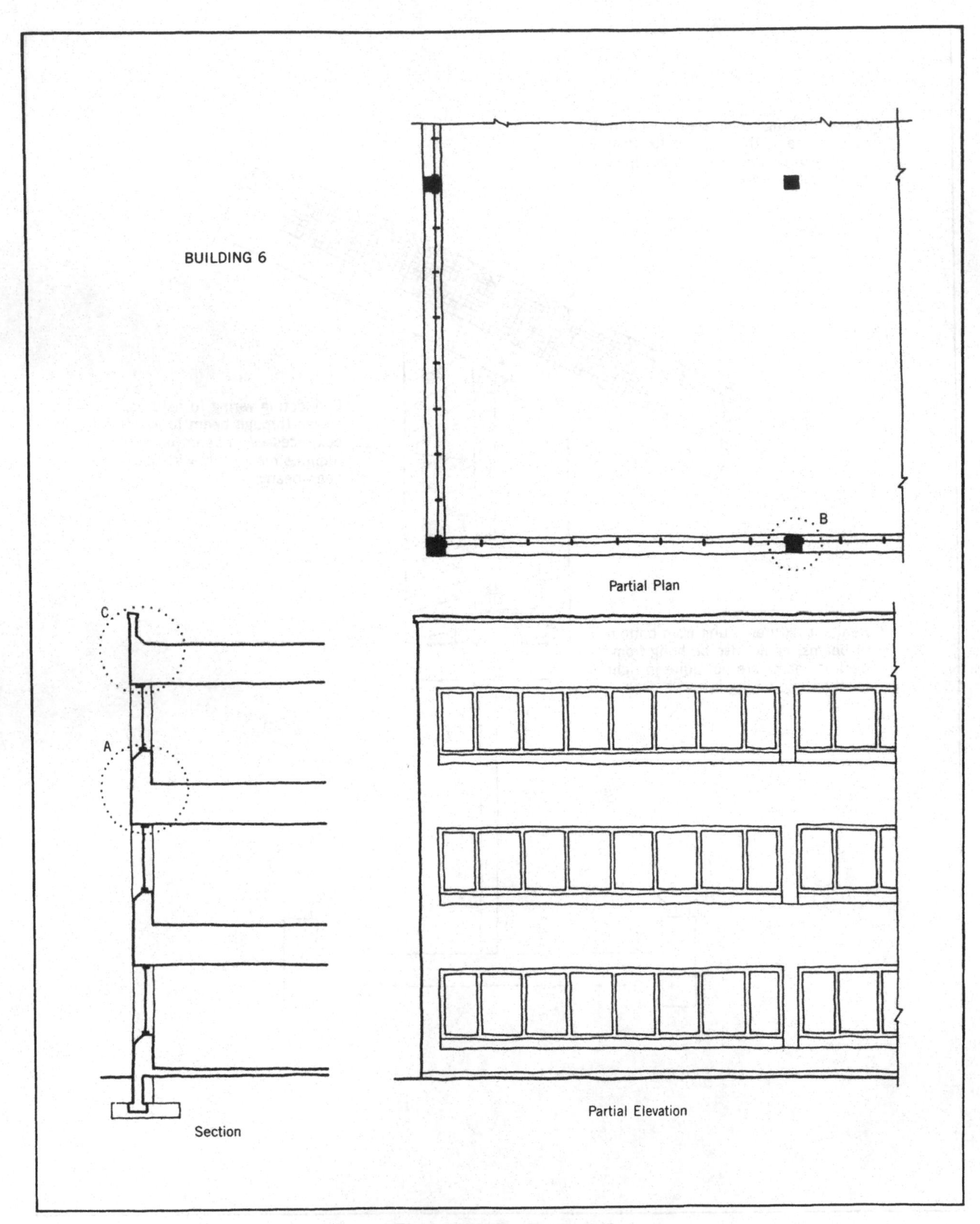

Figure 10.68 Building 6.

values cause realignments in the building design, financing, constructing, and insuring fields. Suddenly, yesterday's favored system is out, and a new champion emerges to reign (for awhile).

Structural options for this example are considerable, including possibly the light wood frame if the total floor area and zoning requirements permit its use. Certainly, all the possible steel frame, concrete frame, and masonry bearing wall systems are feasible. Choice of the structural elements will depend mostly on the desired plan form, type of window arrangements, and clear spans required for the building interior.

Shown here is a steel frame structure with W sections used for columns, girders, and beams and a formed sheet steel deck for the floors and roof. (See Figures 10.69 through 10.71.) With the continuous horizontal strip windows, the exterior walls will not accommodate a perimeter shear wall system, so lateral bracing must be achieved with a core bracing system or a development of the steel frame for rigid bent action. If wind is the critical concern, the core bracing may be selected; if seismic effects are critical, the bent system would probably be favored.

Many variations are possible with this basic structural system, mostly dealing with choices for the deck system and the beams that directly support the deck. Some options are as follows:

Steel open web joists for the deck support. These become more feasible as the span is increased. At a 30-ft span, the W sections are competitive; at 40-ft the joists undoubtedly are favored. Open web joists would most likely be closer spaced, which may affect the choice of the deck.

Sitecast concrete slab with steel beams. This would most likely include the development of the beams in composite action with the slab, using lug connectors on top of the beams to engage the cast concrete.

Plywood deck with nailable joists (fire codes permitting). This is a presently popular system with the use of either light composite wood and steel trusses (wood chords and steel web members) or the various forms of wood I sections, using wood flange members and plywood or particleboard webs. Wood 2 X members are also bolted to the tops of the steel frame members to permit nailing of the plywood at these locations.

Decking must also develop the horizontal diaphragm action required for wind or seismic effects. This can affect the choice of the decking materials as well as the details for installation—in particular, how the deck is attached to the steel frame. The concrete fill on the deck adds considerable weight, so it is sometimes feasible to achieve the weight savings possible by using a lightweight structural aggregate for the fill.

Some modular planning is usually required for this type of building, involving the coordination of dimensions for spacing of columns, window mullions, and interior partitions in the building plan. This modular coordination may also be extended to development of ceiling construction, lighting, ceiling HVAC elements, and the systems for access to electric power, phones, and other signal wiring systems. There is no single magic number for this modular system; all dimensions between 3 and 5 ft have been used and strongly advocated by various designers. Selection of a particular priority system for the curtain wall, interior modular partitioning, or an integrated ceiling system may establish a reference dimension.

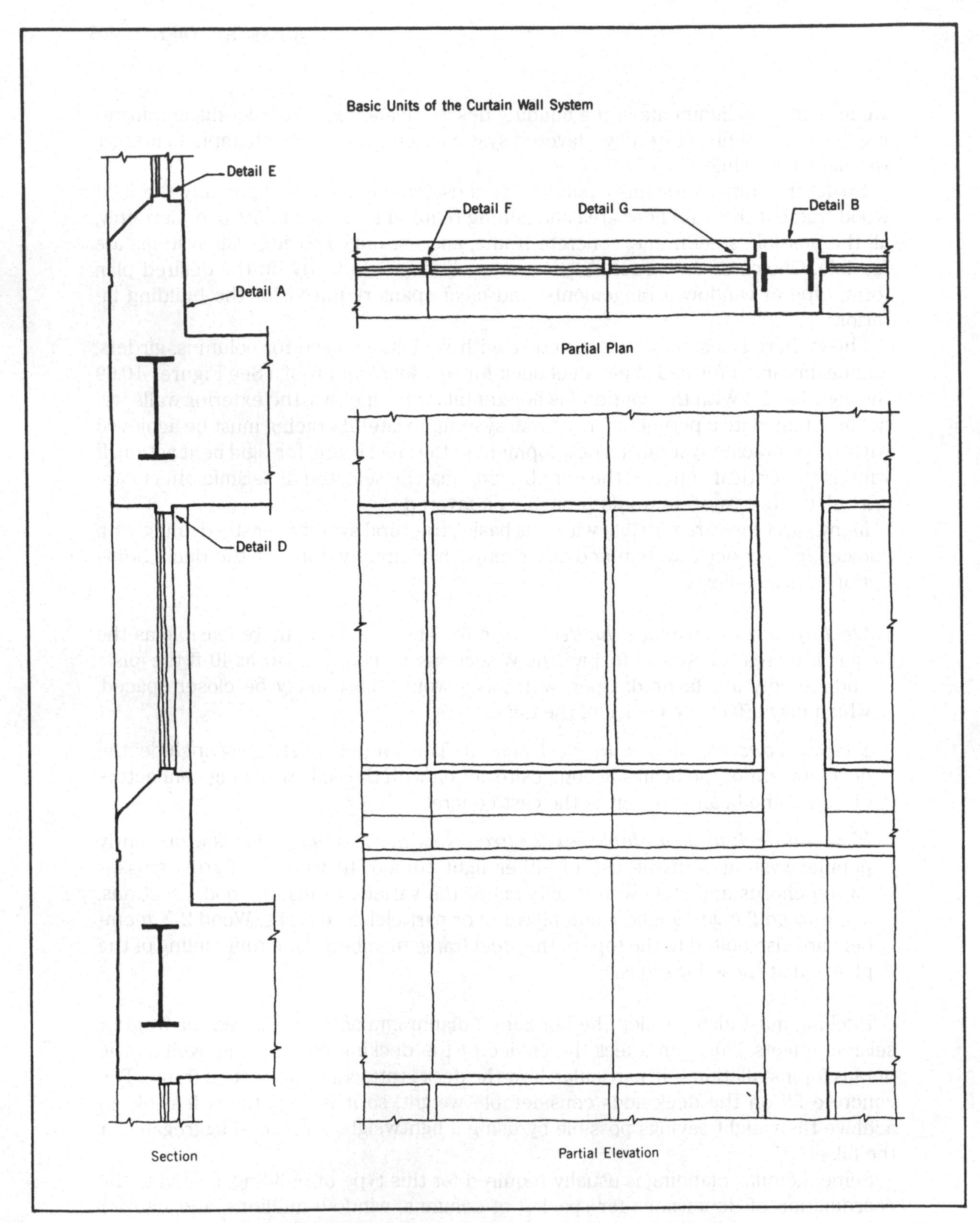

Figure 10.69 Building 6: basic units of the curtain wall system.

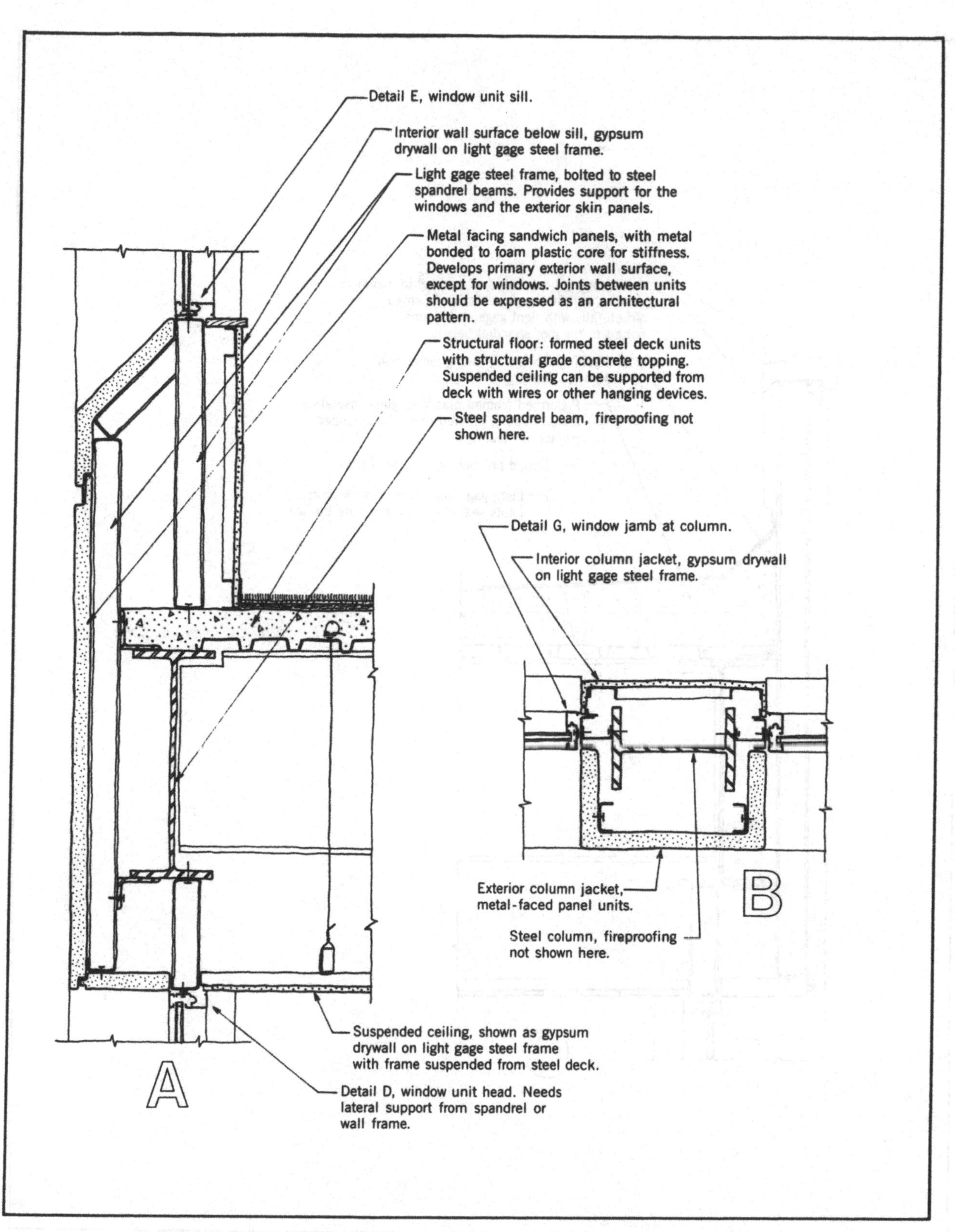
Detail E, window unit sill.
Interior wall surface below sill, gypsum drywall on light gage steel frame.
Light gage steel frame, bolted to steel spandrel beams. Provides support for the windows and the exterior skin panels.
Metal facing sandwich panels, with metal bonded to foam plastic core for stiffness. Develops primary exterior wall surface, except for windows. Joints between units should be expressed as an architectural pattern.
Structural floor: formed steel deck units with structural grade concrete topping. Suspended ceiling can be supported from deck with wires or other hanging devices.
Steel spandrel beam, fireproofing not shown here.
Detail G, window jamb at column.
Interior column jacket, gypsum drywall on light gage steel frame.
Exterior column jacket, metal-faced panel units.
Steel column, fireproofing not shown here.
B
Suspended ceiling, shown as gypsum drywall on light gage steel frame with frame suspended from steel deck.
Detail D, window unit head. Needs lateral support from spandrel or wall frame.
A

Figure 10.70 Building 6: details of the exterior wall and typical floor.

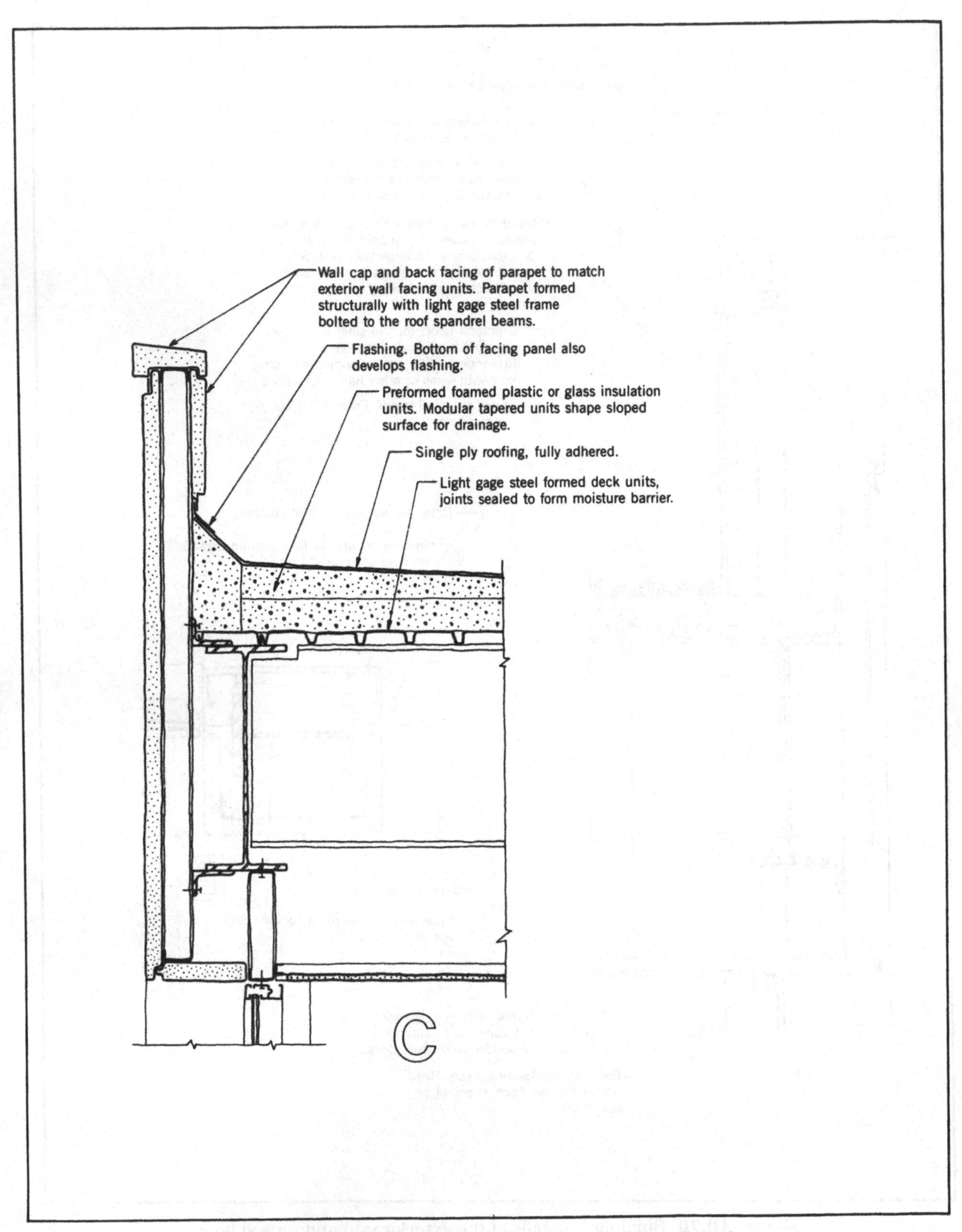

Figure 10.71 Building 6: details of the parapet and roof.

Spacing of columns on the building interior should be as wide as possible, basically to reduce the number of freestanding columns in the building plan. A column-free space for offices may be possible, if the distance from a central core (grouped permanent elements) to the outside walls is not too far for a single span. Spacing of columns at the building perimeter does not affect this issue, so additional columns are sometimes used at this location to reduce their size for gravity loading or to develop a better perimeter rigid frame system for lateral loads.

A major architectural design issue for this building is the choice of basic form of the construction of the exterior walls. For the column-framed structure, there are two elements that must be integrated: the columns and the nonstructural infill wall. In this example the basic form of the construction involves the incorporation of the columns into the wall, with windows developed in horizontal strips between the columns. With the exterior column and spandrel covers developing a general continuous surface, the window units are thus developed as "punched" holes in the wall.

The windows in this example do not exist as parts of a continuous curtain wall system (as in Building 7). They are essentially single individual units, placed in and supported by the general wall system (see Figure 10.72). The general wall is developed as a stud-and-surfacing system not unlike the typical light wood stud wall system in character. The studs in this case are light-gage steel, the exterior covering is a system of metal faced sandwich panel units, and the interior covering, where required, is gypsum drywall, attached to the metal studs with screws.

Detailing of the wall construction (as shown in detail A) results in a considerable interior void space. Although taken up partly with insulation materials, this space may easily contain elements for the electrical system or other services. In cold climates, a perimeter hot water heating system would most likely be used, and it could be incorporated in the wall space shown here.

The general development of this curtain wall construction should be compared with that in Building 7. There are major differences—in exterior appearance, interior planning, generation of interstitial space in the wall, and various design concerns.

For development of the interior, some relationships with the exterior wall and structure are as follows:

1. Interior finish of the exterior wall is part of the curtain wall development, but also an interior surface that must be coordinated with other interior construction. Shown here is a simple gypsum drywall, but many other options are possible, including application of various finishes to the drywall surface.
2. Positioning of the column with relation to the exterior wall and spandrel establishes the nature of the column's presence in the building interior. In this case the wall is thickened to enclose the column, so its only interior presence is in the form of the width of space it occupies in the wall. For comparison, see the situation with the exterior columns in Building 7, as shown in Figures 10.85, 10.86 and 10.87.
3. The locations of vertical window mullions represent potential locations for interior partition walls that intersect the exterior wall. This calls for some design coordination with the planning of interior walls and possibly with any modular ceiling, lighting, and so on.
4. Other details of the windows will also relate to development of interior spaces; such as sill height, head height, presence of horizontal mullions, space for installation of drapes or shades, form of operation of operable sash, and so on.

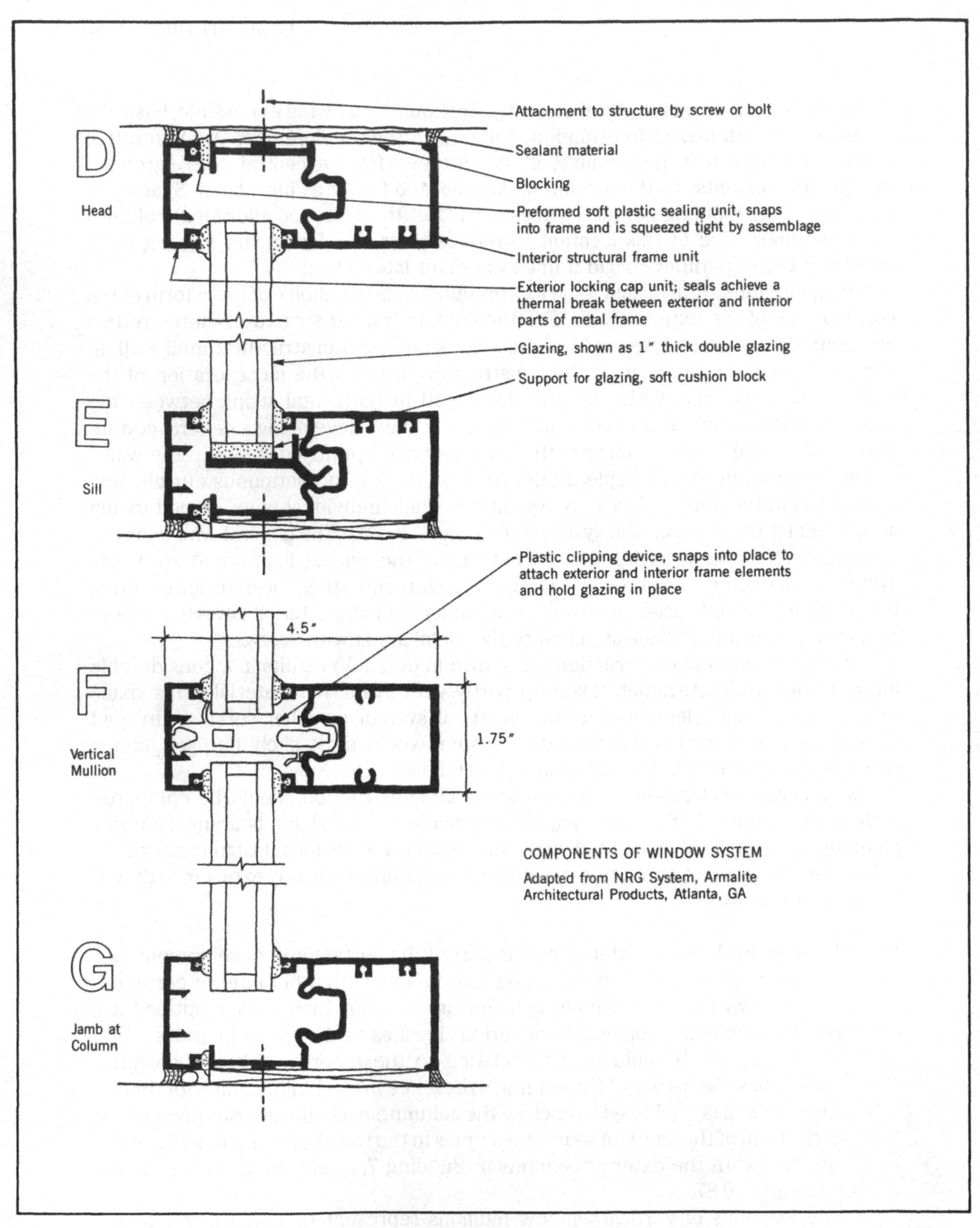

Figure 10.72 Building 6: typical elements of the window system. Adapted from the NRG System, produced by Armalite Architectural Products, Atlanta, Georgia.

It is common to use some basic modular planning in this type of building. Spacing of the building columns constitutes one large unit in this regard, but ordinarily leaves a wide range for smaller divisions. Modular coordination may also be extended to development of ceiling construction, lighting, ceiling HVAC elements, fire sprinklers, and the systems for access to electric power, phones, and other signal wiring systems. There is no single magic number for an interior modular planning system, and all dimensions between 3 and 5 ft have been used for partitions and window mullions. In the United States, common modular units are 4 ft, 12 in., and 4 in. Most priority ceiling systems and corresponding lighting systems use the 1 ft-2-ft-4 ft system for basic elements. However, 4 ft elements can be spaced on 5 ft centers or otherwise utilized in other planning modules. Selection of a particular manufacturer's system for interior development may establish some criteria for this planning.

For buildings built as investment properties, with speculative occupancies that vary over the life of the building, it is usually desirable to accommodate future redevelopment of the building interior with some ease. For the basic construction, this means a design with as few permanent structural elements as possible. At a bare minimum, what is usually required is the construction of the major structure (columns, floors, roof), exterior enclosing walls, and interior walls around stairs, elevators, rest rooms, equipment rooms, and risers for building services. Everything else preferably should be nonstructural and easily demountable, if possible.

The space between the underside of suspended ceilings and the top of floor or roof structures must typically contain many elements besides those of the basic construction. This represents a situation requiring major coordination for the integration of the space needs for the elements of the structural, HVAC, plumbing, electrical, communication, lighting, and fire-fighting systems. A major design decision that must often be made very early in the design process is that of the overall dimension of the space required for this collection of elements. Depth permitted for the spanning structure and the general level-to-level vertical building height will be established—and not easy to change later, if the detailed design of any of the enclosed systems indicates a need for more space.

Generous provision of the space for building elements will make the work of the designers of the various susbsystems easier, but the overall effects on the building design must be considered. Extra height for the exterior walls and permanent interior walls, and for stairs, elevators, and service risers all result in additional cost, making tight control of the level-to-level distance very important.

As previously mentioned, a detail consideration in planning the interior is the manner in which interior partitioning meets the exterior walls. Partitions at right angles to the exterior must join to a column, a solid portion of wall, or—in some manner—to a vertical framing member in the window system. The construction of the partitions and of the columns, windows, and other exterior wall elements must be studied in this regard.

Obviously, many other options are possible for the general form, the materials, and the details of the construction of the exterior walls and the elements of the building structure that occur at the building edge. Figure 10.72 shows some variations on the building edge condition and the ramifications with regard to development of interior partitioning and ceilings.

Alternative details for the windows, consisting of standard curtain wall units, are shown in Figure 10.73. These are units made for the development and not of a general curtain wall system such as that shown for Building 7.

Suspended ceilings—as shown in Figure 10.74—are usually developed with a

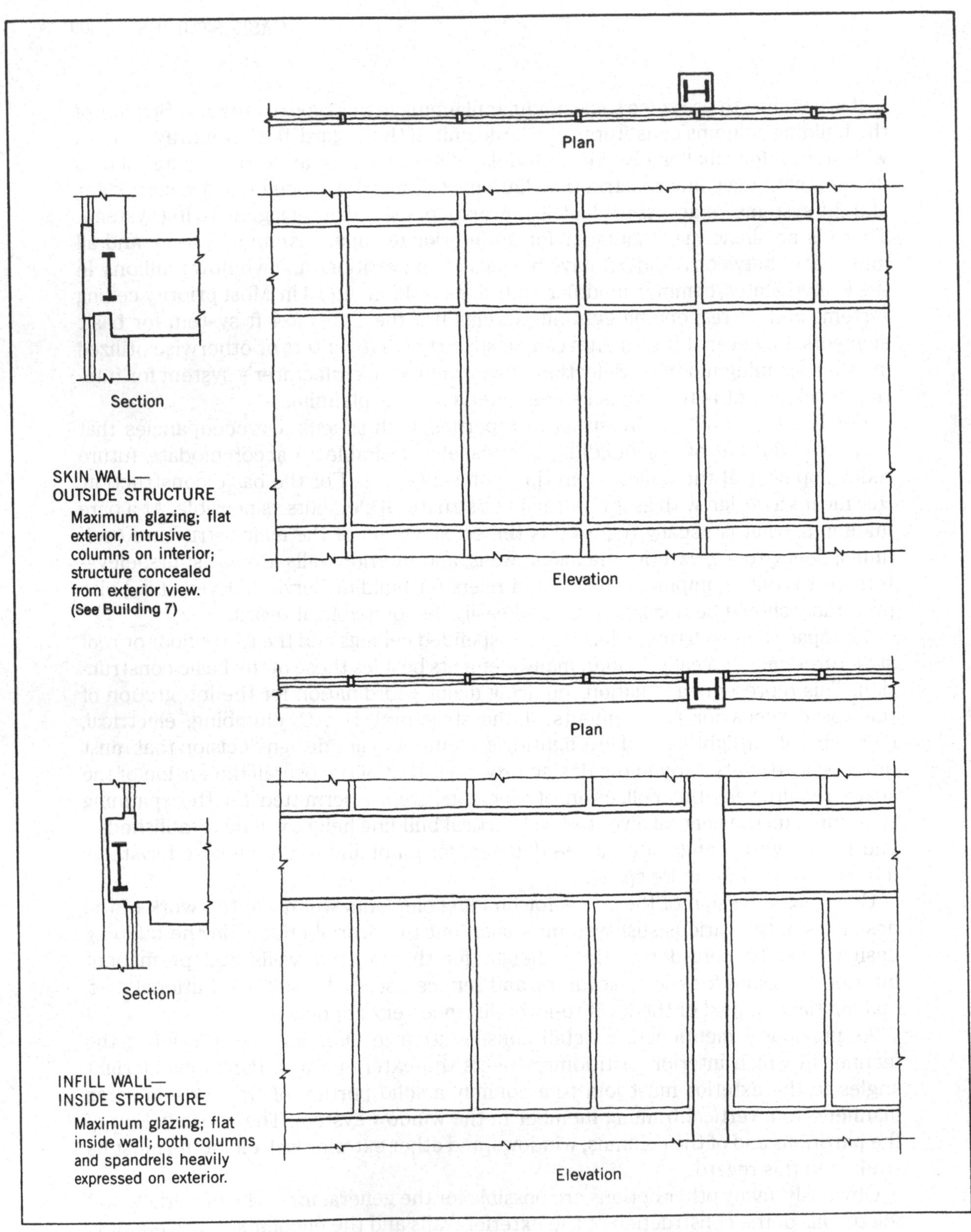

Figure 10.73 Building 6: alternative details for the development of the building exterior walls.

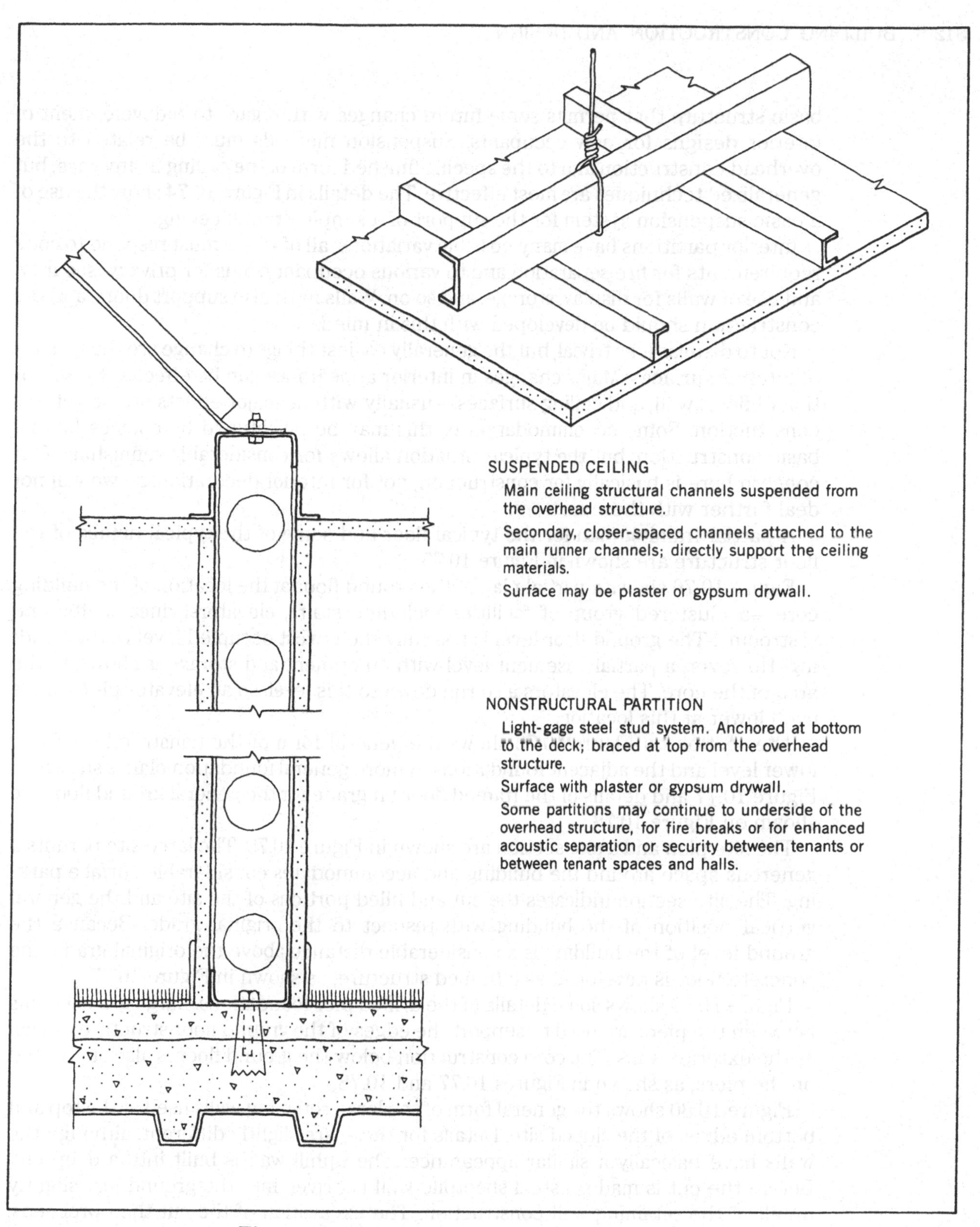

SUSPENDED CEILING
Main ceiling structural channels suspended from the overhead structure.
Secondary, closer-spaced channels attached to the main runner channels; directly support the ceiling materials.
Surface may be plaster or gypsum drywall.
NONSTRUCTURAL PARTITION
Light-gage steel stud system. Anchored at bottom to the deck; braced at top from the overhead structure.
Surface with plaster or gypsum drywall.
Some partitions may continue to underside of the overhead structure, for fire breaks or for enhanced acoustic separation or security between tenants or between tenant spaces and halls.

Figure 10.74 Building 6: ceiling and partition details.

basic structure that permits some future changes with regard to redevelopment of interior designs for new occupants. Suspension methods must be related to the overhead construction and to the specific finished form of the ceiling in any case, but generalized techniques are most effective. The details in Figure 10.74 show the use of a basic suspension system for the support of a simple drywall ceiling.

Interior partitions have many possible variations, all of which must respond to code requirements for fire separation and to various occupant needs for privacy, security, and use of walls for display, storage, and so on. Walls must also support doors, and the construction should be developed with this in mind.

Not to dismiss it as trivial, but the generally easiest things to change are the finishes of interior surfaces. Major changes in interior appearance can be effected by alteration of floor, wall, and ceiling surfaces—usually with no major effects on the general construction. Some accommodation of this may be heightened by choices for the basic construction, but the typical situation allows for considerable refinishing. Our concern here is basically for construction, not for interior decoration; so we will not deal further with these matters.

A partial framing plan for the typical floor and some of the typical details of the floor structure are shown in Figure 10.75.

Figure 10.76 shows a partial plan of the ground floor at the location of the building core—a clustered group of facilities including stairs, elevators, duct shafts, and restrooms. The ground floor level is basically the lowest occupied level of the building. However, a partial basement level with equipment and storage is shown in the area of the core. The elevators also run down to this level, so an elevator pit extends even lower at this location.

The section on Figure 10.76 shows the general form of the construction of this lower level and the adjacent foundations. A more general foundation plan is shown on Figure 10.77 and details of the framed floor on grade for the general ground floor are shown on Figure 10.78.

The site plan and a site section are shown in Figure 10.79. The large site permits a generous space around the building and accommodates considerable surface parking. The site section indicates the cut and filled portions of the site and the general vertical position of the building with respect to the original grade. Because the ground level of the building is a considerable distance above the original grade, the concrete floor is developed as a framed structure, as shown in Figure 10.77.

Figure 10.77 shows some details of the drilled pier foundations. Large grade beams between the piers are used to support the edges of the ground floor structure as well as the exterior walls. The core construction below the ground floor is also supported on the piers, as shown in Figures 10.77 and 10.78.

Figure 10.80 shows the general form of the large retaining walls used at the top and bottom edges of the sloped site. Details for these are slightly different, although the walls have basically a similar appearance. The uphill wall is built into a deep cut. Before the cut is made, a steel sheetpile wall is driven into the ground just slightly uphill of the retaining wall construction. The excavation of the cut then proceeds, with anchors used to brace the sheetpile wall as the cut gets deeper. When the cut is complete to the level of the bottom of the retaining wall footing, construction of the wall begins. When the concrete wall is completed, the steel wall may be removed, or left in place with a draining fill placed between it and the concrete wall.

The downhill wall, on the other hand, penetrates only a short distance into the original grade and serves basically to retain fill. This wall can therefore be built without the sheetpile wall. The positions of the footings for these two walls with

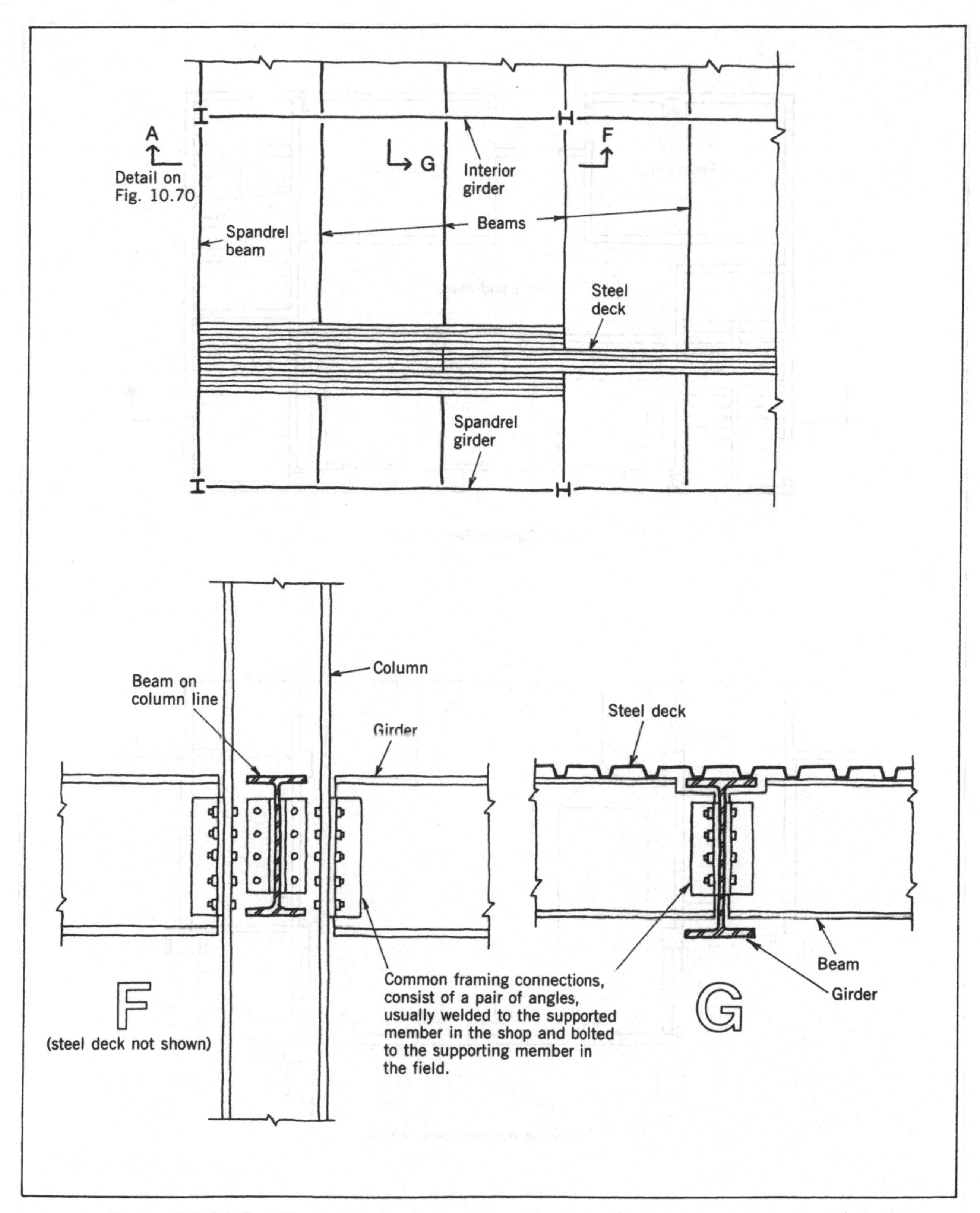

Figure 10.75 Building 6: framing plan and structural details for the typical office floor.

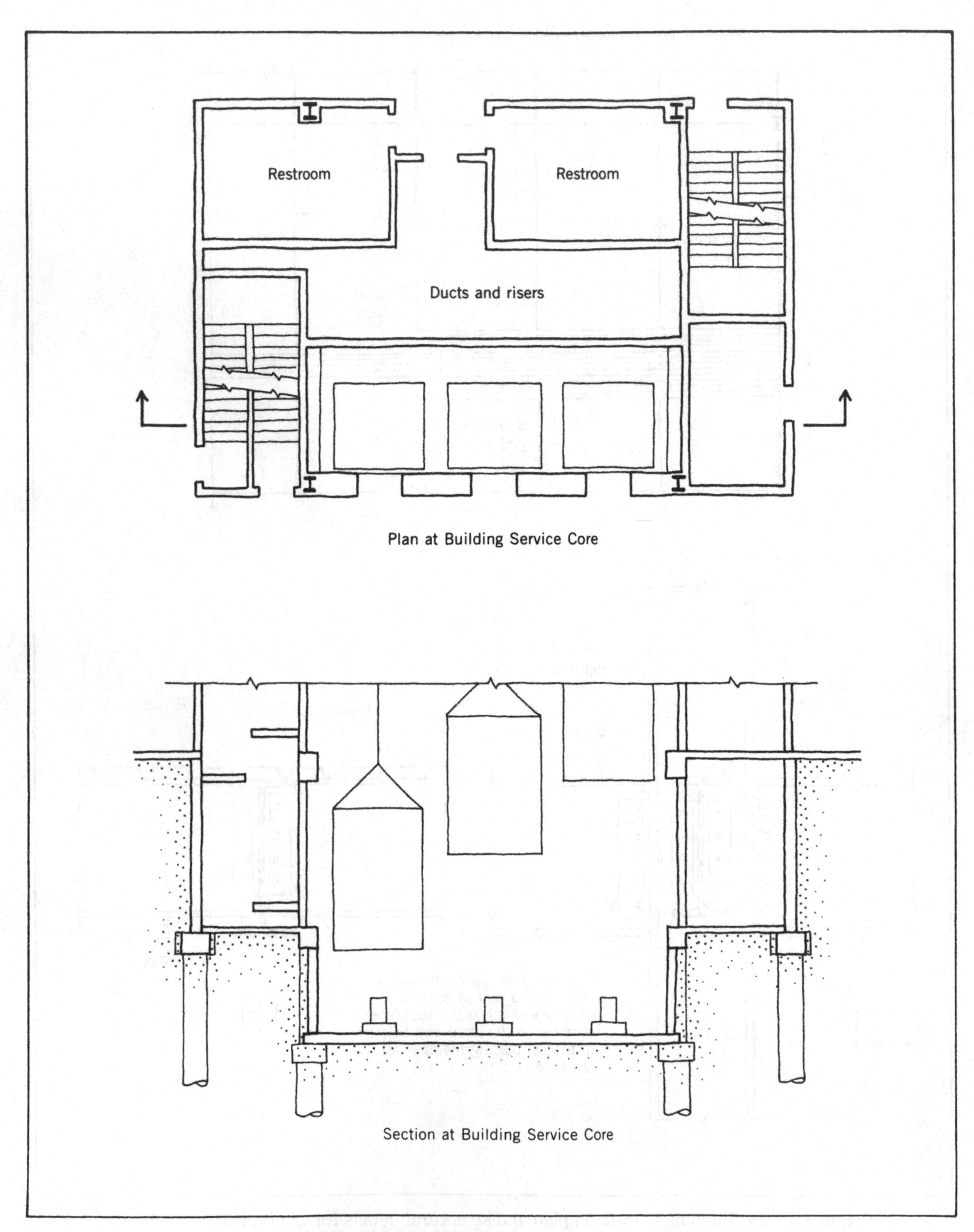

Figure 10.76 Building 6: core plan and sections.

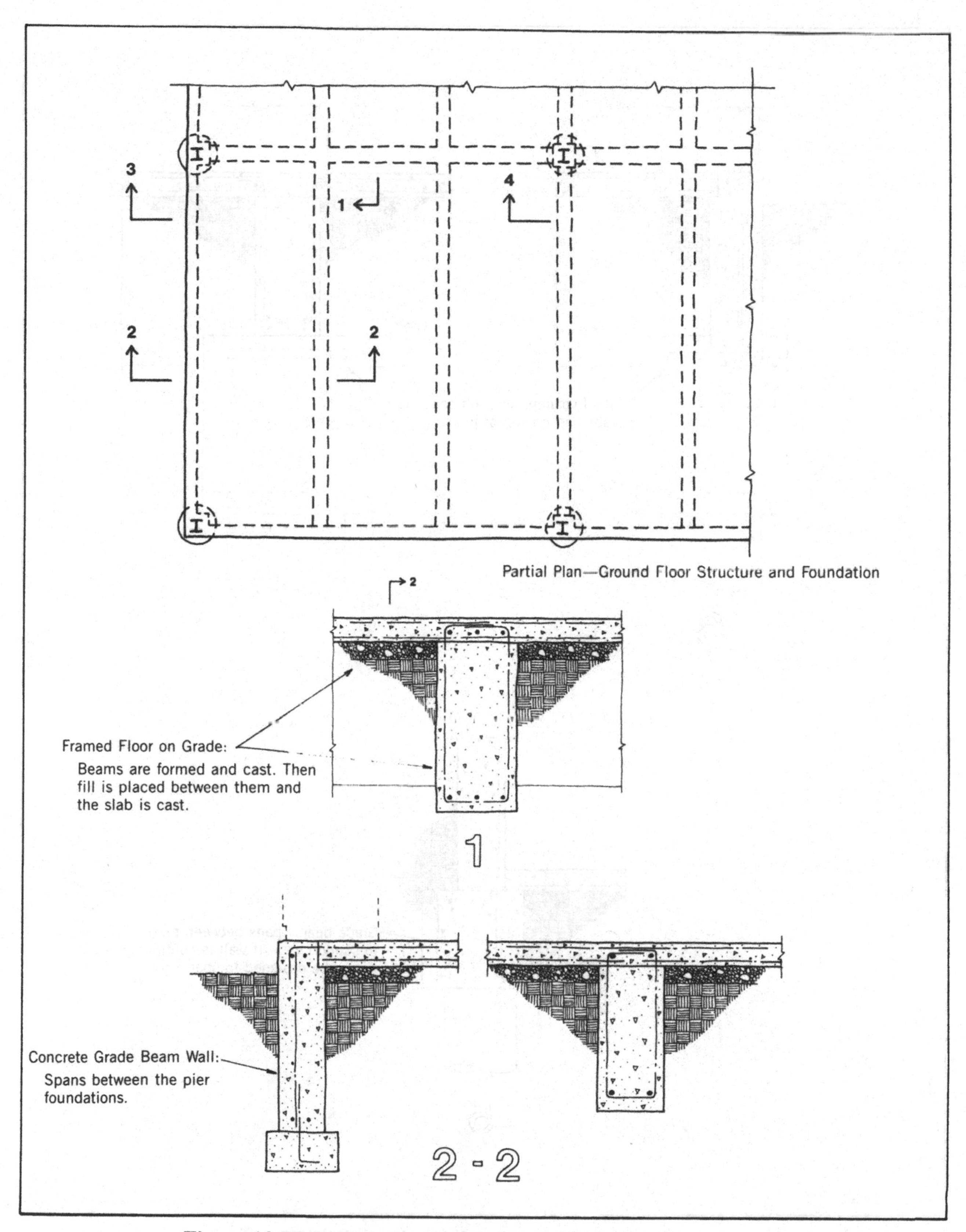
3
1
4
2
2
Partial Plan—Ground Floor Structure and Foundation
2
Framed Floor on Grade:
Beams are formed and cast. Then fill is placed between them and the slab is cast.
1
Concrete Grade Beam Wall:
Spans between the pier foundations.
2 - 2

Figure 10.77 Building 6: ground floor and foundation plan and details.

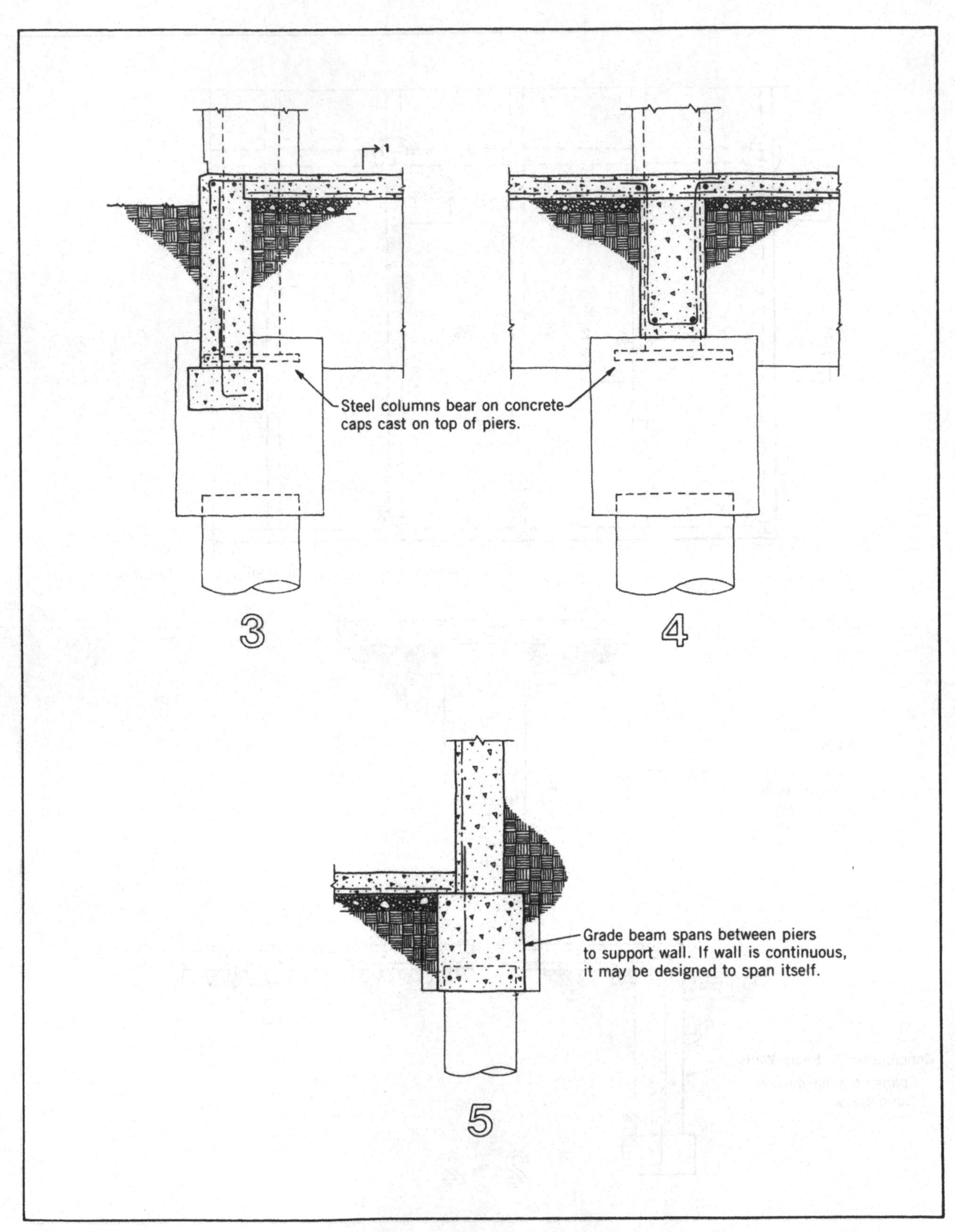

Figure 10.78 Building 6: foundation details.

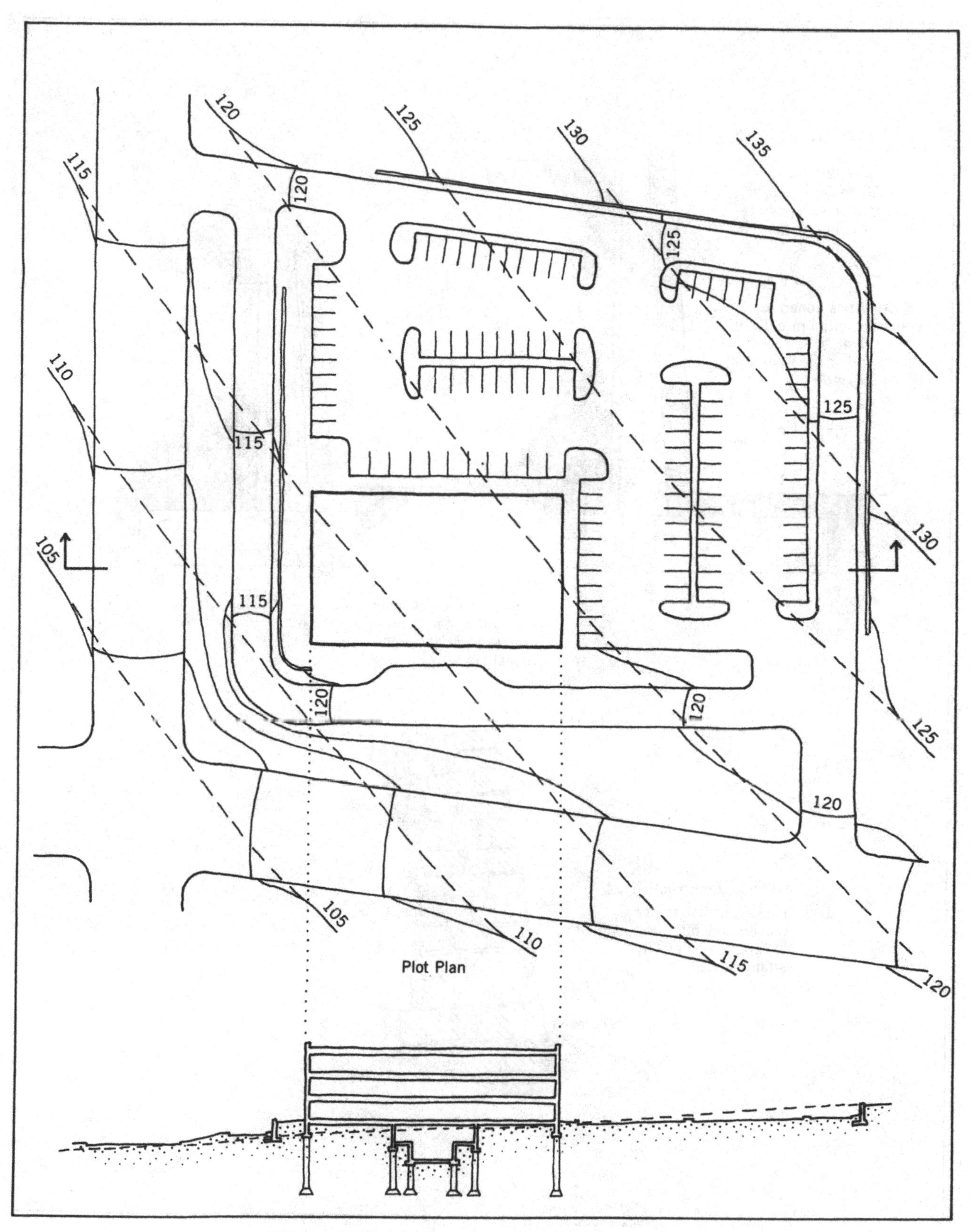

Figure 10.79 Building 6: site plan and section.

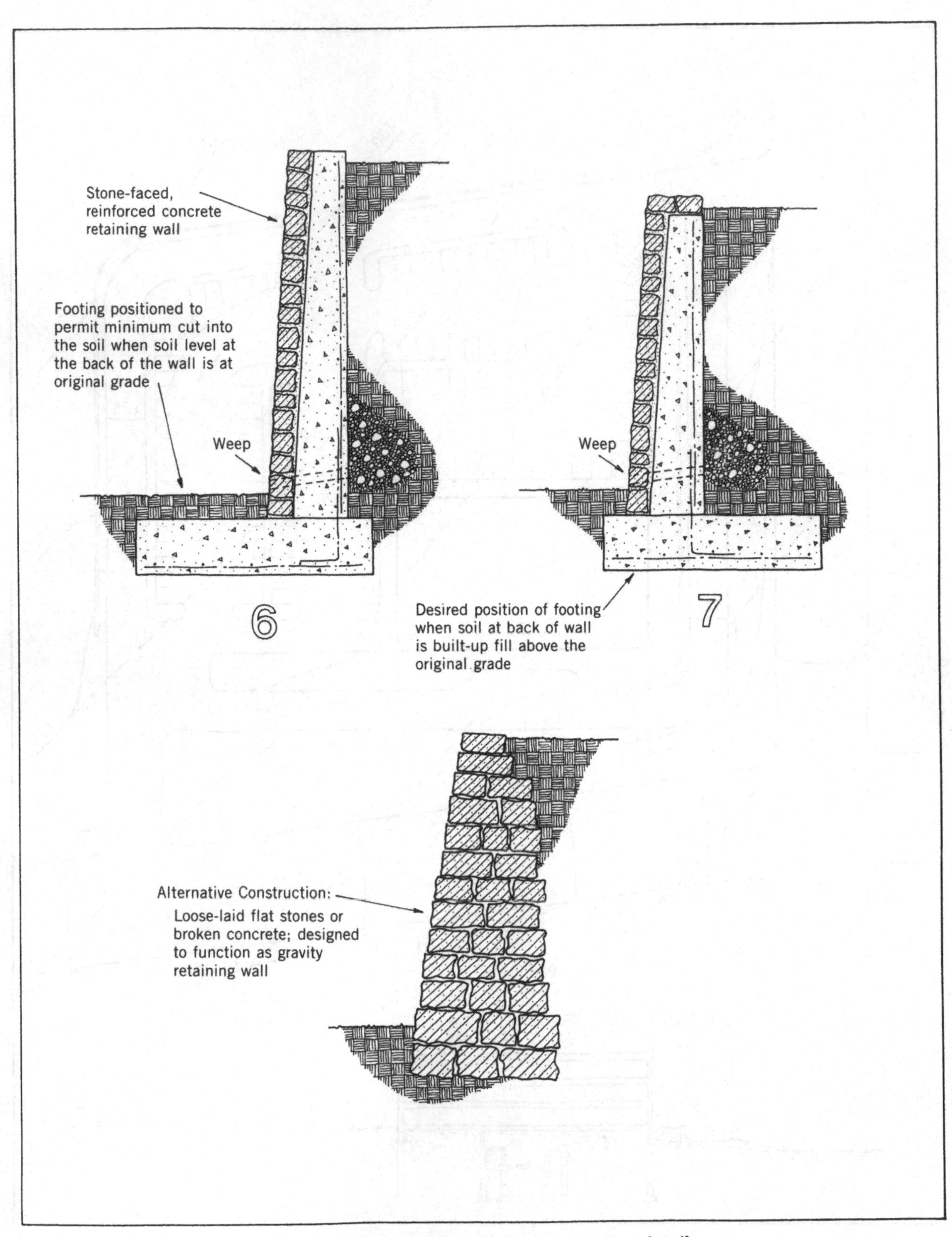

Figure 10.80 Building 6: site construction details.

respect to the wall are slightly different, reflecting the conditions for excavation and the proximity of the site boundaries.

Figure 10.81 shows the general scheme and some details for the HVAC system for the office areas. A ceiling mounted air system delivers air through registers that are integrated with the general modular ceiling system. Return air to the circulating system is pulled through the ceiling, using the enclosed, interstitial space between the ceiling and the structure above as a general plenum space (one big duct). Air also enters the plenum space from grilles in the ceiling of the corridors. Additional details for the ceiling units are shown in Figure 10.82.

The air-handling equipment and heat-dissipating units for the air conditioning system are located on the rooftop. Supply and return air for the offices is fed through a vertical duct shaft in the building core. The air conditioning chiller and the furnace and hot water boiler are in the partial basement at the building core. Hot and cold water is circulated between the basement and the equipment on the rooftop through pipe chases in the core.

The chilled and heated water is also fed to two other systems. One system consists of a series of *zone reconditioners* that are located on the office floors. These serve specific limited areas of the floor and perform minor modification of the temperature of the air generally circulating in the ducts; responding to the varying needs at different points on a single floor. Thus, at any given time, extra heat may be supplied on the shady side of the building at the same time that slightly cooled air is being supplied on the sunny side.

The other system fed by the dual piped water system consists of fin tube radiators along the outside walls. These are installed in the hollow space in the walls beneath the windows and feed up through grilles in the window sills, providing a wash of warm air up the inside surface of the windows. This reduces condensation and frosting on the windows during cold weather and generally adds an extra dose of heat at the cold outside walls.

This dual heating system—perimeter radiators and circulating air—is common in colder climates. The air system is generally necessary for ventilation anyway, and is also most likely the choice for air conditioning (cooling); now provided in just about all climates.

As with other buildings, special systems may be provided for HVAC service of other building spaces. The building rest rooms will most likely have a continuously operating ventilation exhaust system, with undercut or louvered doors pulling air in from corridors to assure a one-way air flow between rest rooms and their connecting corridors.

Figure 10.83 shows some details for a general illumination system for the office areas. This system uses recessed light fixtures integrated with the general ceiling surfacing and other elements in the ceiling location. In addition to the HVAC elements (see Figure 10.82), this may include fire sprinklers, speakers for an audio system, signs, and smoke detectors. Various priority systems are available, with varying degrees of integration of these elements in a total system.

Figure 10.84 shows some details for the development of wiring distribution through a system built into the floor structure. These systems were widely used in the past to provide a form of modular, modifiable attachment for both electrical power and telephones at random locations on the floors. This random attachment is an ongoing necessity, but there are now various other options for responding to the needs. One such option is that shown for Building 7, involving use of an elevated, platform floor structure.

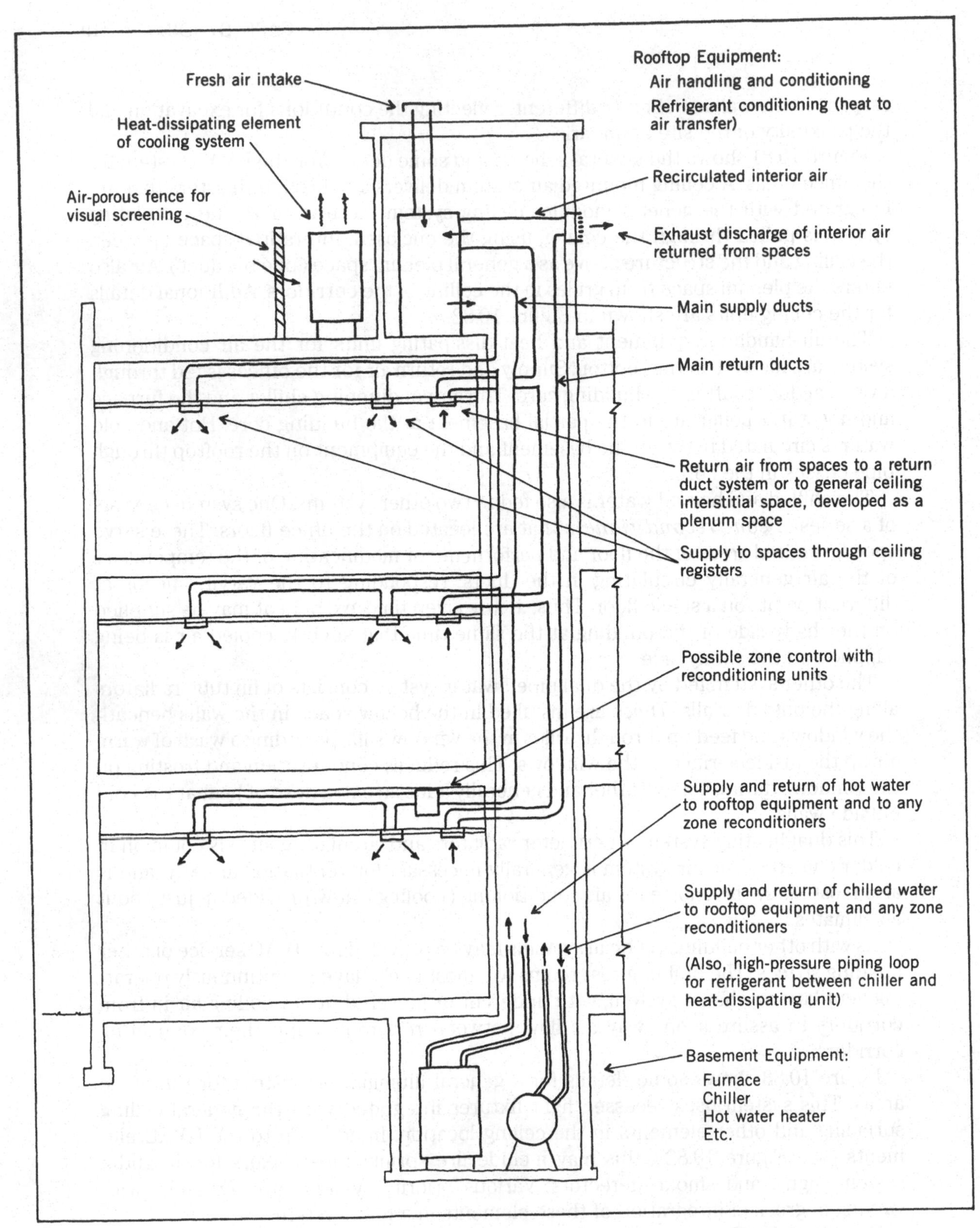

Figure 10.81 Building 6: general scheme and details for the HVAC system.

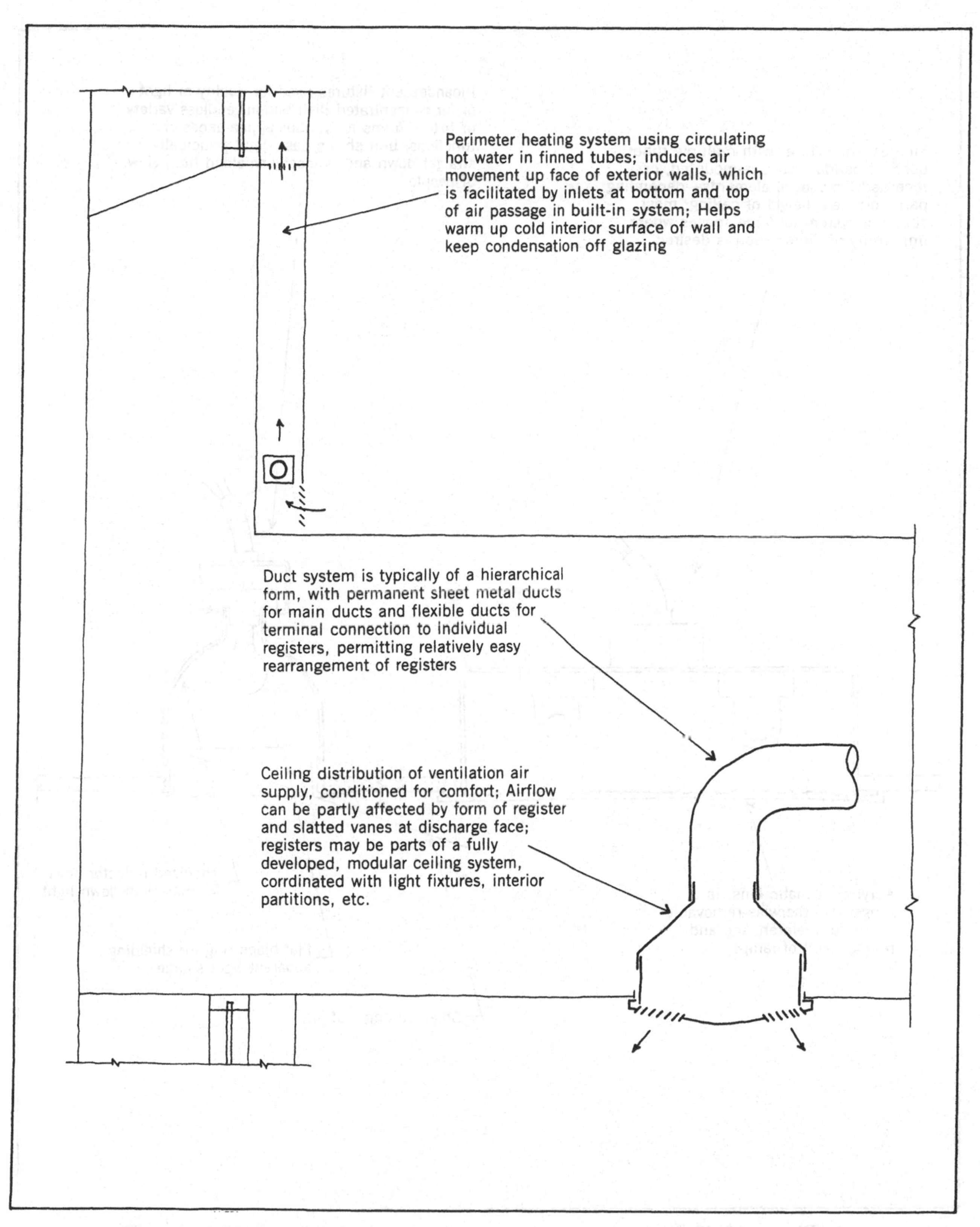

Perimeter heating system using circulating hot water in finned tubes; induces air movement up face of exterior walls, which is facilitated by inlets at bottom and top of air passage in built-in system; Helps warm up cold interior surface of wall and keep condensation off glazing
Duct system is typically of a hierarchical form, with permanent sheet metal ducts for main ducts and flexible ducts for terminal connection to individual registers, permitting relatively easy rearrangement of registers
Ceiling distribution of ventilation air supply, conditioned for comfort; Airflow can be partly affected by form of register and slatted vanes at discharge face; registers may be parts of a fully developed, modular ceiling system, corrdinated with light fixtures, interior partitions, etc.

Figure 10.82 Building 6: details for the HVAC system at the typical office floor.

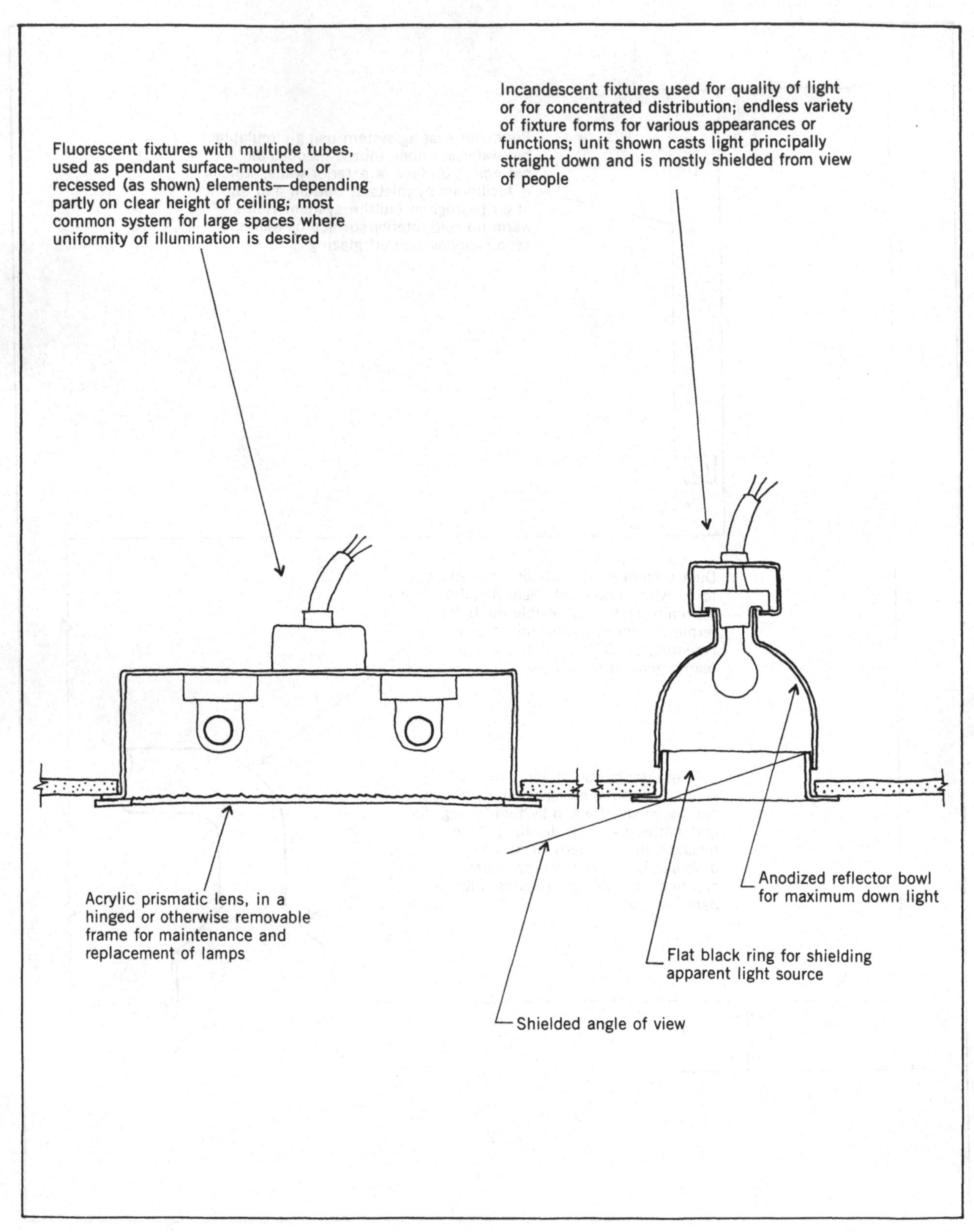
Incandescent fixtures used for quality of light or for concentrated distribution; endless variety of fixture forms for various appearances or functions; unit shown casts light principally straight down and is mostly shielded from view of people
Fluorescent fixtures with multiple tubes, used as pendant surface-mounted, or recessed (as shown) elements—depending partly on clear height of ceiling; most common system for large spaces where uniformity of illumination is desired
Anodized reflector bowl for maximum down light
Acrylic prismatic lens, in a hinged or otherwise removable frame for maintenance and replacement of lamps
Flat black ring for shielding apparent light source
Shielded angle of view

Figure 10.83 Building 6: details for the general illumination system at the typical office floor.

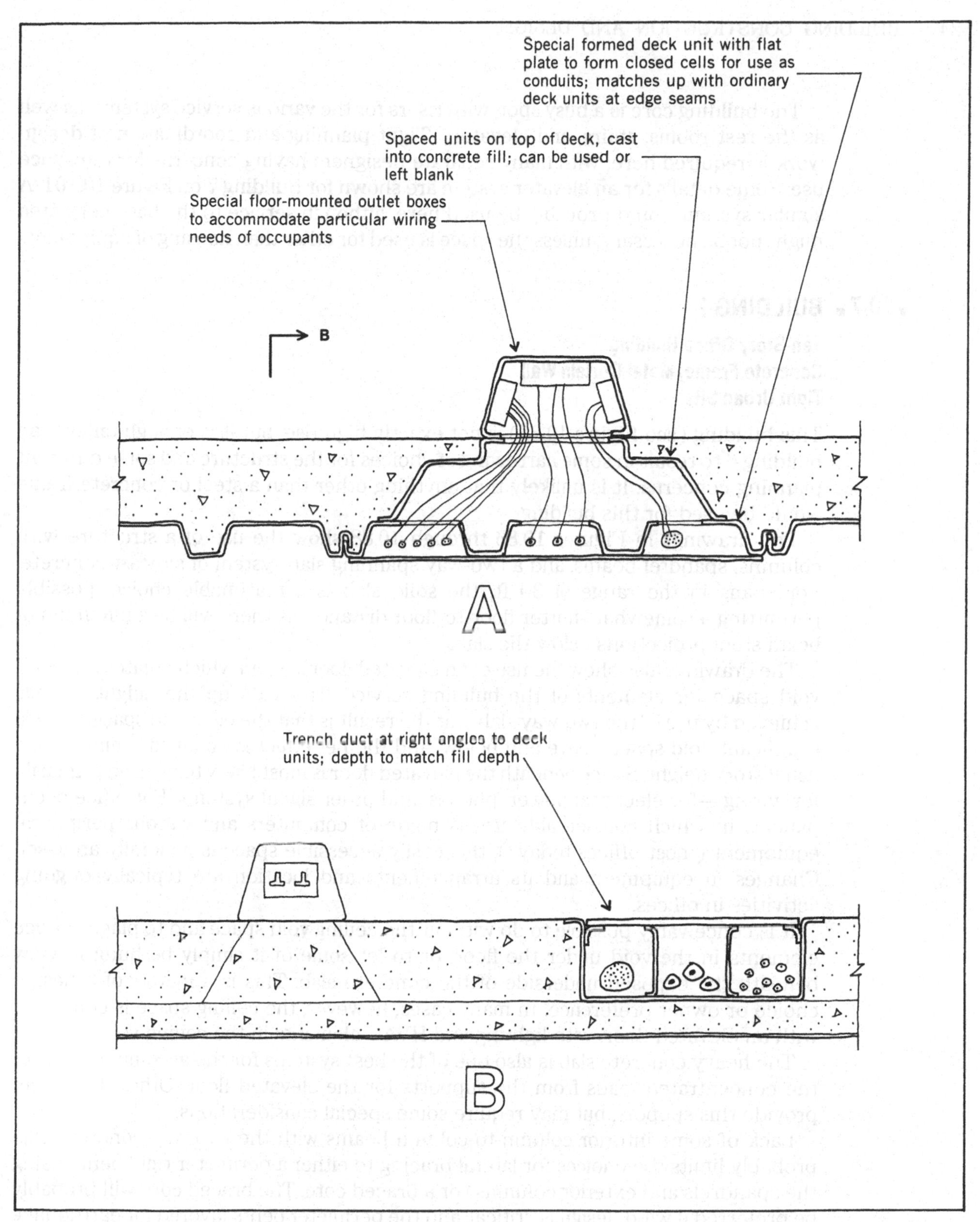

Figure 10.84 Building 6: details for the electrical power system at the office floors.

The building core is a busy spot, with risers for the various service systems, as well as the rest rooms, stairs, and elevators. Tight planning and coordination of design work is required here, with many separate designers having concerns for the space use. Some details for an elevator system are shown for Building 7 on Figure 10.101. A similar system would probably be used here, although service to the basement area might not be necessary, unless the space is used for other than housing of equipment.

■ 10.7 ■ BUILDING 7

Ten-Story Office Building
Concrete Frame, Metal Curtain Wall
Tight Urban Site

This building (see Figure 10.85) is not exactly high rise, but it is enough taller than building 6 to result in some narrowing of choices for the structure and some different planning concerns. It is unlikely that anything other than a steel or concrete frame would be used for this building.

The drawings in Figures 10.86 through 10.88 show the use of a structure with columns, spandrel beams, and a two-way spanning slab system of sitecast concrete. For spans in the range of 30 ft, the solid slab is a reasonable choice, possibly permitting a somewhat shorter floor-to-floor distance, as there will be a minimum of beam stem projections below the slab.

The drawings also show the use of an elevated floor system which creates a second void space for elements of the building service. This eats up the height savings achieved by use of the two-way slab, but the result is that the occupied spaces have a significant void space above and below, and the net effect is about the same as the usual story height. Space beneath the elevated floor is most likely to be used primarily for wiring—for electrical power, phones, and other signal systems. For office occupancies in which considerable use is made of computers and various peripheral equipment (most offices today?), this easily accessible space is generally an asset. Changes in equipment and its arrangements and location are typically ongoing activities in offices.

It is conceivably possible to do without the ceiling void space and to place service elements in the void under the floor, or to let some of it simply be hung in view beneath the exposed underside of the concrete slab. This is a matter of a design choice or owner preference. In many cases, however, the ceiling space is combined with an elevated floor, with lighting and HVAC elements in the ceiling void.

The heavy concrete slab is also one of the best systems for the accommodation of the concentrated loads from the supports for the elevated floor. Other decks can provide this support, but may require some special considerations.

Lack of some interior column-to-column beams with the two-way concrete slab probably limits the choices for lateral bracing to either a perimeter rigid bent (using the spandrels and exterior columns) or a braced core. The braced core will probably be preferred if wind design is critical, and the perimeter bents favored for earthquake resistance.

The form of curtain wall used here is one that creates a continuous skin outside the building structure. The columns are thus almost freestanding inside the wall, a plan intrusion in the occupied space that some designers or owners might find objectionable. (For comparison see the wall and plan for Building 6, in which the columns are enclosed by the thick exterior wall at floor level.)

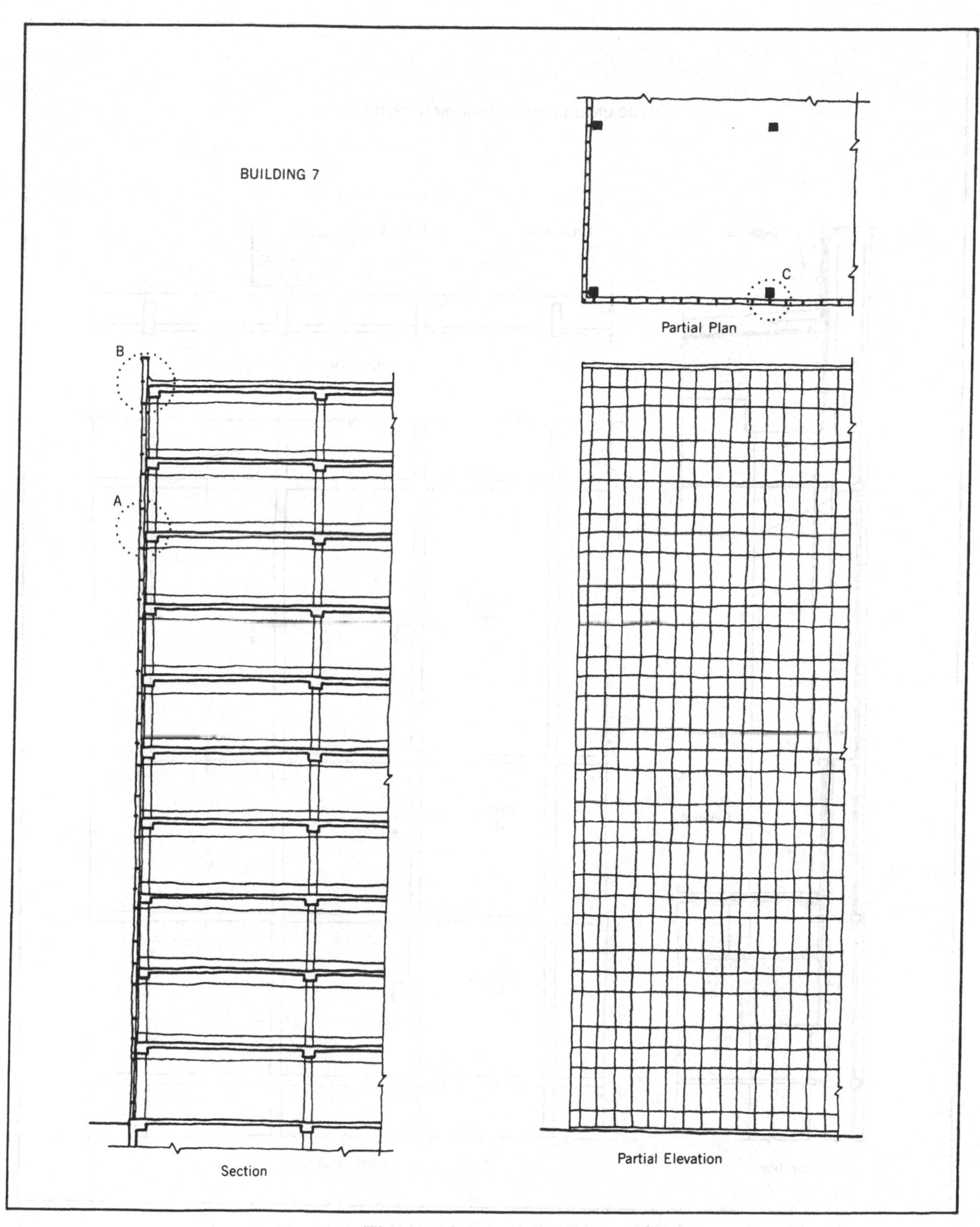

Figure 10.85 Building 7.

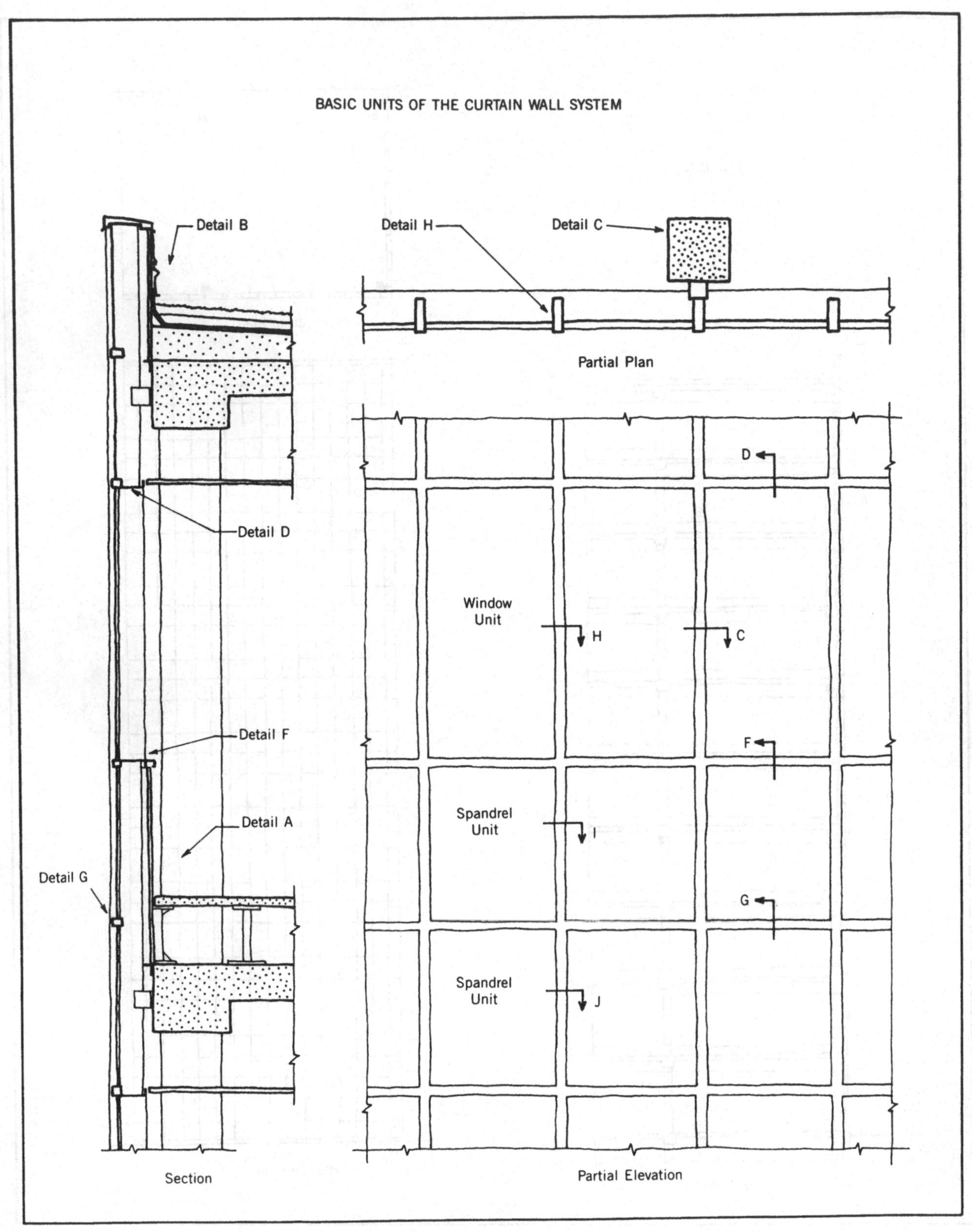

Figure 10.86 Building 7: basic units of the curtain wall system.

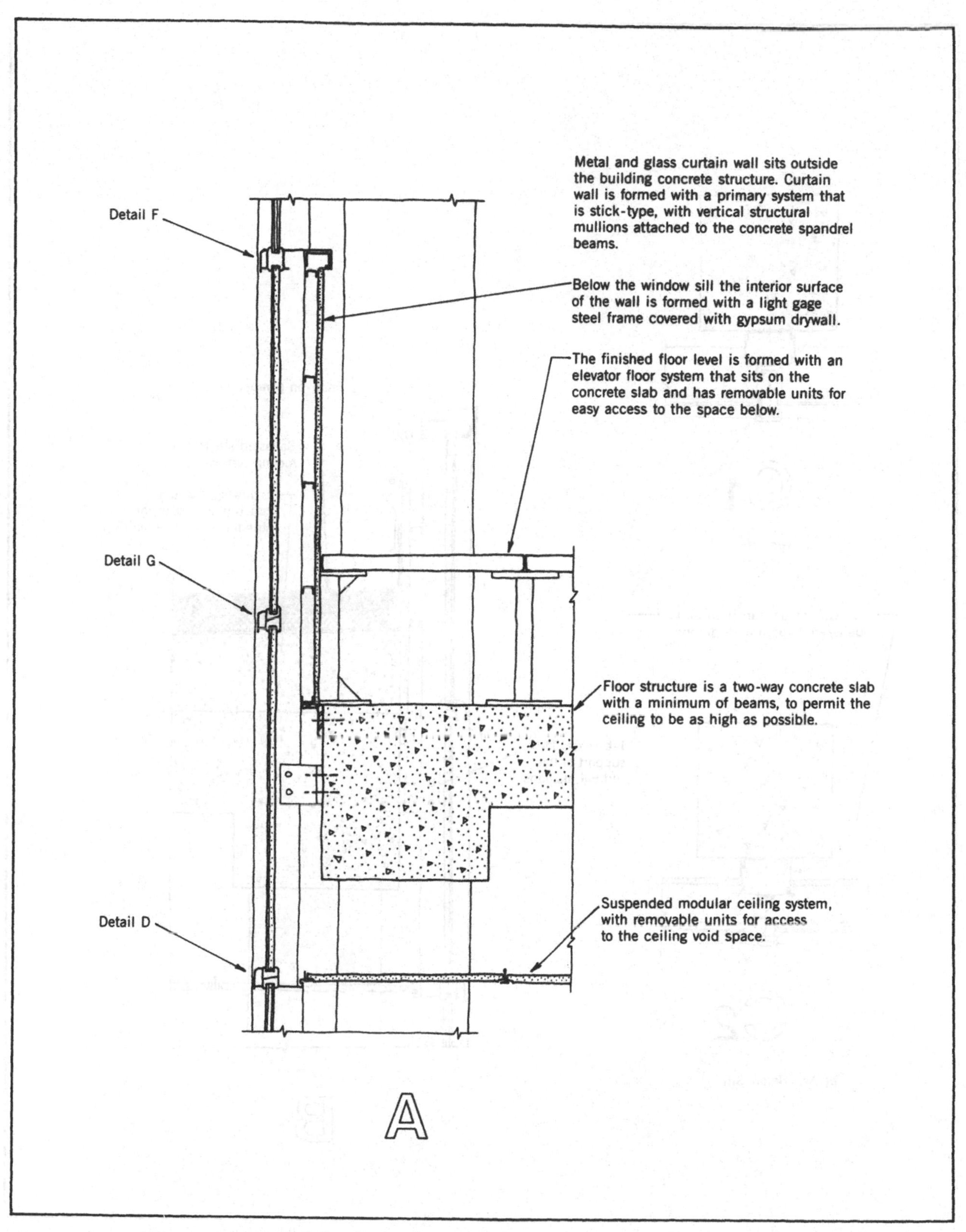

Metal and glass curtain wall sits outside the building concrete structure. Curtain wall is formed with a primary system that is stick-type, with vertical structural mullions attached to the concrete spandrel beams.
Detail F
Below the window sill the interior surface of the wall is formed with a light gage steel frame covered with gypsum drywall.
The finished floor level is formed with an elevator floor system that sits on the concrete slab and has removable units for easy access to the space below.
Detail G
Floor structure is a two-way concrete slab with a minimum of beams, to permit the ceiling to be as high as possible.
Detail D
Suspended modular ceiling system, with removable units for access to the ceiling void space.
A

Figure 10.87 Building 7: details of the exterior wall and typical office floor.

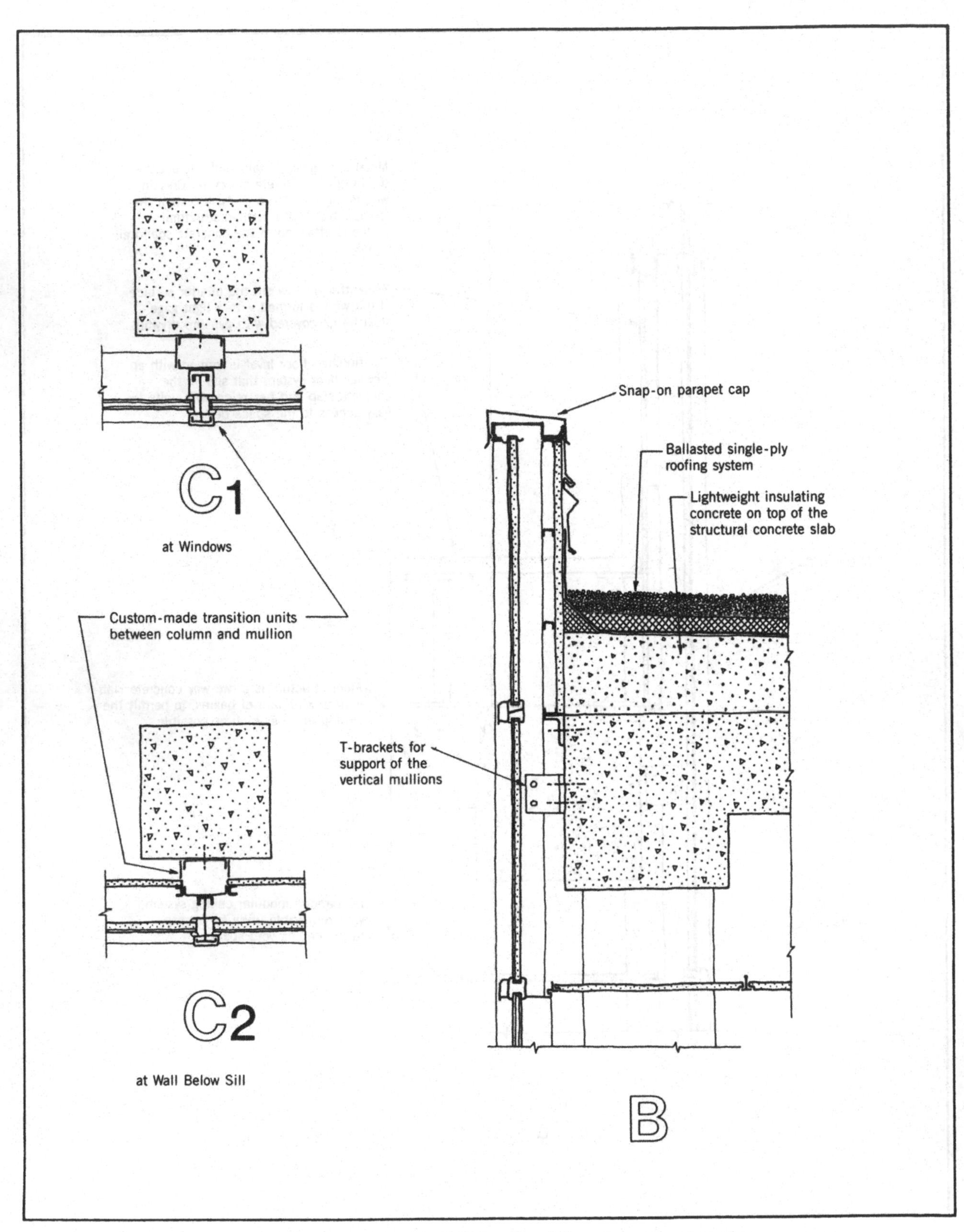

Figure 10.88 Building 7: details of the parapet and roof.

Formation and erection of the curtain wall is achieved using the “stick” system, in which vertical mullions are developed like a stud system to provide basic support for the rest of the wall. The building structure thus supports the entire wall simply by grabbing the vertical mullions, in this case using T-brackets on the face of the spandrels. The mullions are built up from a heavy interior element that is the basic structural member, with additional parts attached to provide exterior finish surfaces and to grab the glazing and spandrel panels.

Horizontal mullions are developed as single-length elements between the continuous vertical mullions. As with the vertical mullions, there is a structural portion with other parts attached to produce the complete mullion.

Windows continue in horizontal rows past the columns, so that the building exterior surface does not reflect the pattern of the column system as it does in Building 6. Spandrel panels are placed in the same plane with the glazing and have glass on the exterior surface. A continuous all-glass wall is thus produced, and if dark or reflective glazing is used, the difference between glazing and spandrel panels may not be apparent in daylight. Thus the spandrels also become obscured, and a very undelineated surface results, with neither vertical nor horizontal structural references visually apparent.

Details of the general construction shown in Figures 10.86 through 10.90 indicate the form of the curtain wall and the possibilities for development of interior partitioning that abuts the exterior wall. As with Building 6, there are two general cases: partitions that meet a column and partitions that meet a vertical mullion in the curtain wall. The difference here is that there is really no solid portion of wall as far as the interior partitions are concerned, as the vertical mullions are essentially continuous from floor to ceiling.

Some possible details for the development of interior construction are shown in Figure 10.91. These indicate the use of a ceiling system with lay-in units supported by a suspended, modular grid of inverted T-shaped elements; a basic system available from many manufactures. Interior partitions are formed with light gage steel studs and surfaced with gypsum drywall screwed to the studs.

The section in Figure 10.91 shows the partitions seated on the concrete slab and running completely to the underside of the concrete slab above. This provides for maximum separation for fire, security, and sound control. The suspended ceiling and elevated floors thus abut the partitions, with possible details as shown in the section.

Figure 10.92 shows some details for the elevated floor system, using the priority elements of one manufacturer. The basic manufactured system provides for the general floor structure and some special conditions, but some expedient detailing or matching construction is usually needed for the full development of the building floor system. In this case, in order to keep the whole floor at one level, a special system is used for the corridors. Similar solutions may be developed for the floors of rest rooms, stair landings, elevator lobbies, and other locations on a typical office floor. There are many options in this situation, and the suppliers of a particular priority system may well have their own solutions for most common occurrences in order to further promote use of their products.

A partial framing plan and some details for the typical tower floor are shown in Figure 10.93.

Figure 10.94 shows a site plan and building/site section, which reveals that the office tower is surrounded by a plaza which is the roof of a two story underground parking garage. A partial framing plan for the plaza is shown in Figure 10.95, together with some typical details of the structural elements.

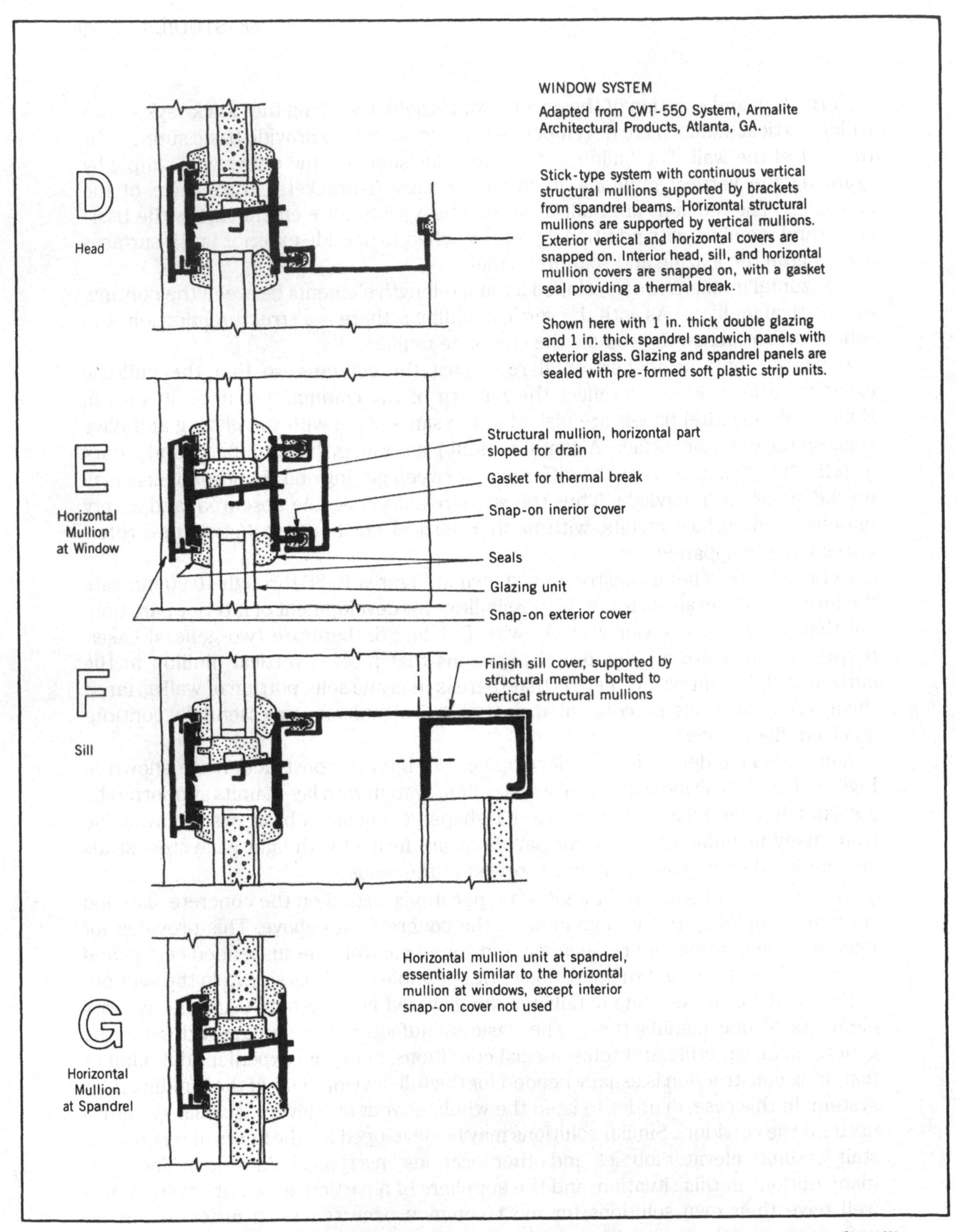

Figure 10.89 Building 7: basic elements of the curtain wall system—horizontal mullions. Details adapted from the CWT-550 System, produced by Armalite Architectural Products, Atlanta, GA.

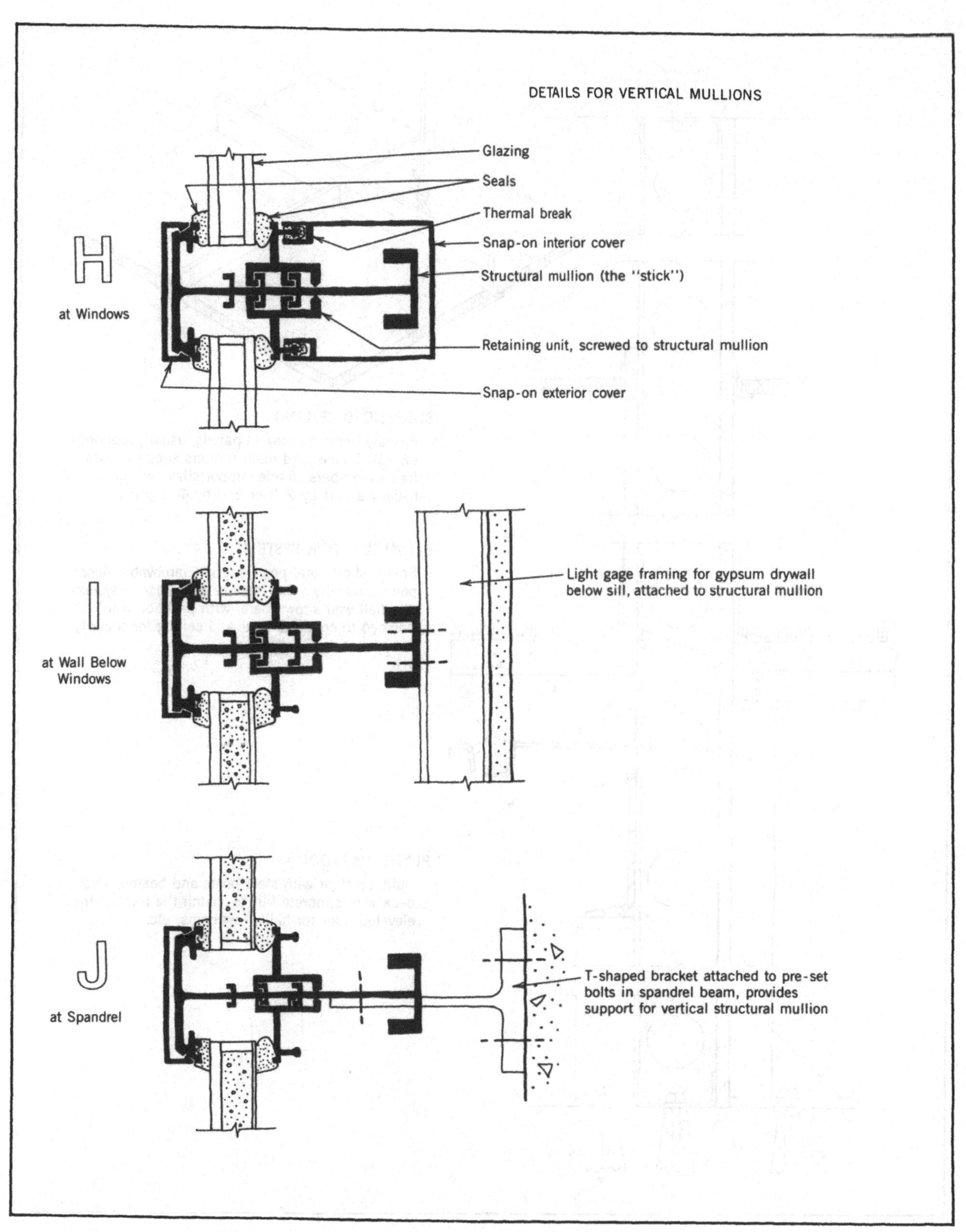

Figure 10.90 Building 7: details for the vertical mullions.

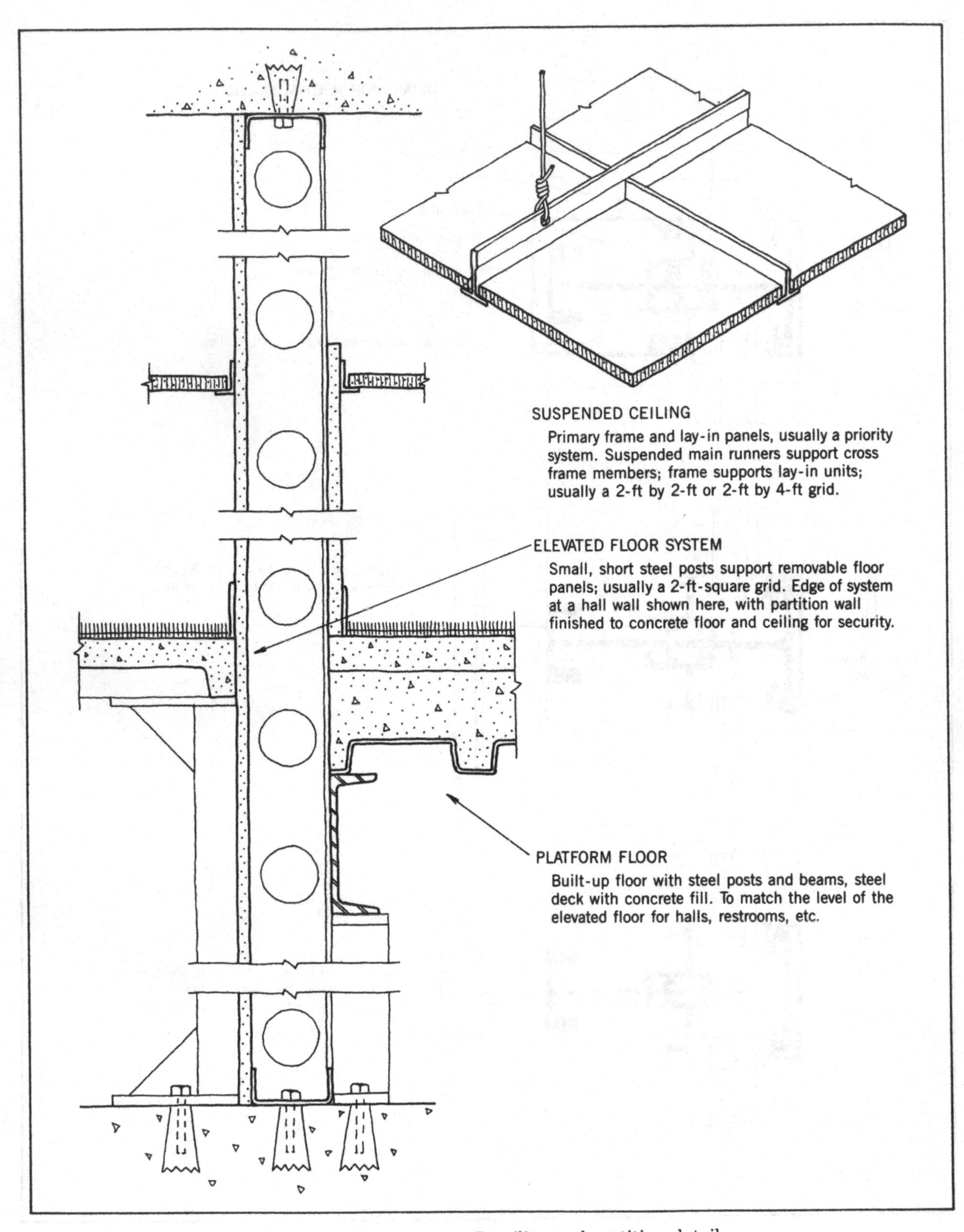

SUSPENDED CEILING
Primary frame and lay-in panels, usually a priority system. Suspended main runners support cross frame members; frame supports lay-in units; usually a 2-ft by 2-ft or 2-ft by 4-ft grid.
ELEVATED FLOOR SYSTEM
Small, short steel posts support removable floor panels; usually a 2-ft-square grid. Edge of system at a hall wall shown here, with partition wall finished to concrete floor and ceiling for security.
PLATFORM FLOOR
Built-up floor with steel posts and beams, steel deck with concrete fill. To match the level of the elevated floor for halls, restrooms, etc.

Figure 10.91 Building 7: ceiling and partition details.

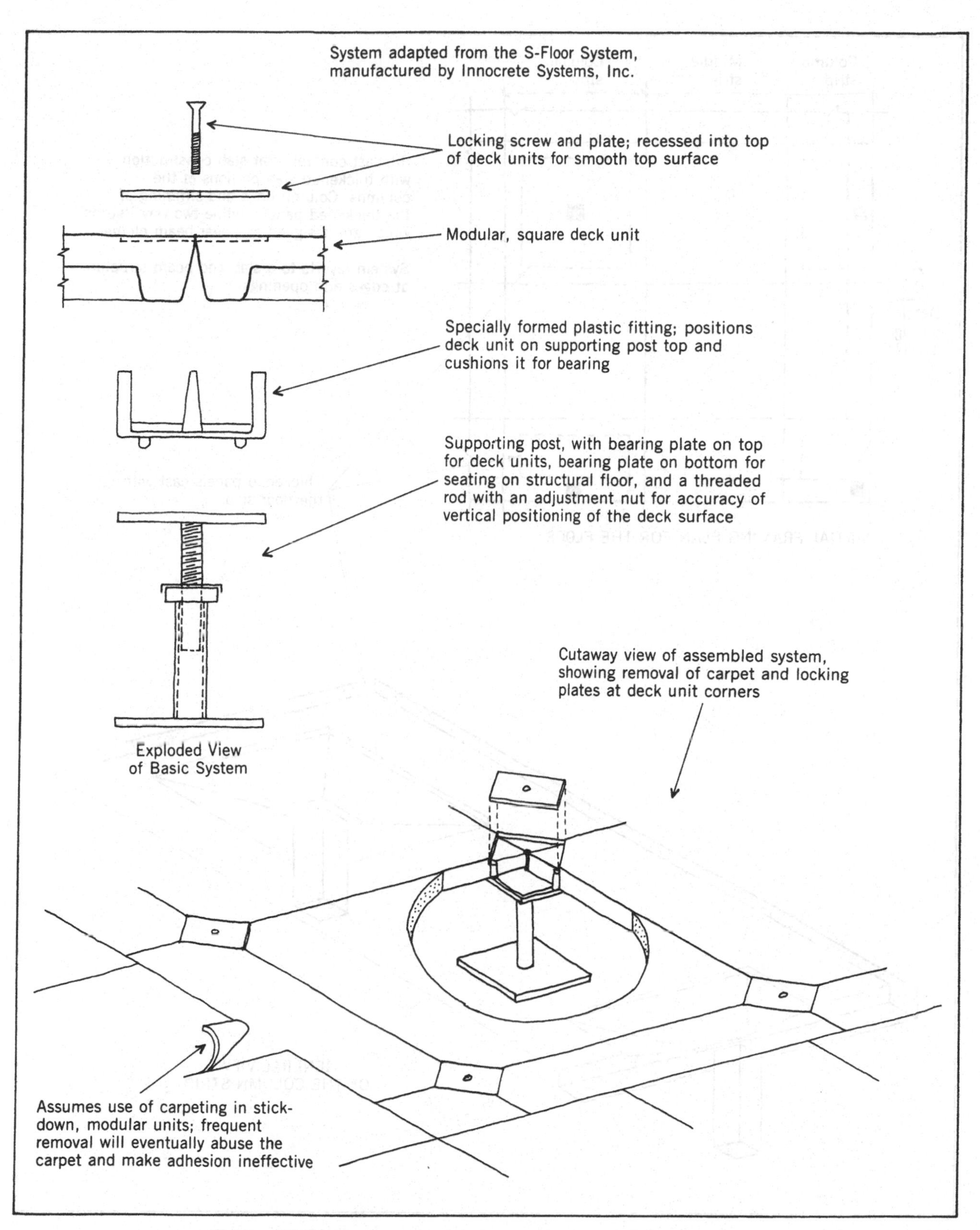
System adapted from the S-Floor System, manufactured by Innocrete Systems, Inc.
Locking screw and plate; recessed into top of deck units for smooth top surface
Modular, square deck unit
Specially formed plastic fitting; positions deck unit on supporting post top and cushions it for bearing
Supporting post, with bearing plate on top for deck units, bearing plate on bottom for seating on structural floor, and a threaded rod with an adjustment nut for accuracy of vertical positioning of the deck surface
Exploded View of Basic System
Cutaway view of assembled system, showing removal of carpet and locking plates at deck unit corners
Assumes use of carpeting in stick-down, modular units; frequent removal will eventually abuse the carpet and make adhesion ineffective

Figure 10.92 Building 7: details for the elevated floor system.

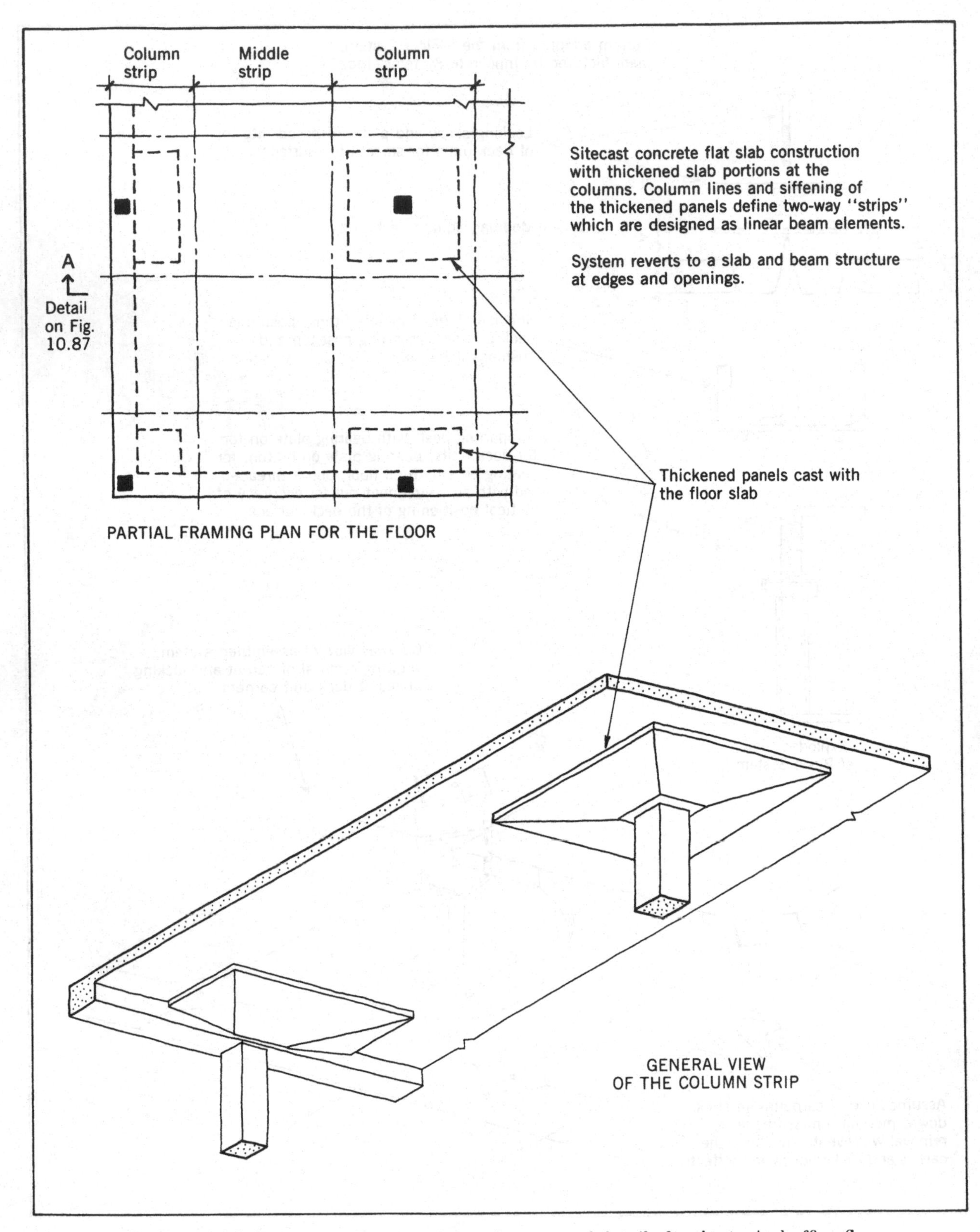
Column strip
Middle strip
Column strip
Sitecast concrete flat slab construction with thickened slab portions at the columns. Column lines and siffening of the thickened panels define two-way "strips" which are designed as linear beam elements.
System reverts to a slab and beam structure at edges and openings.
A
Detail on Fig. 10.87
Thickened panels cast with the floor slab
PARTIAL FRAMING PLAN FOR THE FLOOR
GENERAL VIEW OF THE COLUMN STRIP

Figure 10.93 Building 7: framing plan and structural details for the typical office floor.

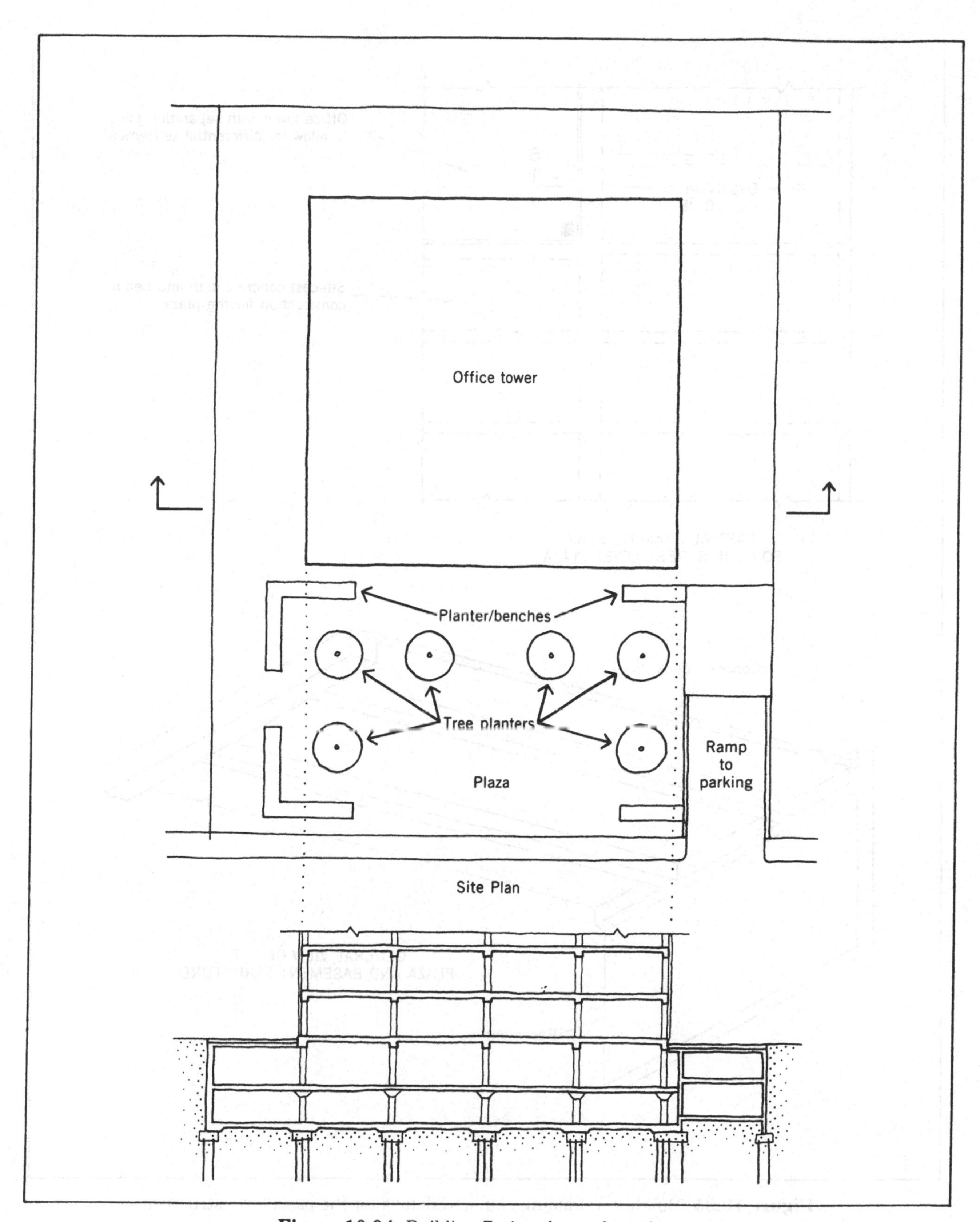

Figure 10.94 Building 7: site plan and section.

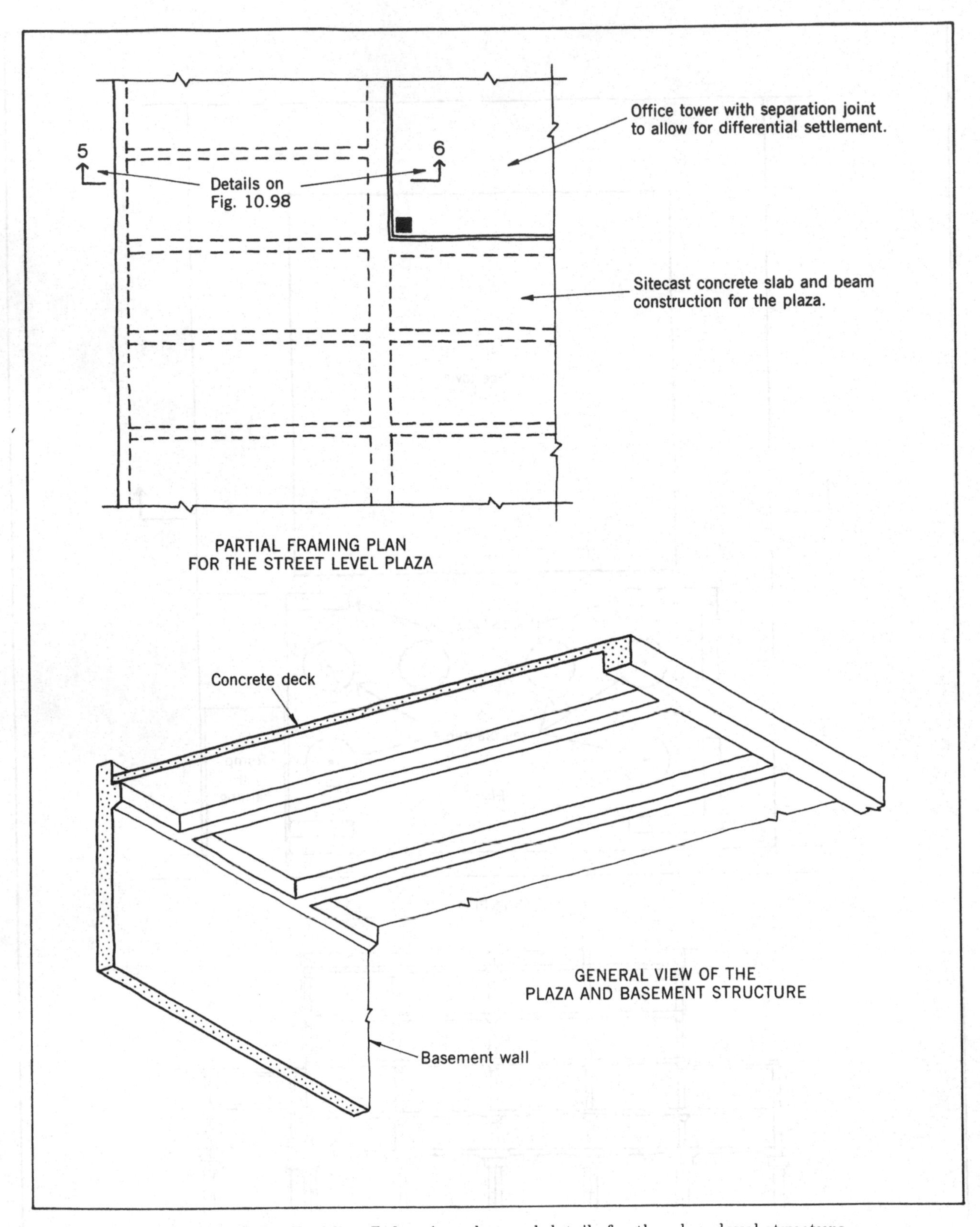

Figure 10.95 Building 7: framing plan and details for the plaza level structure.

Figure 10.96 shows the plan at the lowest level and also indicates the layout of the foundation system, consisting of piles. The piles are driven in clusters to support both the columns and the walls of the building.

Some details of the pile foundations are also shown in Figure 10.96. A problem with piles is the limited accuracy with which the locations of their tops can be controlled. Thus it is unreasonable to place a structure on a single pile, even though its vertical, axial, compression load capacity may be more than sufficient. The plans and details therefore show the use of a minimum of two piles for wall support and three for a single column.

Figure 10.97 shows some of the details for the construction of the basement walls and floors. It is assumed here that the lowest floor level is deep enough to have major water intrusion problems, so that the bottoms of the walls and the lower level floor are designed for a hydrostatic pressure assuming a ground water level at some distance above the lower floor. The construction shown for the floor uses a so-called "mud slab" which is placed on the soil and used as a base for installation of the waterproofing membrane and the structural slab.

The structural slab for the lower basement level must also be designed for the uplift force of the water pressure. This makes it function in a manner similar to that of the tower floor slabs, albeit in a reversed direction, and its design and construction may well be quite similar.

The upper basement floor is simply a spanning floor system, not unlike that for the office tower upper floors. Here also, a concrete flat slab or shallow waffle system may be used as an exposed structure.

Figure 10.98 shows a partial plan of a portion of the office tower and the plaza at the ground floor and plaza level, as well as some of the details for the plaza construction. There are various options for the form of the plaza paving and waterproofing and draining of this structure. All options involve considerable total weight, so that the supporting structure here will be quite heavily loaded.

Figure 10.99 shows some details of additional elements of the plaza level development. Planters may be quite shallow for small plantings, but must be appropriately deep for large shrubs or trees.

Figure 10.100 shows some details for an HVAC system for a typical office floor. Exploiting the potential for an exposed structure here, the interior is developed without the usual suspended ceiling. While the true ceiling is the lightly treated underside of the concrete structure (probably getting just a coat of paint), an implied ceiling is visually created by the level at which most of the suspended fixtures and equipment are placed.

The exposure of the overhead service elements may be visualized as creating a freer condition for their installation. It is true of course that they are not strictly bound by the hard definition of a structured interstitial space and do not need to dodge and fit into a constructed ceiling system. However, as much as their exposure may create a sense of reality, it is quite likely that much greater care will be required for their installation. This is not a basement or a warehouse, where there is virtually no concern for appearance. This is a *high tech* interior design style, with a very controlled appearance; akin to a girl wearing intentionally trashed designer jeans—not K-mart jeans, mind you!

With both the possibilities for interstitial space (the suspended ceiling and the elevated floor) at their disposal, the interior designers and the engineers for the various service systems have some options for location of items. Lighting for general illumination and fire sprinklers must be placed at the usual ceiling locations.

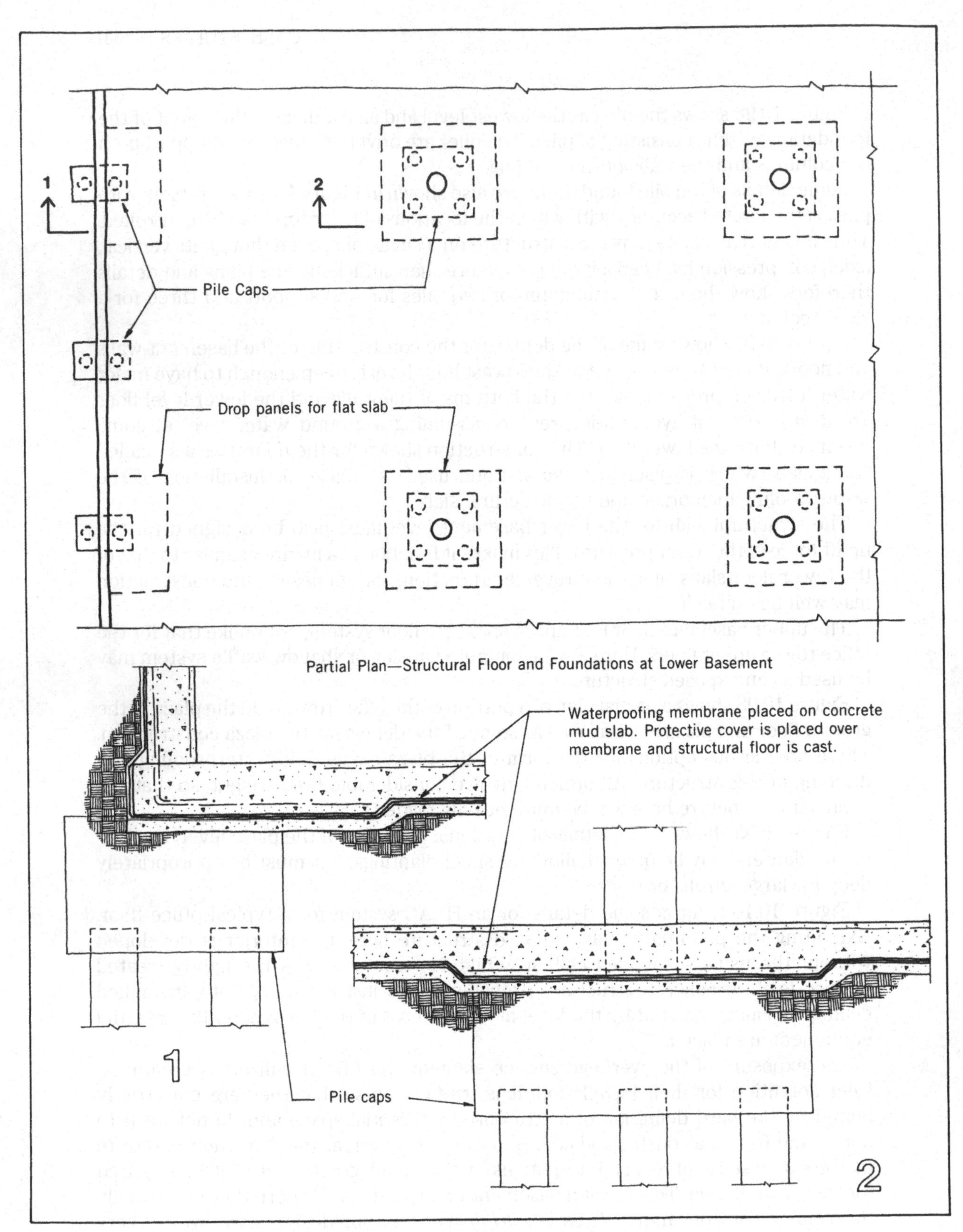

Figure 10.96 Building 7: foundation plan and details, lower basement level.

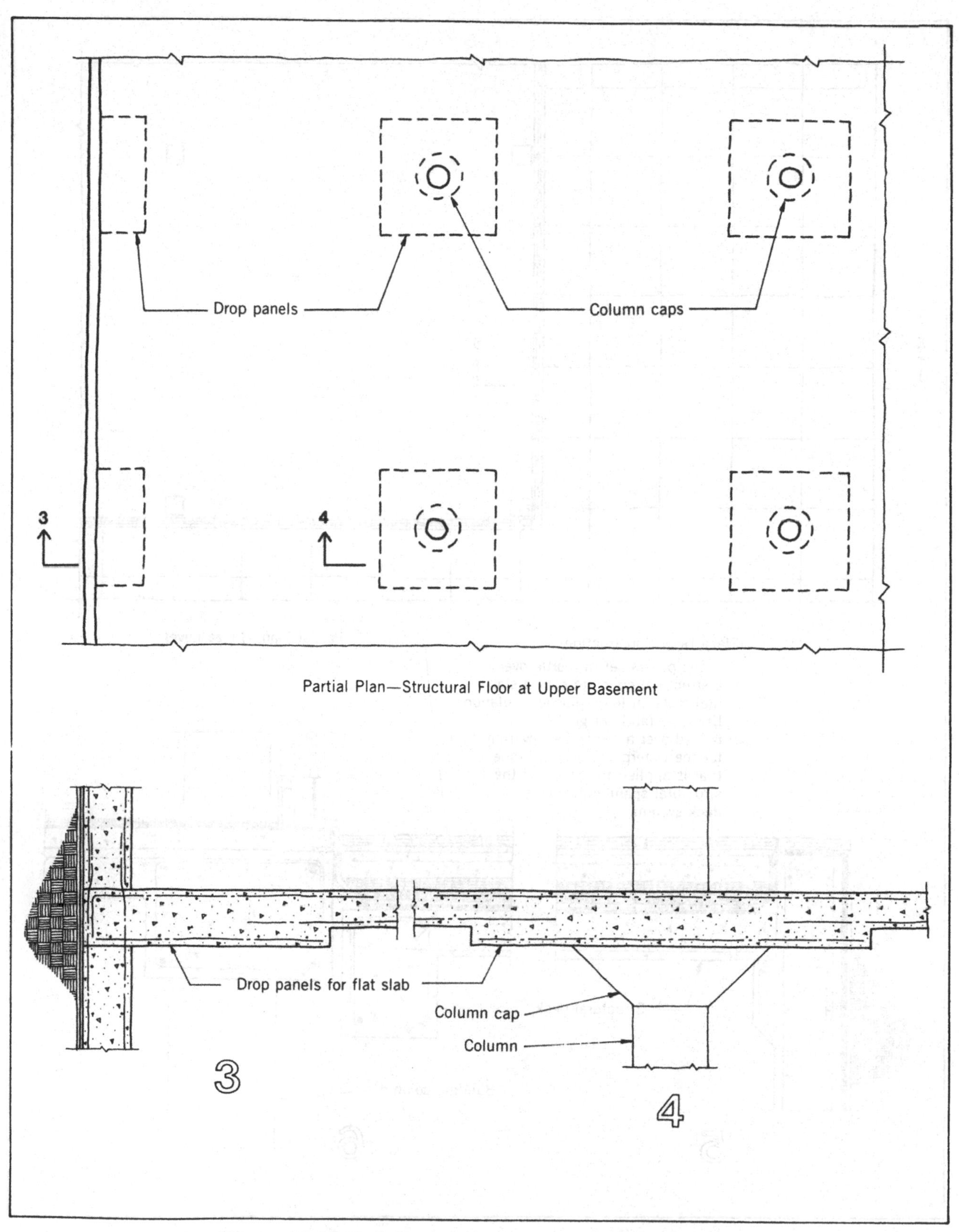

Figure 10.97 Building 7: upper basement plan and details.

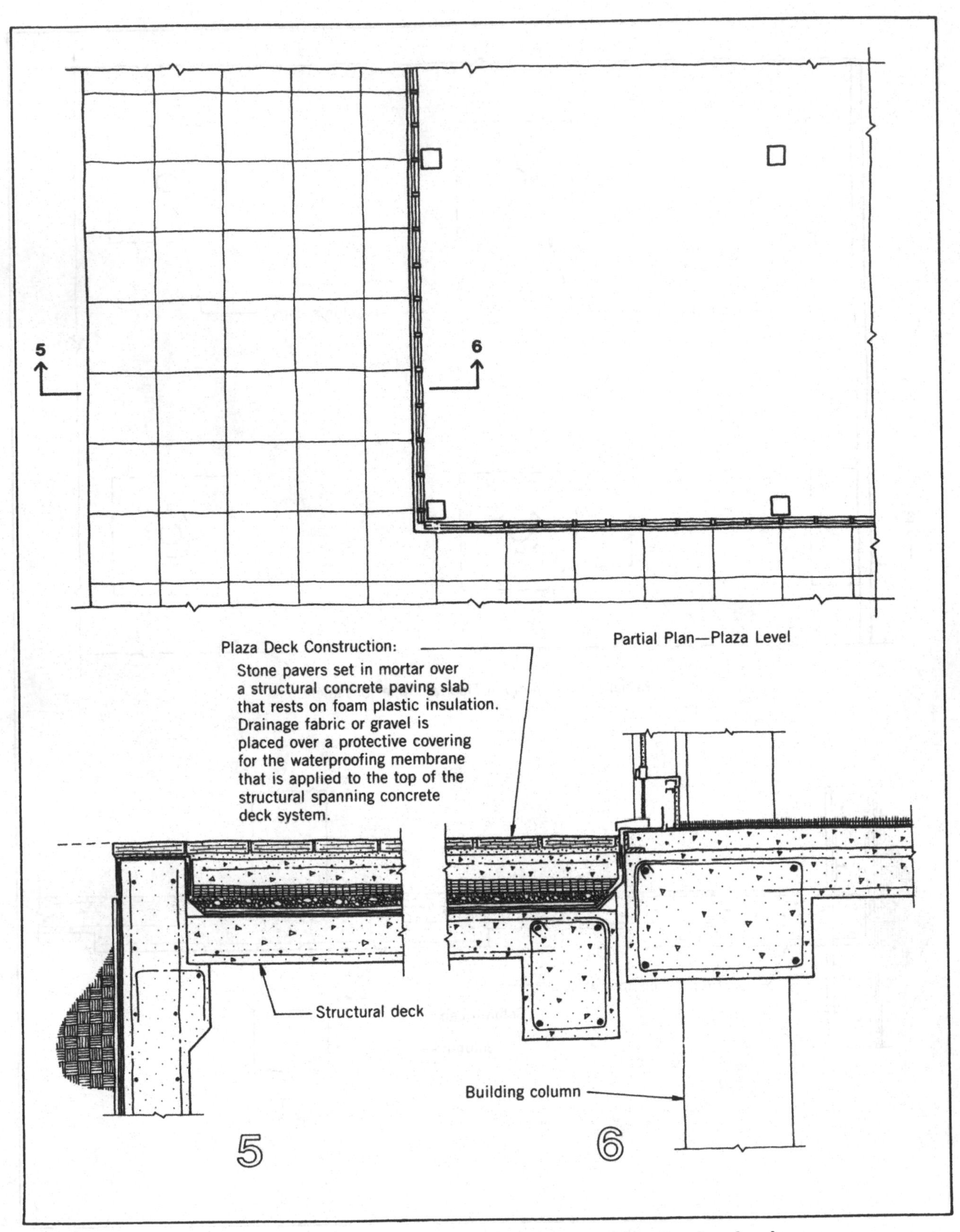
5
6
Partial Plan—Plaza Level
Plaza Deck Construction:
Stone pavers set in mortar over
a structural concrete paving slab
that rests on foam plastic insulation.
Drainage fabric or gravel is
placed over a protective covering
for the waterproofing membrane
that is applied to the top of the
structural spanning concrete
deck system.
Structural deck
Building column
5
6

Figure 10.98 Building 7: ground floor plan and details, plaza level.

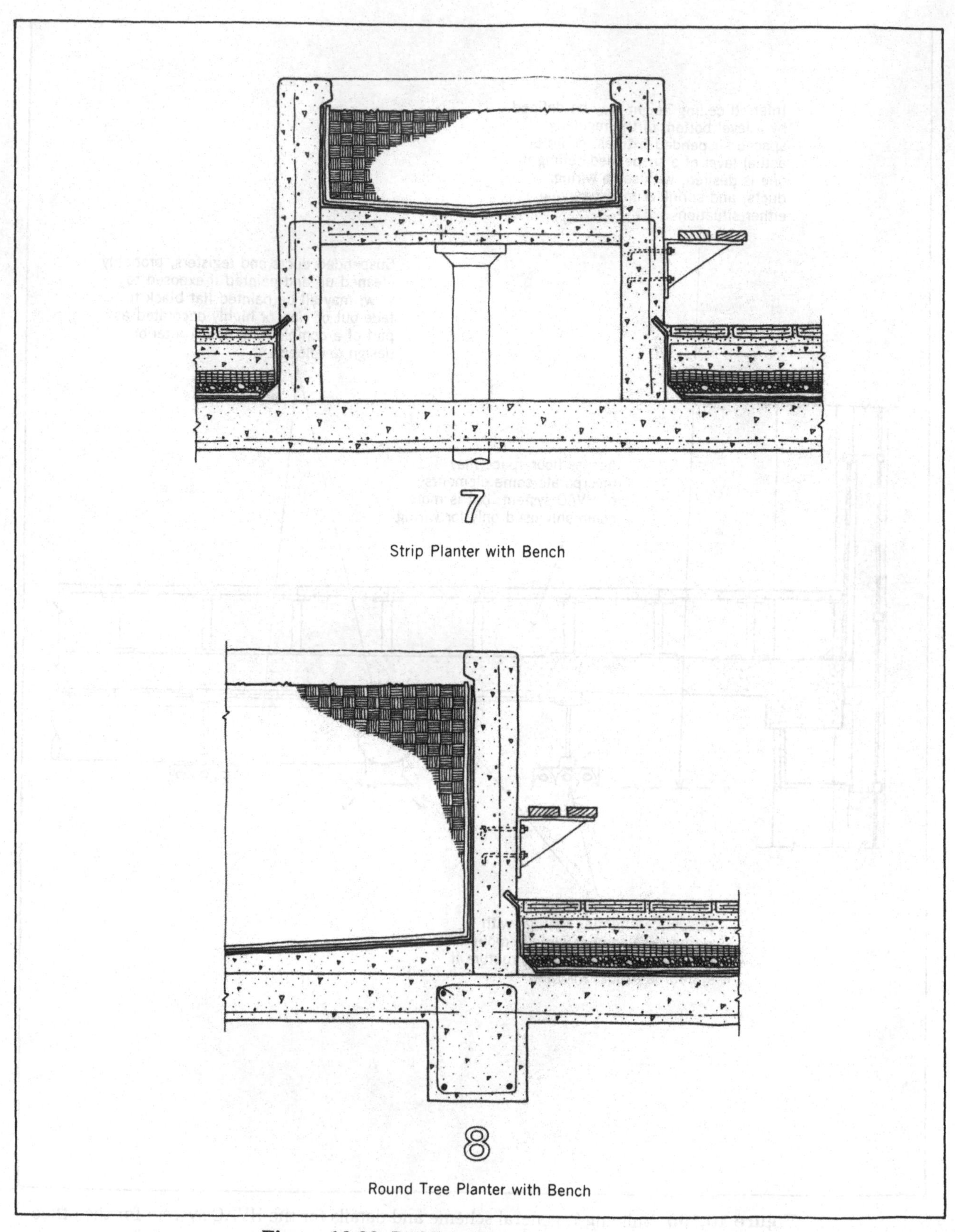

Figure 10.99 Building 7: site construction details.

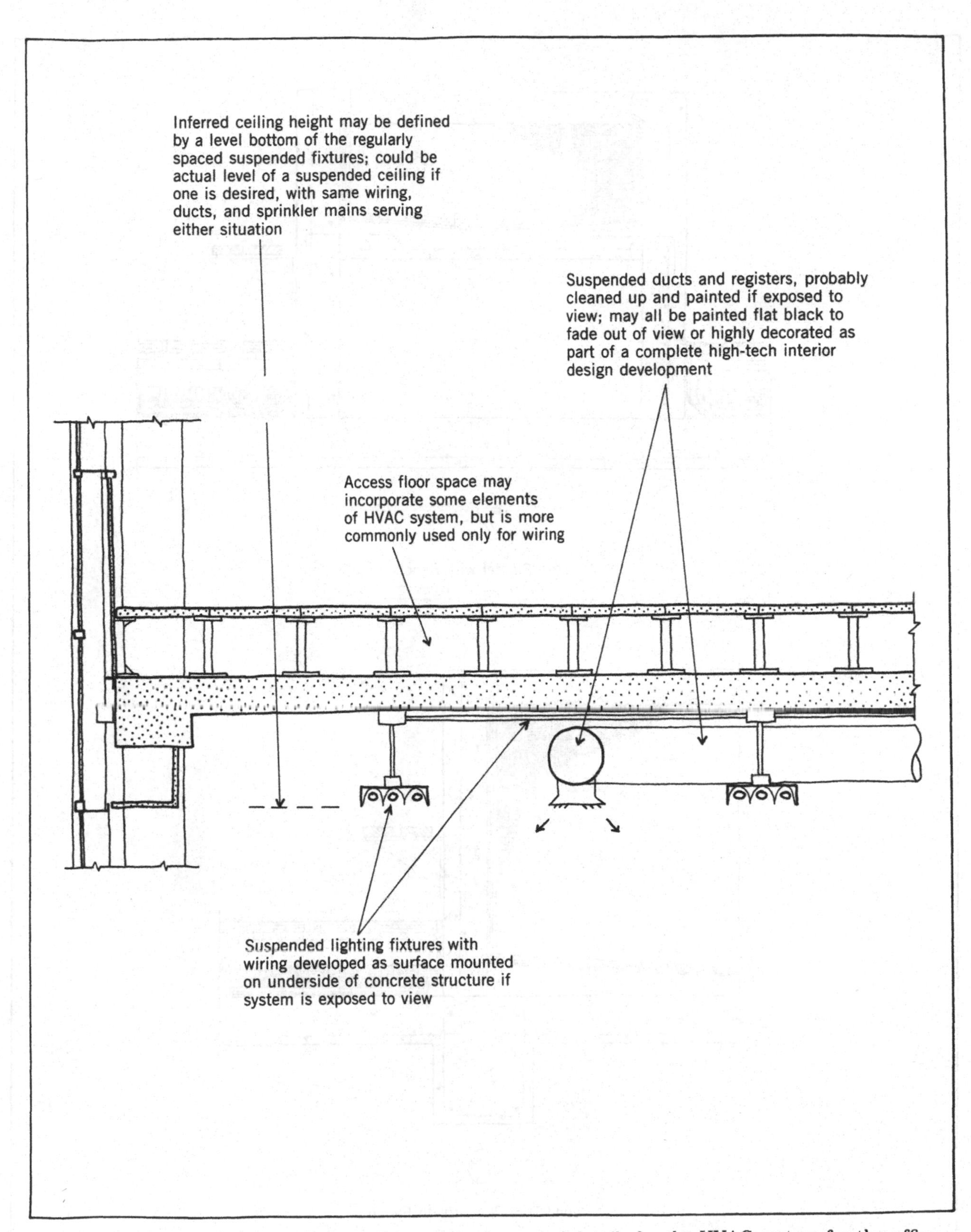
Inferred ceiling height may be defined by a level bottom of the regularly spaced suspended fixtures; could be actual level of a suspended ceiling if one is desired, with same wiring, ducts, and sprinkler mains serving either situation
Suspended ducts and registers, probably cleaned up and painted if exposed to view; may all be painted flat black to fade out of view or highly decorated as part of a complete high-tech interior design development
Access floor space may incorporate some elements of HVAC system, but is more commonly used only for wiring
Suspended lighting fixtures with wiring developed as surface mounted on underside of concrete structure if system is exposed to view

Figure 10.100 Building 7: general scheme and details for the HVAC system for the office tower.

However, wiring, ducts, and various items may be located in either the ceiling or underfloor positions. This flexibility can be exploited for real practical purposes, or may be played with for freedom of expression in the interior design.

Figure 10.101 shows some details for the elevator system, which uses a high-speed, overhead suspended operating system. To serve the lowest levels of the basement parking, the construction must be extended considerably below the level of the floor of the lower basement; requiring some special considerations in development of the building foundations in the immediate vicinity. For the typical installation, there is also the need to extend the top of the building construction almost two stories above the highest level served by the elevators. The latter situation normally produces some form of "penthouse" structure on the building rooftop, unless some considerable attic space is created by the development of the building roof.

When the elevators are placed in a "core" group with other permanent service elements (notably the vertical ducts for the HVAC system), it is common to use a general rooftop structure to enclose some of the necessary rooftop equipment for other services. This is a chronic problem for the flat-roofed building, and design responses vary considerably.

The solutions shown here for Buildings 6 and 7 are not particularly unique, and show some range of possibilities for the typical multistory building. There are, however, many other possibilities that may well be more appropriate for other occupancies or for particular climate conditions, budgetary limitations, local building code restrictions, and so on.

The building shown here is a quite common occurrence in tight urban locations, where site limitations and local zoning requirements conspire to require some underground parking. A service system problem of some consequence in this situation is that of the provision of a ventilating system for the parking facility (see Figure 10.102). The basic concern is for removal of gas fumes, which typically lay at floor level.

While local codes may vary, the usual requirements for an exhaust system for enclosed parking garages include the following provisions.

Positioning of intake grilles for the ducted exhaust system at not more than 18 in. above the floor; at spacing of not more than some distance (50 ft or less) along the entire perimeter of enclosed spaces, and near corners or other spots where gas may become trapped.

Total air change per hour at a rate considerably greater than that for typical occupied spaces.

Air change that is 100% fresh air, with exhaust dumped out at points not likely to permit the toxic fumes to disturb people on or off the site.

The typical result is a system with some very large fans (with enormous noise and vibration) which must usually operate continuously, with large main ducts and good sized intake and exhaust grilles. This results in the strong preference for an open parking structure (no actual enclosure for an HVAC system to have to deal with) whenever possible. If planning of the building and site simply does not allow this, the exhaust system must be paid for and incorporated into the construction and the space planning for the building and site.

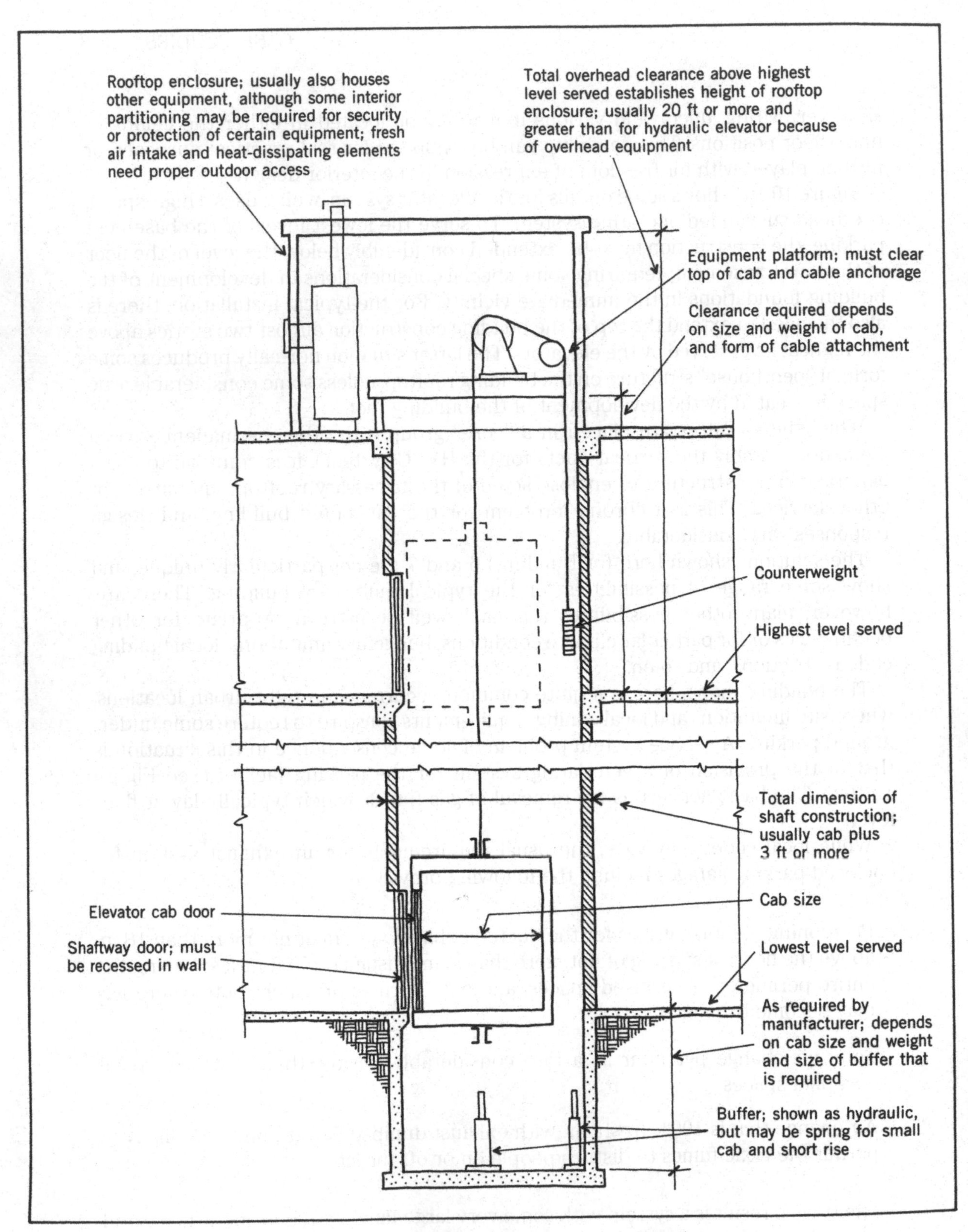

Rooftop enclosure; usually also houses other equipment, although some interior partitioning may be required for security or protection of certain equipment; fresh air intake and heat-dissipating elements need proper outdoor access
Total overhead clearance above highest level served establishes height of rooftop enclosure; usually 20 ft or more and greater than for hydraulic elevator because of overhead equipment
Equipment platform; must clear top of cab and cable anchorage
Clearance required depends on size and weight of cab, and form of cable attachment
Counterweight
Highest level served
Total dimension of shaft construction; usually cab plus 3 ft or more
Cab size
Elevator cab door
Shaftway door; must be recessed in wall
Lowest level served
As required by manufacturer; depends on cab size and weight and size of buffer that is required
Buffer; shown as hydraulic, but may be spring for small cab and short rise

Figure 10.101 Building 7: elevator details.

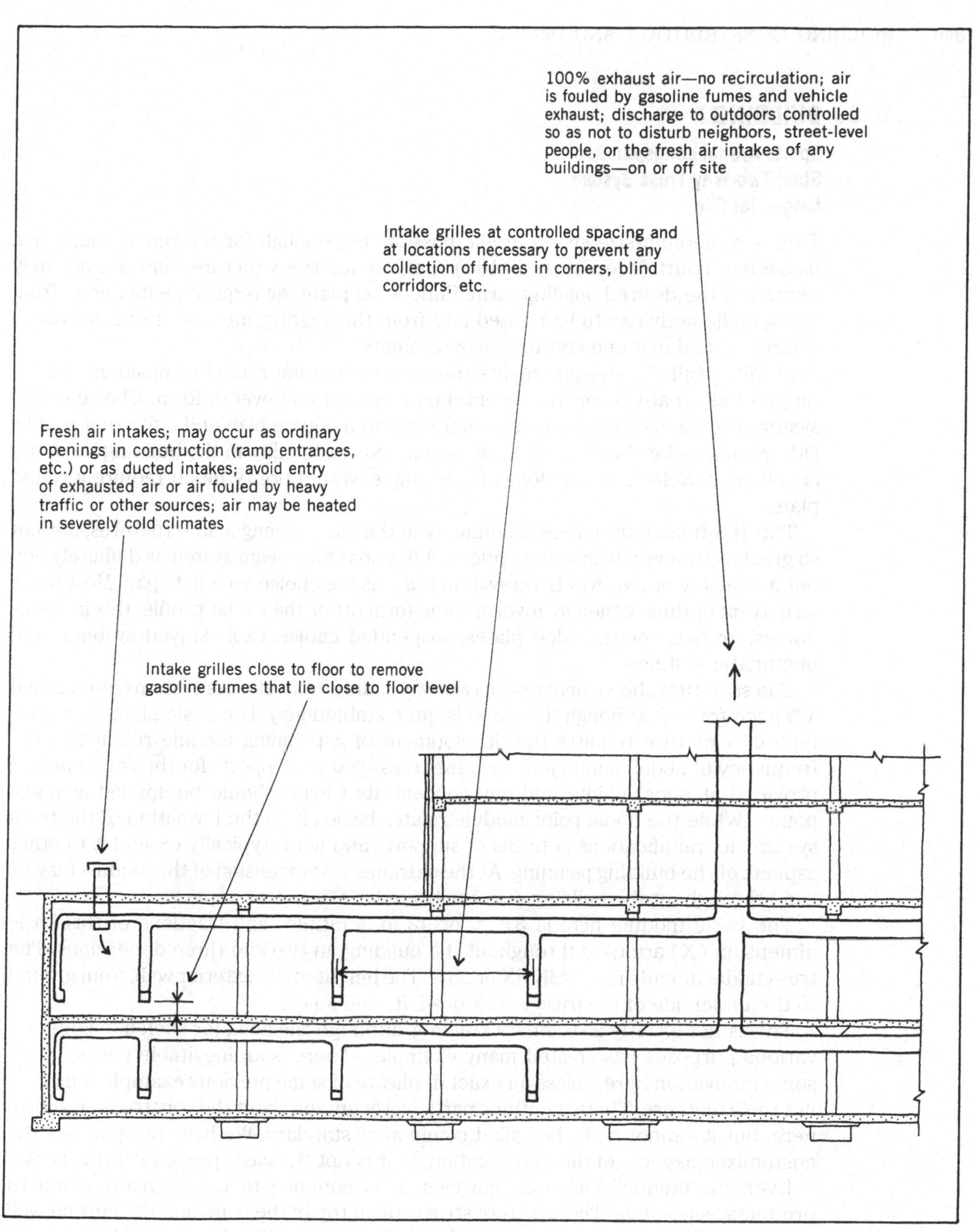
100% exhaust air—no recirculation; air is fouled by gasoline fumes and vehicle exhaust; discharge to outdoors controlled so as not to disturb neighbors, street-level people, or the fresh air intakes of any buildings—on or off site
Intake grilles at controlled spacing and at locations necessary to prevent any collection of fumes in corners, blind corridors, etc.
Fresh air intakes; may occur as ordinary openings in construction (ramp entrances, etc.) or as ducted intakes; avoid entry of exhausted air or air fouled by heavy traffic or other sources; air may be heated in severely cold climates
Intake grilles close to floor to remove gasoline fumes that lie close to floor level

Figure 10.102 Building 7: general scheme and details for the ventilation system for the parking garage.

10.8 BUILDING 8

Sports Arena, Longspan Roof
Steel Two-Way Truss System
Large Flat Site

This is a medium-size sports arena, possibly big enough for a swim stadium or a basketball court (see Figure 10.103). Options for the structure here are strongly related to the desired building form. Functional planning requirements derive from the specific activities to be housed and from the seating, internal traffic, overhead clearance, and exit and entrance arrangements.

In spite of all of the requirements, there is usually some room for consideration of a range of alternatives for the general building plan and overall form. Choice of the system shown here relates to a commitment to a square plan and a flat roof profile. Other choices for the structure may permit more flexibility in the building form and also limit it. Selection of a dome, for example, would pretty much require a round plan.

The 168-ft span used here is definitely in the class of longspan structures, but not so great as to severely limit the options. A flat-spanning beam system is definitely out, but a one-way or two-way truss system is a feasible choice for a flat span. Most other structural options generally involve some form other than a flat profile; this includes domes, arches, shells, folded plates, suspended cables, cable-stayed systems, and pneumatic systems.

The structure shown here uses a two-way spanning steel truss system (often called a "space frame," although the name is quite ambiguous). The basic planning of this type of structure requires the development of a planning module relating to the frequency of nodal points (joints) in the truss system. Supports for the truss must be provided at nodal points, and any concentrated loads should be applied at nodal points. While the nodal point module relates basically to the formation of the truss system, its ramifications in terms of supports and loads typically extend it to other aspects of the building planning. At the extreme, this extension of the module may be used throughout the building—as has been done in this example.

The basic module here is 3.5 ft, or 42 in. Multiples and fractions of this basic dimension (X) are used throughout the building, in two and three dimensions. The truss nodal module is actually 8X or 28 ft. The height of the exterior wall, from ground to the underside of the truss is 12X or 42 ft. And so on.

This is not exactly an ordinary building, although the need for such a building for various purposes has created many examples. There is an inevitable necessity for some innovation here, unless an exact duplicate of some previous example is used—not the usual case. There is nothing particularly unique about the construction shown here, but it cannot really be called common or standard. We have not pursued the customized aspects of this construction, as it is not the basic purpose of this book.

Even in "unique" buildings, however, it is common to use as many standard products as possible. Thus the roof structure on top of the truss and the curtain wall system for the exterior walls use off-the-shelf products. (See Figures 10.104 through 10.106.) The supporting columns and general seating structure would also most likely be of conventional construction.

The truss system shown here could be produced from various available priority systems, manufactured by different producers. It is generally advisable to pursue these when beginning the design of such a structure. A completely custom-designed structure of this kind requires enormous investments of design time for basic devel-

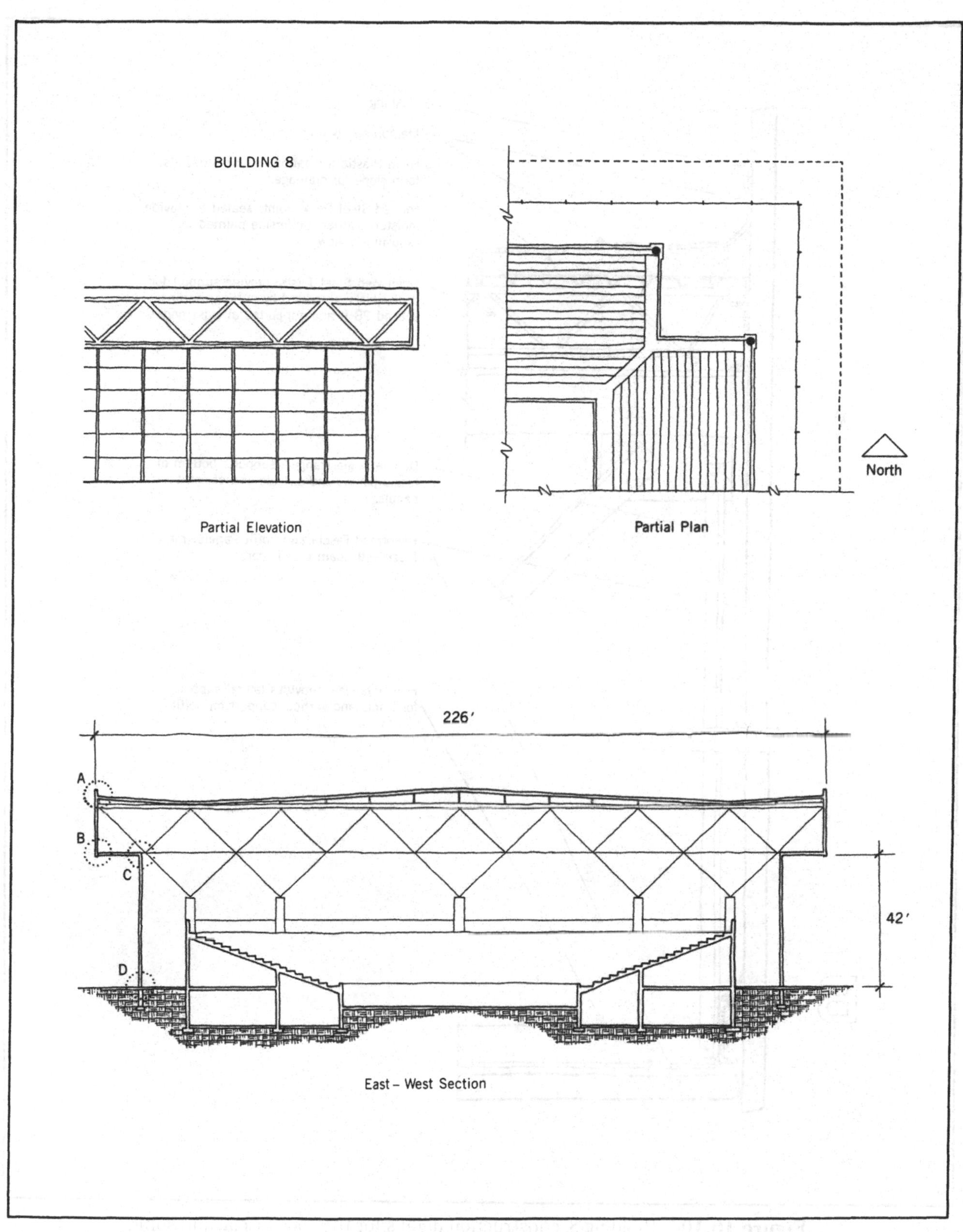

Figure 10.103 Building 8.

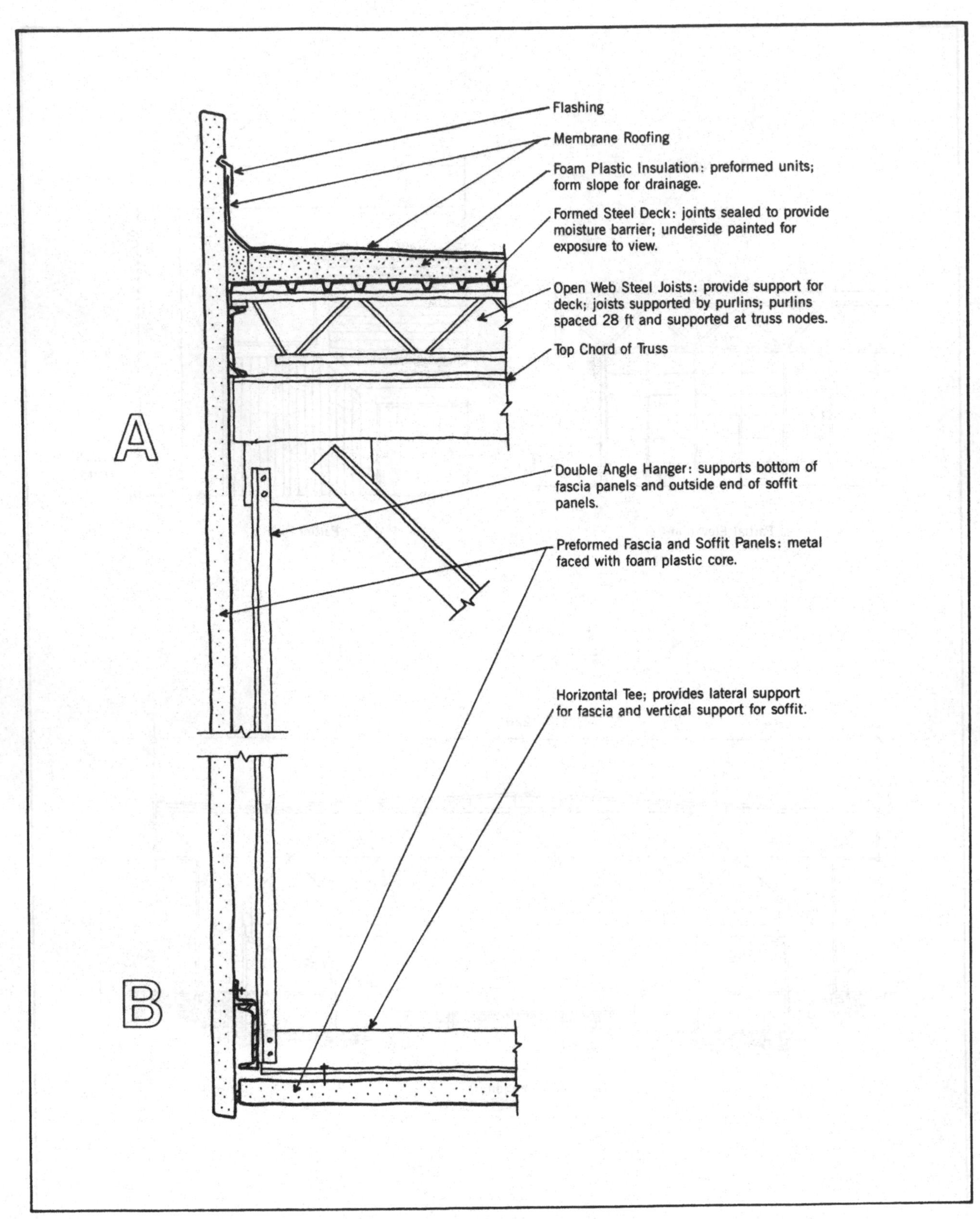

Figure 10.104 Building 8: construction details for the roof, fascia, and soffit.

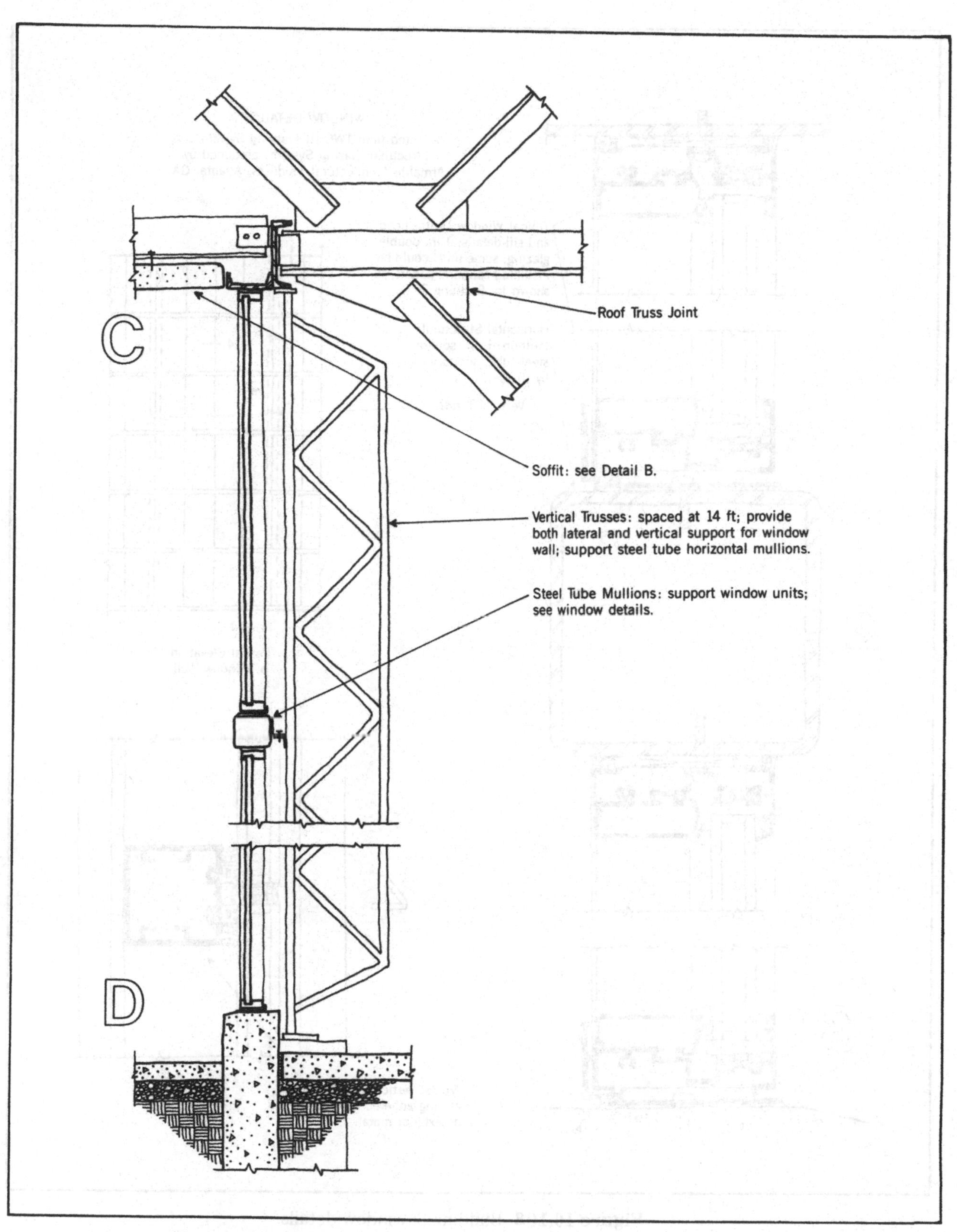
Roof Truss Joint
C
Soffit: see Detail B.
Vertical Trusses: spaced at 14 ft; provide both lateral and vertical support for window wall; support steel tube horizontal mullions.
Steel Tube Mullions: support window units; see window details.
D

Figure 10.105 Building 8: construction details for the window wall.

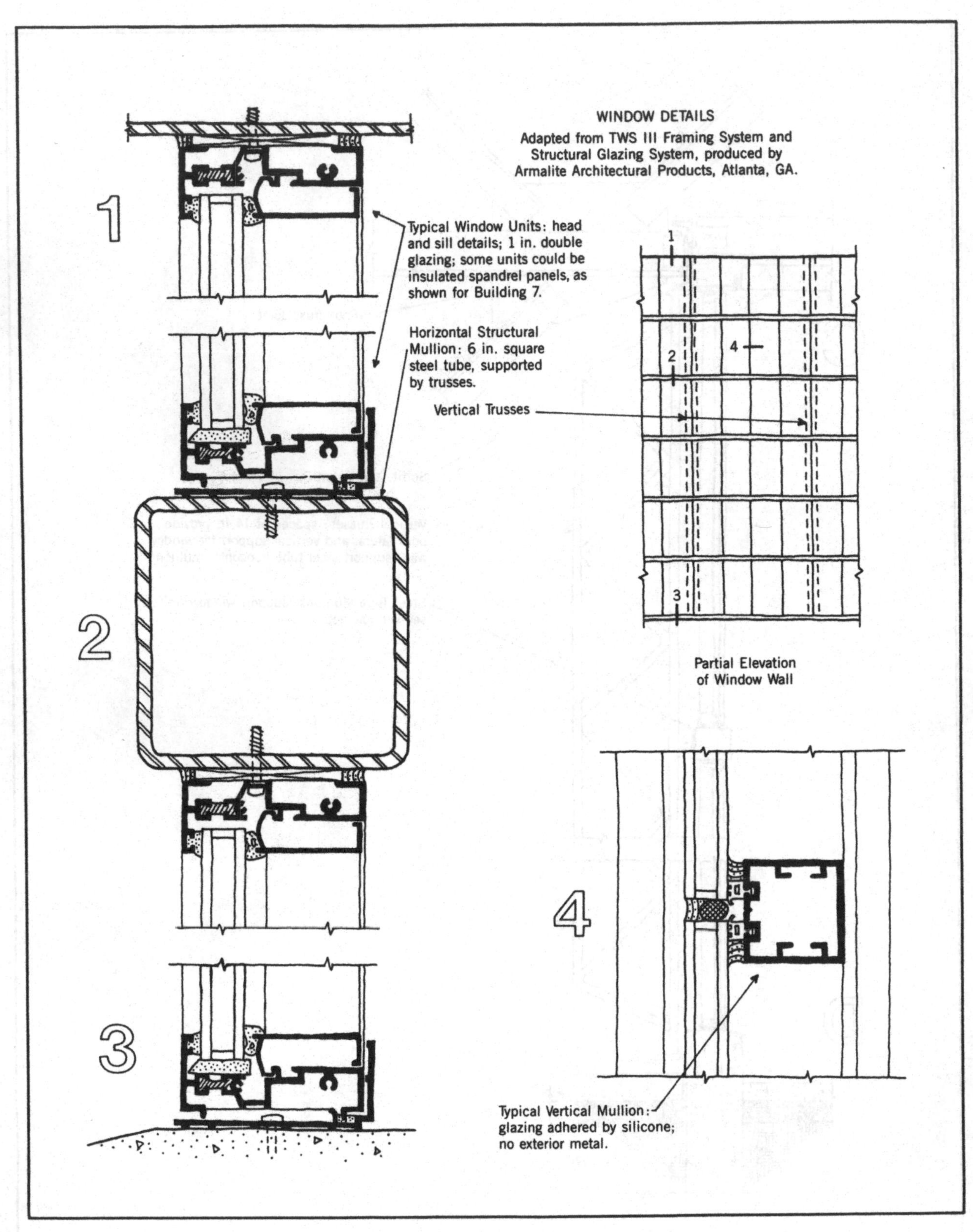
WINDOW DETAILS
Adapted from TWS III Framing System and Structural Glazing System, produced by Armalite Architectural Products, Atlanta, GA.
1
2
3
4
Typical Window Units: head and sill details; 1 in. double glazing; some units could be insulated spandrel panels, as shown for Building 7.
Horizontal Structural Mullion: 6 in. square steel tube, supported by trusses.
Vertical Trusses
Partial Elevation of Window Wall
Typical Vertical Mullion: glazing adhered by silicone; no exterior metal.

Figure 10.106 Building 8: window details.

opment and planning of the structural form, development of nodal joint construction, and reliable investigation of structural behavior of the highly indeterminate structure. Development of assemblage and erection processes are themselves a major design problem. And—of course—all of the design time and effort has to be paid for. If the project deserves this effort, the time can be used, and the budget can absorb the cost, the end result might justify the expenditures. But if what is really desired and needed can be obtained with available products and systems, a lot of time and money can most likely be saved.

In addition to the longspan structure in this case, there is also a major problem in developing the 42-ft-high curtain wall. Braced laterally only at the top and bottom, this is a 42-ft span structure sustaining wind pressure as a loading. With even modest wind conditions, producing pressures in the 20-psf range, this requires some major vertical mullion structure to span the 42 ft.

In this example the fascia of the roof trusses, the soffit of the overhang, and the curtain wall are all developed with products ordinarily available for curtain wall construction. The vertical span of the tall wall is developed by a series of custom-designed trusses that brace the major vertical mullions. These vertical mullions are themselves custom developed to relate to the standard units of the basic wall system, something that may well be done by the supplier of the basic system.

As an exposed structure, the truss system is a major visual element of the building interior. The truss members define a pattern that is orderly and pervading. There will, however, most likely be many additional items overhead, within and possibly beneath the trusses. These may include:

Leaders and other elements of the roof drainage system.

Ducts and registers for the HVAC system.

A general lighting system.

Signs, scoreboards, etc.

Elements of an audio system.

Catwalks for access to the various equipment.

To preserve some design order, these should be related to the truss geometry and detailing, if possible. However, some amount of independence is to be expected, especially with items installed or modified after the building is completed.

The building shell, of course, exists only to house the building activities. In this case, an essentially independent construction is developed under the large overhead enclosure to form the seating and other required facilities. The support columns for the truss are incorporated into this construction, but the separation is otherwise complete and the planning could be totally independent. Some details for the development of the interior construction are shown in Figure 10.107.

A partial framing plan and some details for the truss system are shown in Figure 10.108. Depending on the scale of the truss in terms of spans and dimension of the basic module, it is possible that some priority system may be used for this structure; in which case the manufacturer's details and member forms would be used. The details here show the use of some reasonably conventional forms of steel framing, which are also possible, and may be required if the geometry or scale of the structure cannot be achieved with any available priority system. As in most cases, some

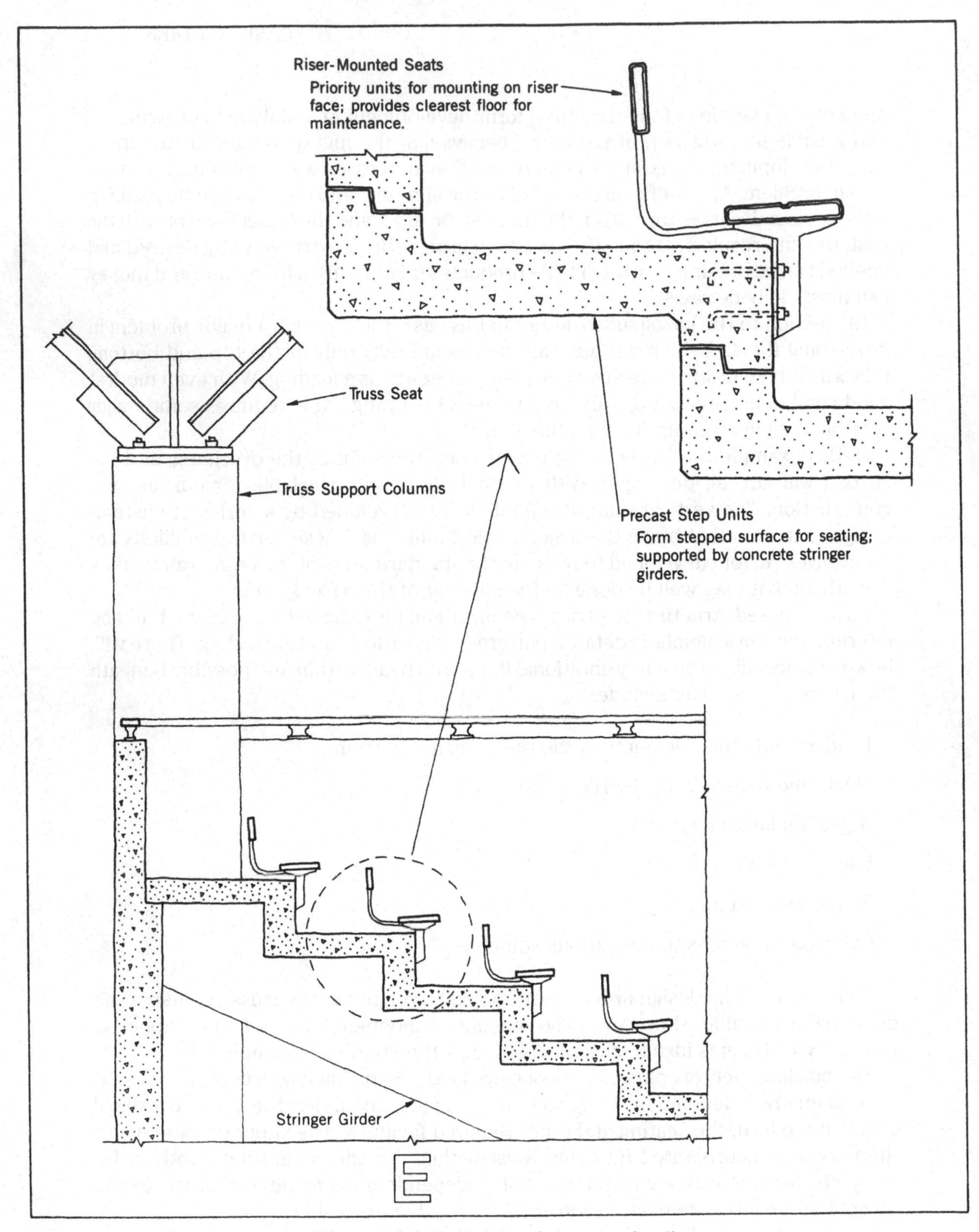

Figure 10.107 Building 8: interior details.

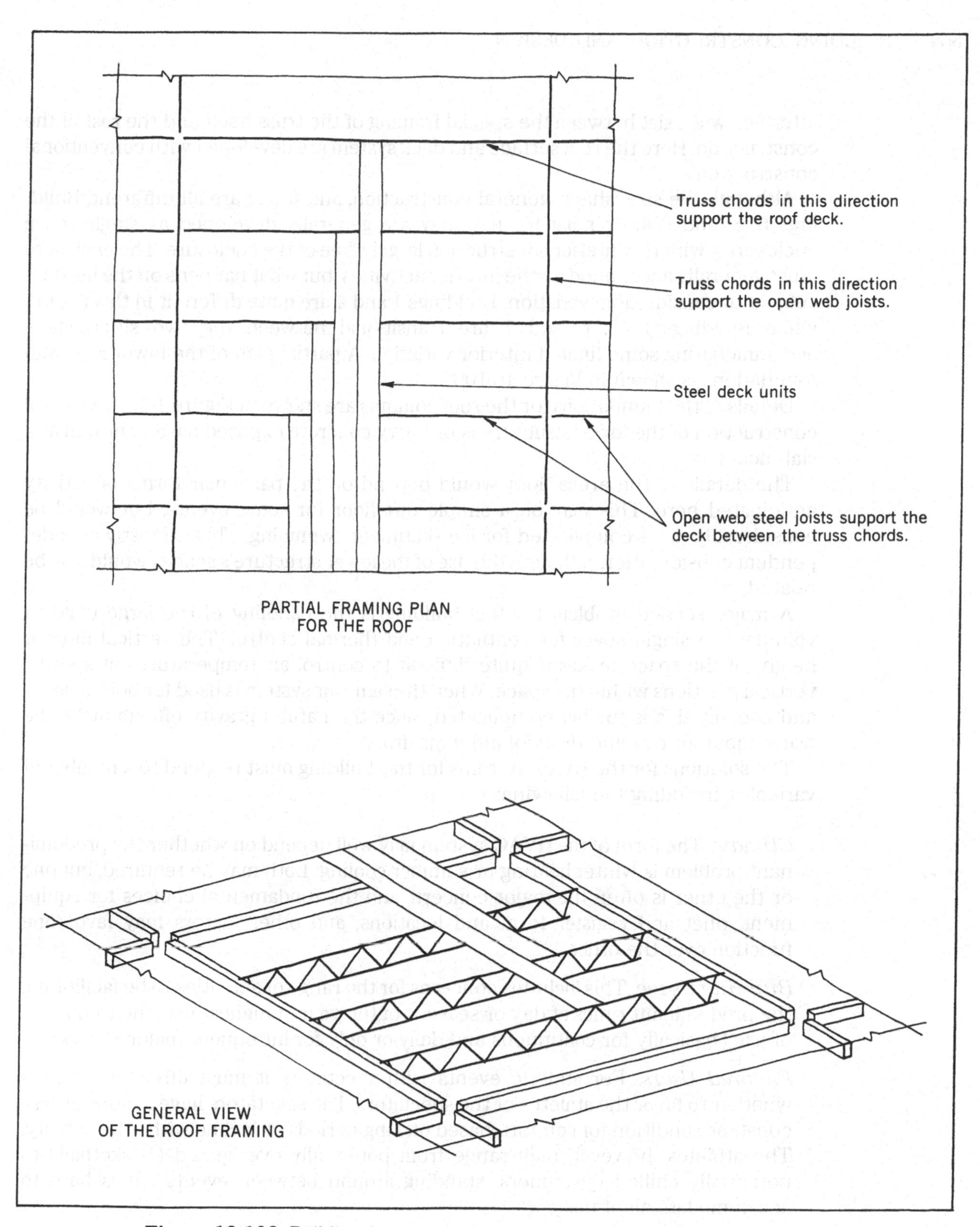

Truss chords in this direction support the roof deck.
Truss chords in this direction support the open web joists.
Steel deck units
Open web steel joists support the deck between the truss chords.
PARTIAL FRAMING PLAN FOR THE ROOF
GENERAL VIEW OF THE ROOF FRAMING

Figure 10.108 Building 8: framing plan and structural details for the roof.

interface will exist between the special framing of the truss itself and the rest of the construction. Here the roof surface and deck system are developed with conventional construction.

Although the size, shape, general construction, and usage are all different, Buildings 5, 8, and 9 are similar in that they are generally developed as single-space enclosures, with the interior construction largely free of the enclosure. The enclosure must generally accommodate the interior activities, but what happens on the inside is subject to considerable variation. Buildings 1 and 4 are quite different in this regard, while Buildings 2, 3, 6, and 7 are transitional between the two situations—accommodating some limited interior variation. A partial plan of the lower level and foundations is shown in Figure 10.109.

Details of the foundations for the roof columns are shown in Figure 10.110. General construction of the lower structure is ordinary concrete exposed basement wall and slab floors.

The details of the arena floor would depend on the particular forms of activity anticipated here. This may be a simple dirt floor for some events, but would be considerably more complicated for ice skating or swimming. This is mostly an independent consideration, although the use of the lower structure's spaces would also be related.

A major service problem for this building is the handling of the large building volume as a single space for ventilation and thermal control. The vertical interior height of the space makes it quite difficult to control air temperatures at specific vertical positions within the space. When the same air system is used for both heating and cooling, this is further complicated; since the natural gravity effects make the warm input air rise and the cool input air drop.

The solutions for the HVAC systems for this building must respond to a number of variables, including the following:

Climate. The form of the HVAC system may well depend on whether the predominant problem is winter heating or summer cooling. Both may be required, but one or the other is often the major concern, and the fundamental choices for equipment, duct and register form and locations, and other factors may favor one function over the other.

Building Usage. This includes concerns for the range of activities to be facilitated, the predominant times of day or seasons of the year of highest use, the frequency of use (basically for continuous and daily or only for infrequent major events).

Favored Users. For athletic events with spectators it must often be decided whether to favor the athletes or the spectators. The spectators have a more-or-less constant condition for comfort, based on long periods of recline with little activity. The athletes, however, may range from potentially overheated (basketball) to potentially chilled (swimmers, standing around between events). It is hard to accommodate all of this.

Adding to the problem here is the need to deal with what amounts to a second building within the larger building, consisting of the enclosed spaces beneath the seating. This space has its own unique problems and particular needs for ventilation and general thermal control (locker rooms, large showers, smelly storage, etc.). This is likely to result in an essentially independent system, with connections bypassing the upper space through tunnels for air intake and exhaust; virtually like a separate,

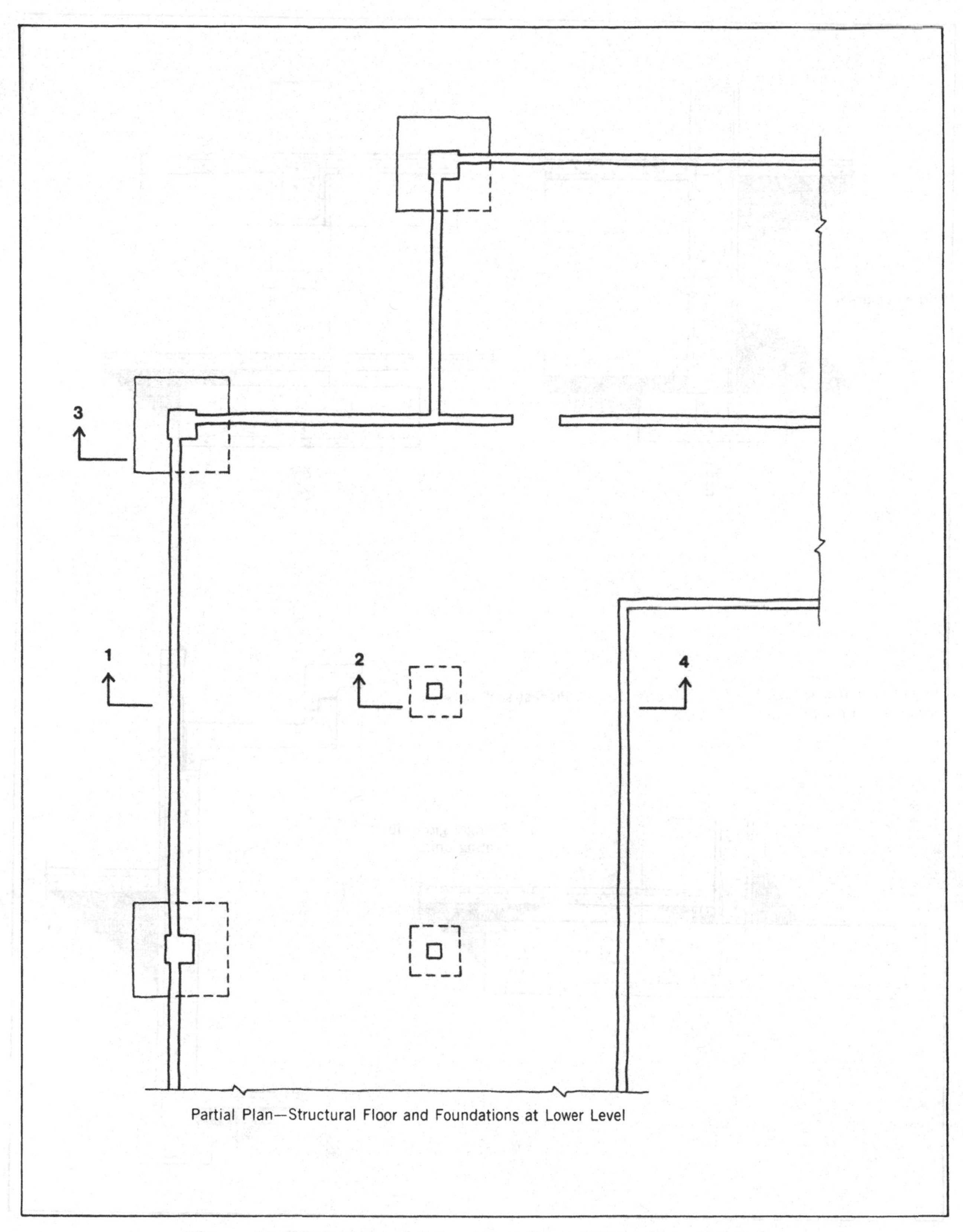

Figure 10.109 Building 8: basement and foundation plan and details.

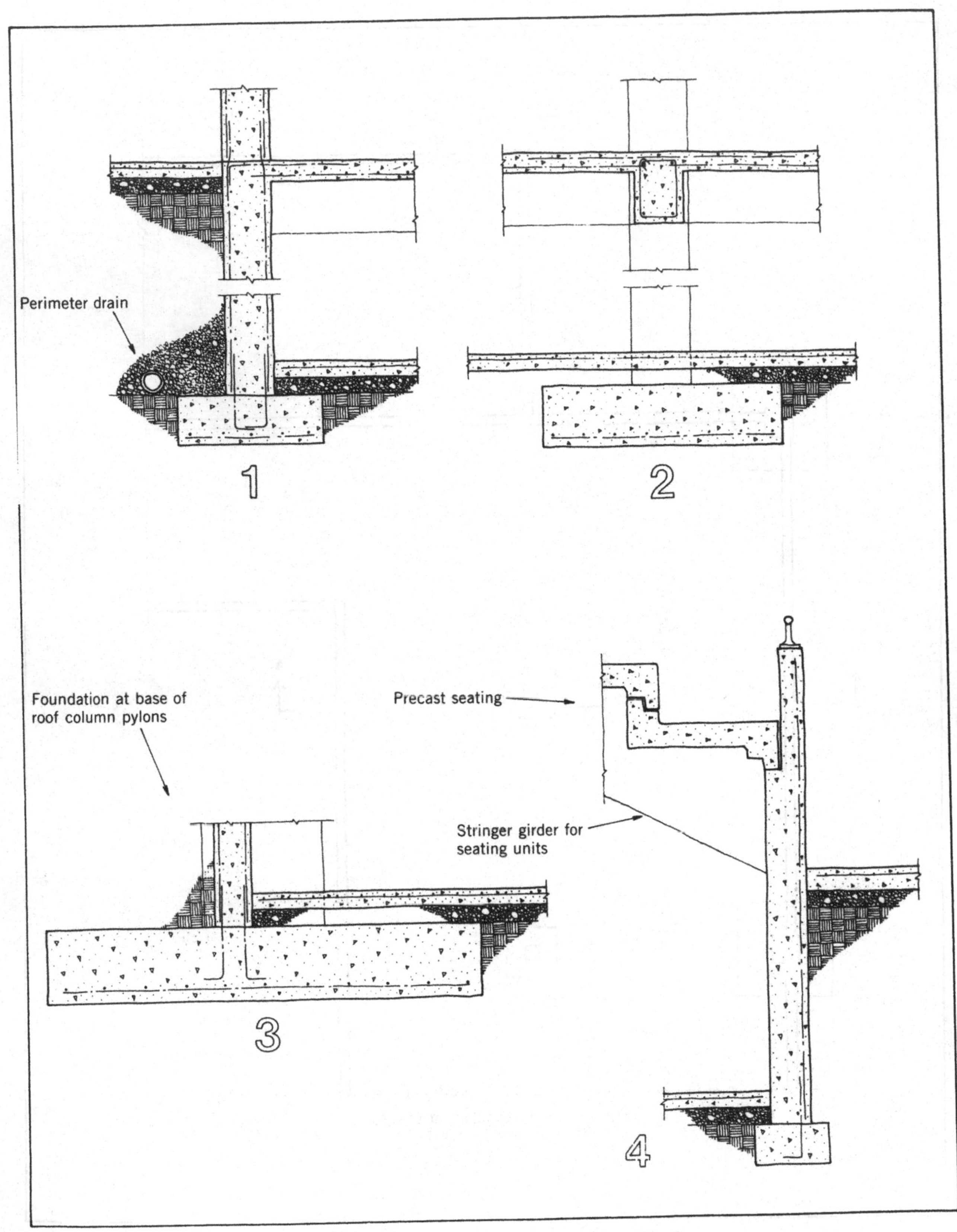

Figure 10.110 Building 8: foundation details.

underground building. However, space for equipment and the fresh air intakes may possibly be shared.

Figure 10.111 shows some details for a system for ventilation of the large arena space, using four separate supply air towers in the building corners, feeding a perimeter duct and overhead register system. Return air is handled through grilles in the seating structure. This system might be coupled with a low pressure system using small registers within the seating to provide better control at the seating locations. Equipment for the system is in the concrete structure, with air intake and exhaust through tunnels connecting to terminals outside the building.

Another major problem for this building is the provision of a general illumination system. If usage is primarily daytime, it is likely that use would be made of natural light as much as possible. This would include input from the large glazed walls and conceivably from some skylights. However, the facility is sure to be used at night as well, and the basic system should be one that permits both conditions.

Figure 10.112 shows some details for a lighting system that uses suspended fixtures with multiple lamps. Individual lamps, as well as individual fixtures, can be separately switched to permit a considerable range for the system. During daytime, for example, the fixtures near the perimeter would probably be switched completely off.

Considerable other lighting would also be necessary, including that for the enclosed spaces beneath the seating. Special lighting for events is also likely, and a catwalk system may well be permanently installed at the level of the bottom of the trusses to facilitate this as well as maintenance of other elements within the truss system.

10.9 BUILDING 9

Open Pavilion
Wood Pole Frame and Timber Truss
Park Site

This is an open-air facility of medium span, consisting only of a large roof over a small amphitheater that is built into the ground (see Figure 10.113). Foundations and vertical supports for the roof structure are provided by wood poles with their bottom ends buried in the ground. The roof structure consists of a two-way spanning timber truss system with timber purlins and a wood plank roof deck. Roofing consists of sheet metal units with standing ribs (see Figure 10.114).

This is a relatively primitive type of structure, with roots in ancient, indigenous construction in many cultures. However, the truss system here represents some clever engineering and some reasonably developed technology in terms of jointing methods.

There is some analogy here with the structure for Building 8, although the different scale and materials and the lack of a wall enclosure creates a much different situation. Appearance is also significantly different here, with a rough-hewn and somewhat crude finish of the timber elements, compared to the relatively high-tech precision of the steel structure.

It is assumed that drainage of the roof is controlled only to the extent of sloping to the roof edges. Some minor diversion of drainage away from the major points of entry to the amphitheater may be achieved; otherwise everybody walks in and out through a waterfall when it rains or the snow melts. However, the major usage of such a building is likely to be on sunny days in times of mild weather, in general.

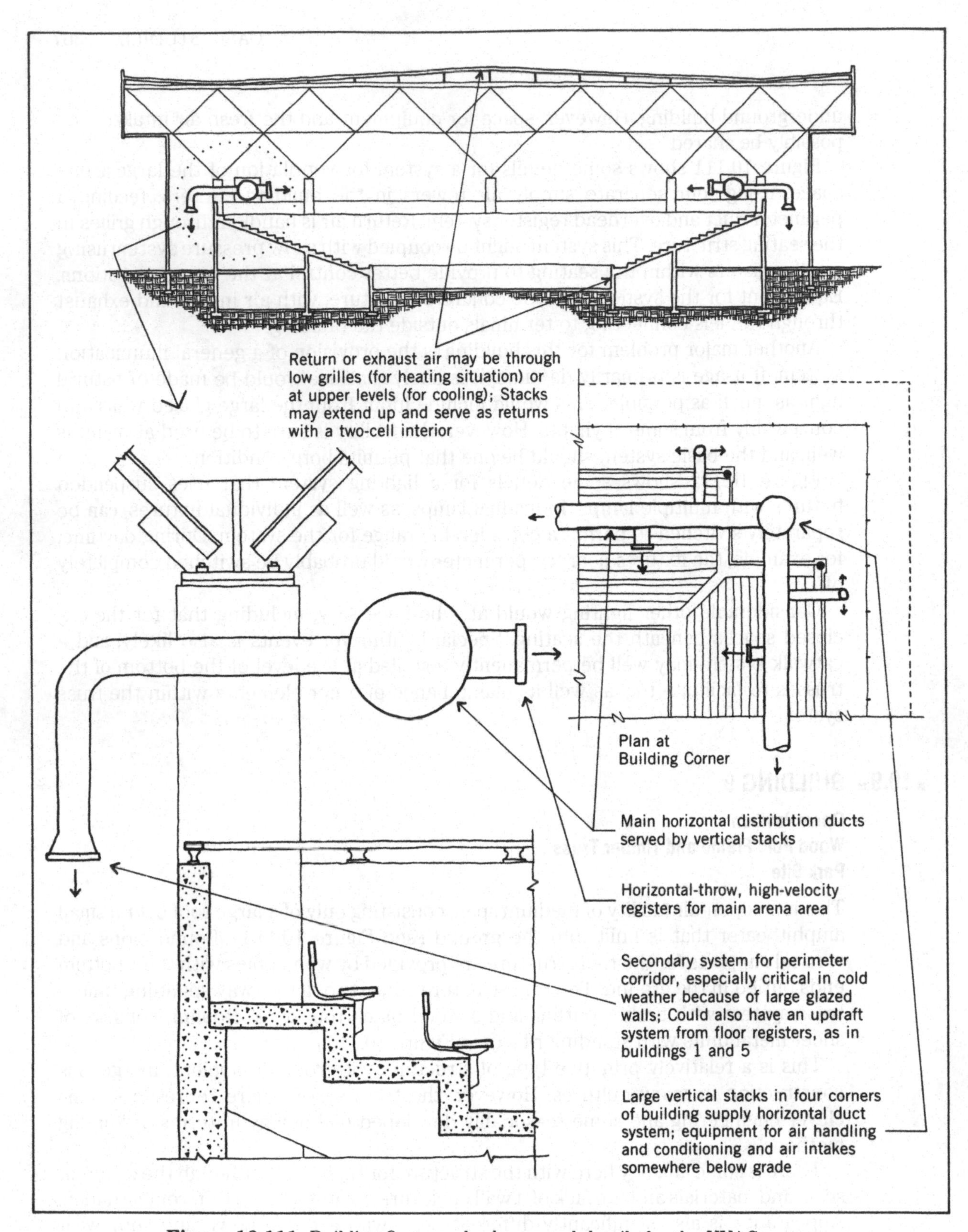

Return or exhaust air may be through low grilles (for heating situation) or at upper levels (for cooling); Stacks may extend up and serve as returns with a two-cell interior
Plan at Building Corner
Main horizontal distribution ducts served by vertical stacks
Horizontal-throw, high-velocity registers for main arena area
Secondary system for perimeter corridor, especially critical in cold weather because of large glazed walls; Could also have an updraft system from floor registers, as in buildings 1 and 5
Large vertical stacks in four corners of building supply horizontal duct system; equipment for air handling and conditioning and air intakes somewhere below grade

Figure 10.111 Building 8: general scheme and details for the HVAC system.

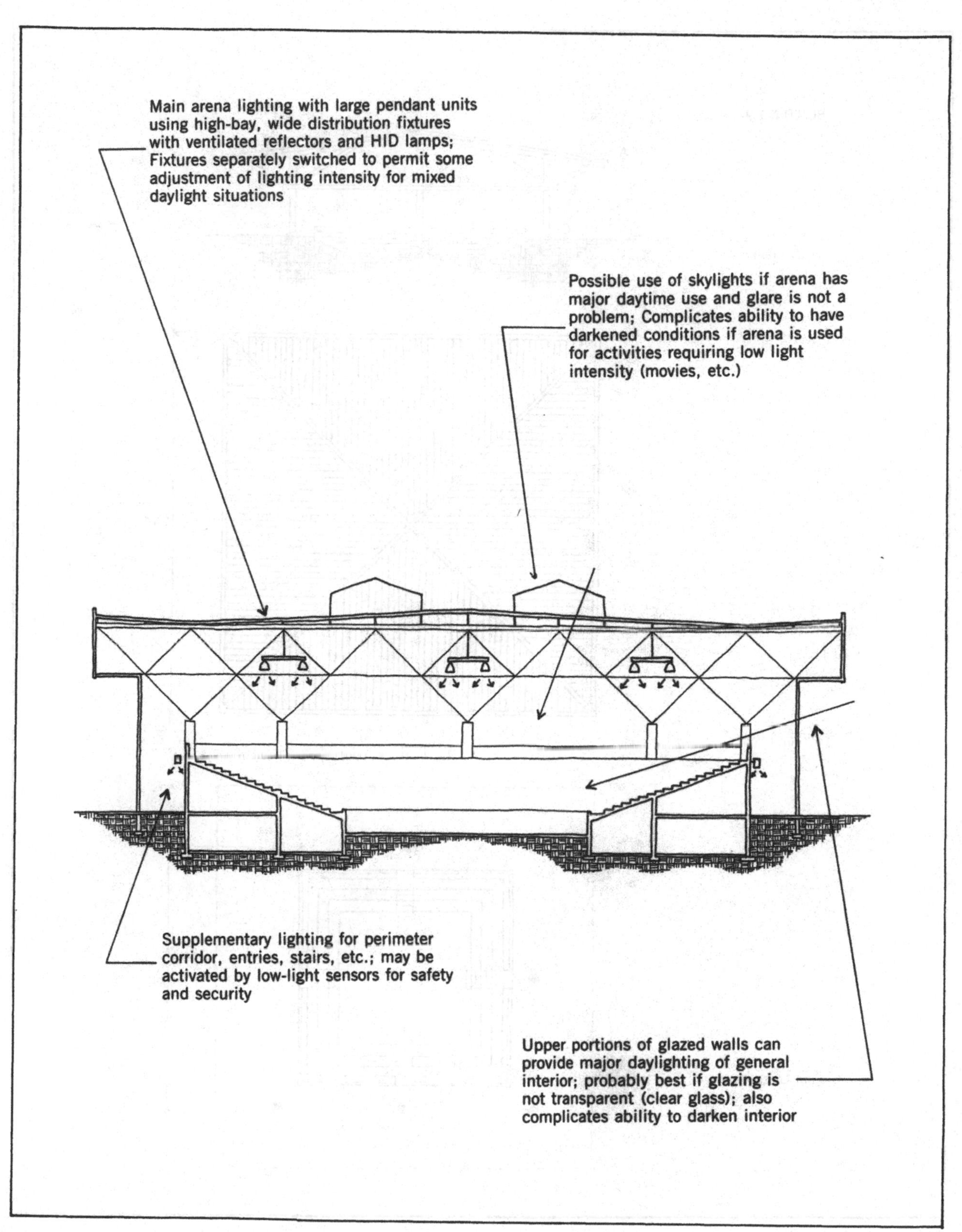
Main arena lighting with large pendant units using high-bay, wide distribution fixtures with ventilated reflectors and HID lamps; Fixtures separately switched to permit some adjustment of lighting intensity for mixed daylight situations
Possible use of skylights if arena has major daytime use and glare is not a problem; Complicates ability to have darkened conditions if arena is used for activities requiring low light intensity (movies, etc.)
Supplementary lighting for perimeter corridor, entries, stairs, etc.; may be activated by low-light sensors for safety and security
Upper portions of glazed walls can provide major daylighting of general interior; probably best if glazing is not transparent (clear glass); also complicates ability to darken interior

Figure 10.112 Building 8: lighting details.

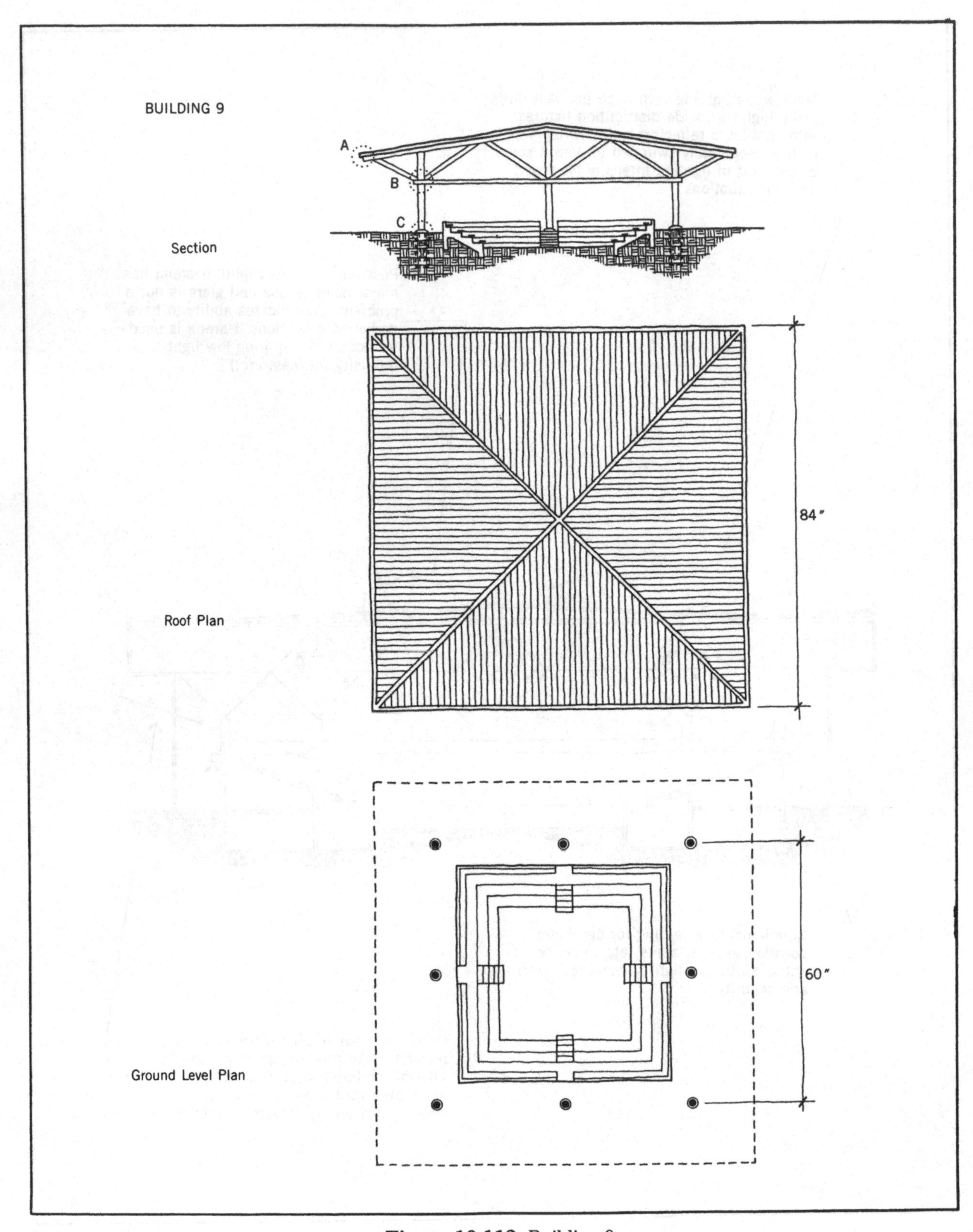

Figure 10.113 Building 9.

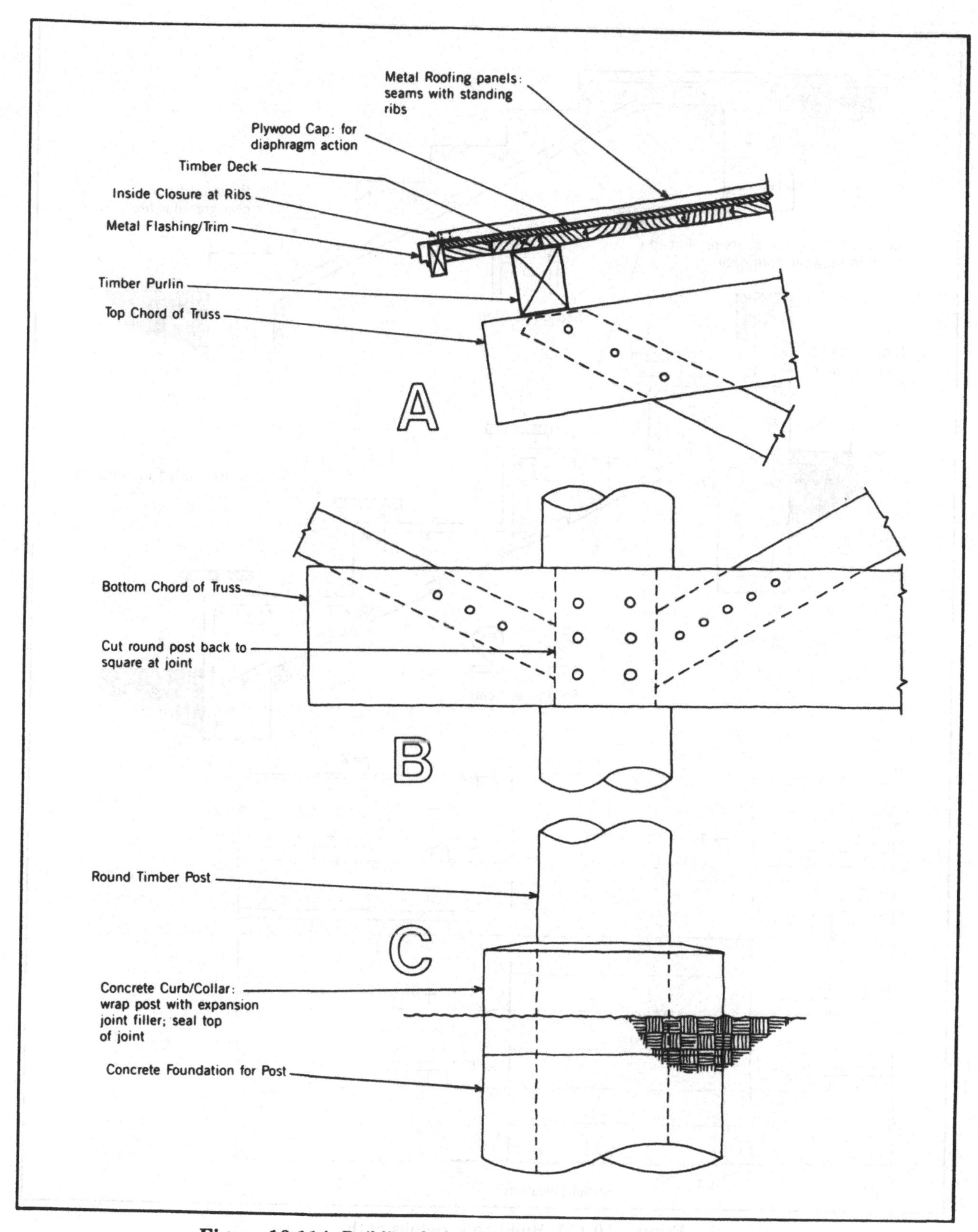

Figure 10.114 Building 9: details of the wood structure.

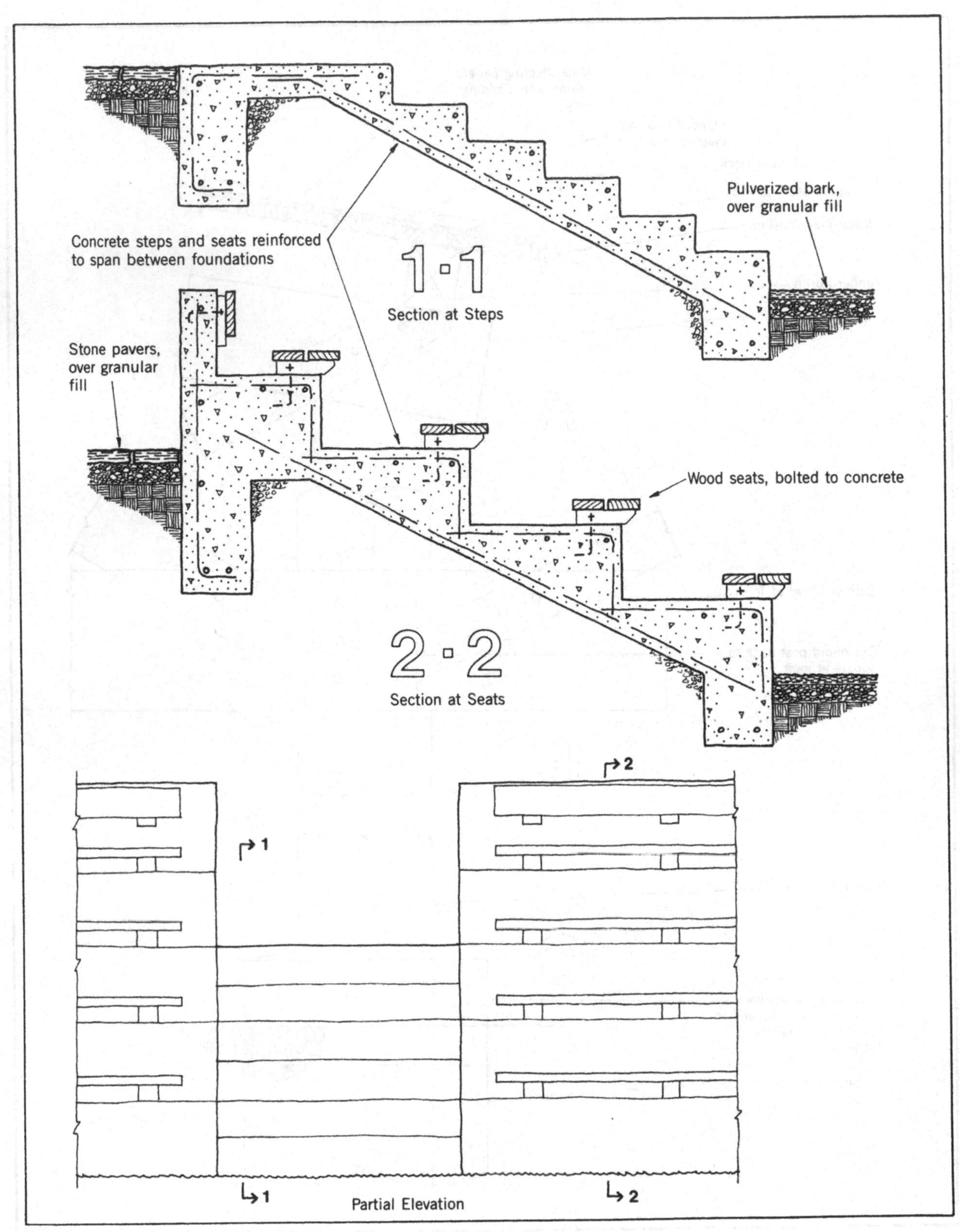

Figure 10.115 Building 9: details of the seating.

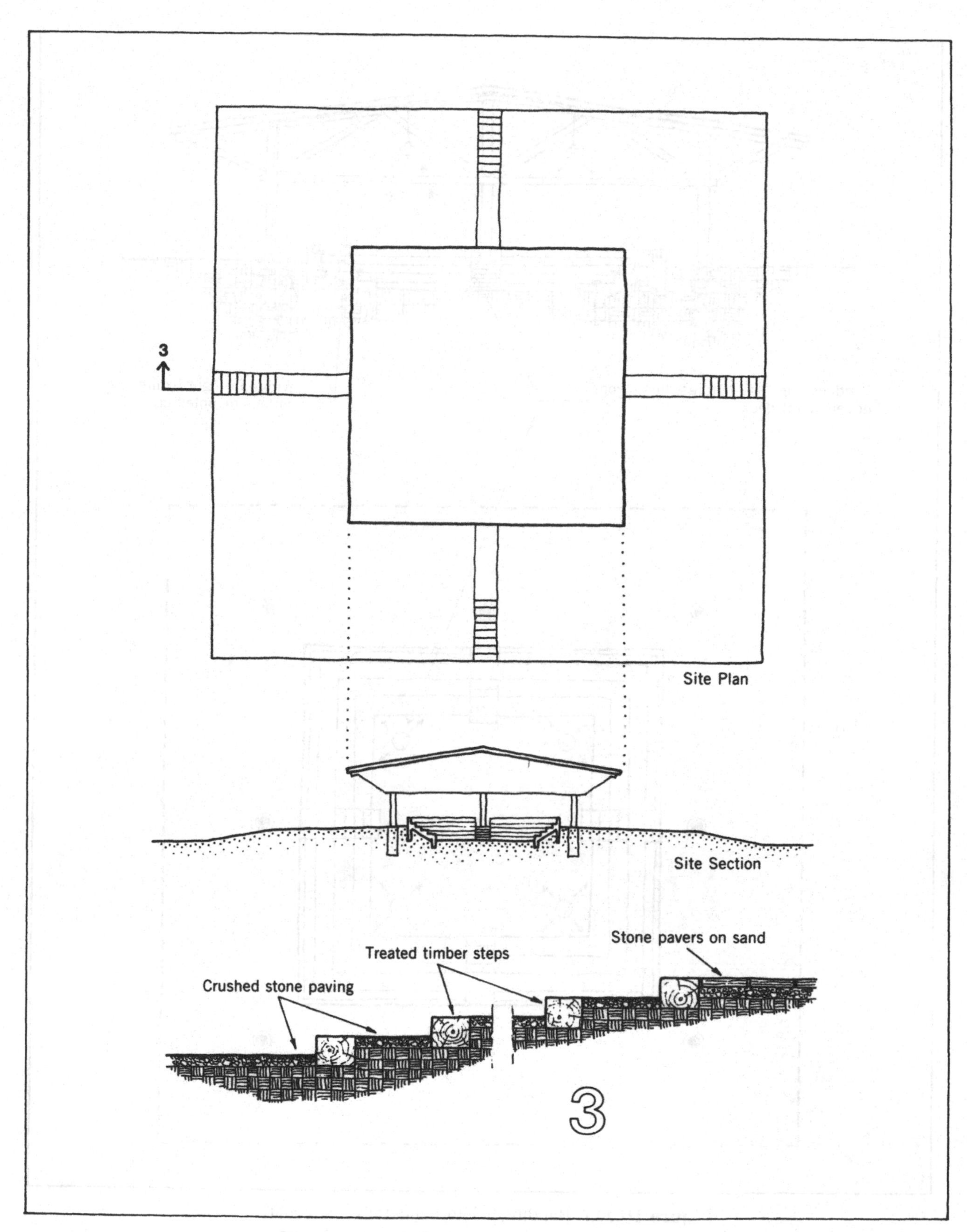

Figure 10.116 Building 9: site plan and details.

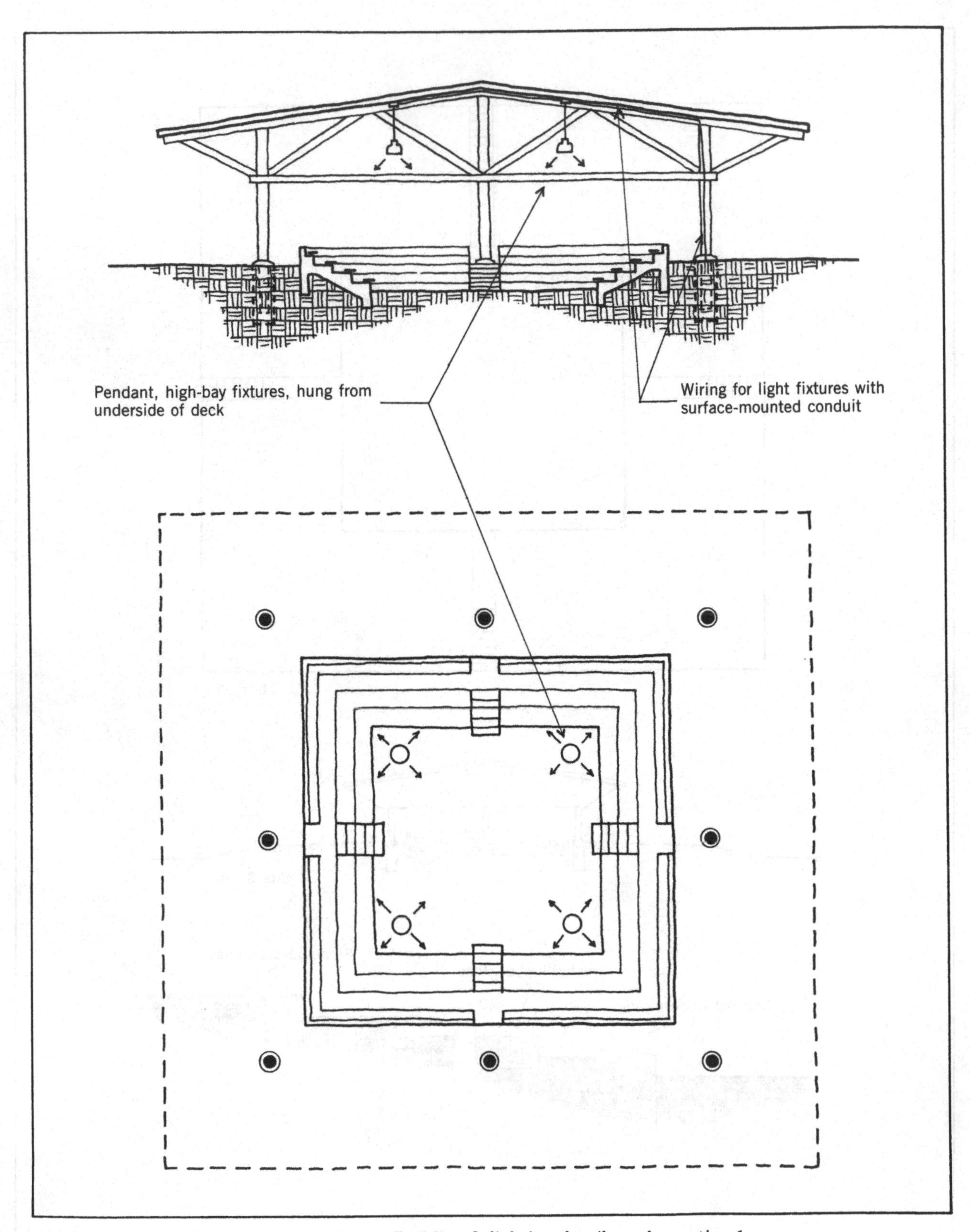

Figure 10.117 Building 9: lighting details—alternative 1.

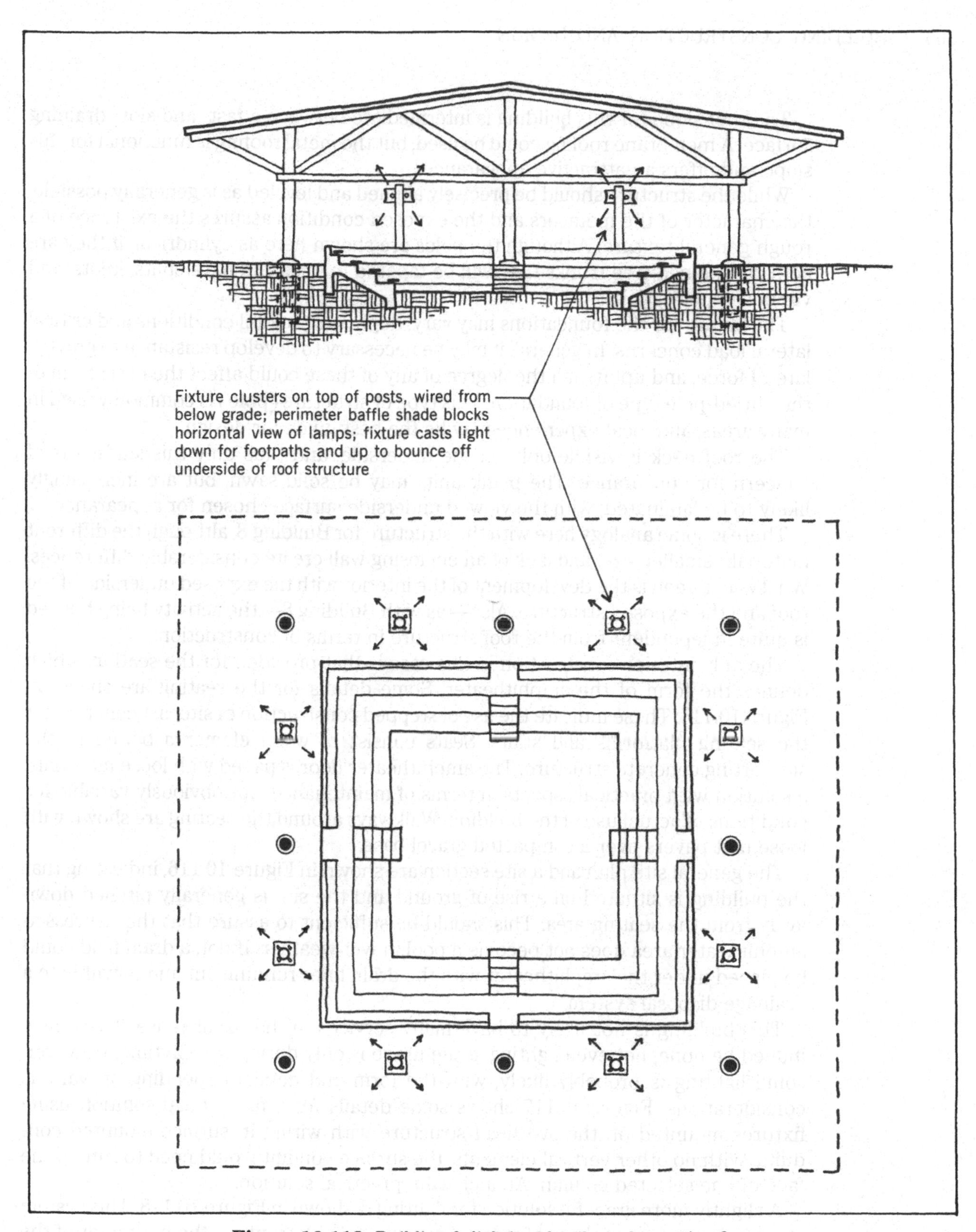
Fixture clusters on top of posts, wired from below grade; perimeter baffle shade blocks horizontal view of lamps; fixture casts light down for footpaths and up to bounce off underside of roof structure

Figure 10.118 Building 9: lighting details—alternative 2.

The roof slope for this building is intermediate between a fast- and slow-draining surface. A membrane roofing could be used, but the metal roofing is functional for this slope and offers an attractive appearance.

While the structure should be precisely aligned and leveled as is generally possible, the character of the members and the exposed condition assures the existence of a rough general texture. Although the poles are shown here as cylindrical, if they are actually skinned tree trunks they will be tapered in form and have splits, knots, and various imperfections.

The details for the foundations may vary, depending on soil conditions and critical lateral load concerns. In general, it may be necessary to develop resistance to gravity, lateral force, and uplift, and the degree of any of these could affect the basic form of the buried-pole type of foundation. This type of structure (pole) is commonly used in many areas, and local experience may be the best guide for design.

The roof deck is visible only on the underside, and thus only this surface is of concern for appearance. The plank units may be solid sawn, but are increasingly likely to be laminated, with the viewed underside surface chosen for appearance.

There is some analogy here with the structure for Building 8, although the different materials, smaller size, and lack of an enclosing wall create considerable differences. What *is* common is the development of the interior with the exposed underside of the roof and the exposed structure. Also—as with Building 8—the activity being housed is quite independent from the roof structure in terms of construction.

The only interior construction in this case is that provided for the seating, which defines the form of the amphitheater. Some details for the seating are shown in Figure 10.115. These indicate the use of stepped construction of sitecast concrete for the seating platforms and stairs. Seats consist of wood elements bolted to the supporting concrete structure. The amphitheater floor is paved with loose materials; a solution with practical aspects in terms of maintenance, but obviously variable for conditions of actual use of the building. Walkways around the seating are shown with loose unit pavers over a compacted gravel base.

The general site plan and a site section are shown in Figure 10.116, indicating that the building is situated on a rise of ground and the site is generally pitched down away from the seating area. This should be sufficient to assure that the depressed amphitheater area does not become a pool in wet weather. If not, a drain field could be placed under the amphitheater with the drain lines running out and downhill to a drainage disposal system.

This building is not likely to have many services in the usual sense. There may indeed be none; not even lighting, if the usage is only during the daytime. However, some lighting is probably likely, with the form and details depending on various considerations. Figure 10.117 shows some details for a fast, cheap solution using fixtures mounted on the overhead structure with wiring in surface-mounted conduits. With no other vertical elements, the surface conduit would need to run up the face of one selected column. An inelegant, practical solution.

A slightly more graceful solution for lighting is shown in Figure 10.118. This system stays independent of the roof structure; using pole elements at the perimeter of the seating. The fixtures used in this case throw their light mostly up and down—not sideways. Fixtures on the opposite side would thus be less disturbing as glare spots to viewers. The downward cast light will generally illuminate the seating, while the upward cast light will light the underside of the roof and bounce some light into the amphitheater.

Glossary

Access Flooring A floor structure built up as a platform above another floor to create an interstitial space for building services.

Acoustical ceiliing A ceiling surface created with materials intended to absorb sound.

Awning window A window that pivots and cantilevers outward, in the manner of an awning.

Backfill Soil deposited in the excessive part of an excavation after completion of the construction.

Bay A structural planning unit, defined by spaced walls or columns; a portion of an exterior wall that extends beyond the general plane of the wall.

Bent A unit of a framework defined by the frame members in a single vertical plane.

Blocking Short pieces of wood framing placed between the main framing members; as firestops, bracing, etc.

Bridging Bracing elements placed between rows of trusses or beams to provide for lateral stability and some load sharing.

Butt-joint glazing Multi-paned glazing in which the vertical edges of glass units are joined with sealant, rather than being inserted into mullions.

Caisson A vertical shaft of concrete cast in the ground for a foundation; more properly called a *Pier.*

Cant strip A filler strip, cut at an angle, and used to ease the bend from horizontal to vertical for a waterproofing membrane.

Casement window A window that pivots or is hinged to swing like a door.

Cast-in-place Refers to concrete that is formed and cast in its intended position; also called *Sitecast.*

Cavity wall A multiple layered wall in which an outer wythe of masonry is separated from the rest of the construction by an air space.

Cladding Material used to produce the exterior surface of a building; other than glazing for windows.

Concrete masonry unit (CMU) Precast block of concrete, usually with hollow cores, for use in masonry construction.

Conduit A tube of metal or plastic through which electric wiring is run.

Control joint General term for a joint used to control some phenomenon; usually movement due to temperature change, moisture change, or structural action. Sometimes refers specifically to joints constructed in continuous elements (slabs, walls) to provide a location for cracking in a controlled manner.

Coping A cap on a wall.

Cost estimate Informed, educated guess at the final cost of the whole construction or some part of it, before the work is actually done and paid for.

Counterflashing A flashing element from above that laps over one turned up from below.

Course A horizontal strip or layer of masonry units or shingles.

Crawlspace Space between the ground surface and the bottom of a floor that is framed over the ground.

Curtain wall An exterior wall that does not provide support for other elements, but rather depends on the building structure for support.

Cut section An orthographic drawing of what is seen on a plane sliced through the building with the portion on one side of the plane removed. A building plan is a horizontal cut section; the term "cut section" usually refers to vertical cut sections.

Daylighting Use of sunlight for illumination of a building interior.

Decking Structural surfacing used over spanning elements to produce a floor or roof.

Deep foundation Foundation established in the ground some distance below the bottom of the building construction, usually consisting of piles or piers. See also Shallow Bearing Foundation.

Doorway Opening in a wall permitting passage through the wall at floor level; may provide the frame for a door.

Dormer A construction built out from a sloping roof plane, usually with its own separate roof.

Double-hung window Window with two sash that slide vertically past each other.

Duct Usually refers to a large conduit for moving of air.

Eave The bottom horizontal edge of a sloping roof plane.

Expansion joint A control joint for accommodation of movements due to temperature change.

Fascia The exposed vertical face of a roof edge.

Fireproofing Material used to protect steel structural members by forming of thermal insulation.

Firestop Baffle placed in a hollow space in a light wood frame to prevent the spread of fire inside the space.

Fixed window A window that cannot be opened except by removing the glazing from the sash.

Flame spread rating Rating of surfacing materials with regard to hazard for spread of fire on the surface.

Flashing Sheet material of metal, plastic, rubber, or waterproof paper; used to seal water out of a construction joint.

Flooring Finish material used to achieve the exposed, wearing surface of a floor.

Flush door A door with smooth, undetailed faces.

Footing A foundation element designed to spread a compression force out over a large soil surface.

French door A pair of glazed doors in a single opening, hinged at the sides of the opening and meeting at the center.

Frost line The depth to which frost penetrates below the ground surface during a severe winter.

Furring strip Strip material of wood or metal, attached to a surface to allow for attachment of surfacing material to the surface by nails or screws that are driven into the strips.

Gable roof A roof with two opposed, sloping planes that meet at a ridge at the top. A *Gable* is the triangular-shaped section of wall at the end of a gable roof.

Glazing Describes the infill surface material (usually glass) for windows or door openings (noun); the work of installing the material (verb); or some forms of sealing compound used to fix the glass into the window sash (adjective).

Glued Laminated Timber Large wood structural members, laminated from stock lumber of one or two inch nominal thickness.

Gutter A trough or channel, situated to collect water runoff from a sloping roof surface.

Gypsum A natural cement: calcium sulfate, mixed with water to produce gypsum plaster or gypsum concrete.

Gypsum board A facing panel consisting of two faces of paper with a gypsum plaster core; also called *Drywall*, or *Plasterboard*.

Gypsum plaster Plaster made with gypsum as the cementing substance.

Hopper window A window that pivots on a horizontal axis, opening to the interior at its top.

Interstitial space Literally: the spaces in between. Used to describe the spaces *inside* the building construction that are not occupied by the users of the building.

Jamb The vertical edge of a window.

Joist One of a set of closely-spaced beams, used to frame a floor or a flat roof.

Laminated glass Glazing material with two outer faces of glass laminated to one or more inner plies of plastic.

Landing The platform at the top, bottom, or intermediate height of a stair.

Lateral bracing Structural elements of the building construction that are involved in resistance to the horizontal force effects of wind and earthquakes.

Leader Vertical element in a drainage system.

Let-in bracing Thin boards placed diagonally in notches in the face of studs, so as not to increase the wall thickness.

Light wood frame Wood structure achieved primarily with framing members of nominal two inch thickness (2 by 4 studs, etc.).

Lintel Horizontal framing member at the top of an opening in a wall to provide for support of the wall above the opening.

Masonry Construction produced by cementing together of loose units.

Masonry unit The loose units that are cemented together to produce masonry construction; usually brick, tile, stone, or precast concrete.

Membrane A sheet or film that is used to resist the passage of water or water vapor.

Moisture barrier A membrane placed to resist the passage of moisture by migration through a construction element—usually a floor or wall.

Mullion Framing element used between glazing units in a window or door with multiple units of glazing.

Nominal dimension An approximate reference dimension, used to identify an element; actual dimensions may be larger or smaller.

Nosing The leading, outer edge of a step; usually projected (cantilevered) beyond the riser face.

One-way action Spanning nature of a planar system that is supported on two opposite edges, creating bending of a single curvature form.

Open-web steel joist Light, prefabricated steel truss; usually produced as a priority product by an individual manufacturer.

Operable window Window with a moveable sash that permits the window to be opened.

Oriented strand board (OSB) Wood fiber product, produced by bonding of long shredded fibers that simulate the oriented wood grain character.

Parapet Top portion of an exterior wall that projects above the roof.
Particle board Wood fiber product consisting of panel material formed with wood fibers bonded together under pressure.
Partition. An interior wall that serves no structural purposes of support or bracing.
Pilaster A column or stiffening rib formed by a vertical strip projected from the face of a concrete or masonry wall.
Pile A linear form element driven into the ground to provide foundation support.
Precast concrete Concrete construction that is formed and cast in a position other than that of its intended use.
Priority product Product that is copy protected by its manufacturer.
Rafter Inclined framing member for a sloping roof.
Raised access flooring See *Access flooring.*
Reinforced concrete Structural concrete that has steel rods cast into the concrete to develop increased tensile resistance.
Retaining wall A wall that assists in effecting an abrupt change in ground surface elevation.
Riser The unit height dimension of stair steps; the face of a step; the vertical elements in a piping or wiring system.
Roof window An operable window in a roof, or a skylight that can be opened. The term skylight is now used to describe a *fixed* window in a roof.
Roofing Construction material used to achieve the exposed surface of a roof and usually to affect the shedding of water from the roof surface.
Rough opening The clear dimensioned opening that is required for the installation of some element in the basic construction, such as a window in a wall or a skylight in a roof.
Safing A fire resistant plug that helps prevent passage of fire vertically through the construction, such as at the spandrel in curtain walls.
Sash A frame that holds glazing.
Scupper A roof drain consisting of a hole through a wall.
Shallow bearing foundation Foundation established at a short distance below the bottom of the building construction, typically resolving vertical loads by simple contact bearing with the soil.
Sheathing Structural surfacing of a framed wall, floor, or roof.
Shim Thin elements, sometimes wedge-shaped, used to tighten the fit of or more precisely locate elements of the construction; the placing of shims.
Siding Exterior finish material, usually as applied to a light frame structure.
Sill Bottom framing element for a stud wall; bottom edge of a window; the sloped, outer portion of the bottom of a window, intended to drain water away from the window.
Single hung window Two-element window in which only the lower sash slides up and down.
Sitecast (concrete) Concrete which is formed and cast in its desired position; also called *Cast-in-place, In situ,* or *Poured-in-place.*
Skylight Roof window which is not operable (can't be opened).
Slab on grade Sitecast concrete slab resting on and supported by the ground.
Sliding window Window in which one or more sash moves by sliding; usually refers to horizontal sliding, as the terms *Double-hung* or *Single-hung* are used to describe windows that slide vertically.

Soffit The underside surface of some construction, such as a roof overhang, a stairs, a balcony, a canopy, or a bridge.

Spandrel The portion of an exterior wall that is above the head of one window and below the sill of one above it.

Spandrel beam Framing member at the outside edge of a floor or roof.

Specifications Written documents, describing the construction in sufficient detail so as to establish the basis for a binding contract.

Specs Short for specifications.

Stick system Describes the curtain wall structure that uses the vertical mullions (the "sticks") for basic support of the rest of the wall construction.

Storm window A separate sash added to the outside of a window during cold weather, to increase thermal insulation and reduce air infiltration.

Stucco Plaster with portland cement, used for exterior wall cladding.

Stud Columns used in closely-spaced sets to produce light framed walls.

Subfloor Structural decking for a floor; floor *Sheathing.*

Suspended ceiling A ceiling whose structure is supported by being hung from a roof or floor structure that is overhead.

Terrazzo Floor surfacing consisting of concrete made with marble chips, with the surface ground and polished to expose the stone.

Thermal break A nonconductive element used in an assemblage to reduce conductive flow of heat through the construction.

Thermal bridge An element of high conductivity in an insulated wall, such as the studs in a wall with insulation in the wall cavities.

Tilt-up construction Wall construction utilizing precast panels that are cast flat on the site and then moved into their desired positions.

Tread The horizontal portion of a step.

Two-way action Spanning action resulting when a flat spanning system is supported in a way that results in bending deformation that is doubly-curved (domed, dished, etc.). See also *One-way action.*

Underlayment Panel material laid over a rough subfloor to provide a smooth surface (and possibly attachment) for a finish flooring material.

Unreinforced Concrete or masonry construction without steel reinforcement. Term is commonly used for this, but the construction is more correctly described as nonreinforced.

Vapor barrier Term widely used to describe a membrane or some other layer of material which is incorporated into roof or wall construction to retard the flow of airborne moisture (water vapor) through the construction. More precise term is *Vapor retarder.*

Vapor retarder See *Vapor barrier.*

Veneer An outer layer of finish material in a multi-layered assemblage, such as the face plies in a plywood panel.

Waffle slab A concrete joist system with two-way joists.

Water table The level to which water will rise in a hole dug in the ground. *High water table* refers to the highest expected seasonal level of the water table in the ground that is likely to occur at a given location.

Wythe (Pronounced why-th). A single vertical layer of masonry units.

Soil Properties and Behaviors

The materials in this appendix present a general summary of concerns for the soils that constitute the surface and subsurface ground mass of most sites. This material has been abstracted from various references, but mostly from *Simplified Design of Building Foundations,* by James Ambrose, Wiley, 1988.

Information about the materials that constitute the earth's surface is forthcoming from a number of sources. Persons and agencies involved in fields such as agriculture, landscaping, highway and airport paving, waterway and dam construction, and the basic earth sciences of geology, mineralogy, and hydrology have generated research and experience that is useful to those involved in general site development.

A.1 SOIL CONSIDERATIONS RELATED TO SITE DEVELOPMENT

Some of the fundamental properties and behaviors of soils related to concerns in site development are the following:

Soil Strength. For bearing-type foundations a major concern is the soil resistance to vertical compression. Resistance to horizontal pressure and to sliding friction are also of concern for situations involving the lateral (horizontally-directed) effects due to wind, earthquakes, or retained soil.

Dimensional Stability. Soil volumes are subject to change; principal sources being changes in stress or water content. This effects settlement of foundations and pavements, swelling or shrinking of graded surfaces, and general movements of site structures.

General Relative Stability. Frost actions, seismic shock, organic decomposition, and disturbance during site work can also produce changes in physical conditions of soils. The degree of sensitivity of soils to these actions is called their relative stability. Highly sensitive soils may require modification as part of the site development work.

Uniformity of Soil Materials. Soil masses typically occur in horizontally stratified layers. Individual layers vary in their composition and their thickness. Conditions can also vary considerably at different locations on a site. A major early investigation that must precede any serious engineering design is that for the soil profiles and properties of individual soil components at the site. Depending on the site itself and the nature of the work proposed for the site, this investigation may need to proceed to considerable depth below grade.

Ground Water Conditions. Various conditions, including local climate, proximity to large bodies of water, and the relative porosity (water penetration resistance) of soil layers, effect the presence of water in soils. Water conditions may affect soil stability, but also relate to soil drainage, excavation problems, need for irrigation, and so on.

Sustaining of Plant Growth. Where site development involves considerable new planting, the ability of surface soils to sustain plant growth and to respond to irrigation systems must be considered. Existing surface soils must often be modified or replaced to provide the necessary conditions.

The discussions that follow present various issues and relationships that affect these and other concerns for site development.

A.2 GROUND MATERIALS

A major purpose for early investigations of a site is to identify the nature of the materials that constitute the surface and near-surface ground mass. These are typically stratified mixtures of the following basic materials.

Soil and Rock. The two basic solid materials that constitute the earth's crust are soil and rock. At the extreme, the distinction between the two is clear; loose sand versus solid granite, for example. A precise division is somewhat more difficult, however, since some highly compressed soils may be quite hard, while some types of rock are quite soft or contain many fractures, making them relatively susceptible to disintegration. For practical use in engineering, soil is generally defined as any material consisting of discrete particles that are relatively easy to separate, while rock is any material that requires considerable brute force for its excavation.

Fill. In general, fill is material that has been deposited on the site, to build up the ground surface. Many naturally-occurring soil deposits are of this nature, but for engineering purposes, the term fill is mostly used to describe *man-made fill* or other deposits of fairly recent origin. The issue of concern for fill is primarily its recent origin and the lack of stability that this represents. Continuing consolidation, decomposition, and other changes are likely. The uppermost soil materials on a site are likely to have the character of fill—man-made or otherwise.

Organic Materials. Organic materials near the ground surface occur mostly as partially decayed plant materials. These are highly useful for sustaining new plant growth, but generally represent undesirable stability conditions for various engineering purposes. Organically-rich surface soils (generally called *topsoil*) are a valuable resource for landscaping—and indeed may have to be imported to the site where they do not exist in sufficient amounts. For support of pavements, site structures, or building foundations, however, they are generally undesirable.

Investigation of site conditions is done partly to determine the general inventory of these and other existing materials, with a view towards the general management of the site materials for the intended site development.

A.3 SOIL PROPERTIES AND IDENTIFICATION

The following material deals with the definition of various soil properties and their significance to the identification of soils and determination of soil behaviors.

Soil Composition

A typical soil mass is visualized as consisting of three parts, as shown in Figure A.1. The total soil volume is taken up partly by the solid particles and partly by the open spaces between the particles, called the *void.* The void space is typically filled by some combination of gas (usually

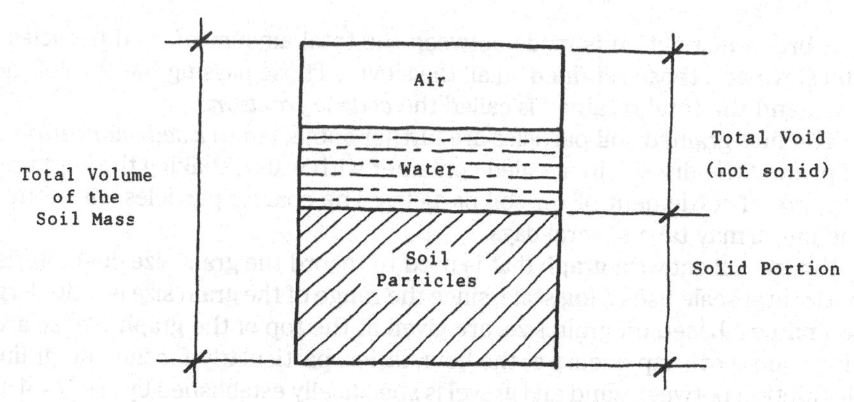

Figure A.1 Typical three-part composition of a soil mass.

air) and liquid (usually water). There are several soil properties that relate to this composition, such as the following:

Soil Weight. Most of the materials that constitute the solid particles in ordinary soils have a unit density that falls within a narrow range. Expressed as specific gravity (the ratio of the unit density to that of water), the range is from 2.6 to 2.75, or from about 160 to 170 lb/cu. ft. Sands typically average about 2.65; clays about 2.70. Notable exceptions are soils containing large amounts of organic (mostly plant) materials. If the unit weight of a dry soil sample is determined, the amount of void can thus be closely approximated.

Void Ratio. The amount of the soil void can be expressed as a percentage of the total volume. However, in engineering work, the void is usually expressed as the ratio of the volume of the void to that of the solid. Thus a soil with 40% void would be said to have a void ratio of 40/60 = 0.67. Either means of expression can be used for various computations of soil properties.

Porosity. The actual percentage of the void is expressed as the porosity of the soil, which in coarse-grained soils (sand and gravel) is generally an indication of the rate at which water flows through or drains from the soil. The actual water flow is determined by standard tests, however, and is described as the relative *permeability* of the soil.

Water Content. The amount of water in a soil sample can be expressed in two ways: by the *water content* and by the *saturation.* They are defined as follows:

$$\text{Water Content} = (\text{weight of water}/\text{weight of solids}) \times 100$$

$$\text{Saturation} = \text{volume of water}/\text{volume of void}$$

Full saturation occurs when the void is totally filled with water. Oversaturation is possible when the water literally floats or suspends the solid particles, increasing the void. Muddy water is really very oversaturated soil.

Size and Gradation of Soil Particles

The size of the discrete particles that constitute the solids in a soil is significant with regard to the identification of the soil and the evaluation of many of its physical characteristics. Most soils have a range of particles of various sizes (expressed as the size *gradation*), so the full identification of the soil typically consists of determining the percentages of particles of particular size categories.

The two common means for measuring grain size are by sieve and sedimentation. The sieve method consists of passing the pulverized dry soil sample through a series of sieves of increasingly smaller openings. The percentage of the total original sample that is retained on each sieve is recorded. The finest sieve is a No. 200, with openings of approximately 0.003 in.

A broad distinction is made between the total amount of solid particles that pass the No. 200 sieve and those retained on all the sieves. Those passing the No. 200 sieve are called the *fines* and the total retained is called the *coarse fraction.*

The fine-grained soil particles are further subjected to a *sedimentation test.* This consists of placing the dry soil in a sealed container with water, shaking the container, and measuring the rate of settlement of the soil particles. The coarser particles will settle in a few minutes; the finest may take several days.

Figure A.2 shows a graph that is used to record the grain size characteristics for soils. The horizontal scale uses a log scale, since the range of the grain size is quite large. Some common soil names, based on grain size, are given at the top of the graph. These are approximations, since some overlap occurs at the boundaries, particularly for the fine-grained materials. The distinction between sand and gravel is specifically established by the No. 4 sieve, although the actual materials that constitute the coarse fraction are often the same across the grain size range.

The curves shown on the graph in Figure A.23 are representative of some particularly characteristic soils, described as follows:

A *well-graded* soil consists of some significant percentages of a wide range of soil particle sizes.

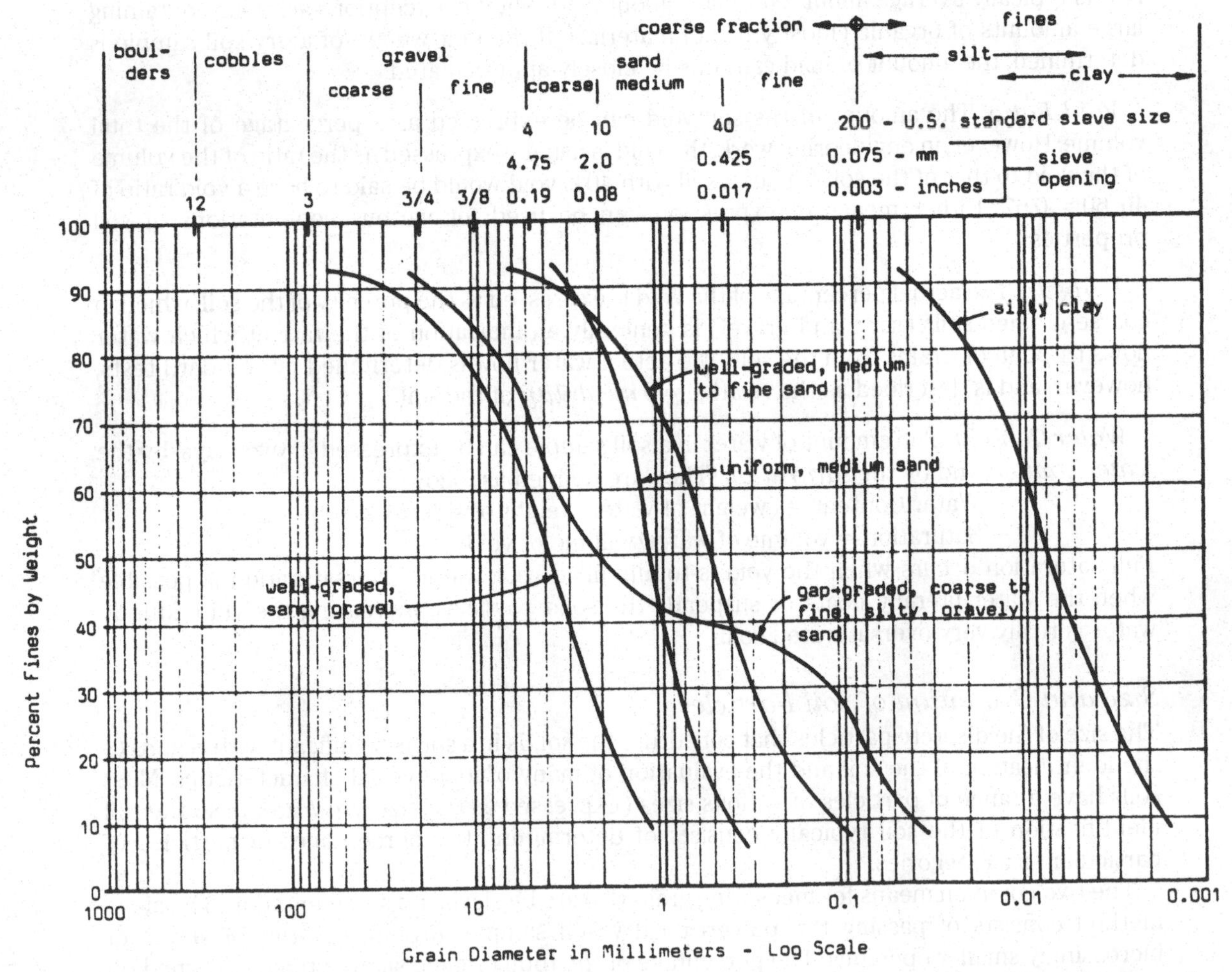

Figure A.2 Grain size range for typical soils. (Reproduced from *Simplified Design of Building Foundations,* with permission of the publisher, John Wiley & Sons.

A *uniform* soil has a major portion of the particles grouped in a small size range.

A *gap-graded* soil has a wide range of sizes, but with some concentrations of single sizes and small percentages over some ranges.

Shape of Soil Particles

The shape of soil particles is also significant for some soil properties. The three major classes of shape are bulky, flaky, and needlelike; the latter being quite rare. Sand and gravel are typically bulky, with a further distinction being made as to the degree of roundness of the particle shape—ranging from angular to well rounded.

Bulky-grained soils are usually quite strong in resisting static loads, especially when the grain shape is quite angular, as opposed to well rounded. Unless a bulky-grained soil is well-graded or contains some significant amount of fine-grained material, however, it tends to be subject to displacement and consolidation (reduction in volume) due to shock or vibration.

Flaky-grained soils tend to be easily deformable or highly compressible, similar to the action of randomly thrown loose sheets of paper or dry leaves in a container. A relatively small percentage of flaky grained particles can give the character of a flaky soil to an entire soil mass.

Effects of Water

Water has various effects on soils, depending on the proportions of water and on the particle size, shape, and chemical properties. A small amount of water tends to make sand particles stick together somewhat, generally aiding the excavation and handling of the sand. When saturated, however, most sands behave like highly viscous fluids, moving easily under stress due to gravity or other sources.

The effect of the variation of water content is generally most dramatic in fine-grained soils. These will change from rocklike solids when totally dry to virtual fluids when supersaturated.

Another water-related property is the relative ease with which water flows through or can be extracted from the soil mass—called the permeability. Coarse-grained soils tend to be rapid-draining and highly permeable. Fine-grained soils tend to be nondraining or impervious, and may literally seal out flowing water.

Plasticity of Fine-Grained Soils

Table A.1 describes for fine-grained soils the Atterberg limits. These are the water content limits, or boundaries, between four stages of structural character of the soil. An important property for such soils is the *plasticity index,* which is the numeric difference between the liquid limit and the plastic limit.

A major physical distinction between clays and silts is the range of the plastic state, referred to as the relative plasticity of the soil. Clays typically have a considerable plastic range while silts generally have practically none—going almost directly from the semisolid state to the liquid state. The plasticity chart, shown in Figure A.3, is used to classify clays and silts on the basis of two properties: liquid limit and plasticity. The diagonal line on the chart is the classification boundary between the two soil types.

Soil Structures

Soil structures may be classified in many ways. A major distinction is that made between soils considered to be *cohesive* and those considered to be *cohesionless.* Cohesionless soils consist primarily of sand and gravel with no significant bonding of the discrete soil particles. The addition of a small amount of fine-grained material will cause a cohesionless soil to form a weakly bonded mass when dry, but the bonding will virtually disappear with a small percentage of moisture. As the percentage of fine materials is increased, the soil mass becomes progressively more cohesive, tending to retain some defined shape right up to the fully saturated, liquid consistency.

TABLE A.1 Atterberg Limits for Water Content in Fine-Grained Soils

Description of Structural Character of the Soil Mass	Analogous Material and Behavior	Water Content Limit
Liquid	Thick soup; flows or is very easily deformed	
		Liquid limit: w_L
Plastic	Thick frosting or toothpaste; retains shape, but is easily deformed without cracking	Magnitude of range is *plasticity index:* I_p
		Plastic limit: w_P
Semisolid	Cheddar cheese or hard caramel candy; takes permanent deformation but cracks	
		Shrinkage limit: w_S
Solid	Hard cookie; crumbles up if deformed	(Least volume attained upon drying out)

The extreme cases of cohesive and cohesionless soils are personified by a pure clay and a pure, or clean, sand respectively. Typically, soils range between these extremes, making the extremes useful mostly for the defining of boundary conditions for classification.

Sands

For a clean sand the structural nature of the soil mass will be largely determined by three properties: the particle shape (well rounded versus angular), the nature of size gradation (well-graded, gap-graded, or uniform), and the density or degree of *compaction* of the soil mass.

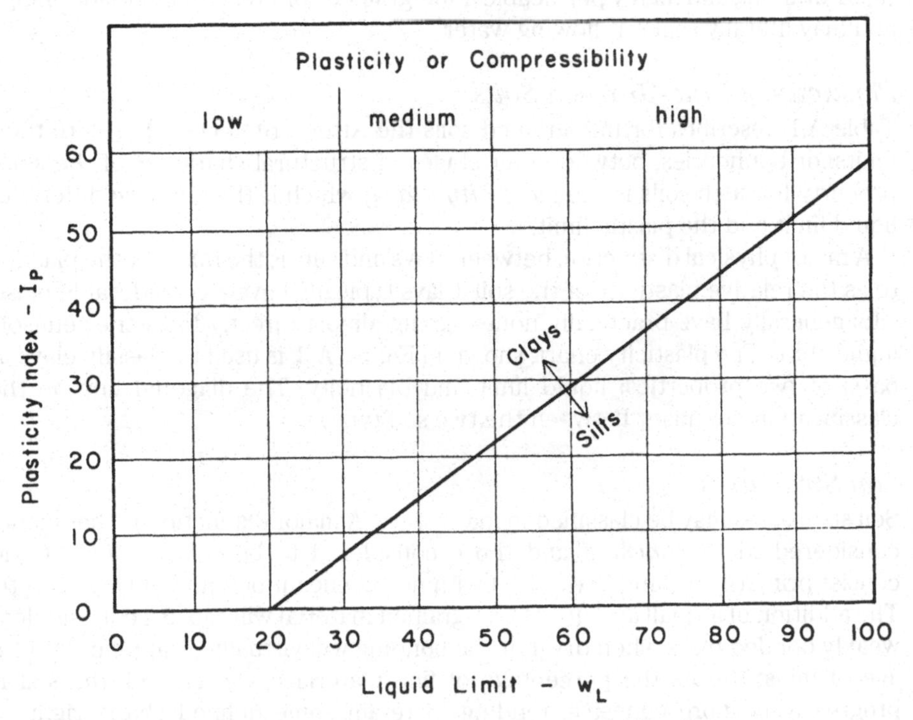

Figure A.3 Plasticity of fine-grained soils.

The density of a sand deposit is related to how closely the particles are fit together and is essentially measured by the void ratio or percentage. The actions of water, vibration, shock, or compressive force will tend to pack the particles into tighter (more dense) arragements. Thus the same sand particles may produce strikingly different soil deposits as a result of density variation.

Table A.2 gives the range of density classifications that are commonly used in describing sand deposits, ranging from very loose to very dense. The general character of the deposit and the typical range of usable bearing strength are shown as they relate to density. As mentioned previously, however, the effective nature of the soil depends on additional considerations, principally the particle shape and the size gradation.

Clays and Silts

Many physical and chemical properties affect the structural character of clays. Major considerations are the particle size, the particles shape, and whether the particles are organic or inorganic. The amount of water in a clay deposit has a very significant effect on its structural nature, changing it from a rocklike material when dry to a viscous fluid when saturated.

The property of a clay corresponding to the density of sand is its consistency, varying from very soft to very hard. The general nature of clays and their typical usable bearing strengths as they relate top consistency are shown in Table A.3.

Another major property of fine-grained soils (clays and silts) is their relative *plasticity.* This was discussed previously in terms of the Atterberg limits and the classification was made using the plasticity chart shown in Figure A.3.

Most fine-grained soils contain both silt and clay, and the predominant character of the soil is evaluated in terms of various measured properties, most significant of which is the plasticity index. Thus an identification as "silty" usually indicates a lack of plasticity (crumbly, friable, etc.), while that of "claylike" or "clayey" usually indicates some significant degree of plasticity (moldable without fracture, even when only partly wet).

Special Soil Structures

Various special soil structures are formed by actions that help produce the original soil deposit or work on the deposit after it is in place. Coarse-grained soils with a small percentage of fine-grained material may develop arched arrangements of the cemented coarse particles, resulting in a soil structure that is called *honeycombed.* Organic decomposition, electrolytic action, or other factors can cause soils consisting of mixtures of bulky and flaky particles to form highly voided soils that are called *flocculent.* The nature of formation of these soils is shown in Figure A.4. Water deposited silts and sands, such as those found at the bottom of dry streams or ponds, should be suspected of this condition if the tested void ratio is found to be quite high.

TABLE A.2 Average Properties of Cohesionless Soils

Relative Density	Blow Count, N (blows/ft)	Void Ratio, e	Simple Field Test with ½-in. Diameter Rod	Usable Bearing Strength	
				(k/ft^2)	(kPa)
Loose	<10	0.65–0.85	Easily pushed in by hand	0–1.0	0–50
Medium	10–30	0.35–0.65	Easily driven in by hammer	1.0–2.0	50–100
Dense	30–50	0.25–0.50	Driven in by repeated hammer blows	1.5–3.0	75–150
Very dense	>50	0.20–0.35	Barely penetrated by repeated hammer blows	2.5–4.0	125–200

TABLE A.3 Average Properties of Cohesive Soils

Consistency	Unconfined Compressive Strength (k/ft²)	Simple Field Test by Handling of an Undisturbed Sample	Usable Bearing Strength k/ft²	kPa
Very soft	<0.5	Oozes between fingers when squeezed	0	0
Soft	0.5–1.0	Easily molded by fingers	0.5–1.0	25–50
Medium	1.0–2.0	Molded by moderately hard squeezing	1.0–1.5	50–75
Stiff	2.0–3.0	Barely molded by strong squeezing	1.0–2.0	50–100
Very stiff	3.0–4.0	Barely dented by very hard squeezing	1.5–3.0	75–150
Hard	4.0 or more	Dented only with a sharp instrument	3.0+	150+

Honeycombed and flocculent soils may have considerable static strength and be quite adequate for foundation purposes as long as no unstabilizing effects are anticipated. A sudden, unnatural increase in the water content (such as that due to introduction of continuous irrigation) or some significant shock or vibration may disturb the fragile bonding, however, resulting in major consolidation of the soil deposit. This can produce major settlement of ground surfaces, pavements, or structures if the affected soil mass is extensive.

Structural Behavior of Soils

Behavior under stress is usually quite different for the two basic soil types: sand and clay. Sand has little resistance to compressive stress unless it is confined. Consider the difference in behavior of a handful of dry sand and sand rammed into a strong container. Clay, on the other hand, has some resistance to both compression and tension in an unconfined condition, all the way up to its liquid consistency. If a hard, dry clay is pulverized, however, it becomes similar to loose sand until some water is added.

In summary, the basic nature of structural behavior and the significant properties that affect it for the two extreme soil types are as follows:

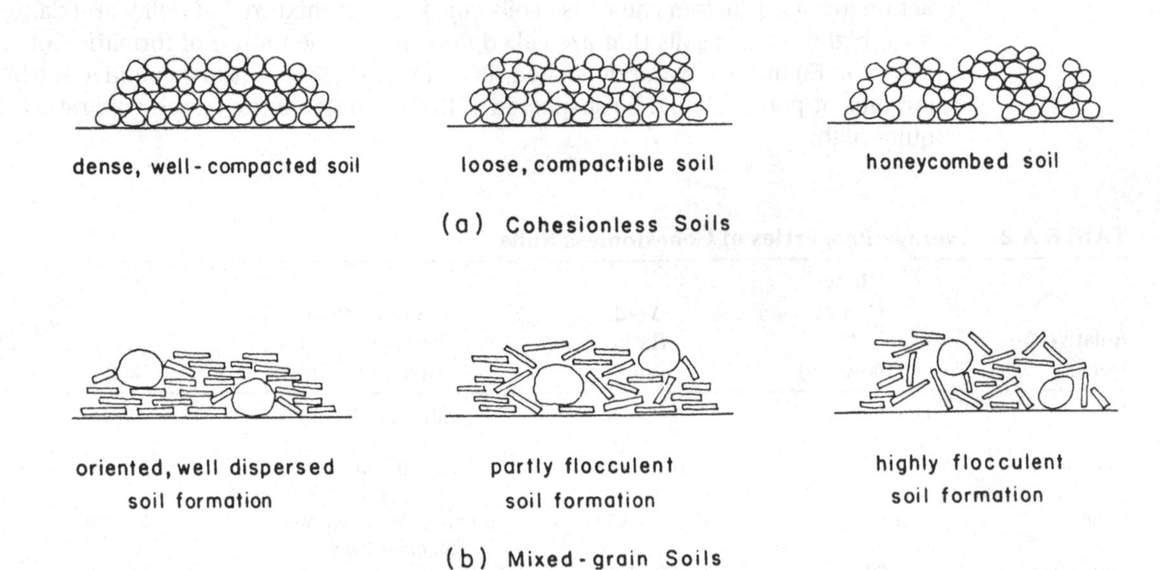

Figure A.4 Arrangements of particles in various soil structures: (a) In bulky-grained soils. (b) In soils with mixtures of bulky and flaky grains.

Sand. Has little compression resistance without some confinement. Principal stress mechanism is shear resistance (interlocking particles grinding together). Important properties are penetration resistance to a driven object, unit density, grain shape, predominant grain size and size gradation, and a derived property called the *angle of internal friction.* Some reduction in capacity with high (saturated or supersaturated) water content.

Clay. Principal stress resistance in tension (cohesive tendency). Confinement generally of concern only in very soft, wet clays which may ooze or flow if compressed or relieved of confinement. Important properties are the tested unconfined compressive strength, liquid limit, plasticity index, and relative consistency (soft to hard).

These are, of course, the extreme limits or outer boundaries in terms of soil types. Most soils are neither pure clays nor clean sands and thus possess some characteristics of both of the basic extremes.

A.4 SOIL CLASSIFICATION AND IDENTIFICATION

Soil classification and identification must deal with a number of properties for precise categorization of a particular soil sample. Many systems exist and are used by various groups with different concerns. The three most widely used systems are the triangular textural system used by the U.S. Department of Agriculture; the AASHO system, named for its developer, the American Association of State Highway Officials; and the unified system which is used in foundation engineering.

The unified system relates to properties that are of major concern in stress and deformation behaviors, excavation and dewatering problems, stability under loads, and other issues of concern to foundation designers. The triangular textural system relates to problems of erosion, water retention, ease of cultivation, and so on. The AASHO system relates primarily to the effectiveness of soils for use as base materials for pavements, both as natural deposits and as fill materials.

Figure A.5 shows the triangular textural system, which is given in graphic form and permits easy identification of the limits used to distinguish the named soil types. The property used is strictly grain size percentage, which makes the identification somewhat approximate since there are potential overlaps between fine sand and silt and between silt and clay. For agricultural purposes this is of less concern than it may be for foundation engineering.

Use of the textural graph consists of finding the percentage of the three basic soil types in a sample and projecting the three edge points to an intersection point that falls in one of the named groups. For example, a soil having 46% sand (possibly including some gravel), 21% silt, and 33% clay would fall in the group called "sandy clay loam." However, it would be close to the border of "clay loam" and would in reality be somewhere between the two soil types in actual nature.

One use of the triangular graph is to observe the extent to which percentages of the various materials affect the essential nature of a soil. A sand, for example, must be relatively clean (free of fines) to be considered as such. With as much as 60% sand and only 40% clay a soil is considered essentially a clay.

The AASHO system is shown in Table A.4. The three basic items of data used are the grain size analysis, the liquid limit, and the plasticity index; the latter two properties relating only to fine-grained soils. On the basis of this data the soil is located by group, and its general usefulness as a base for paving is rated.

The unified system is shown in Figure A.6. It consists of categorizing the soil into one of 15 groups, each identified by a two-letter symbol. As with the AASHO system, primary data used includes the grain size analysis, liquid limit, and plasticity index. It is thus not significantly superior to that system in terms of its data base, but it provides more distinct identification of soils pertaining to their general structural behavior.

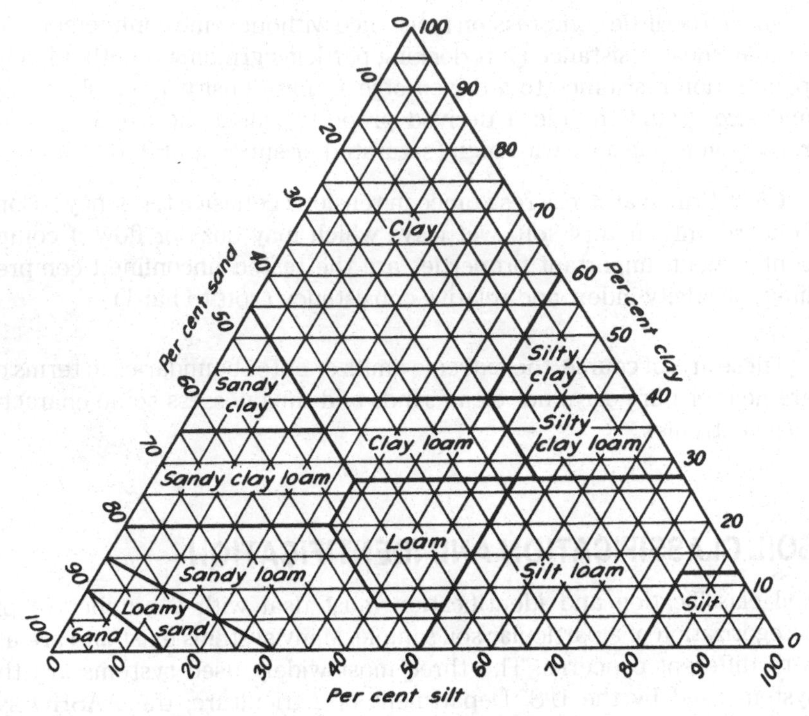

Figure A.5 Triangular textural classification chart used by the U.S. Department of Agriculture.

TABLE A.4 American Association of State Highway Officials Classification of Soils and Soil Aggregate Mixtures—AASHO Designation M-145

General Classification[a]	Granular Materials (35% or Less Passing No. 200)							Silt-Clay Materials (More than 35% Passing No. 200)			
											A-7
Group Classification	*A*-1		*A*-3	*A*-2				*A*-4	*A*-5	*A*-6	*A*-7-5, *A*-7-6
	A-1-*a*	*A*-1-*b*		*A*-2-4	*A*-2-5	*A*-2-6	*A*-2-7				
Sieve analysis percent passing:											
No. 10	50 max										
No. 40	30 max	50 max	51 min								
No. 200	15 max	25 max	10 max	35 max	35 max	35 max	35 max	36 min	36 min	36 min	36 min
Characteristics of fraction passing No. 40:											
Liquid limit				40 max	41 min	40 max	41 min	40 max	41 min	40 max	41 min
Plasticity index	6 max		N.P.[b]	10 max	10 max	11 min	11 min	10 max	10 max	11 min	11 min
Usual types of significant constituent materials	Stone fragments—gravel and sand		Fine sand	Silty or clayey gravel and sand				Silty soils		Clayey soils	
General rating as subgrade	Excellent to good							Fair to poor			

[a] Classification procedure: With required test data in mind, proceed from left to right in chart; correct group will be found by process of elimination. The first group from the left consistent with the test data is the correct classification. The *A*-7 group is subdivided into *A*-7-5 or *A*-7-6 depending on the plastic limit. For $wp \ngtr 30$, the classification is *A*-7-6; for $wp > 30$, *A*-7-5.

[b] N.P. denotes nonplastic.

Major Divisions			Group Symbols	Descriptive Names
Coarse-Grained Soils More than 50% retained on No. 200 sieve	Gravels 50% or more of coarse fraction retained on No. 4 sieve	Clean Gravels	GW	Well-graded gravels and gravel-sand mixtures, little or no fines
			GP	Poorly graded gravels and gravel-sand mixtures, little or no fines
		Gravels with Fines	GM	Silty gravels, gravel-sand-silt mixtures
			GC	Clayey gravels, gravel-sand-clay mixtures
	Sands More than 50% of coarse fraction passes No. 4 sieve	Clean Sands	SW	Well-graded sands and gravelly sands, little or no fines
			SP	Poorly graded sands and gravelly sands, little or no fines
		Sands with Fines	SM	Silty sands, sand-silt mixtures
			SC	Clayey sands, sand-clay mixtures
Fine-Grained Soils 50% or more passes No. 200 sieve	Silts and Clays Liquid limit 50% or less		ML	Inorganic silts, very fine sands, rock flour, silty or clayey fine sands
			CL	Inorganic clays of low to medium plasticity, gravelly clays, sandy clays, silty clays, lean clays
			OL	Organic silts and organic silty clays of low plasticity
	Silts and Clays Liquid limit greater than 50%		MH	Inorganic silts, micaceous or diatomaceous fine sands or silts, elastic silts
			CH	Inorganic clays of high plasticity, fat clays
			OH	Organic clays of medium to high plasticity
Highly Organic Soils			Pt	Peat, muck and other highly organic soils

Figure A.6 Unified system for classification of soils for engineering design, ASTM designation D2487.5.

Building codes and engineering handbooks often use some simplified system of grouping soil types for the purpose of regulating foundation design and construction.

A.5 SPECIAL SOIL PROBLEMS

A great number of special soil situations can be of major concern in site and foundation construction. Some of these are predictable on the basis of regional climate and geological conditions. A few common problems of special concern are discussed in the following material.

Expansive Soils

In climates with long dry periods, fine-grained soils often shrink to a minimum volume, sometimes producing vertical cracking in the soil masses that extends to considerable depths. When significant rainfall occurs, two phenomena occur that can produce problems for structures. The first is the rapid seepage of water into lower soil strata through the vertical cracks. The second is the swelling of the ground mass as water is absorbed, which can produce considerable upward or sideways pressures on structures.

The soil swelling can produce major stresses in foundations, especially when it occurs nonuniformly, which is the general case because of ground coverage by paving, buildings, and landscaping. Local building codes usually have provisions for design with these soils in regions where they are common.

Collapsing Soils

Collapsing soils are in general soils with large voids. The collapse mechanism is essentially one of rapid consolidation (volume reduction) as whatever tends to maintain the soil structure with the large void condition is removed or altered. Very loose sands may display such behavior when they experience drastic changes in water content or are subjected to shock or vibration.

The most common cases of soil collapse, however, are those involving soil structures in which fine-grained materials achieve a bonding or molding of cellular voids. These soils may be quite strong when relatively dry but the bonds may dissolve when the water content is significantly raised. Weakly bonded structures may also be collapsed by shock or simply by excessive compression.

The two ordinary methods of dealing with collapsing soils are to stabilize the soil by introducing materials to partly fill the void and substantially reduce the potential degree of collapse, or to use some means to cause the collapse to occur prior to construction. Infusion with bentonite slurry may be possible to achieve the reduction of the void without collapse. Water saturation, vibration, or overloading of the soil with fill are methods that can be used to induce collapse.

A.6 SOIL MODIFICATION

In connection with site and building construction, modifications of existing soil deposits occur in a number of ways. The recontouring of the site, placing of the building on a site, covering the ground surface with large areas of paving, installation of extensive plantings, and development of a continuous irrigation system all represent major changes in the site environment.

Major changes, done in a few months or years, are equivalent to ones that may take thousands of years to occur by natural causes, and the sudden disruption of the equilibrium of the geological environment is likely to have an impact that results in readjustments that will eventually cause major changes in soil conditions—particularly those near the ground surface. The single *major* modification of site conditions, therefore, is the construction work.

In some cases, in order to achieve the construction work, to compensate for disruptions caused by construction activity, or to protect against future potential problems, it is necessary to make deliberate modifications of existing soils. The most common type of modification is that undertaken to stabilize or densify some soil deposit by compaction, preconsolidation, cementation, or other means. This type of modification is usually done either to improve bearing capacity and reduce settlements, or to protect against future failures in the form of surface subsidence, slippage of slopes, or erosion.

Another common modification relates to changes in the groundwater and moisture conditions. Recontouring for channelled drainage, covering the ground with buildings or paving, and major irrigation or dewatering cause changes of this type.

Deliberate changes may consist of altering the soil to change its degree of permeability or its water-retaining properties. Fine-grained materials may be leached from a mixed grain soil to make it more cohesionless and permeable. Conversely, fine materials may be introduced to make a coarse-grained soil less permeable or to produce a more stable soil mass.

Except for relatively simple compaction of surface materials, all soil modifications should be undertaken only with the advice of an experienced soils consultant.

Study Aids

The materials in this section are given to provide the readers with some means to measure their comprehension with regard to the book presentations. It is recommended that upon completion of an individual chapter, the materials given here for that chapter be utilized to find out what has been accomplished. Answers to the questions are given at the end of this section.

WORDS AND TERMS

Using the glossary, the index, and the text of each chapter indicated, review the meaning of the following words and terms.

Chapter 1

Prime designer
Participants—in building design/build process
Structural engineer
Civil engineer
Landscape architect
Mechanical engineer
Electrical engineer
General contractor
Subcontractor
Specifications
Building code
Zoning ordinance
CSI
ACI
AISC
UBC
BOCA
ASTM
ANSI
NFPA

Chapter 2

Budget (for a building)
Real costs (for construction)
Cost estimates

Fast-track method (building design/build process)
Safety (in buildings)
Design criteria (as standards)
Minimal criteria
Optimal criteria
Hazardous materials (building construction)
Energy use (in buildings)
Design stages: preliminary, definitive, final

Chapter 3

Bias (of information sources)
Access (to design information)

Chapter 4

Common practice
Standards (for design)
Legally-enforceable code
Permit (for construction, occupancy)
Specs

Chapter 5

Building enclosure
Controlled barrier
Selective filter
Shear wall
Seismic effect
Fenestration
Flame spread
Nonstructural wall
Partition
Curtain wall
Slow draining roof
Fast draining roof
Roofing
Waterproof membrane
Roof drainage elements: saddle, gutter, scupper, vertical leader, area drain
Window elements: rough opening, frame, sash, glazing, sill, jamb, mullion
Operable window
Double glazing
Doorway
Flooring
Suspended ceiling
Stair elements: tread, riser, flight, landing
Shallow bearing foundation
Footing
Deep foundation
Pile

Pier (as foundation)
Caisson
Easement
HVAC system
Air conditioning
Ventilation
Illumination
Lighting system
Daylighting
Hard-wired (electrical devices)

Chapter 6

Wood:
- Solid-sawn products
- Glued laminated products
- Wood fiber products
- Green wood
- Species

Metals:
- Rolled steel products
- Light-Gauge steel products
- Extruded products

Concrete:
- Structural concrete
- Insulating concrete
- Fiber concrete
- Gypsum concrete
- Bituminous concrete
- Reinforced concrete
- Prestressed concrete
- Curing

Masonry:
- Structural masonry
- CMU
- Veneer
- Mortar
- Reinforced masonry

Miscellaneous:
- Foamed plastic
- Laminated products
- Laminated glazing
- Spandrel glass
- Waterproof membrane
- Moisture-resistive membrane
- Composite material

Chapter 7

Light wood frame
Shear wall
Lateral bracing
Fire blocking (in wall)
Furring
Modular system
Prefabrication
Site construction
Priority product
Natural surface (of construction)
Applied surfacing
Reflective properties (of surfaces)
Sheathing
Surface structure
Shingle
Roofing tile
Built-up roofing
Single-ply roofing
Ballasted roofing
Flashing
Standing rib metal roofing
Sealants
Slab on grade
Pavement base
Control joint
Slip-resistant surface
Framed construction on grade
Hydrostatic uplift
Heavy timber construction
One-way spanning system
Two-way spanning system
Bearing wall
Column
Set (of concrete surface)
Underlayment
Thin tile
Terrazzo
Interstitial space
Modular ceiling system
Drywall
Adobe
Stucco
Wet wall (finish)
Integrated ceiling system
Skylight
Sky window
Door swing
Wall footing
Column footing

Pedestal
Combined footing
Pole foundation
Friction pile
End-bearing pile
Belled pier (foundation)
Socketted caisson
Earth sheltering
Grade wall
Ground water
Loose pavers
Asphalt
Soil structure
Erosion
Stability of soil slopes
Sheetpiling
Tieback
Soldier beam
Lagging
Raker strut
Drilled-in anchor
Curb
Loose-laid retaining wall
Cantilever retaining wall

Chapter 8

Thermal flow (through construction)
Infiltration
Superinsulation
Moisture barrier
Vapor barrier
Condensation
Noise
Barrier-free environment (access)

QUESTIONS

The following questions serve as a review of principal issues covered in the text. Answers to the questions follow the end of the question list.

Chapter 1

1. Common practices in building design and construction in the United States are largely based on experience. What considerations favor this?
2. What are the fundamental activities in the building design and production process?
3. What is the meaning of the statement that building construction is an architectural design problem?

Chapter 2

1. What are some of the major sources for the derivation of basic design criteria?
2. What elements essentially represent the completion of the design for building construction?
3. What aspects of the completed building construction are primarily controlled by the written specifications?
4. What are the three basic forms of cut section drawings used extensively for construction documentation?

Chapter 3

1. With regard to information sources for design of construction, what attitude of the sources is most important to be cautious of?
2. Why is access to information about building construction not always easy to obtain for those who are in a basic learning situation?
3. With regard to applications to design work, what are some cautions to exercise in using information about construction?

Chapter 4

1. What are the two major concerns for use of any design standard?
2. Why are building codes of limited use as standards for good design?
3. What is the primary concern of those who enforce building codes?
4. What does a building permit actually permit?

Chapter 5

1. Why is the development of the construction for the exterior walls of such critical concern to architects?
2. What basic situation is achieved by the barrier and selective filter functions of the building enclosure?
3. What portion of the construction is most affected by the slope of a roof surface?
4. Where are roof leaks most likely to occur?
5. What two sources of movements of the construction can result in fracture of roofing materials?
6. What are the major reasons for avoiding the use of area drains for flat roofs?
7. Why are high slope roofs of major concern to architects?
8. What problems do windows create for the walls that they occur in?
9. What dimensional relationship must occur between the sash for an operable window and its supporting frame?
10. What type of building activity mostly affects the physical requirements for a door?
11. What are the two basic forms of floor structures?
12. Other than achieving the necessary horizontal span, what functions are required of a floor framed over an occupied space below?
13. Why is there no such thing as a fully enclosed room with no ceiling?
14. What major functions for exterior walls are seldom critical for interior walls?
15. What are the major functional physical requirements for interior walls?

16. From what is a suspended ceiling suspended?
17. What major features of a door are of concern when the door occurs in the required exit path in case of a fire?
18. Since people almost always use the elevators for vertical transportation in multistory buildings, why are stairs necessary?
19. Why are shallow bearing foundation systems almost always preferred over deep foundations when an option for choice exists?
20. The major portion of site construction work involves dealing with what material?
21. The term "air conditioning" is commonly used to mean only cooling of the indoor air. What does it include in building design work?
22. In modern commercial buildings, what are the usual primary required elements of a complete HVAC system?
23. What are the major considerations for accommodating HVAC systems that must commonly be dealt with in planning of the building construction?
24. Despite all the efforts to utilize natural light (sunlight), what usually makes it necessary to use some artificial (electrical) light in most buildings?
25. Besides the utilization of basic lighting sources, what aspects of the construction affect the conditions for illumination in building interiors?
26. With regard to use of electrical power in buildings, what are the major concerns for safety of the building occupants?

Chapter 6

1. What form of wood product is steadily replacing solid sawn elements and ordinary plywood in building construction?
2. Even though steel is not combustible (won't burn), steel structures are highly vulnerable to fire. Why is this?
3. Why is aluminum used in relatively small total volume in buildings, even though it is used for a great number of products?
4. What basic forming process, used extensively with aluminum and hardly ever with steel, makes aluminum a favored material for window frames and various forms of trim?
5. Other than strength, what single physical property basically distinguishes structural concrete from insulating concrete?
6. What major ingredient in the concrete mix distinguishes ordinary concrete from bituminous concrete?
7. Concrete as a material has a relatively low bulk cost. What makes concrete structures expensive?
8. What are the two primary ingredients of structural masonry?
9. Besides the glass, what is a basic ingredient in double glazing?
10. Besides the glass, what is almost always included in laminated glazing?
11. What is the difference in performance required for construction described as waterproof as opposed to that which is merely moisture-resistive?

Chapter 7

1. Surfacing materials on light wood frame walls are not assumed to add to resistance to vertical loads, but still have some other structural tasks. Describe these potential tasks.

2. Hollow spaces in framed walls are useful for many purposes, but dangerous in what one major regard?
3. What type of production process is especially facilitated by use of a panelized wall system?
4. Roofs present the opportunity for use of a greater range of structural systems than floors. What are some reasons for this?
5. Selection of roof construction must relate to many concerns. Describe some major ones.
6. What is the basic task of roofing?
7. What general character of the roofing is not required when the roof slope is fast-draining?
8. Why is appearance generally of greater concern for shingles than for built-up roofing?
9. What is built up in a built-up roof?
10. Why are paving slabs of concrete typically cast in units with relatively closely-spaced joints?
11. Why may it not be desireable to have the smoothest finish possible on a concrete slab on grade?
12. What is the usual primary reason for using a framed floor on grade?
13. What usual geometric requirements most limit the choices for structures for framed floors?
14. Concrete fill is often used on top of steel or wood decks. What are some functional purposes of the fill?
15. What two functions does underlayment usually provide with carpetted floors?
16. What are the two primary reasons for using pedestals with columns supported by footings?
17. What does a retaining wall retain?

Chapter 8

1. Other than general potential for loss in a fire, what is critical about having combustible materials in the building construction?
2. Why do present building code requirements not necessarily relate to optimal life safety concerns?
3. Development of the building enclosure for thermal resistance (heat loss or gain of the interior) affects what concerns for the construction?
4. With regard to roofs and exterior walls, what is the major concern for migration of airborne water through the construction?
5. Where should a required vapor barrier be placed in the construction?
6. What generally qualifies a building environment as barrier-free?

Chapter 9

1. How does the hierarchy of a system relate to modifications of the system?
2. Why do most buildings consist of mixtures of systems, rather than whole, predetermined systems?
3. What is the difference between designing and choosing?

ANSWERS TO THE QUESTIONS

Chapter 1

1. Other than old habits, the desire for assurance of performance by all concerned: designers, builders, owners, financiers, insurers, code agencies, and so on. Tried and true is the name of the game.
2. 1: someone wants/needs a building and will pay for it; 2: someone designs it; 3: someone supplies the materials to build it; 4: someone builds it.
3. If "design" means to contrive to make, then planning for the actual making of the building is part of the design.

Chapter 2

1. Personal goals and philosophy of the designer; dollar cost of construction; time required for construction; safety for everybody—users, general public, fire fighters, etc.; absence of hazardous/toxic materials; energy conservation; etc.
2. The documents that describe the complete construction of the building (typically drawings and specifications).
3. General quality and soundness of the finished work.
4. Plans, full building sections (cut on vertical planes), and selected detailed views of pieces of the construction.

Chapter 3

1. Vested interest in the outcome of design choices.
2. First, they don't know what to look for because they don't know what they don't know. Second, they can't get much information about current products and materials because it is supplied mostly only to potential purchasers, working designers, spec writers, etc.
3. Neutral sources (unbiased); timeliness (up to date); regional applicability (local weather, codes, common practices, etc.).

Chapter 4

1. Reliability of the source and obligation to use the standard (legally enforceable, common practice, etc.).
2. They mostly set minimal, barely adequate limits; not the highest standards or optimal conditions.
3. Protection of the public health, welfare, and interests.
4. Construction of the building—but only in accordance with the design documents submitted for the permit review.

Chapter 5

1. They are the most generally viewed parts of the buildings; what is seen by the most people (and cameras).
2. Maintaining of desired interior conditions.
3. Roofing; choices for roofing materials are strongly limited by the slope.
4. At edges, valleys, and other discontinuities of the roof surface.

5. Thermal change (expansion and contraction) and movements of the supporting structure (deflections, settlements, etc.).
6. Cost; likelihood of clogging with dirt and debris; the extended problems of dealing with the water brought into the building.
7. They are prominently in view.
8. Discontinuity or disruption of the wall construction; need for support; need to span the wall over the opening; loss of barrier functions.
9. The sash must be smaller or it won't be able to move.
10. Traffic.
11. Slab on grade (pavement) and framed (spanning) floor.
12. Development of the ceiling; barrier functions for the separation of the two spaces (above and below); incorporation of necessary service elements for the two spaces.
13. By definition a ceiling is the upper bounding surface, usually the underside of the overhead construction. It may not be specially treated, but it is always there.
14. Thermal barrier; air and water tightness; incorporation of windows.
15. Resistance to fire and transmission of sound.
16. The structure (roof, floor) overhead.
17. Size, operating hardware, direction of swing, and maybe the fire resistance of the door itself.
18. For use in fires and when the elevators break down.
19. They cost less, and almost always take less time and effort to build.
20. Soil.
21. Heating and the general handling and treatment of the air—for humidity, freshness (oxygen content), odors, dust, etc.
22. Heater, chiller, air movement system, general control system.
23. Space required, noise and vibration, properties of the construction affecting the interior conditions, planning considerations (form, size, etc.), sun orientation.
24. Night use of buildings; interior spaces away from windows.
25. Size and shape of rooms; reflective character of interior surfaces.
26. Possibility for electrical shock and the hazard of ignition of fires.

Chapter 6

1. Products consisting primarily of reconstituted wood fiber.
2. Steel heats rapidly and softens (losing strength and stiffness) at low temperatures.
3. Because of its cost per unit volume.
4. Extrusion, producing complex forms and/or cross sections with sharp corners.
5. Unit density (weight).
6. The binder—portland cement in ordinary concrete; coal tar or asphalt in bituminous concrete.
7. Costs for transporting, forming, placing, finishing and curing of the concrete, and for reinforcing—especially with steel rods.
8. Mortar and masonry units.
9. Trapped air between the glass panes.
10. Plastic film.
11. Waterproof means resistance to the penetration of flowing water, standing water or water under pressure. Moisture-resistance means interruption of moisture mi-

gration—as airborne moisture for roofs or walls or as absorbed or capillary-action moisture movement for concrete slabs.

Chapter 7

1. Developing resistance to lateral loads: as shear walls or for wind pressure on the wall surface; generally spanning between the wall frame members for any forces applied to the wall surface.
2. Growth and spread of fires inside the wall; dangerously out of view of occupants or fire fighters.
3. Off-site prefabrication.
4. No need for flat surface; less concern for bounciness; less restriction on depth; possible positive use (drainage) for non-flat surface.
5. The rest of the building construction; span; need for particular form; nature of supports; penetrations; people or things on the roof; concerns for movement of the roof structure; architectural impact—exterior and interior.
6. To help keep water out of the building.
7. That it be waterproof.
8. They are usually much more in view from outside the building.
9. Multiple plies of the roof membrane.
10. Because of the high potential for cracking due to shrinkage and temperature changes.
11. It is too slippery for people to walk on or too smooth for adhering of applied finishes.
12. The supporting ground is likely to settle.
13. Need for a flat surface; need to restrict overall depth.
14. Added resistance to fire and to sound transmission; increased stiffness and general solidity; smoother surface for flooring; possible inclusion of wiring in the concrete.
15. A smoother surface and possibility for mechanical attachment of carpet over concrete floors.
16. To raise the bottom of the column above the ground; to ease transfer of the highly concentrated column load to the footing.
17. Soil; at the face of a vertical cut, where the soil wants to move out from the face.

Chapter 8

1. Their contribution to the fire load (amount of total combustible material in the building) and the potential release of toxic gasses during combustion.
2. They were originally mostly developed under the sponsorship of insurance companies with major concern for property loss, which occurs more frequently in building fires than loss of life.
3. Resistance to thermal flow; air tightness (leaks); affects on interior air movements and radiant affects; ability to accommodate building service elements for the HVAC system.
4. Possibility of condensation into liquid water within the construction.
5. Where it will stay the warmest.
6. Ability to accommodate building users with limited physical abilities (deaf, blind, aged, very young, illiterate, foreign, very short, wheelchair-bound, etc.)

Chapter 9

1. Modification of primary elements in the system may alter the basic character of the system; modification of minor elements generally has little effect on the basic nature of the system.

2. Few manufacturers produce whole building systems; most produce small fractional portions of whole buildings.

3. Designing means planning for a purpose; typically involving some choosing between alternate solutions. Choosing may be done for good reasons and for a purpose in mind, but can also be arbitrary or without purpose.

Index